STUDENT'S
SOLUTIONS MANUAL

JOHN R. MARTIN
Tarrant County College, Northeast Campus

STUDENT'S
SOLUTIONS MANUAL

JOHN R. MARTIN

Tarrant County College, Northeast Campus

TO ACCOMPANY

BASIC TECHNICAL
MATHEMATICS

SEVENTH EDITION

BASIC TECHNICAL
MATHEMATICS WITH
CALCULUS

SEVENTH EDITION

Allyn J. Washington

ADDISON-WESLEY

An imprint of Addison Wesley Longman, Inc.

Reading, Massachusetts • Menlo Park, California • New York • Harlow, England
Don Mills, Ontario • Sydney • Mexico City • Madrid • Amsterdam

Acknowledgments

The author gratefully acknowledges the contributions of the following individuals: Judy Martinez for the word processing, Jim McLaughlin for the calculator screens, John Garlow for accuracy checking, and Kathie Morrison for the art work. Their attention to detail and cooperation were of great assistance in preparing this manual.

John R. Martin
Tarrant County College, NE Campus
Hurst, Texas
1999

CONTENTS

CHAPTER 4 THE TRIGONOMETRIC FUNCTIONS

CHAPTER 5 SYSTEMS OF LINEAR EQUATIONS; DETERMINANTS

CHAPTER 6 FACTORING AND FRACTIONS

CHAPTER 11 EXPONENTS AND RADICALS

CHAPTER 12 COMPLEX NUMBERS

CHAPTER 13 EXPONENTIAL AND LOGARITHMIC FUNCTIONS

CHAPTER 22 INTRODUCTION TO STATISTICS

CHAPTER 23 THE DERIVATIVE

CHAPTER 24 APPLICATIONS OF THE DERIVATIVE

CHAPTER 29 EXPANSION OF FUNCTIONS IN SERIES

CHAPTER 30 DIFFERENTIAL EQUATIONS

Basic Algebraic Operations

1.1 Numbers

1. 3 is an integer (whole number); 3 is rational (may be written as a ratio of integers, $3/1$); 3 is real (not the square root of a negative number). $-\pi$ is irrational; it is not a ratio of integers; $-\pi$ is real.

5. $|3| = 3;\ \left|\dfrac{7}{2}\right| = \dfrac{7}{2}$

9. $6 < 8$

13. $-|-3| = -3 \Rightarrow -4 < -3 = -|-3|$

17. The reciprocal of $3 = \dfrac{1}{3}$. The reciprocal of $-\dfrac{1}{3} = \dfrac{1}{-\dfrac{1}{3}} = -3$.

21.

		$-\frac{1}{2}$				2.5		
-2	-1		0	1	2		3	4

25.

-18	$-|-3|$	-1	$\sqrt{5}$	π	$|-8|$	9
-18	-3	-1	2.24	3.14	8	9

29. (a) Let x be a positive integer, then $|x| = x$ which is positive integer and thus an integer. Let x be a negative integer, then $|x| = -x$ which is a positive integer and thus an integer. Yes.
(b) Let x be a positive or negative integer, then the reciprocal is $\frac{1}{x}$ which is the ratio of two integers, 1 and x, and is thus a rational number. Yes.

33. If $x > 1$, then $\frac{1}{x}$ is a positive number less than 1.
(x is between 0 and 1)

37. $N = \dfrac{a \text{ bits}}{\text{byte}} \cdot \dfrac{1000 \text{ bytes}}{\text{kilobytes}} \cdot n \text{ kilobytes} = 1000 \cdot a \cdot n \text{ bits}$

1.2 Fundamental Laws and Operations of Algebra

1. $8 + (-4) = 8 - 4 = 4$

5. $-19 - (-16) = -19 + 16 = -3$

9. $-7(-5) = 35$

13. $-2(4)(-5) = (-8)(-5) = 40$

17. $9 - 0 = 9$

21. $8 - 3(-4) = 8 - (-12) = 8 + 12 = 20$

25. $\dfrac{3(-6)(-2)}{0 - 4} = \dfrac{-18(-2)}{-4} = \dfrac{36}{-4} = -9$

29. $-7 - \dfrac{-14}{2} - 3(2) = -7 - (-7) - 6 = -7 + 7 - 6 = 0 - 6 = -6$

33. $6(7) = 7(6)$ demonstrates the commutative law of multiplication.

37. $3 + (5 + 9) = (3 + 5) + 9$ demonstrates the associative law of addition.

41. $-a + (-b) = -a - b$ which is expression (d).

45. (a) The product of an even number of negative numbers is positive.

    (b) The product of an odd number of negative numbers is negative.

49. 100 m + 200 m = 200 m + 100 m illustrates the commutative law of addition.

1.3   Calculators and Approximate Numbers

1. 8 cylinders is exact because they can be counted. 55 mi/h is approximate since it is measured.

5. 107 has 3 significant digits. 3004 has 4 significant digits.

9. 3000 has 1 significant digit. 3000.1 has 5 significant digits.

13. (a) Both numbers have the same precision with digits in the tenths place.

    (b) 78.0 with 3 significant digits is more accurate than 0.1 with 1 significant digit.

17. (a) 4.936 = 4.94 rounded to 3 significant digits.

    (b) 4.936 = 4.9 rounded to 2 significant digits.

21. (a) 9549 = 9550 rounded to 3 significant digits.

    (b) 9549 = 9500 rounded to 2 significant digits.

25. (a) Calculator: 3.8 + 0.154 + 47.26 = 51.2

    (b) Estimate: 4 + 0 + 47 = 51

29. (a) Calculator: $0.0350 - \dfrac{0.0450}{1.909} = 0.0114$ (b) Estimate: $0.04 - \dfrac{0.05}{2} = 0.015$

33. (a) Calculator: $\dfrac{23.962 \times 0.01537}{10.965 - 8.249} = 0.1356$ (b) Estimate: $\dfrac{20 \times 0.02}{11 - 8} = 0.1$

37. 0.9788 + 14.9 = 15.8788 since 4 is the number of decimal places in the least precise number.

41. 2.745 MHz and 2.755 MHz are the least possible and greatest possible frequencies respectively.

45. (a) $2.2 + 3.8 \times 4.5 = 19.3$ (b) $(2.2 + 3.8) \times 4.5 = 27$

49. (a) $\dfrac{8}{33} = 0.242424 \cdots = 0.\overline{24}$ (b) $\pi = 3.141592653589873 \cdots$

53. 1 K = 1024 bytes
    256 K = 256 × 1024 = 262,144 bytes

1.4   Exponents

1. $x^3 \cdot x^4 = x^{3+4} = x^7$

5. $\dfrac{m^5}{m^3} = m^{5-3} = m^2$

9. $(a^2)^4 = a^{2 \cdot 4} = a^8$

13. $(2n)^3 = 2^3 \cdot n^3 = 8n^3$

17. $\left(\dfrac{2}{b}\right)^3 = \dfrac{2^3}{b^3} = \dfrac{8}{b^3}$

21. $7^0 = 1$

25. $6^{-1} = \dfrac{1}{6^1} = \dfrac{1}{6}$

29. $(-t^2)^7 = -t^{2 \cdot 7} = -t^{14}$

33. $(4xa^{-2})^0 = 1$

37. $\dfrac{2a^4}{(2a)^4} = \dfrac{2a^4}{2^4 a^4} = \dfrac{1}{2^3} = \dfrac{1}{8}$

41. $(5^0 x^2 a^{-1})^{-1} = 1 \cdot x^{2 \cdot (-1)} \cdot a^{(-1)(-1)} = x^{-2} \cdot a^1 = \dfrac{a}{x^2}$

45. $(-8gs^3)^2 = (-8)^2 \cdot g^2 \cdot (s^3)^2 = 64g^2 s^{3 \cdot 2} = 64g^2 s^6$

49. $7(-4) - (-5)^2 = -28 - 25 = -53$

53. $\dfrac{3.07(-1.86)}{(-1.86)^4 + 1.596} = \dfrac{-5.71}{11.97 + 1.596} = \dfrac{-5.71}{13.57} = -0.421$

57. $\pi \left(\dfrac{r}{2}\right)^3 \left(\dfrac{4}{3\pi r^2}\right) = \pi \cdot \dfrac{r^3}{8} \cdot \dfrac{4}{3\pi r^2} = \dfrac{r}{6}$

1.5   Scientific Notation

1. $4.5 \times 10^4 = 45{,}000$

5. $3.23 \times 10^0 = 3.23$

9. $40{,}000 = 4 \times 10^4$

13. $6 = 6 \times 10^0$

17. $28{,}000(2{,}000{,}000{,}000) = 2.8 \times 10^4 (2 \times 10^9) = 5.6 \times 10^{13}$

21. $1280(865{,}000)(43.8) = 4.85 \times 10^{10}$

25. $(3.642 \times 10^{-8})(2.736 \times 10^5) = 9.965 \times 10^{-3}$

29. $6{,}500{,}000 \text{ kW} = 6.5 \times 10^6 \text{ kW}$

33. $10^{30}{}^\circ\text{C} = 1{,}000{,}000{,}000{,}000{,}000{,}000{,}000{,}000{,}000{,}000{}^\circ\text{C}$

37. $\dfrac{7.5 \times 10^{-15} \text{ seconds}}{\text{addition}} \cdot 5.6 \times 10^6 \text{ additions} = 4.2 \times 10^{-8} \text{ seconds}$

41. $730{,}000 \text{ km}^2 \cdot \dfrac{1000^2 \text{ m}^2}{\text{km}^2} \cdot \dfrac{100^2 \text{ cm}^2}{\text{m}^2} = 7.3 \times 10^{15} \text{ cm}^2$

1.6   Roots and Radicals

1. $\sqrt{25} = 5$

5. $-\sqrt{49} = -7$

9. $\sqrt[3]{125} = 5$

13. $\left(\sqrt{5}\right)^2 = 5$

17. $\left(-\sqrt{18}\right)^2 = 18$

21. $2\sqrt{84} = 2\sqrt{4 \cdot 21} = 2 \cdot 2\sqrt{21} = 4\sqrt{21}$

25. $\sqrt{36 + 64} = \sqrt{100} = 10$

29. $\sqrt{85.4} = 9.24$

33. (a) $\sqrt{1296 + 2304} = \sqrt{3600} = 60.00$

(b) $\sqrt{1296} + \sqrt{2304} = 36.00 + 48.00 = 84.00$

37. $v = \sqrt{64.4 \cdot 2.75} = \sqrt{177.1} = 13.3 \text{ ft/s}$

41. $d = \sqrt{w^2 + h^2} = \sqrt{15.2^2 + 11.4^2} = \sqrt{361} = 19.0 \text{ in}$

1.7   Addition and Subtraction of Algebraic Expressions

1. $5x + 7x - 4x = 12x - 4x = 8x$

5. $2a - 2c - 2 + 3c - a = 2a - a + 3c - 2c - 2 = a + c - 2$

9. $s + (4 + 3s) = s + 4 + 3s = 4s + 4$

13. $2 - 3 - (4 - 5a) = -1 - 4 + 5a = -5 + 5a = 5a - 5$

17. $-(t - 2u) + (3u - t) = -t + 2u + 3u - t = 5u - 2t$

21. $-7(6 - 3j) - 2(j + 4) = -42 + 21j - 2j - 8 = 19j - 50$

25. $2[4 - (t^2 - 5)] = 2[4 - t^2 + 5] = 2[9 - t^2] = 18 - 2t^2$

29. $\begin{aligned} a\sqrt{LC} - [3 - (a\sqrt{LC} + 4)] &= a\sqrt{LC} - [3 - a\sqrt{LC} - 4] \\ &= a\sqrt{LC} - [-1 - a\sqrt{LC}] \\ &= a\sqrt{LC} + 1 + a\sqrt{LC} \\ &= 2a\sqrt{LC} + 1 \end{aligned}$

33. $\begin{aligned} 5p - (q - 2p) - [3q - (p - q)] &= 5p - q + 2p - [3q - p + q] \\ &= 7p - q - [4q - p] \\ &= 7p - q - 4q + p \\ &= 8p - 5q \end{aligned}$

37. $\begin{aligned} 5V^2 - (6 - (2V^2 + 3)) &= 5V^2 - (6 - 2V^3 - 3) \\ &= 5V^2 - (3 - 2V^2) \\ &= 5V^2 - 3 + 2V^2 \\ &= 7V^2 - 3 \end{aligned}$

41. $3D - (D - d) = 3D - D + d = 2D + d$

1.8   Multiplication of Algebraic Expressions

1. $(a^2)(ax) = a^{2+1}x = a^3x$

5. $(2ax^2)^2(-2ax) = 4a^2x^4(-2ax) = -8a^3x^5$

9. $i^2(R+r) = i^2R + i^2r$

13. $5m(m^2n + 3mn) = 5m^3n + 15m^2n$

17. $ab^2c^4(ac - bc - ab) = a^2b^2c^5 - ab^3c^5 - a^2b^3c^4$

21. $(x-3)(x+5) = x^2 + 5x - 3x - 15 = x^2 + 2x - 15$

25. $(2a - b)(3a - 2b) = 6a^2 - 4ab - 3ab + 2b^2 = 6a^2 - 7ab + 2b^2$

29. $(x^2 - 1)(2x + 5) = 2x^3 + 5x^2 - 2x - 5$

33. $(x + 1)(x^2 - 3x + 2) = x^3 - 3x^2 + 2x + x^2 - 3x + 2 = x^3 - 2x^2 - x + 2$

37. $2(a + 1)(a - 9) = 2(a^2 - 8a - 9) = 2a^2 - 16a - 18$

41. $(2x - 5)^2 = 4x^2 - 20x + 25$

45. $(xyz - 2)^2 = x^2y^2z^2 - 4xyz + 4$

49. $(2 + x)(3 - x)(x - 1)$
$= (6 + x - x^2)(x - 1)$
$= 6x - 6 + x^2 - x - x^3 + x^2$
$= -x^3 + 2x^2 + 5x - 6$

53. $(n + 100)^2 = n^2 + 200n + 100^2$
$= n^2 + 200n + 10{,}000$

1.9   Division of Algebraic Expressions

1. $\dfrac{8x^3y^2}{-2xy} = -4x^2y$

5. $\dfrac{(15x^2)(4bx)(2y)}{30bxy} = 4x^2$

9. $\dfrac{a^2x + 4xy}{x} = \dfrac{a^2x}{x} + \dfrac{4xy}{x}$
$= a^2 + 4y$

13. $\dfrac{4pq^3 + 8p^2q^2 - 16pq^5}{4pq^2} = \dfrac{4pq^3}{4pq^2} + \dfrac{8p^2q^2}{4pq^2} - \dfrac{16pq^5}{4pq^2}$
$= q + 2p - 4q^3$

17. $\dfrac{3ab^2 - 6ab^3 + 9a^2b^2}{9a^2b^2} = \dfrac{3ab^2}{9a^2b^2} - \dfrac{6ab^3}{9a^2b^2} + \dfrac{9a^2b^2}{9a^2b^2} = \dfrac{1}{3a} - \dfrac{2b}{3a} + 1$

21.

$$
\begin{array}{r}
2x + 1 \\
x + 3\overline{)2x^2 + 7x + 3} \\
\underline{2x^2 + 6x} \\
x + 3 \\
\underline{x + 3} \\
0
\end{array}
$$

25.

$$
\begin{array}{r}
4x^2 - \ x - 1 \\
2x - 3\overline{)8x^3 - 14x^2 + \ x + 0} \\
\underline{8x^3 - 12x^2} \\
-2x^2 + \ x \\
\underline{-2x^2 + 3x} \\
-2x + 0 \\
\underline{-2x + 3} \\
-3
\end{array}
$$

29.
$$
\begin{array}{r}
x^2 + x - 6 \\
x + 2{\overline{\smash{\big)}\,x^3 + 3x^2 - 4x - 12}} \\
\underline{x^3 + 2x^2} \\
x^2 - 4x \\
\underline{x^2 + 2x} \\
-6x - 12 \\
\underline{-6x - 12} \\
0
\end{array}
$$

33.
$$
\begin{array}{r}
x^2 - 2x + 4 \\
x + 2{\overline{\smash{\big)}\,x^3 + 0x^2 + 0x + 8}} \\
\underline{x^3 + 2x^2} \\
-2x^2 + 0x \\
\underline{-2x^2 - 4x} \\
4x + 8 \\
\underline{4x + 8} \\
0
\end{array}
$$

37. $\dfrac{8A^5 + 4A^3\mu^2 E^2 - A\mu^4 E^4}{8A^4} = \dfrac{8A^5}{8A^4} + \dfrac{4A^3\mu^2 E^2}{8A^4} - \dfrac{A\mu^4 E^4}{8A^4} = A + \dfrac{\mu^2 E^2}{2A} - \dfrac{\mu^4 E^4}{8A^3}$

1.10   Solving Equations

1. $x - 2 = 7$
$x = 7 + 2$
$x = 9$

5. $\dfrac{t}{2} = 5$
$t = 2 \cdot 5$
$t = 10$

9. $3t + 5 = -4$
$3t = -4 - 5$
$3t = -9$
$t = -3$

13. $3x + 7 = x$
$3x - x = -7$
$2x = -7$
$x = \dfrac{-7}{2}$

17. $6 - (r - 4) = 2r$
$6 - r + 4 = 2r$
$-3r = -10$
$r = \dfrac{10}{3}$

21. $x - 5(x - 2) = 2$
$x - 5x + 10 = 2$
$-4x = -8$
$x = 2$

25. $5.8 - 0.3(x - 6.0) = 0.5x$
$5.8 - 0.3x + 1.8 = 0.5x$
$7.6 = 0.8x$
$x = 9.5$

29. $\dfrac{x}{2.0} = \dfrac{17}{6.0}$
$x = \dfrac{34}{6.0}$
$x = 5.7$

33. $15(5.5 + v) = 24(5.5 - v)$
$82.5 + 15v = 132 - 24v$
$39v = 49.5$
$v = 1.3 \text{ mi/h}$

37. $\dfrac{x}{150} = \dfrac{2}{5}$
$x = \dfrac{300}{5}$
$x = 60 \text{ mg}$

1.11   Formulas and Literal Equations

1. $ax = b$
$x = \dfrac{b}{a}$

5. $ax + 6 = 2ax - c$
$ax = 6 + c$
$x = \dfrac{6 + c}{a}$

9. $\theta = kA + \lambda$
$\lambda = \theta - kA$

13. $P = 2\pi T f$

$$T = \frac{P}{2\pi f}$$

17. $A = \dfrac{Rt}{PV}$

$$Rt = APV$$

$$t = \frac{APV}{R}$$

21. $C_0^2 = C_1^2(1 + 2V)$

$$C_0^2 = C_1^2 + 2C_1^2 V$$

$$V = \frac{C_0^2 - C_1^2}{2C_1^2}$$

25.
$$Q_1 = P(Q_2 - Q_1)$$
$$Q_1 = PQ_2 - PQ_1$$
$$Q_1(1 + P) = PQ_2$$
$$Q_2 = \frac{Q_1 + PQ_1}{P}$$

29.
$$L = \pi(r_1 + r_2) + 2x_1 + x_2$$
$$L = \pi r_1 + \pi r_2 + 2x_1 + x_2$$
$$\pi r_1 = L - \pi r_2 - 2x_1 - x_2$$
$$r_1 = \frac{L - \pi r_2 - 2x_1 - x_2}{\pi}$$

33. $p = p_0 + kh \Rightarrow h = \dfrac{p - p_0}{k} = \dfrac{205 - 101}{9.80} = 10.6$ meters

---

1.12 Applied Word Problems

1. $x + (x + 114) = 390 \Rightarrow 2x = 276 \Rightarrow x = 138$ and $x + 114 = 252$ where $x =$ cost of first program

The programs cost $138 and $252.

5. Let $x =$ number of acres @ $20,000

$$20{,}000 \cdot x + 10{,}000 \cdot (70 - x) = 900{,}000$$
$$2x + 70 - x = 90$$
$$x = 20 \text{ acres @ } \$20{,}000/\text{acre}$$
$$70 - x = 50 \text{ acres @ } \$10{,}000/\text{acre}$$

9. $x + 2x + (x + 9.2) = 0$
$4x = -9.2 \Rightarrow x = -2.3$ μA for the first current
$2x = -4.6$ μA for the second current
$x + 9.2 = 6.9$ μA for the third current

13. Let $x =$ amount invested at 9.00%
$0.0700 \cdot 8000 + 0.0900x = 1550 \Rightarrow 0.0900x = 990 \Rightarrow x = 11{,}000$

$11,000 must be invested at 9.00% to have a total interest of $1550.

17. Let $v =$ speed of French train

$$v \cdot \frac{17}{60} + (v - 8) \cdot \frac{17}{60} = 50 \Rightarrow 17v + 17v - 136 = 3000 \Rightarrow 34v = 3136 \Rightarrow v = 92.2$$

$v - 8 = 84.2$.

The speed of the France train is 92.2 km/h. The speed of the English train is 84.2 km/h.

21. Assume the customer is located between A and B at a distance x from A.

$$0.0005 \cdot x + 0.85 = 0.0005(228 - x) + 0.80$$
$$0.0005x + 0.85 = 0.114 - 0.0005x + 0.80$$
$$0.001x = 0.064$$
$$x = 64 \text{ mi}$$

The customer is 64 mi from A and 164 mi from B.

Chapter 1 Review Exercises

1. $(-2) + (-5) - 3 = -7 - 3$
$$= -10$$

5. $-5 - |2(-6)| + \dfrac{-15}{3} = -5 - |-12| + (-5)$
$$= -5 - 12 - 5$$
$$= -17 - 5$$
$$= -22$$

9. $\sqrt{16} - \sqrt{64} = 4 - 8 = -4$

13. $(-2rt^2)^2 = 4r^2(t^2)^2 = 4r^2t^4$

17. $\dfrac{-16s^{-2}(st^2)}{-2st^{-1}} = 8s^{-2}t^{2-(-1)}$
$$= \dfrac{8t^3}{s^2}$$

21. (a) 8840 has 3 significant digits
(b) 8840 rounded to 2 significant digits is 8800

25. $37.3 - 16.92(1.067)^2 = 18.03676612$ on a calculator; 18.0

29. $a - 3ab - 2a + ab = a - 2a - 3ab + ab = -a - 2ab$

33. $(2x - 1)(x + 5) = 2x^2 + 10x - x - 5$
$$= 2x^2 + 9x - 5$$

37. $\dfrac{2h^3k^2 - 6h^4k^5}{2h^2k} = \dfrac{2h^3k^2}{2h^2k} - \dfrac{6h^4k^5}{2h^2k}$
$$= hk - 3h^2k^4$$

41. $2xy - \{3z - [5xy - (7z - 6xy)]\} = 2xy - \{3z - [5xy - 7z + 6xy[\}$
$$= 2xy - \{3z - [11xy - 7z]\}$$
$$= 2xy - \{3z - 11xy + 7z\}$$
$$= 2xy - \{10z - 11xy\}$$
$$= 2xy - 10z + 11xy$$
$$= 13xy - 10z$$

45. $-3y(x - 4y)^2 = -3y(x^2 - 8xy + 16y^2)$
$$= -3x^2y + 24xy^2 - 48y^3$$

49. $\dfrac{12p^3q^2 - 4p^4q + 6pq^5}{2p^4q} = \dfrac{12p^3q^2}{2p^4q} - \dfrac{4p^4q}{2p^4q} + \dfrac{6pq^5}{2p^4q}$
$$= \dfrac{6q}{p} - 2 + \dfrac{3q^4}{p^3}$$

53.
$$
\begin{array}{r}
x^2 - 2x + 3 \\
3x - 1\overline{)3x^3 - 7x^2 + 11x - 3} \\
\underline{3x^3 - x^2} \\
-6x^2 + 11x \\
\underline{-6x^2 + 2x} \\
9x - 3 \\
\underline{9x - 3}
\end{array}
$$

57. $-3\{(r + s - t) - 2[(3r - 2s) - (t - 2s)]\}$
$$= -3\{(r + s - t) - 2[3r - 2s - t + 2s]\}$$
$$= -3\{r + s - t - 2[3r - t]\}$$
$$= -3\{r + s - t - 6r + 2t\}$$
$$= -3\{-5r + s + t\}$$
$$= 15r - 3s - 3t$$

61. $3x + 1 = x - 8$
$3x - x = -8 - 1$
$2x = -9$
$x = \dfrac{-9}{2}$

65. $6x - 5 = 3(x - 4)$
$6x - 5 = 3x - 12$
$3x = -7$
$x = \dfrac{-7}{3}$

69. $3t - 2(7 - t) = 5(2t + 1)$
$3t - 14 + 2t = 10t + 5$
$5t - 14 = 10t + 5$
$-5t = 19$
$t = \dfrac{-19}{5}$

73. $25,000 \text{ mi/h} = 2.5 \times 10^4 \text{ mi/h}$

77. $0.0000012 \text{ cm}^2 = 1.2 \times 10^{-6} \text{ cm}^2$

81. $3s + 2 = 5a$
$3s = 5a - 2$
$s = \dfrac{5a - 2}{3}$

85. $R = n^2 Z$
$Z = \dfrac{R}{n^2}$

89. $m = dV(1 - e)$
$\dfrac{m}{dV} = 1 - e$
$e = 1 - \dfrac{m}{dV}$

93. $R = \dfrac{A(T_2 - T_1)}{H}$
$RH = AT_2 - AT_1$
$AT_2 = RH + AT_1$
$T_2 = \dfrac{RH + AT_1}{A}$

97. $0.553 \text{ km} \cdot \dfrac{1000 \text{ m}}{\text{km}} - 443 \text{ m} = 553 \text{ m} - 443 \text{ m} = 110 \text{ m higher.}$

101. $2V(r - a) - V(b - a) = 2Vr - 2Va - Vb + Va$
$= 2Vr - Va - Vb$

105. $x + 1.5x = 10.5 \Rightarrow 2.5x = 10.5 \Rightarrow x = 4.2 \text{ and } 1.5x = 6.3$

The disks have 4.2 gigabytes and 6.3 gigabytes of memory.

109. $\dfrac{5.0 \text{ ft}}{1.7 \text{ lb}} = \dfrac{27 \text{ ft}}{x} \Rightarrow 5.0x = 46 \Rightarrow x = 9.2 \text{ lb}$

113. Let $x =$ the number of liters of 0.50% mixture.
$100 - x =$ the number of liters of 0.75% mixture.

$$0.50\% \cdot x + 0.75\%(1000 - x) = 0.65\% \cdot 1000$$
$$50x + 75(1000 - x) = 65,000$$
$$50x + 75,000 - 75x = 65,000$$
$$25x = 10,000$$
$$x = 400 \text{ liters of the 0.50\% mixture}$$
$$1000 - x = 600 \text{ liters of the 0.75\% mixture}$$

117. Answers will vary.

Chapter 2

GEOMETRY

2.1 Lines and Angles

1. $\angle EBD$ and $\angle DBC$ are acute angles.

5. The complement of $\angle CBD = 65°$ is $25°$.

9. $\angle AOB = 90° + 50° = 140°$

13. $\angle 1$ and $62°$ are vertical angles $\Rightarrow \angle 1 = 62°$

17. $\angle ABF = 136°$ and $\angle FCE$ are supplementary $\Rightarrow \angle FCE = 44°$

21. $\dfrac{a}{4.75} = \dfrac{3.05}{3.20} \Rightarrow a = 4.75 \cdot \dfrac{3.05}{3.20} = 4.53$ in.

25. $\angle BCD = 180° - 47° = 133°$

2.2 Triangles

1. $\angle A = 180° - 84° - 40° = 56°$

5. $p = 3.5 + 2.3 + 4.1 = 9.9$ ft

9. $A = \dfrac{1}{2}bh = \dfrac{1}{2}(7.6)(2.2) = 8.4$ ft^2

13. $A = \dfrac{1}{2}bh = \dfrac{1}{2}(3.46)(2.55) = 4.41$ ft^2

17. $c = \sqrt{13.8^2 + 22.7^2} = 26.6$ ft

21. $\angle B = 90° - 23° = 67°$

25. $\angle LMK$ and $\angle OMN$ are vertical angles and thus equal $\Rightarrow \angle KLM = \angle MON$. The corresponding angles are equal and the triangles are similar.

29. Each base angle is $\dfrac{180° - 38°}{2} = 71°$

33. $A = \dfrac{1}{2}bh = \dfrac{1}{2}(8.0)(15) = 60$ ft^2

37. $d = \sqrt{18^2 + 12^2 + (8.0)^2} = 23$ ft

2.3 Quadrilaterals

1. $p = 4s = 4 \cdot 0.65 = 2.6$ m

5. $p = 2\ell + 2w = 2(3.7) + 2(2.7) = 12.8 = 13$ m to two significant digits

9. $A = s^2 = 2.7^2 = 7.3$ mm^2

13. $A = bh = 3.7(2.5) = 9.3$ m^2

17. $p = 2b + 4a$

21. $p = 6(2) = 12$ cm

25. For the courtyard: $s = \dfrac{p}{4} = \dfrac{320}{4} = 80$. For the outer edge of the walkway:

$p = 4s = 4(80 + 6) = 344$ m.

2.4   Circles

1. (a) AD is a secant line.   (b) AEF is a tangent line.

5. $c = 2\pi r = 2\pi \cdot 2.75 = 17.3$ ft

9. $A = \pi r^2 = \pi \cdot 0.0952^2 = 0.0285$ yd^2

13. $\angle CBT = 90° - \angle ABC = 90° - 65° = 25°$

17. ARC BC $= 2(60°) = 120°$

21. $22.5° \cdot \dfrac{\pi}{180°} = 0.393$ rad

25. $P = \dfrac{1}{4} \cdot 2\pi r + 2r = \dfrac{1}{2} \cdot \pi r + 2r$

29. $C = 2\pi r = 2\pi \cdot 3960 = 24{,}900$ mi

33. $A = 2 \cdot \dfrac{\pi D^2}{4} + s^2 = \dfrac{\pi \cdot 4.50^2}{2} + 4.50^2 = 52.1$ m^2

2.5   Measurement of Irregular Areas

1. $A_{\text{trap}} = \dfrac{2.0}{2} [0.0 + 2(6.4) + 2(7.4) + 2(7.0) + 2(6.1) + 2(5.2) + 2(5.0) + 2(5.1) + 0.0]$

$A_{\text{trap}} = 84.4 = 84$ m^2 to two significant digits

5. $A_{\text{trap}} = \dfrac{0.5}{2} [0.6 + 2(2.2) + 2(4.7) + 2(3.1) + 2(3.6) + 2(1.6) + 2(2.2) + 2(1.5) + 0.8]$

$A_{\text{trap}} = 9.8$ mi^2

9. $A_{\text{trap}} = \dfrac{45}{2} [170 + 2(360) + 2(420) + 2(410) + 2(390) + 2(350) + 2(330) + 2(290) + 230]$

$A_{\text{trap}} = 120{,}000$ ft^2

13. $A_{\text{trap}} = \dfrac{0.500}{2} [0.0 + 2(1.732) + 2(2.000) + 2(1.732) + 0.0] = 2.73$ in^2

This value is less than 3.14 in^2 because all of the trapezoids are inscribed.

2.6   Solid Geometric Figures

1. $V = e^3 = 7.15^3 = 366$ ft^3

5. $V = \dfrac{4}{3}\pi r^3 = \dfrac{4}{3}\pi \cdot 0.877^3 = 2.83$ yd^3

9. $V = \dfrac{1}{3}Bh = \dfrac{1}{3} \cdot 16^2 \cdot 13 = 1100$ in^3

13. $V = \dfrac{1}{2} \cdot \dfrac{4}{3}\pi r^3 = \dfrac{2}{3}\pi \cdot \left(\dfrac{0.83}{2}\right)^3 = 0.15$ yd^3

17. $A = 2\ell h + 2\ell w + 2wh = 2 \cdot 12.0 \cdot 8.75 + 2 \cdot 12.0 \cdot 9.50 + 2 \cdot 9.50 \cdot 8.75$
$A = 604$ in^2

21. $V = \frac{1}{3}BH = \frac{1}{3} \cdot 250^2 \cdot 160 = 3,300,000 \text{ yd}^3$

25. $A = l^2 + \frac{1}{2}ps = 16.0^2 + \frac{1}{2} \cdot (4 \cdot 16) \cdot \sqrt{8.0^2 + 40.0^2} = 1560 \text{ mm}^2$

Chapter 2 Review Exercises

1. $\angle CGE = 180° - 148° = 32°$

5. $c = \sqrt{9^2 + 40^2} = 41$

9. $c = \sqrt{6.30^2 + 3.80^2} = 7.36$

13. $P = 3s = 3 \cdot 8.5 = 25.5 \text{ mm}$

17. $C = \pi d = \pi \cdot 98.4 = 309 \text{ mm}$

21. $V = Bh = \frac{1}{2} \cdot 26.0 \cdot 34.0 \cdot 14.0 = 6190 \text{ cm}^3$

25. $A = 6e^2 = 6 \cdot 5.20^2 = 162 \text{ m}^2$

29. $\angle BTA = \frac{50°}{2} = 25°$

33. $\angle ABE = 90° - 37° = 53°$

37. $P = b + \sqrt{b^2 + (2a)^2} + \frac{1}{2}\pi \cdot 2a = b + \sqrt{b^2 + 4a^2} + \pi a$

41. A square is a rectangle with four equal sides and a rectangle is a parallelogram with perpendicular intersecting sides so a square is a parallelogram. A rhombus is a parallelogram with four equal sides and since a square is a parallelogram, a square is a rhombus.

45. $L = \sqrt{1.2^2 + 7.8^2} = 7.9 \text{ m}$

49. $s = \frac{18.0 + 15.5 + 7.50}{2} = 20.5, A = \ell w + \sqrt{s(s-a)(s-b)(s-c)}$

$A = 18.0 \cdot 8.75 + \sqrt{20.5(20.5 - 18.0)(20.5 - 15.5)(20.5 - 7.50)} = 215 \text{ ft}^2$

53. $C = \pi D = \pi \cdot (7920 + 2 \cdot 210) = 26,200 \text{ mi}$

57. $A = \frac{250}{3}[220 + 4 \cdot 530 + 2 \cdot 480 + 4 \cdot (320 + 190 + 260) + 2 \cdot 510 + 4 \cdot 350 + 2 \cdot 730 + 4 \cdot 560 + 240]$

$A = 1,000,000 \text{ m}^2$

61. $d = \sqrt{2.4^2 + 3.7^2} = 4.4 \text{ mi}$

65. Label the vertices of the pentagon ABCDE. The area is the sum of the areas of three triangles, one with sides 921, 1490, and 1490 and two with sides 921, 921, and 1490. The semi-perimeters are given by

$$s_1 = \frac{921 + 921 + 1490}{2} = 1666 \text{ and } s_2 = \frac{921 + 1490 + 1490}{2} = 1950.5.$$

$A = 2\sqrt{1666(1666 - 921)(1666 - 921)(1666 - 1490)} + \sqrt{1950.5(1950.5 - 1490)(1950.5 - 1490)(1950.5 - 921)}$
$= 1,460,000 \text{ ft}^2$

FUNCTIONS AND GRAPHS

3.1 Introduction to Functions

1. (a) $A(r) = \pi r^2$ (b) $A(D) = \pi \cdot \left(\dfrac{D}{2}\right)^2 = \pi \cdot \dfrac{D^2}{4}$ **5.** $A(l) = lw = 5l$

9. $f(x) = 2x + 1; f(1) = 2 \cdot 1 + 1 = 3;$
$f(-1) = 2(-1) + 1 = -1$

13. $\phi(x) = \dfrac{6 - x^2}{2x}; \phi(1) = \dfrac{6 - 1^2}{2 \cdot 1} = \dfrac{5}{2};$
$\phi(-2) = \dfrac{6 - (-2)^2}{2 \cdot (-2)} = \dfrac{2}{-4} = -\dfrac{1}{2}$

17. $K(s) = 3s^2 - s + 6; K(-s) = 3(-s)^2 - (-s) + 6 = 3s^2 + s + 6$
$K(2s) = 3(2s)^2 - 2s + 6 = 12s^2 - 2s + 6$

21. $f(x) = 5x^2 - 3x; f(3.86) = 5 \cdot 3.86^2 - 3 \cdot 3.86 = 62.9$
$f(-6.92) = 5 \cdot (-6.92)^2 - 3 \cdot (-6.92) = 260$

25. $f(x) = x^2 + 2$: square x and add 2 to the result.

29. Take 3 times the sum of twice the independent variable and 5, then subtract 1.

33. $y = x^2; f(x) = x^2$

37. $s = f(t) = 17.5 - 4.9t^2; f(1.2) = 17.5 - 4.9 \cdot 1.2^2 = 10.4$ m

3.2 More About Functions

1. The domain and range of $f(x) = x + 5$ is all real numbers.

5. The domain of $f(s) = \dfrac{2}{s^2}$ is all real numbers except zero since it gives a division by zero. The range is all positive real numbers because $\dfrac{2}{s^2}$ is always positive.

9. The domain of $Y(y) = \dfrac{y+1}{\sqrt{y-2}}$ is $y > 2$ because the square root requires $y - 2 \geq 0$ or $y \geq 2$ and to avoid a division by zero, $y > 2$ is required.

13. $F(t) = 3t - t^2$ for $t \leq 2; F(2) = 3 \cdot 2 - 2^2 = 2; F(3)$ does not exist.

17. $d(t) = 40 \cdot 2 + 55t = 80 + 55t$

21. $m(h) = \begin{cases} 110 & \text{for } h \leq 1000 \\ 110 + 0.5(h - 1000) & \text{for } h > 1000 \end{cases}$

25. (a) $0.1x + 0.4y = 1200 \Rightarrow y(x) = \dfrac{1200 - 0.1x}{0.4}$ (b) $y(400) = \dfrac{1200 - 0.1 \cdot 400}{0.4} = 2900$ L

29. $A = f(x) = \pi(6 - x)^2$ with domain $0 \leq x \leq 6$ since x represents the radius and it must be greater than or equal to zero and less than or equal to six. Using the end point values for the radius gives a range $0 \leq A \leq 36\pi$.

33. The domain of $f = \dfrac{1}{2\pi\sqrt{C}}$ is $C > 0$ because C must be ≥ 0 to avoid taking the square root of a negative and > 0 to prevent division by zero.

3.3   Rectangular Coordinates

1. $A(2,1)$; $B(-1,2)$; $C(-2,-3)$

5. Joining the points in the order ABCA form an isosceles triangle.

9. The coordinates of the fourth vertex, V, are $(5, 4)$.

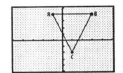

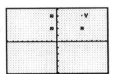

13. All points with abscissas of 1 are on a vertical line through $(1, 0)$. The equation of this vertical line is $x = 1$.

17. All points whose abscissas equal their ordinates are on a $45°$ line through the origin. The equation of this line is $y = x$.

21. All points for which $x > 0$ are in QI and QIV to the right of the y-axis.

25. The ratio $\dfrac{y}{x}$ is positive in QI and QIII.

3.4   The Graph of a Function

1. $y = 3x$

x	y
-1	-3
0	0
1	3

5. $y = 7 - 2x$

x	y
-1	9
0	7
1	5

9. $y = x^2$

x	y
-2	4
-1	1
0	0
1	1
2	4

13. $y = \dfrac{1}{2}x^2 + 2$

x	y
-4	10
-2	4
0	2
2	4
4	10

17. $y = x^2 - 3x + 1$

x	y
3	1
2	-1
1.5	-1.25
1	-1
0	1

21. $y = x^3 - x^2$

x	y
-2	-12
-1	-2
0	0
$\frac{2}{3}$	$-\frac{4}{27}$
1	0
2	4

25. $P = \dfrac{1}{V}$

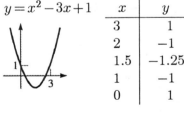

V	-3	-2	-1	1	2	3
P	$-\frac{1}{3}$	$-\frac{1}{2}$	-1	1	$\frac{1}{2}$	$\frac{1}{3}$

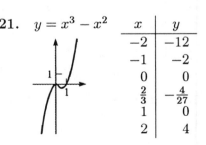

29. $y = \sqrt{x}$

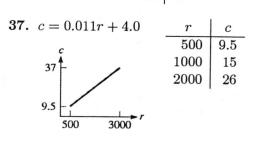

x	y
0	0
1	1
4	2
9	3
16	4

33. $n = 0.40m$

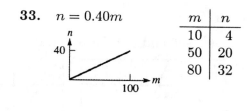

m	n
10	4
50	20
80	32

37. $c = 0.011r + 4.0$

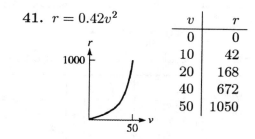

r	c
500	9.5
1000	15
2000	26

41. $r = 0.42v^2$

v	r
0	0
10	42
20	168
40	672
50	1050

45. $P = 2l + 2w = 200 \Rightarrow l = 100 - w$
$A = lw = (100 - w)w = 100w - w^2$
for $30 \le w \le 70$

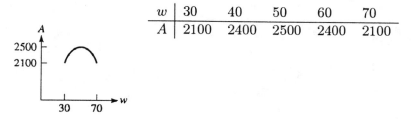

w	30	40	50	60	70
A	2100	2400	2500	2400	2100

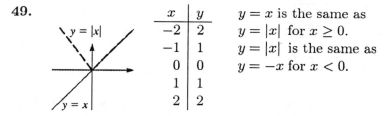

49.

x	y
-2	2
-1	1
0	0
1	1
2	2

$y = x$ is the same as
$y = |x|$ for $x \ge 0$.
$y = |x|$ is the same as
$y = -x$ for $x < 0$.

For negative values of x, $y = |x|$ becomes $y = -x$.

53. The graph passes the vertical line test and is, therefore, a function.

3.5   Graphs on the Graphing Calculator

1. $7x - 5 = 0$

Graph $y = 7x - 5$ and use the zero feature to solve.

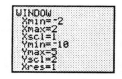 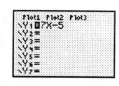 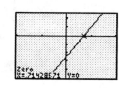

$x = 0.7$

5. $x^2 + x - 5 = 0$

Graph $y = x^2 + x - 5$ and use the zero feature to solve.

 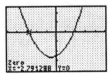 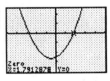 $x = -2.8, \ 1.8$

9. $x^3 = 4x \Rightarrow x^3 - 4x = 0$

Graph $y = x^3 - 4x$ and use the zero feature to solve.

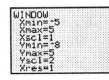 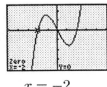 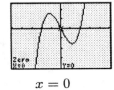 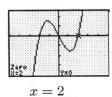

$\qquad\qquad\qquad x = -2 \qquad\qquad x = 0 \qquad\qquad x = 2$

13. $\sqrt{5R + 2} = 3 \Rightarrow \sqrt{5R + 2} - 3 = 0$

Graph $y = \sqrt{5x + 2} - 3$ and use zero feature to solve.

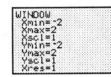 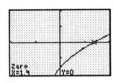 $R = 1.4$

17. From the graph, $y = \dfrac{4}{x^2 - 4}$ has range $y \leq -1$ or $y > 0$.

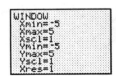 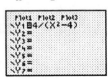 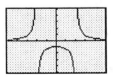

21. Graph $y = \dfrac{x + 1}{\sqrt{x - 2}}$ on graphing calculator and use the minimum feature, then from the graph,

$Y(y) = \dfrac{y + 1}{\sqrt{y - 2}}$ has range $Y \geq 3.464$.

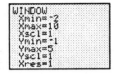 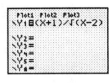 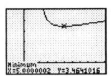

25. From the graph, $v = 6.0$ for $i = 0$.

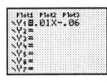 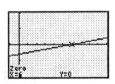

29. $A = 520 = lw = (w + 12)w = w^2 + 12w \Rightarrow w^2 + 12w - 520 = 0$

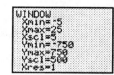 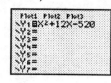 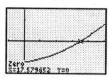

The approximate dimensions, in cm, to 2 significant digits are $w \approx 18$, $l \approx 30$.

33. $x^2 - 2x = 1 \Rightarrow x^2 - 2x - 1 = 0$

 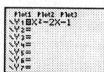 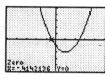 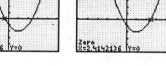

$\qquad\qquad\qquad\qquad\qquad x = -0.4142 \qquad\qquad x = 2.4142$

3.6   Graphs of Functions Defined by Tables of Data

1.

5.

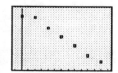

9. (a) Reading from the graph, $T = 132°C$ for $t = 4.3$ min.

(b) Reading from the graph, $t = 0.7$ min for $T = 145.0°C$.

13.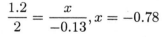

$\dfrac{1.2}{2} = \dfrac{x}{-0.13}, x = -0.78$

Therefore $M = 0.38 - 0.078 = 0.30$ H

17. (a) From the graph, $H \approx 3.4$ when $R = 20$

(b) From the graph, $R \approx 17$ when $H = 2.5$

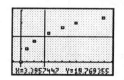 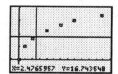

21. $10 \begin{bmatrix} 6 \begin{bmatrix} 30 & 0.30 \\ 46 & ? \\ 40 & 0.37 \end{bmatrix} x \end{bmatrix} 0.07$

$\dfrac{6}{10} = \dfrac{x}{0.07}, x = 0.042$

Therefore, $f = 0.30 + 0.042 = 0.34$

25. From the graph, $T \approx 130.3°C$ for $t = 5.3$ min

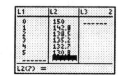 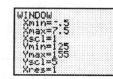 

Chapter 3 Review Exercises

1. $A = \pi r^2 = \pi(2t)^2$
   $A = 4\pi t^2$

5. $f(x) = 7x - 5$
   $f(3) = 7 \cdot 3 - 5 = 21 - 5 = 16$
   $f(-6) = 7(-6) - 5 = -42 - 5 = -47$

9. $f(x) = 3x^2 - 2x + 4$

$$\begin{aligned}
f(x+h) - f(x) &= 3(x+h)^2 - 2(x+h) + 4 - (3x^2 - 2x + 4)\\
&= 3(x^2 + 2xh + h^2) - 2x - 2h + 4 - 3x^2 + 2x - 4\\
&= 3x^2 + 6xh + 3h^2 - 2x - 2h + 4 - 3x^2 + 2x - 4\\
&= 6xh + 3h^2 - 2h
\end{aligned}$$

13. $f(x) = 8.07 - 2x$
   $f(5.87) = 8.07 - 2 \cdot 5.87 = -3.67$
   $f(-4.29) = 8.07 - 2(-4.29) = 16.65 \approx 16.7$

17. The domain of $f(x) = x^4 + 1$ is $-\infty < x < \infty$. The range is $y \geq 1$.

21. The graph of $y = 4x + 2$ is

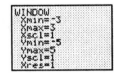 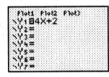 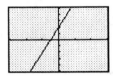

25. The graph of $y = 3 - x - 2x^2$ is

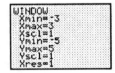  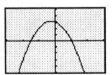

29. The graph of $y = 2 - x^4$ is

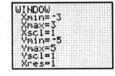 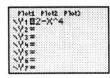 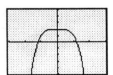

33. $7x - 3 = 0$

Graph $y = 7x - 3$ and use the zero feature to solve.

 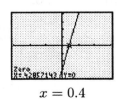

$$x = 0.4$$

37. $x^3 - x^2 = 2 - x \Rightarrow x^3 - x^2 + x - 2 = 0$

Graph $y = x^3 - x^2 + x - 2$ use the zero feature to solve.

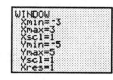 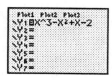 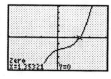 $x = 1.4$

41. Graph $y = x^4 - 5x^2$ and use the minimum feature, then from the graph, the range is $y \geq -6.25$.

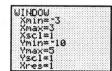 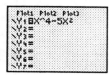 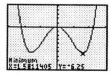

45. $A(a, b) = A(2, -3)$ is in QIV while $B(b, a) = B(-3, 2)$ is in QII.

49. $I = f(m) = 12.5\sqrt{1 + 0.5m^2}$
$F(0.55) = 12.5\sqrt{1 + 0.5(0.55)^2} = 13.41203983\cdots \approx 13.4$

53. $T = f(t) = 28.0 + 0.15t,\ 0 \leq t \leq 30$

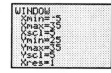 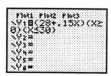 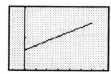

57. $P = f(i) = 1.5 \times 10^{-6}i^3 - 0.77,\ 80 \leq i \leq 140$

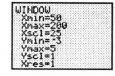 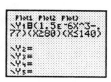 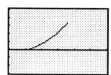

61. $d = f(D)$

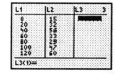 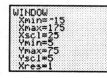 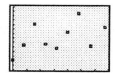

65. $d = f(t) = 250 - 60y,\ 0 \leq t \leq 2;\ d = f(t) = 130 - 40(t - 2),\ 2 < t < 5.25$

 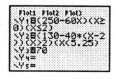 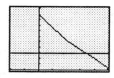 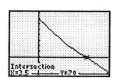

The person is 70 mi from home after 3.5 hours.

69. $v = 7.6x - 2.1x^2$, $0 \leq x \leq 1.75$

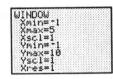 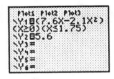 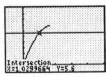

$x = 1.0299664 \approx 1.03$ ft for $v = 5.6$ ft/s

73. $V = \pi r^2 h = 250.0 \Rightarrow h = \dfrac{250.0}{\pi r^2}$

$A = A_{\text{base}} + A_{\text{side}} = \pi r^2 + 2\pi rh = \pi r^2 + 2\pi r \cdot \dfrac{250.0}{\pi r^2} = \pi r^2 + \dfrac{500.0}{r}$

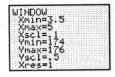 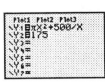 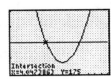 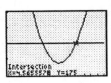

THE TRIGONOMETRIC FUNCTIONS

4.1 Angles

1. (a) 60°

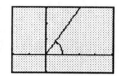

(b) 120°

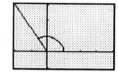

(c) −90°

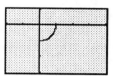

5. positive: $45° + 360° = 405°$
   negative: $45° − 360° = −315°$

9. positive: $70°30' + 360 = 430°30'$
   negative: $70°30' − 360 = −289°30'$

13. To change 0.265 rad to degrees multiply by $\dfrac{180}{\pi}$, $0.265 \text{ rad} \cdot \dfrac{180°}{\pi \text{ rad}} \approx 15.18°$

17. $0.329 \text{ rad} \approx 18.85°$

 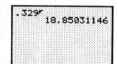

21. $56.0° = 0.977 \text{ rad}$ to three significant digits

 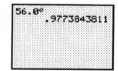

25. $47° + 0.5° \cdot \dfrac{60'}{1°} = 47° + 30' = 47°30'$

29. $15°12' = 15° + 12' \cdot \dfrac{1°}{60'} = 15.2°$

33. Angle in standard position terminal side passing through $(4, 2)$.

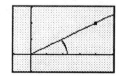

37. Angle in standard position terminal side passing through $(-7, 5)$.

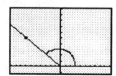

41. $21°42'36'' = 21° + 42' \cdot \dfrac{1°}{60'} + 36'' \cdot \dfrac{1°}{60''} \cdot \dfrac{1°}{60'} = 21.710°$

4.2   Defining the Trigonometric Functions

1. $r = \sqrt{6^2 + 8^2} = \sqrt{36 + 64} = \sqrt{100} = 10$

$\sin\theta = \dfrac{y}{r} = \dfrac{8}{10} = \dfrac{4}{5}$

$\cos\theta = \dfrac{x}{r} = \dfrac{6}{10} = \dfrac{3}{5}$

$\tan\theta = \dfrac{y}{x} = \dfrac{8}{6} = \dfrac{4}{3}$

$\csc\theta = \dfrac{r}{y} = \dfrac{10}{8} = \dfrac{5}{4}$

$\sec\theta = \dfrac{r}{x} = \dfrac{10}{6} = \dfrac{5}{3}$

$\cot\theta = \dfrac{x}{y} = \dfrac{6}{8} = \dfrac{3}{4}$

5. $r = \sqrt{9^2 + 40^2} = 41$

$\sin\theta = \dfrac{y}{r} = \dfrac{40}{41}$

$\cos\theta = \dfrac{x}{r} = \dfrac{9}{41}$

$\tan\theta = \dfrac{y}{x} = \dfrac{40}{9}$

$\csc\theta = \dfrac{r}{y} = \dfrac{41}{40}$

$\sec\theta = \dfrac{r}{x} = \dfrac{41}{9}$

$\cot\theta = \dfrac{x}{y} = \dfrac{9}{40}$

9. $r = \sqrt{1^2 + 1^2} = \sqrt{2}$

$\sin\theta = \dfrac{y}{r} = \dfrac{1}{\sqrt{2}} = \dfrac{\sqrt{2}}{2}$

$\cos\theta = \dfrac{x}{r} = \dfrac{1}{\sqrt{2}} = \dfrac{\sqrt{2}}{2}$

$\tan\theta = \dfrac{y}{x} = \dfrac{1}{1} = 1$

$\csc\theta = \dfrac{r}{y} = \dfrac{\sqrt{2}}{1} = \sqrt{2}$

$\sec\theta = \dfrac{r}{x} = \dfrac{\sqrt{2}}{1} = \sqrt{2}$

$\cot\theta = \dfrac{x}{y} = \dfrac{1}{1} = 1$

13. $r = \sqrt{3.25^2 + 5.15^2} = \sqrt{37.085} \approx 6.09$

to three significant ditits

$\sin\theta = \dfrac{y}{r} = \dfrac{5.15}{6.09} = 0.846$

$\cos\theta = \dfrac{x}{r} = \dfrac{3.25}{6.09} = 0.534$

$\tan\theta = \dfrac{y}{x} = \dfrac{5.15}{3.25} = 1.58$

$\csc\theta = \dfrac{r}{y} = \dfrac{6.09}{5.15} = 1.18$

$\sec\theta = \dfrac{r}{x} = \dfrac{6.09}{3.25} = 1.87$

$\cot\theta = \dfrac{x}{y} = \dfrac{3.25}{5.15} = 0.631$

17. $\cos\theta = \dfrac{12}{13} \Rightarrow x = 12$ and $r = 13$ with θ in QI.

$r^2 = x^2 + y^2 \Rightarrow 169 = 144 + y^2 \Rightarrow y^2 = 25$
$$y = 5$$

$\sin\theta = \dfrac{y}{r} = \dfrac{5}{13}$, $\cot\theta = \dfrac{x}{y} = \dfrac{12}{5}$.

21. $\sin\theta = 0.750 \Rightarrow y = 0.750$ and $r = 1$ with θ in QI.
$r^2 = x^2 + y^2 \Rightarrow 1^2 = x^2 + 0.750^2 \Rightarrow x^2 = 0.4375 \Rightarrow x = 0.661$

$\cot\theta = \dfrac{x}{y} = \dfrac{0.661}{0.750} = 0.881$,

$\csc\theta = \dfrac{r}{y} = \dfrac{1}{0.750} = 1.333$.

25. For $(3, 4)$, $r = 5$, $\sin\theta = \dfrac{y}{r} = \dfrac{4}{5}$ and $\tan\theta = \dfrac{y}{x} = \dfrac{4}{3}$.

For $(6, 8)$, $r = 10$, $\sin\theta = \dfrac{y}{r} = \dfrac{8}{10} = \dfrac{4}{5}$ and $\tan\theta = \dfrac{y}{x} = \dfrac{8}{6} = \dfrac{4}{3}$.

For $(4.5, 6)$, $r = 7.5$, $\sin\theta = \dfrac{y}{r} = \dfrac{6}{7.5} = \dfrac{4}{5}$ and $\tan\theta = \dfrac{y}{x} = \dfrac{6}{4.5} = \dfrac{4}{3}$.

29. $\sin^2\theta + \cos^2\theta = \left(\dfrac{3}{5}\right)^2 + \left(\dfrac{4}{5}\right)^2 = \dfrac{9}{25} + \dfrac{16}{25} = \dfrac{25}{25} = 1$

4.3   Values of the Trigonometric Functions

1. Answers may vary. One set of measurements gives $x = 7.6$ and $y = 6.5$.

$\sin 40° = \dfrac{6.5}{10} = 0.65$ $\qquad\qquad$ $\csc 40° = \dfrac{10}{6.5} = 1.54$

$\cos 40° = \dfrac{7.6}{10} = 0.76$ $\qquad\qquad$ $\sec 40° = \dfrac{10}{7.6} = 1.32$

$\tan 40° = \dfrac{6.5}{7.6} = 0.86$ $\qquad\qquad$ $\cot 40° = \dfrac{7.6}{6.5} = 1.17$

5. $\sin 22.4° = 0.381$

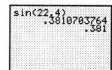

9. $\cos 15.71° = 0.9626$

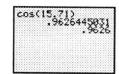

13. $\cot 67.78° = 0.4085$

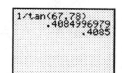

17. $\csc 49.3° = 1.32$

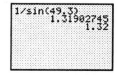

21. $\cos 70.97° = 0.3261$

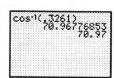

25. $\tan 11.7° = 0.207$

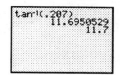

29. $\csc 53.44° = 1.245$

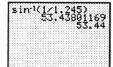

33. $\sec 74.1° = 3.65$

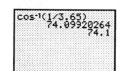

37.

$\dfrac{\sin 43.7°}{\cos 43.7°} = \tan 43.7° = 0.96$

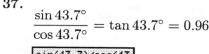

41. Given $\tan\theta = 1.936$, $\sin\theta = 0.8885$

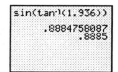

45. $d = 87$ dB

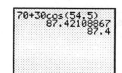

4.4   The Right Triangle

1. A 60° angle between sides of 3 in and 6 in determines the unique triangle shown in the figure below.

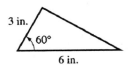

5. $\sin 77.8° = \dfrac{6700}{c} \Rightarrow c = 6850$, $\angle B = 90° - 77.8° = 12.2°$

$\tan 77.8° = \dfrac{6700}{b} \Rightarrow b = 1450$

9. $\angle A = 90° - 32.1° = 57.9°$, $\sin 32.1° = \dfrac{b}{23.8} \Rightarrow b = 12.6$

$\cos 32.1° = \dfrac{a}{23.8} \Rightarrow a = 20.2$

13. $\angle B = 90° - 32.1° = 57.9°$, $\sin 32.1° = \dfrac{a}{56.85} \Rightarrow a = 30.21$

$\cos 32.1° = \dfrac{b}{56.85} \Rightarrow b = 48.16$

17. $\angle A = 90° - 37.5° = 52.5°$, $\tan 37.5° = \dfrac{b}{0.862} \Rightarrow b = 0.661$

$\cos 37.5° = \dfrac{0.862}{c} \Rightarrow c = 1.09$

21. $\tan A = \dfrac{591.87}{264.93} \Rightarrow A = 65.89°$ $\tan B = \dfrac{264.93}{591.87} \Rightarrow B = 24.11°$

$c = \sqrt{264.93^2 + 591.87^2} = 648.46$

25. $\sin 61.7° = \dfrac{3.92}{x} \Rightarrow x = \dfrac{3.92}{\sin 61.7°} = 4.45$

29. $\angle B = 90° - \angle A$, $\sin A = \dfrac{a}{c} \Rightarrow a = c \sin A$, $\cos A = \dfrac{b}{c} \Rightarrow b = c \cos A$

4.5   Applications of Right Triangles

1. $\sin 54° = \dfrac{h}{120} \Rightarrow h = 120 \sin 54° = 97$ ft

5. $25.0 \text{ ft} \cdot \dfrac{12 \text{ in}}{1 \text{ ft}} = 300 \text{ in}$, $\theta = \tan^{-1} \dfrac{2.00}{300} = 0.4°$

9. $\sin 6.0° = \dfrac{2.65}{x} \Rightarrow x = \dfrac{2.65}{\sin 6.0°} = 25.4$ ft

13. $\theta = \tan^{-1} \dfrac{6.0}{100} = 3.4°$

17. $\theta = \sin^{-1} \dfrac{12.0}{85.0} = 8.1°$

21. Each angle of the pentagon is $\frac{360°}{5} = 72°$. Radii drawn from the center of the pentagon (which is also the center of the circle) through adjacent vertices of the pentagon outward to the fence form an isosceles triangle with base 92.5 and equal sides x. A $\perp$ bisector from the center of the pentagon to the base this isosceles triangle forms a right triangle with hypotenuse x and base 46.25. The base angle of this right triangle is $\frac{180° - 72°}{2} = 54°$. Thus,

$$\cos(54°) = \frac{46.25}{x} \Rightarrow x = \frac{46.25}{\cos 54°} \text{ and } C = 2\pi(x + 25)$$

$$C = 2\pi\left(\frac{46.25}{\cos 54°} + 25\right) = 651 \text{ ft}$$

25. $\frac{1}{2}C = \frac{1}{2}[2\pi(11.8 + 1)] = 12.8\pi \qquad \cos 31.8° = \frac{12.8\pi}{l} \Rightarrow l = 47.3 \text{ m}$

Chapter 4 Review Exercises

1. $17.0° + 360.0° = 377.0°, \quad 17.0 - 360.0° = -343.0°$ **5. and 9.** $31°54' = 31.9°, 17.5° = 17°30'$

```
31°54'
17.5▶DMS      31.9
          17°30'0"
```

13. $r = \sqrt{x^2 + y^2} = \sqrt{24^2 + 7^2} = \sqrt{625} = 25$

$$\sin\theta = \frac{y}{r} = \frac{7}{25} \qquad\qquad \csc\theta = \frac{r}{y} = \frac{25}{7}$$

$$\cos\theta = \frac{x}{r} = \frac{24}{25} \qquad\qquad \sec\theta = \frac{r}{x} = \frac{25}{24}$$

$$\tan\theta = \frac{y}{x} = \frac{7}{24} \qquad\qquad \cot\theta = \frac{x}{y} = \frac{24}{7}$$

17. $r^2 = x^2 + y^2 \Rightarrow 13^2 = x^2 + 5^2 \Rightarrow x = 12$

$$\cos\theta = \frac{x}{r} = \frac{12}{13} = 0.923$$

$$\cot\theta = \frac{x}{y} = \frac{12}{5} = 2.40$$

21. $\sin 72.1° = 0.952$ **25.** $\sec 18.4° = 1.05$ **29.** $\cos 18.2° = 0.950$ **33.**

$$\csc 12.25° = 4.713$$

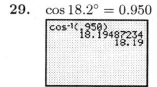

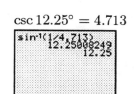

```
sin(72.1)
       .9515944039
             .952
```

```
1/cos(18.4)
       1.053878471
             1.05
```

```
cos⁻¹(.950)
       18.19487234
             18.19
```

```
sin⁻¹(1/4.713)
       12.25008249
             12.25
```

37. $\angle B = 90.0° - 17.0° = 73.0°$ $\qquad \tan 17.0° = \frac{a}{6.00} \Rightarrow a = 1.83$

$$\cos 17.0° = \frac{6.00}{c} \Rightarrow c = \frac{6.00}{\cos 17.0°} = 6.27$$

41. $\angle B = 90.0° - 37.5° = 52.5°$ $\tan 37.5° = \dfrac{12.0}{b} \Rightarrow b = 15.6$

$\sin 37.5° = \dfrac{12.0}{c} \Rightarrow c = \dfrac{12.0}{\sin 37.5°} = 19.7$

45. $\angle B = 90.00° - 49.67° = 40.33°$ $\sin 49.67° = \dfrac{a}{0.8253} \Rightarrow a = 0.6292$

$\cos 49.67° = \dfrac{b}{0.8253} \Rightarrow b = 0.5341$

49. $e = E \cos \alpha \Rightarrow 56.9 = 339 \cos \alpha \Rightarrow \alpha = \cos^{-1} \dfrac{56.9}{339} = 80.3°$

53. (a) A triangle with angle θ included between sides a and b, the base, has an altitude of $a \sin \theta$. The area, A, is $A = \dfrac{1}{2} \cdot b \cdot a \sin \theta$.

(b) The area of the tract is $A = \dfrac{1}{2} \cdot 31.96 \cdot 47.25 \sin 64.09° = 679.2$ m^2.

57. $\tan(90° - 65°) = \dfrac{2.0}{2.5 + x} \Rightarrow x = \dfrac{2.0}{\tan 25°} - 2.5 = 1.8$

$\dfrac{1.8}{3.2} = 0.5625$, 56% of the window is shaded.

61. $d = a + b = \dfrac{1.85}{\tan 28.3°} + \dfrac{1.85}{\tan(90.0° - 28.3°)} = 4.43$ m

65. $\sin 31.0° = \dfrac{d}{x} \Rightarrow \sin 31.0° = \dfrac{14.2 \sin 21.8°}{x} \Rightarrow x = \dfrac{14.2 \sin 21.8°}{\sin 31.0°} = 10.2$ in

69. Each angle of a regular pentagon is $\dfrac{(5-2)\cdot 180}{5} = 108°$. A regular pentagon with a side of 45.0 mm consists of 5 triangles of base 45.0 and altitude of 22.5 tan 54°. The area of 12 such pentagons is $12 \cdot 5 \cdot \frac{1}{2} \cdot 45.0 \cdot 22.5 \tan 54°$ or 41,807.60083 mm^2. Each angle of a regular hexagon is $\dfrac{(6-2)\cdot 180}{6} = 120°$. A regular hexagon with a side of 45.0 mm consists of 6 triangles of base 45.0 and an altitude of 22.5 tan 60°. The area of 20 such hexagons is

$$20 \cdot 6 \cdot \dfrac{1}{2} \cdot 45 \cdot 22.5 \tan 60° = 105,222.0866 \text{ mm}^2.$$

Thus, the area of 12 regular pentagons of side 45.0 mm and 20 regular hexagons of side 45.0 mm is 147,029.6874 mm^2 (147,000 mm^2 rounded off). Since this is the area of a flat surface it approximates the area of the spherical soccer ball which is given by

$$4\pi r^2 = 4\pi \cdot \left(\dfrac{222}{2}\right)^2 = 154,830.2523 \text{ mm}^2 \text{ (155,000 mm}^2 \text{ rounded off)}.$$

73. $h = a + b = 375 \sin 25° + d \tan 42°$
$\qquad\qquad\quad = 375 \sin 25° + 375 \cos 25° \tan 42°$
$\qquad\qquad\quad = 464 \text{ m}$

77. For small angles, such as 2.3°, the sine and tangent are approximately equal.

$$\sin \, 2.3° = 0.0401317925 \approx \tan \, 2.3° = 0.040164149.$$

SYSTEMS OF LINEAR EQUATIONS; DETERMINANTS

5.1 Linear Equations

1. The coordinates of the point $(3, 1)$ do not satisfy the equation since $2(3) + 3(1) = 6 + 3 = 9$.

   The coordinates of the point $(5, 1/3)$ do not satisfy the equation since $2(5) + 3\left(\dfrac{1}{3}\right) = 10 + 1 = 11 \neq 9$.

5. $5(1) - y = 6,\ y = 5 - 6 = -1$
   $5(-2) - y = 6,\ y = -10 - 6 = -16$

9. If the values $x = 4$ and $y = -1$ satisfy both equations, they are a solution.

   $4 - (-1) = 4 + 1 = 5$
   $2(4) + (-1) = 8 - 1 = 7$

   Therefore the given values are a solution.

13. If the values $x = 1/2, y = -1/5$ satisfy both equations, they are a solution.

    $2\left(\dfrac{1}{2}\right) - 5\left(\dfrac{-1}{5}\right) = 1 + 1 = 2 \neq 0.$

    Therefore, the given values are not a solution.

17. If $p = 260$ mi/h and $w = 40$ mi/h, then

    $p + w = 260 + 40 = 300$ mi/h
    $p - w = 260 - 40 = 220$ mi/h

    The speeds are 260 mi/h and 40 mi/h.

5.2 Graphs of Linear Functions

1. By taking $(3, 8)$ as (x_2, y_2) and $(1, 0)$ as (x_1, y_1)

   $m = \dfrac{8 - 0}{3 - 1} = \dfrac{8}{2} = 4$

5. By taking $(-2, -5)$ as (x_2, y_2) and $(5, -3)$ as (x_1, y_1)

   $m = \dfrac{-5 - (-3)}{-2 - 5} = \dfrac{-5 + 3}{-7} = \dfrac{2}{7}$

9. $m = 2,\ (0, -1)$ Plot the y-intercept point $(0, -1)$. Since the slope is $2/1$, from this point, go over 1 unit and up 2 units, and plot a second point. Sketch the line between the 2 points.

13. $m = \dfrac{1}{2}$, $(0,0)$

Plot the y-intercept point $(0,0)$. Since the slope is $1/2$, from this point, go over 2 units and up 1 unit, and plot a second point. Sketch the line between the 2 points.

17. $y = -2x + 1$, $m = -2$, $b = 1$

Plot the y-intercept point $(0,1)$. Since the slope is $-2/1$, from this point, go over 1 unit and down 2 units, and plot a second point. Sketch the line between the 2 points.

21. $5x - 2y = 40 \Rightarrow y = \dfrac{5}{2}x - 20$, $m = \dfrac{5}{2}$, $b = -20$

Plot the y-intercept point $(0, -20)$. Since the slope is $5/2$, from this point, go over 2 units and up 5 units, and plot a second point. Sketch the line between these 2 points.

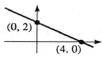

25. $x + 2y = 4|_{x=0} \Rightarrow 0 + 2y = 4 \Rightarrow y\text{-int} = 2$

$x + 2y = 4|_{y=0} \Rightarrow x + 2 \cdot 0 = 4 \Rightarrow x\text{-int} = 4$

Plot the y-intercept point $(4,0)$ and the y-intercept point $(0,2)$. Sketch the line between these 2 points. A third point is found as a check. Let $x = -2$, $-2 + 2y = 4$, $2y = 6$, $y = 3$. Therefore the point $(-2, 3)$ should lie on the line.

29. $y = 3x + 6|_{x=0} \Rightarrow y = 3 \cdot 0 + 6 \Rightarrow y\text{-int} = 6$
$y = 3x + 6|_{y=0} \Rightarrow 0 = 3x + 6 \Rightarrow x\text{-int} = -2$

Plot the x-intercept point $(-2,0)$ and the y-intercept point $(0,6)$. Sketch the line between these 2 points. A third point is found as a check. Let $x = 1$, $y = 3(1) + 6 = 9$. Therefore the point $(1,9)$ should lie on the line.

33. $d = 0.2l + 1.2$

"The d-intercept is $(0, 1.2)$. Since the slope is $2/10$, from the d-intercept go over 10 units and up 2 units the point $(10, 3.2)$ and plot a second point. Sketch the line between these 2 points."

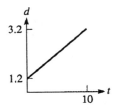

5.3   Solving Systems of Two Linear Equations in Two Unknowns Graphically

1. $y = -x + 4$ and $y = x - 2$.

"The slope of the first line is -1, and the y-intercept is 4. The slope of the second line is 1 and the y-intercept is -2. From the graph, the point of intersection is $(3.0, 1.0)$. Therefore, the solution of the system of equations is $x = 3.0, y = 1.0$."

5. $3x + 2y = 6, x - 3y = 3$

"The intercepts of the first line are $(0, 3)$, $(2, 0)$. A third point is $(4, -3)$. The intercepts of the second line are $(0, -1)$, $(3, 0)$. A third point is $(6, 1)$. From the graph, the point of intersection is $(2.2, -0.3)$. Therefore, the solution of the system of equations is $x = 2.2$, $y = -0.3$."

9. $s - 4t = 8 \Rightarrow s = 4t + 8$ and $2s = t + 4 \Rightarrow s = \dfrac{1}{2}t + 2$

"The slope of the first line is 4 and the y-intercept is 8. The slope of the second line is $1/2$ and the slope is 2. From the graph, the point of intersection is $(-1.7, 1.1)$. Therefore, the solution of the system of equations is $t = -1.7$, $s = 1.1$."

13. $x - 4y = 6, 2y = x + 4$

"The intercepts of the first line are $(0, -1.5)$, $(6, 0)$. A third point is $(2, -1)$. The intercepts of the second line are $(0, 2)$, $(-4, 0)$. A third point is $(2, 3)$. From the graph, the point of intersection is $(-14.0, -5.0)$. Therefore, the solution of the system of equations is $x = -14.0$, $y = -5.0$."

17. $x = 4y + 2 \Leftrightarrow y = \dfrac{x - 2}{4}$ and

$3y = 2x + 3 \Leftrightarrow y = \dfrac{2x + 3}{3}$

On a graphing calculator let $y_1 = \dfrac{x - 2}{4}$ and $y_2 = \dfrac{2x + 3}{3}$. Using the intersect feature, the point of intersection is $(-3.6, -1.4)$, and the solution of the system of equations is $x = -3.6$, $y = -1.4$.

21. $x = 4y + 2 \Leftrightarrow y = \dfrac{x-2}{4}$ and

$3y = 2x + 3 \Leftrightarrow y = \dfrac{2x+3}{3}$

On a graphing calculator let $y_1 = \dfrac{x-2}{4}$ and $y_2 = \dfrac{2x+3}{3}$. Using the intersect feature, the point of intersection is $(-3.6, -1.4)$, and the solution of the system of equations is $x = -3.6$, $y = -1.4$.

25. $5x - y = 3 \Leftrightarrow y = 5x - 3$ and

$4x = 2y - 3 \Leftrightarrow y = \dfrac{4x+3}{2}$

On a graphing calculator let $y_1 = 5x - 3$ and $y_2 = \dfrac{4x+3}{2}$. Using the intersect feature, the point of intersection is $(1.5, 4.5)$, and the solution to the system of equations is $x = 1.5$, $y = 4.5$.

29. $0.8T_1 - 0.6T_2 = 12 \Leftrightarrow$

$$T_2 = \frac{0.8T_1 - 12}{0.6} \Rightarrow y = \frac{0.8x - 12}{0.6}$$

$0.6T_1 + 0.8T_2 = 68 \Leftrightarrow$

$$T_2 = \frac{68 - 0.6T_1}{0.8} \Rightarrow y = \frac{68 - 0.6x}{0.8}$$

On a graphing calculator let $y_1 = \dfrac{0.8x - 12}{0.6}$ and $y_2 = \dfrac{68 - 0.6x}{0.8}$. Using the intersect feature, the point of intersection is $(50, 47)$. The tensions are $T_1 = 50$ N, $T_2 = 47$ N.

5.4   Solving Systems of Two Linear Equations in Two Unknowns Algebrically

1. (1) $x = y + 3$
  (2) $x - 2y = 5$

$(y + 3) - 2y = 5$ substitute x from
                        (1) into (2)

$\qquad -y = 2$
$\qquad\quad y = -2$ substitute -2 for
                        y in (1)
$\quad x = -2 + 3 = 1$

5. (1) $x + y = -5, y = -x - 5$
  (2) $2x - y = 2$

$2x - (-x - 5) = 2$ substitute y from
                        (1) into (2)

$\qquad 3x = -3$
$\qquad\ x = -1$
$-1 + y = -5$ substitute -1
                        for x in (1)
$\qquad y = -4$

9. (1) $3x + 2y = 7, y = \dfrac{7 - 3x}{2}$

(2) $2y = 9x + 11$

$$2\frac{7 - 3x}{2} = 9x + 11 \quad \text{substitute } y \text{ from (1) into (2)}$$
$$7 - 3x = 9x + 11$$
$$12x = -4$$
$$x = -\frac{1}{3}$$

$$3\left(-\frac{1}{3}\right) + 2y = 7 \qquad \text{substitute } -\frac{1}{3} \text{ for } x \text{ in (1)}$$
$$2y = 8$$
$$y = 4$$

13. $x + 2y = 5$
$\underline{x - 2y = 1}$
$2x = 6$
$x = 3$
$3 + 2y = 5$
$2y = 2$
$y = 1$
$(3, 1)$

17. $2x + 3y = 8 \Rightarrow \quad -2x - 3y = -8$
$x = 2y - 3 \Rightarrow \quad \underline{2x - 4y = -6}$
$\qquad\qquad\qquad\qquad -7y = -14$
$\qquad\qquad\qquad\qquad\quad y = 2$
$x = 2 \cdot 2 - 3$
$x = 4 - 3$
$x = 1$
$(1, 2)$

21. (1) $2x - 3y - 4 = 0$
(2) $3x + 2 = 2y$
(3) $2x - 3y = 4$ put (1) in standard form
(4) $3x - 2y = -2$ put (2) in standard form
(5) $6x - 9y = 12$ (3) multiplied by 3
(6) $\underline{-6x + 4y = 4} \quad$ (4) multiplied by -2
$\qquad -5y = 16$

$$y = -\frac{16}{5}$$

from (1) $x = \dfrac{3y + 4}{2} = \dfrac{3\left(-\dfrac{16}{5}\right) + 4}{2} = -\dfrac{14}{5}, \left(-\dfrac{14}{5}, -\dfrac{16}{5}\right)$ is the solution.

25. (1) $2x - y = 5, y = 2x - 5$
(2) $6x + 2y = -5$
$6x + 2(2x - 5) = -5 \quad \text{substitute } y \text{ from (1) into (2)}$
$10x = 5$

$$x = \frac{1}{2}$$

$$y = 2\left(\frac{1}{2}\right) - 5 = -4 \qquad \text{substitute } \frac{1}{2} \text{ for } x \text{ in (1)}$$

$x = \dfrac{1}{2}$, from (1) $y = 2x - 5 = 2 \cdot \dfrac{1}{2} - 5 = 1 - 5 = -4$

$$\left(\frac{1}{2}, -4\right)$$

29. Multiplying $3x - 6y = 15$ by 4 gives $\qquad 12x - 24y = 60$

Multiplying $4x - 8y = 20$ by -3 gives $\quad \underline{-12x + 24y = -60}$

$$0 = 0$$

The system is dependent and has an infinite number of solution.

33. $V_1 + V_2 = 15$

$\underline{V_1 - V_2 = 3}$

$2V_1 = 18$

$V_1 = 9$

$9 + V_2 = 15 \Rightarrow V_2 = 6$, the solultion is $V_1 = 9$ and $V_2 = 6$

37. (1) $2000t_1 = 3200t_2$

(2) $t_1 - 12 = t_2$

$2000t_1 = 3200(t_1 - 12) \Rightarrow 2000t_1 = 3200t_1 - 38,400 \Rightarrow -1200t_1 = -38,400$

$t_1 = 32 \ s$ and $t_2 = 32 - 12 = 20 \ s$

41. Let n be the number of turns the large pulley, of radius R, makes for one revolution of the belt. Let r and $\frac{r}{2}$ be the radii of the small pulleys, then

(1) $n(2\pi R) = 40 \Leftrightarrow n\pi R = 20$

(2) $(n + 1)(2\pi r) = 40 \Leftrightarrow (n + 1)\pi r = 20$

(3) $(n + 6)\left(2\pi \cdot \dfrac{r}{2}\right) = 40 \Leftrightarrow (n + 6)\pi r = 40$. From (2) and (3) $\dfrac{n + 1}{n + 6} = \dfrac{1}{2}$ from which $n = 4$ and

from (1), $4 \cdot (2\pi R) = 40 \Rightarrow (2\pi R) = 10$. From (2), $(4 + 1)(2\pi r) = 40 \Rightarrow 2\pi r = 8$ and from (3)

$(4 + 6)\left(2\pi \cdot \dfrac{r}{2}\right) = 40 \Rightarrow \left(2\pi \cdot \dfrac{r}{2}\right) = 4$. The circumferences are 10 ft, 8 ft, and 4 ft.

5.5   Solving Systems of Two Linear Equations in Two Unknowns by Determinants

1. $\begin{vmatrix} 2 & 4 \\ 3 & 1 \end{vmatrix} = (2)(1) - (3)(4) = 2 - 12 = -10$ $\qquad$ **5.** $\begin{vmatrix} 8 & -10 \\ 0 & 4 \end{vmatrix} = (8)(4) - (0)(-10) = 32 - 0 = 32$

9. $\begin{vmatrix} 0.7 & -1.3 \\ 0.1 & 11 \end{vmatrix} = (0.7)(1.1) - (0.1)(-1.3) = 0.77 + 0.13 = 0.9$

13. $x + 2y = 5$

$x - 2y = 1$

$$x = \frac{\begin{vmatrix} 5 & 2 \\ 1 & -2 \end{vmatrix}}{\begin{vmatrix} 1 & 2 \\ 1 & -2 \end{vmatrix}} = \frac{(5)(-2) - (1)(2)}{(1)(-2) - (1)(2)} = \frac{-10 - 2}{-2 - 2} = 3$$

$$y = \frac{\begin{vmatrix} 1 & 5 \\ 1 & 1 \end{vmatrix}}{-4} = \frac{(1)(1) - (1)(5)}{-4} = \frac{1 - 5}{-4} = 1$$

17. Rewrite the system with both equations in standard form.

$$2x + 3y = 8$$
$$x - 2y = -3$$

$$x = \frac{\begin{vmatrix} 8 & 3 \\ -3 & -2 \end{vmatrix}}{\begin{vmatrix} 2 & 3 \\ 1 & -2 \end{vmatrix}} = 1,$$

$$y = \frac{\begin{vmatrix} 2 & 8 \\ 1 & -3 \end{vmatrix}}{\begin{vmatrix} 2 & 3 \\ 1 & -2 \end{vmatrix}} = 2$$

25. $2.1x - 1.0y = 5.2$
$5.8x + 1.6y = -5.4$

$$x = \frac{\begin{vmatrix} 5.2 & -1.0 \\ -5.4 & 1.6 \end{vmatrix}}{\begin{vmatrix} 2.1 & -1.0 \\ 5.8 & 1.6 \end{vmatrix}} = 0.32$$

$$y = \frac{\begin{vmatrix} 2.1 & 5.2 \\ 5.8 & -5.4 \end{vmatrix}}{\begin{vmatrix} 2.1 & -1.0 \\ 5.8 & 1.6 \end{vmatrix}} = -4.5$$

33. Rewrite the system with both equations in standard form.

$$F_1 + F_2 = 21$$
$$2F_1 - 5F_2 = 0$$

$$F_1 = \frac{\begin{vmatrix} 21 & 1 \\ 0 & -5 \end{vmatrix}}{\begin{vmatrix} 1 & 1 \\ 2 & -5 \end{vmatrix}} = 15 \text{ lb}$$

$$F_2 = \frac{\begin{vmatrix} 1 & 21 \\ 2 & 0 \end{vmatrix}}{\begin{vmatrix} 1 & 1 \\ 2 & -5 \end{vmatrix}} = 6 \text{ lb}$$

21. Rewrite the system with both equations in standard form.

$$2x - 3y = 4$$
$$3x - 2y = -2$$

$$x = \frac{\begin{vmatrix} 4 & -3 \\ -2 & -2 \end{vmatrix}}{\begin{vmatrix} 2 & -3 \\ 3 & -2 \end{vmatrix}} = -2.8,$$

$$y = \frac{\begin{vmatrix} 2 & 4 \\ 3 & -2 \end{vmatrix}}{\begin{vmatrix} 2 & -3 \\ 3 & -2 \end{vmatrix}} = -3.2$$

29. $301x - 529y = 1520$
$385x - 741y = 2540$

$$x = \frac{\begin{vmatrix} 1520 & -529 \\ 2540 & -741 \end{vmatrix}}{\begin{vmatrix} 301 & -529 \\ 385 & -741 \end{vmatrix}} = -11.2$$

$$y = \frac{\begin{vmatrix} 301 & 1520 \\ 385 & 2540 \end{vmatrix}}{\begin{vmatrix} 301 & -529 \\ 385 & -741 \end{vmatrix}} = -9.26$$

37. Convert 24 minutes to hours.
24 min = 0.4 h

$$t_2 = t_1 - 0.4 \Leftrightarrow t_1 - t_2 = 0.4$$

$$42t_1 = 50t_2 \Leftrightarrow 42t_1 - 50t_2 = 0$$

$$t_1 = \frac{\begin{vmatrix} 0.4 & -1 \\ 0 & -50 \end{vmatrix}}{\begin{vmatrix} 1 & -1 \\ 42 & -50 \end{vmatrix}} = 2.5 \text{ h}$$

$$t_2 = \frac{\begin{vmatrix} 1 & 0.4 \\ 42 & 0 \end{vmatrix}}{\begin{vmatrix} 1 & -1 \\ 42 & -50 \end{vmatrix}} = 2.1 \text{ h}$$

41. $0.03x + 0.08y = 0.06(2) = 0.12$
$x + y = 2$

$$x = \frac{\begin{vmatrix} 0.12 & 0.08 \\ 2 & 1 \end{vmatrix}}{\begin{vmatrix} 0.03 & 0.08 \\ 1 & 1 \end{vmatrix}} = 0.8 \text{ L} \qquad y = \frac{\begin{vmatrix} 0.03 & 0.12 \\ 1 & 2 \end{vmatrix}}{\begin{vmatrix} 0.03 & 0.08 \\ 1 & 1 \end{vmatrix}} = 1.2 \text{ L}$$

5.6 Solving Systems of Three Linear Equations in Three Unknowns Algebraically

1. (1) $x + y + z = 2$ $\qquad$ $x + y + z = 2$
(2) $x - z = 1$
(3) $x + y = 1$ $\qquad$ (4) $\qquad$ $\underline{x - z = 1}$
(4) $\underline{2x + y = 3}$ $\qquad\qquad$ $2x + y = 3$ adding (1) and (2)
(5) Subtracting (4) from (3). $-x = -2 \Rightarrow x = 2$, from (4) $y = 3 - 2 \cdot 2 = -1$ and from
(1) $z = 2 - 2 - (-1) = 1$.
The solution is $x = 2, y = -1, z = 1$

5. (1) $5l + 6w - 3h = 6$ multiplied by 2, $\qquad$ (4) $\qquad$ $10l + 12w - 6h = 12$
(2) $4l - 7w - 2h = -3$ multiplied by -3, $\qquad$ (5) $\qquad$ $\underline{-12l + 21w + 6h = 9}$ adding
(3) $3l + w - 7h = 1$ $\qquad\qquad\qquad\qquad\quad$ (6) $\qquad\qquad$ $-2l + 33w = 21$
(7) $28l - 49w - 14h = -21$ $\quad$ (2) multiplied by 7
(8) $\underline{-6l - 2w + 14h = -2}$ $\quad$ (3) multiplied by -2, and adding
(9) $\qquad 22l - 51w = -23$
(10) $\qquad \underline{-22l + 363w = 231}$ $\quad$ (6) multiplied by 11, and adding
(11) $\qquad 312w = 208 \Rightarrow w = \dfrac{2}{3}$, from (6) $-2l + 33 \cdot \dfrac{2}{3} = 21 \Rightarrow l = \dfrac{1}{2}$
From (1) $5 \cdot \dfrac{1}{2} + 6 \cdot \dfrac{2}{3} - 3h = 6 \Rightarrow h = \dfrac{1}{6}$ The solution is $x = \dfrac{1}{2}, y = \dfrac{2}{3}, z = \dfrac{1}{6}.$

9. (1) $3x - 7y + 3z = 6$ multiplied by -2, $\qquad$ (4) $\qquad$ $-6x + 14y - 6z = -12$
(2) $3x + 3y + 6z = 1$ $\qquad\qquad\qquad\qquad\quad$ (2) $\qquad$ $\underline{3x + 3y + 6z = 1}$ adding
(3) $5x - 5y + 2z = 5$ $\qquad\qquad\qquad\qquad\quad$ (5) $\qquad$ $-3x + 17y \qquad = -11$
(2) $\qquad 3x + 3y + 6z = 1$
(6) $-15x + 15y - 6z = -15$ $\quad$ (3) multiplied by -3 and adding
(7) $\underline{-12x + 18y \qquad = -14}$
(8) $\qquad 12x - 68y \qquad = 44$ $\quad$ (5) multiplied by -4 and adding
(9) $-50y = 30 \Rightarrow y = -\dfrac{3}{5}$, from (5) $-3x + 17\left(-\dfrac{3}{5}\right) = -11 \Rightarrow x = \dfrac{4}{15}$

From (1) $3 \cdot \dfrac{4}{15} - 7\left(-\dfrac{3}{5}\right) + 3z = 6 \Rightarrow z = \dfrac{1}{3}$

The solution is $x = \dfrac{4}{15}, y = -\dfrac{3}{5}, z = \dfrac{1}{3}.$

13. (1) $2x + 3y - 5z = 7$ (1) $2x + 3y - 5z = 7$
   (2) $4x - 3y - 2z = 1$ (2) $\underline{4x - 3y - 2z = 1}$ adding
   (3) $8x - y + 4z = 3$ (4) $6x \qquad - 7z = 8$
   (2) $4x - 3y - 2z = 1$
   (5) $\underline{-24x + 3y - 12z = -9}$, (3) multiplied by -3 and adding
   (6) $-20x - 14z = -8$
   (7) $\underline{-12x + 14z = -16}$, (4) multiplied by -2 and adding

(8) $-32x = -24 \Rightarrow x = \dfrac{3}{4}$, from (4) $6 \cdot \dfrac{3}{4} - 7z = 8 \Rightarrow z = -\dfrac{1}{2}$.

From (1) $2 \cdot \dfrac{3}{4} + 3y - 5\left(-\dfrac{1}{2}\right) = 7 \Rightarrow y = 1$

The solution is $x = \dfrac{3}{4}, y = 1, z = -\dfrac{1}{2}$.

17. (1) $P + M + I = 1150$, solved for I, (4) $I = 1150 - M - P$
(2) $P = 4I - 100$ with I from (4), (5) $P = 4(1150 - M - P) - 100$
(3) $P = 6M + 50$. Substituting P from (3) into (5) gives
(6) $6M + 50 = 4(1150 - M - (6M + 50)) - 100 \Rightarrow 6M + 50 = 4(1150 - M - 6M - 50) - 100$
$6M + 50 = 4600 - 28M - 200 - 100$
$34M = 4250 \Rightarrow M = 125$, from (3) $P = 6 \cdot 125 + 50 = 800$.
From (1) $800 + 125 + I = 1150 \Rightarrow I = 225$.
$P = 800$ h, $M = 125$ h, $I = 225$ h is the solution.

21. Letting $t = 1, 3, 5$ and $\theta = 19, 30.9, 19.8$ in $at^3 + bt^2 + ct = \theta$ gives

(1) $a + b + c = 19$
(2) $27a + 9b + 3c = 30.9$
(3) $125a + 25b + 5c = 19.8$. Solving (1) for $c = 19 - a - b$ and substituting into (2) and (3) gives
(4) $24a + 6b = -26.1$ and (5) $120a + 20b = -75.2$. From (4), (6) $b = \dfrac{-26.1 - 24a}{6}$.

Substituting in (5) gives

(7) $120a + 20\left(\dfrac{-26.1 - 24a}{6}\right) = -75.2 \Rightarrow a = 0.295$

(6) $b = \dfrac{-26.1 - 24(0.295)}{6} = -5.53$

From (1) $c = 19 - 0.295 - (-5.53) = 24.235$.
The solution is $a = 0.295$, $b = -5.53$, and $c = 24.235$.

25. (1) $x - 2y - 3z = 2 \Rightarrow x = 2y + 3z + 2$ substituted in (2) and (3) gives
(2) $x - 4y - 13z = 14 \Rightarrow 2y + 3z + 2 - 4y - 13z = 14 \Rightarrow -2y - 10z = 12 \Rightarrow$
(4) $y + 5z = -6$
(3) $-3x + 5y + 4z = 0 \Rightarrow -3(2y + 3z + 2) + 5y + 4z = 0 \Rightarrow y + 5z = -6$ which is (4).
Thus, $y = -6 - 5z$ and from (1) $x - 2(-6 - 5z) - 3z = 2 \Rightarrow x = -7z - 10$.
The solution is $x = -72 - 10, y = -5z - 6, z = z$. Letting $z = 0, x = -10, y = -6, z = 0$.

5.7 Solving Systems of Three Linear Equations in Three Unknowns by Determinants

1. $\begin{vmatrix} 5 & 4 & -1 \\ -2 & -6 & 8 \\ 7 & 1 & 1 \end{vmatrix} \begin{matrix} 5 & 4 \\ -2 & -6 \\ 7 & 1 \end{matrix}$

$= (-30) + (224) + (2) - (42) - (40) - (-8) = 122$

5. $\begin{vmatrix} -3 & -4 & -8 \\ 5 & -1 & 0 \\ 2 & 10 & -1 \end{vmatrix} \begin{matrix} -3 & -4 \\ 5 & -1 \\ 2 & 10 \end{matrix}$

$= (-3) + (0) + (-400) - (16) - (0) - (20) = -139$

9. $\begin{vmatrix} 5 & 4 & -5 \\ -3 & 2 & -1 \\ 7 & 1 & 3 \end{vmatrix} \begin{matrix} 5 & 4 \\ -3 & 2 \\ 2 & 1 \end{matrix}$

$= (30) + (-28) + (15) - (-70) - (-5) - (-36) = 128$

13. $x = \dfrac{\begin{vmatrix} 4 & 3 & 1 \\ -3 & 0 & -1 \\ -5 & -2 & 2 \end{vmatrix} \begin{matrix} 4 & 3 \\ -3 & 0 \\ -5 & -2 \end{matrix}}{\begin{vmatrix} 2 & 3 & 1 \\ 3 & 0 & -1 \\ 1 & -2 & 2 \end{vmatrix} \begin{matrix} 2 & 3 \\ 3 & 0 \\ 1 & -2 \end{matrix}}$

$= \dfrac{(0) + (15) + (6) - (0) - (8) - (-18)}{(0) + (-3) + (-6) - (0) - (4) - (18)}$

$= \dfrac{31}{-31} = -1$

$y = \dfrac{\begin{vmatrix} 2 & 4 & 1 \\ 3 & -3 & -1 \\ 1 & -5 & 2 \end{vmatrix} \begin{matrix} 2 & 4 \\ 3 & -3 \\ 1 & -5 \end{matrix}}{-31}$

$= \dfrac{(-12) + (-4) + (-15) - (-3) - (10) - (24)}{-31}$

$= \dfrac{-62}{-31} = 2$

$z = \dfrac{\begin{vmatrix} 2 & 3 & 4 \\ 3 & 0 & -3 \\ 1 & -2 & -5 \end{vmatrix} \begin{matrix} 2 & 3 \\ 3 & 0 \\ 1 & -2 \end{matrix}}{-31}$

$= \dfrac{(0) + (-9) + (-24) - (0) - (12) - (-45)}{-31}$

$= \dfrac{0}{-31} = 0$

17. $x = \dfrac{\begin{vmatrix} 2 & 3 & 1 \\ -1 & 2 & 3 \\ 0 & -3 & 1 \end{vmatrix} \begin{matrix} 2 & 3 \\ -1 & 2 \\ 0 & -3 \end{matrix}}{\begin{vmatrix} 2 & 3 & 1 \\ -1 & 2 & 3 \\ -3 & -3 & 1 \end{vmatrix} \begin{matrix} 2 & 3 \\ -1 & 2 \\ -3 & -3 \end{matrix}}$

$= \dfrac{(4) + (0) + (3) - (0) - (-18) - (-3)}{(4) + (-27) + (3) - (-6) - (-18) - (-3)}$

$= \dfrac{28}{7} = 4$

$y = \dfrac{\begin{vmatrix} 2 & 2 & 1 \\ -1 & -1 & 3 \\ -3 & 0 & 1 \end{vmatrix} \begin{matrix} 2 & 2 \\ -1 & -1 \\ -3 & 0 \end{matrix}}{7}$

$= \dfrac{(-2) + (-18) + (0) - (3) - (0) - (-2)}{7}$

$= \dfrac{-21}{7} = -3$

$z = \dfrac{\begin{vmatrix} 2 & 3 & 2 \\ -1 & 2 & -1 \\ -3 & -3 & 0 \end{vmatrix} \begin{matrix} 2 & 3 \\ -1 & 2 \\ -3 & -3 \end{matrix}}{7}$

$= \dfrac{(0) + (9) + +(6) - (-12) - (6) - (0)}{7}$

$= \dfrac{21}{7} = 3$

21.

$$x = \frac{\begin{vmatrix} 5 & -2 & 3 \\ -1 & 1 & -2 \\ 0 & -1 & -3 \end{vmatrix}\begin{matrix} 5 & -2 \\ -1 & 1 \\ 0 & -1 \end{matrix}}{\begin{vmatrix} 2 & -2 & 3 \\ 2 & 1 & -2 \\ 4 & -1 & -3 \end{vmatrix}\begin{matrix} 2 & -2 \\ 2 & 1 \\ 4 & -1 \end{matrix}}$$

$$= \frac{(-15)+(0)+(3)-(0)-(10)-(-6)}{(-6)+(16)+(-6)-(12)-(4)-(12)}$$

$$= \frac{-16}{-24} = \frac{2}{3}$$

$$y = \frac{\begin{vmatrix} 2 & 5 & 3 \\ 2 & -1 & -2 \\ 4 & 0 & -3 \end{vmatrix}\begin{matrix} 2 & 5 \\ 2 & -1 \\ 4 & 0 \end{matrix}}{-24}$$

$$= \frac{(6)+(-40)+(0)-(-12)-(0)-(-30)}{-24}$$

$$= \frac{8}{-24} = -\frac{1}{3}$$

$$z = \frac{\begin{vmatrix} 2 & -2 & 5 \\ 2 & 1 & -1 \\ 4 & -1 & 0 \end{vmatrix}\begin{matrix} 2 & -2 \\ 2 & 1 \\ 4 & -1 \end{matrix}}{-24}$$

$$= \frac{(0)+(8)+(-10)-(20)-(2)-(0)}{-24}$$

$$= \frac{-24}{-24} = 1$$

25.

$$p = \frac{\begin{vmatrix} 0 & 2 & 2 \\ -1 & 6 & -3 \\ -8 & -3 & 6 \end{vmatrix}\begin{matrix} 0 & 2 \\ -1 & 6 \\ -8 & -3 \end{matrix}}{\begin{vmatrix} 1 & 2 & 2 \\ 2 & 6 & -3 \\ 4 & -3 & 6 \end{vmatrix}\begin{matrix} 1 & 2 \\ 2 & 6 \\ 4 & -3 \end{matrix}}$$

$$= \frac{(0)+(48)+(6)-(-96)-(0)-(-12)}{(36)+(-24)+(-12)-(48)-(9)-(24)}$$

$$= \frac{162}{-81} = -2$$

$$q = \frac{\begin{vmatrix} 1 & 0 & 2 \\ 2 & -1 & -3 \\ 4 & -8 & 6 \end{vmatrix}\begin{matrix} 1 & 0 \\ 2 & -1 \\ 4 & -8 \end{matrix}}{-81}$$

$$= \frac{(-6)+(0)+(-32)-(-8)-(24)-(0)}{-81}$$

$$= \frac{-54}{-81} = \frac{2}{3}$$

$$r = \frac{\begin{vmatrix} 1 & 2 & 0 \\ 2 & 6 & -1 \\ 4 & -3 & -8 \end{vmatrix}\begin{matrix} 1 & 2 \\ 2 & 6 \\ 4 & -3 \end{matrix}}{-81}$$

$$= \frac{(-48)+(-8)+(0)-(0)-(3)-(-32)}{-81}$$

$$= \frac{-27}{-81} = \frac{1}{3}$$

29.

$$A = \frac{\begin{vmatrix} 80 & 0 & -0.60 \\ 0 & 1 & -0.80 \\ 0 & 0 & -10 \end{vmatrix}\begin{matrix} 80 & 0 \\ 0 & 1 \\ 0 & 0 \end{matrix}}{\begin{vmatrix} 1 & 0 & -0.60 \\ 0 & 1 & -0.80 \\ 6.0 & 0 & -10 \end{vmatrix}\begin{matrix} 1 & 0 \\ 0 & 1 \\ 6.0 & 0 \end{matrix}} = \frac{(-800)+(0)+(0)-(0)-(0)-(0)}{(-10)+(0)+(0)-(-3.6)-(0)-(0)} = 125 \text{ N}$$

$$B = \frac{\begin{vmatrix} 1 & 80 & -0.60 \\ 0 & 0 & -0.80 \\ 6.0 & 0 & -10 \end{vmatrix}\begin{matrix} 1 & 80 \\ 0 & 0 \\ 6.0 & 0 \end{matrix}}{-6.4} = \frac{(0)+(-384)+(0)-(0)-(0)-(0)}{-6.4} = 60 \text{ N}$$

$$F = \frac{\begin{vmatrix} 1 & 0 & 80 \\ 0 & 1 & 0 \\ 6.0 & 0 & 0 \end{vmatrix}\begin{matrix} 1 & 0 \\ 0 & 1 \\ 6.0 & 0 \end{matrix}}{-6.4}$$

$$= \frac{(0)+(0)+(0)-(480)-(0)-(0)}{-6.4}$$

$$= 75 N$$

Chapter 5 Review Exercises

1. $\begin{vmatrix} -2 & 5 \\ 3 & 1 \end{vmatrix} = (-2)(1) - (3)(5) = -2 - 15 = -17$ **5.** $m = \dfrac{y_2 - y_1}{x_2 - x_1} = \dfrac{-8 - 0}{4 - 2} = \dfrac{-8}{2} = -4$

9. Comparing $y = -2x + 4$ to $y = mx + b$ gives a slope of -2 and y-int of 4.

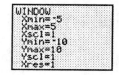 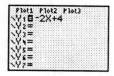 

13.

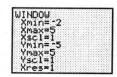  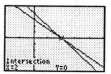

17. $7x = 2y + 14 \Rightarrow y = \dfrac{7x - 14}{2}$

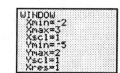 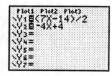 

21. (1) $x + 2y = 5 \Rightarrow x = 5 - 2y$ which substitutes into (2) $x + 3y = 7$ to give
$5 - 2y + 3y = 7 \Rightarrow y = 2$. From (1) $x = 5 - 2 \cdot 2 = 1$. The solution is $(1, 2)$.

25. (1) $3i + 4v = 6 \Rightarrow v = \dfrac{6 - 3i}{4}$ which substitutes into (2) $9i + 8v = 11$ to give

$9i + 8 \cdot \dfrac{6 - 3i}{4} = 11 \Rightarrow i = -\dfrac{1}{3}$. From (1) $v = \dfrac{6 - 3\left(-\frac{1}{3}\right)}{4} = \dfrac{7}{4}$. The solution is $\left(-\dfrac{1}{3}, \dfrac{7}{4}\right)$.

29. (1) $0.9x - 1.1y = 0.4 \Rightarrow x = \dfrac{11y + 4}{9}$ which substitutes into (2) $0.6x - 0.3y = 0.5$ to give

$6 \cdot \dfrac{11y + 4}{9} - 3y = 5 \Rightarrow y = \dfrac{7}{13}$. From (1) $x = \dfrac{11 \cdot \frac{7}{13} + 4}{9} = \dfrac{43}{39}$. The solution is $\left(\dfrac{43}{39}, \dfrac{7}{13}\right)$.

33. (1) $y = 2x - 3$ substitute into (2) $4x + 3y = -4$ gives $4x + 3(2x - 3) = -4 \Rightarrow x = \dfrac{1}{2}$. From (1)
$y = 2 \cdot \dfrac{1}{2} - 3 = -2$. The solution is $\left(\dfrac{1}{2}, -2\right)$.

37. (1) $7x = 2y - 6 \Rightarrow x = \dfrac{2y - 6}{7}$ which substitutes into (2) $7y = 12 - 4x$ to give

$7y = 12 - 4 \cdot \dfrac{2y - 6}{7} \Rightarrow y = \dfrac{36}{19}$. From (1) $x = \dfrac{2 \cdot \frac{36}{19} - 6}{7} = -\dfrac{6}{19}$. The solution is $\left(-\dfrac{6}{19}, \dfrac{36}{19}\right)$.

41. Exercise 33 is most easily solved by substitution because the second equation is already solved for y.

45. $\begin{vmatrix} 4 & -1 & 8 \\ -1 & 6 & -2 \\ 2 & 1 & -1 \end{vmatrix} = 4(-6+2) + (1+4) + 8(-1-12) = -115$

49. (1) $2x + y + z = 4$ mulltiplied by 2 $4x + 2y + 2z = 8$
      (2) $x - 2y - z = 3$ adding   (3) $\underline{3x + 3y - 2z = 1}$ adding
      (4) $3x - y = 7$ (5) $7x + 5y = 9$
                    (4) multiplied by 5 $\underline{15x - 5y = 35}$ adding
                                $22x = 44 \Rightarrow x = 2$

From (4) $y = 3 \cdot 2 - 7 = -1$. From (1) $z = 4 - (-1) - 2(2) = 1$. The solution is $(2, -1, 1)$.

53. Multiply both sides of all three equations by 10 to clear decimals.

(1) $36x + 52y - 10z = -22$ solved for z : (4) $z = \dfrac{36x + 52y + 22}{10}$

(2) $32x - 48y + 39z = 81$

(3) $64x + 41y + 23z = 51$

(5) $32x - 48y + 39 \cdot \dfrac{36x + 52y + 22}{10} = 81$ (2) with z from (4) which simplifes to

(5) $1724x + 1548y = -48 \Rightarrow y = \dfrac{-48 - 1724x}{1548}$

(6) $64x + 41y + 23 \cdot \dfrac{36x + 52y + 22}{10} = 51$ (3) with z from (4) which simplites to

(6) $1468x + 1606 \cdot \dfrac{-48 - 1724x}{1548} = 4 \Rightarrow x = -0.1678084952$. From (5)

$y = \dfrac{-48 - 1724x}{1548} = 0.1558797453$. From (4)

$z = \dfrac{36x + 52y + 22}{10} = 2.406464093$. The solution is $(-0.17, 0.16, 2.4)$.

57. $r = \dfrac{\begin{vmatrix} 8 & 1 & 2 \\ 5 & -2 & -4 \\ -3 & 3 & 4 \end{vmatrix}}{\begin{vmatrix} 2 & 1 & 2 \\ 3 & -2 & -4 \\ -2 & 3 & 4 \end{vmatrix}} = \dfrac{42}{14} = 3,$ $s = \dfrac{\begin{vmatrix} 2 & 8 & 2 \\ 3 & 5 & -4 \\ -2 & -3 & 4 \end{vmatrix}}{14} = \dfrac{-14}{14} = -1$

$t = \dfrac{\begin{vmatrix} 2 & 1 & 8 \\ 3 & -2 & 5 \\ -2 & 3 & -3 \end{vmatrix}}{14} = \dfrac{21}{14} = \dfrac{3}{2}$. The solution is $\left(3, -1, \dfrac{3}{2}\right)$.

61. $\begin{vmatrix} 2 & 5 \\ 1 & x \end{vmatrix} = 3 \Rightarrow 2x - 5 = 3 \Rightarrow 2x = 8 \Rightarrow x = 4$

65. (1) $\dfrac{1}{x} - \dfrac{1}{y} = \dfrac{1}{2} \Rightarrow u - v = \dfrac{1}{2}$

(2) $\dfrac{1}{x} + \dfrac{1}{y} = \dfrac{1}{4} \Rightarrow \underline{u + v = \dfrac{1}{4}}$

Adding (1), (2) $2u = \dfrac{3}{4} \Rightarrow u = \dfrac{3}{8} \Rightarrow x = \dfrac{8}{3}$, from (2) $v = \dfrac{1}{4} - \dfrac{3}{8} = -\dfrac{1}{8}$, $y = -8$. The solution is $\left(\dfrac{8}{3}, -8\right)$.

69. (1) $3x - ky = 6$

(2) $x + 2y = 2$. Multiplying (2) by 3 gives $3x + 6y = 6$ which is (1) with $k = -6$. A k-value of -6 makes the system dependent.

73. $F_1 = \dfrac{\begin{vmatrix} 280 & 2.0 & 0 \\ 0 & 0 & -1 \\ 600 & -4.0 & 0 \end{vmatrix}}{\begin{vmatrix} 1 & 2.0 & 0 \\ 0.87 & 0 & -1 \\ 3.0 & -4.0 & 0 \end{vmatrix}} = \dfrac{-2320}{-10} = 232 = 230 \text{ lb}$

$F_2 = \dfrac{\begin{vmatrix} 1 & 280 & 0 \\ 0.87 & 0 & -1 \\ 3.0 & 600 & 0 \end{vmatrix}}{-10} = \dfrac{-240}{-10} = 24 \text{ lb}$

$F_3 = \dfrac{\begin{vmatrix} 1 & 2.0 & 280 \\ 0.87 & 0 & 0 \\ 3.0 & -4 & 600 \end{vmatrix}}{-10} = \dfrac{-2018.4}{-10} = 201.84 = 200 \text{ lb}$

77. (1) $V + v = 24,200$

(2) $\underline{V - v = 21,400}$ adding

$2V = 45,600 \Rightarrow V = 22,800$ km/h is the speed of the shuttle.

(1) $22,800 + v = 24,200 \Rightarrow v = 1400$ km/h is the speed of the satellite.

81. (1) $2w = 8L \Rightarrow w = 4L$ which may be subsituted into (2)

(2) $4L \cdot 8 = 2w + 20 \cdot 12$ to obtain $32L = 2 \cdot 4L + 240 \Rightarrow L = 10$ lb

$w = 4 \cdot 10 = 40$ lb

85. Reasons for choosing a particular method may vary.

Let Q_1, Q_2 be the ft^3/h removed by each pump. Then $2.2Q_1 + 2.7Q_2 = 1100$ and $1.4Q_1 + 2.5Q_2 = 840$ describes the situation.

The solution is $Q_1 = 280$ ft^3/h and $Q_2 = 180$ ft^3/h.

FACTORING AND FRACTIONS

6.1 Special Products

1. $40(x - y) = 40x - 40y$

5. $(y + 6)(y - 6) = y^2 - 6^2 = y^2 - 36$

9. $(4x - 5y)(4x + 5y) = (4x)^2 - (5y)^2$
$$= 16x^2 - 25y^2$$

13. $(5f + 4)^2 = (5f)^2 + 2(5f)(4) + 4^2$
$$= 25f^2 + 40f + 16$$

17. $(L^2 - 1)^2 = (L^2)^2 - 2 \cdot L^2 \cdot 1 + 1^2$
$$= L^4 - 2L^2 + 1$$

21. $(4x - 2y)^2 = (4x)^2 - 2(4x)(2y) + (2y)^2$
$$= 16x^2 - 16xy + 4y^2$$

25. $(x + 1)(x + 5) = x^2 + 6x + 5$

29. $(3x - 1)(2x + 5) = 6x^2 + 13x - 5$

33. $(5v - 3)(4v + 5) = 20v^2 + 13v - 15$

37. $2(x - 2)(x + 2) = 2(x^2 - 4) = 2x^2 - 8$

41. $6a(x + 2b)^2 = 6a(x^2 + 4bx + 4b^2)$
$$= 6ax^2 + 24abx + 24ab^2$$

45. $[(2R + 3r)(2R - 3r)]^2 = [4R^2 - 9r^2]^2$
$$= 16R^4 - 72R^2r^2 + 81r^4$$

49. $[3 - (x + y)^2] = 9 - 6(x + y) + (x + y)^2 = 9 - 6x - 6y + x^2 + 2xy + y^2$

53. $(2x + 5t)^3 = 8x^3 + 3 \cdot 4x^2 \cdot 5t + 3 \cdot 2x \cdot 25t^2 + 125t^3 = 8x^3 + 60x^2t + 150xt^2 + 125t^3$

57. $(x + 2)(x^2 - 2x + 4) = x^3 - 2x^2 + 4x + 2x^2 - 4x + 8$
$$= x^3 + 8$$

61. $P_1(P_oc + G) = P_1P_oc + P_1G$

65. $\dfrac{1}{2}\pi(R + r)(R - r) = \dfrac{1}{2}\pi(R^2 - r^2) = \dfrac{1}{2}\pi R^2 - \dfrac{1}{2}\pi r^2$

69.

$A = (2x + 3)(2x - 3) = 4x^2 - 9$

6.2 Factoring: Common Factor and Difference of Squares

1. $6x + 6y = 6(x + y)$

5. $3x^2 - 9x = 3x(x - 3)$

9. $12n^2 + 6n = 6n(2n + 1)$

13. $3ab^2 - 6ab + 12ab^3 = 3ab(b - 2 + 4b^2)$

17. $2a^2 - 2b^2 + 4c^2 - 6d^2 = 2(a^2 - b^2 + 2c^2 - 3d^2)$

21. $100 - 9A^2 = (10 - 3A)(10 + 3A)$

25. $81s^2 - 25t^2 = (9s - 5t)(9s + 5t)$

29. $(x + y)^2 - 9 = (x + y - 3)(x + y + 3)$

33. $3x^2 - 27z^2 = 3(x^2 - 9z^2)$
$$= 3(x - 3z)(x + 3z)$$

37. $x^4 - 16 = (x^2 - 4)(x^2 + 4)$
$$= (x - 2)(x + 2)(x^2 + 4)$$

41. Solve $2a - b = ab + 3$ for a.
$$2a - ab = b + 3$$
$$a(2 - b) = b + 3$$
$$a = \frac{b + 3}{2 - b}$$

45. $3x - 3y + bx - by = 3(x - y) + b(x - y)$
$$= (x - y)(b + 3)$$

49. $x^3 + 3x^2 - 4x - 12 = x^2(x + 3) - 4(x + 3)$
$$= (x + 3)(x^2 - 4)$$
$$= (x + 3)(x + 2)(x - 2)$$

53. $2\pi rh + 2\pi r^2 = 2\pi r(h + r)$

57.   Solve $i_3 = (1 + a)i_1 - ai_2$ for a.
$$i_3 = i_1 + ai_1 - ai_2$$
$$a(i_1 - i_2) = i_3 - i_1$$
$$a = \frac{i_3 - i_1}{i_1 - i_2}$$

6.3   Factoring Trinomials

1. $x^2 + 5x + 4 = (x + 1)(x + 4)$

5. $t^2 + 5t - 24 = (t + 8)(t - 3)$

9. $x^2 - 4xy + 4y^2 = (x - 2y)^2$

13. $3y^2 - 8y - 3 = (3y + 1)(y - 3)$

17. $3f^4 - 16f^2 + 5 = (3f^2 - 1)(f^2 - 5)$

21. $3t^2 - 7tu + 4u^2 = (3t - 4u)(t - u)$

25. $9x^2 + 7xy - 2y^2 = (x + y)(9x - 2y)$

29. $4x^2 - 12x + 9 = (2x - 3)^2$

33. $8b^6 + 31b^3 - 4 = (8b^3 - 1)(b^3 + 4)$

37. $12x^2 + 47xy - 4y^2 = (12x - y)(x + 4y)$

41. $4x^2 + 14x - 8 = 2(2x^2 + 7x - 4)$
$$= 2(2x - 1)(x + 4)$$

45. $a^2 + 2ab + b^2 - 4 = (a + b)^2 - 4$
$$= (a + b + 2)(a + b - 2)$$

49. $4s^2 + 16s + 12 = 4(s^2 + 4s + 3)$
$$= 4(s + 3)(s + 1)$$

53. $wx^4 - 5wLx^3 + 6wL^2x^2 = wx^2(x^2 - 5Lx + 6L^2)$
$$= wx^2(x - 3L)(x - 2L)$$

6.4   The Sum and Difference of Cubes

1. $x^3 + 1 = (x + 1)(x^2 - x + 1)$

5. $27x^3 - 8a^3 = (3x)^3 - (2a)^3$
$$= (3x - 2a)(9x^2 + 6ax + 4a^2)$$

9. $6A^4 + 6A = 6A(A^3 + 1)$
$$= 6A(A + 1)(A^2 - A + 1)$$

13. $x^6y^3 + x^3y^6 = x^3y^3(x^3 + y^3)$
$$= x^3y^3(x + y)(x^2 - xy + y^2)$$

17. $0.001R^3 - 0.064r^3 = 0.001(R^3 - 64r^3)$
$$= 0.001(R - 4r)(R^2 + 4Rr + 16r^2)$$

21. $(a+b)^3 - 64 = (a+b)^3 - 4^3$
$$= (a+b-4)((a+b)^2 + 4(a+b) + 4^2)$$
$$= (a+b-4)(a^2 + 2ab + b^2 + 4a + 4b + 16)$$

25.

$$
\require{enclose}
\begin{array}{r}
x^4 \;\; + x^3y \;\; + x^2y^2 + \;\; xy^3 + \;\; y^4 \\[2pt]
x-y\,\overline{)\,x^5 \hspace{6.5cm} -\,y^5} \\[2pt]
\underline{x^5 - x^4y} \hspace{5.5cm} \\[2pt]
x^4y \hspace{5cm} \\[2pt]
\underline{x^4y - x^3y^2} \hspace{4cm} \\[2pt]
x^3y^2 \hspace{3.5cm} \\[2pt]
\underline{x^3y^2 - x^2y^3} \hspace{2.5cm} \\[2pt]
x^2y^3 \hspace{2cm} \\[2pt]
\underline{x^2y^3 - xy^4} \hspace{1cm} \\[2pt]
xy^4 - y^5 \\[2pt]
\underline{xy^4 - y^5}
\end{array}
$$

$$(x^5 - y^5) \div (x-y) = x^4 + x^3y + x^2y^2 + xy^3 + y^4$$
$$x^5 - y^5 = (x-y)(x^4 + x^3y + x^2y^2 + xy^3 + y^4)$$

$$
\begin{array}{r}
x^6 + x^5y + x^4y^2 + x^3y^3 + x^2y^4 + xy^5 + \;\; y^6 \\[2pt]
x-y\,\overline{)\,x^7 \hspace{8cm} -\,y^7} \\[2pt]
\underline{x^7 - x^6y} \hspace{7cm} \\[2pt]
x^6y \hspace{6.5cm} \\[2pt]
\underline{x^6y - x^5y^2} \hspace{5.5cm} \\[2pt]
x^5y^2 \hspace{5cm} \\[2pt]
\underline{x^5y^2 - x^4y^3} \hspace{4cm} \\[2pt]
x^4y^3 \hspace{3.5cm} \\[2pt]
\underline{x^4y^3 - x^3y^4} \hspace{2.5cm} \\[2pt]
x^3y^4 \hspace{2cm} \\[2pt]
\underline{x^3y^4 - x^2y^5} \hspace{1cm} \\[2pt]
x^2y^5 - xy^6 \\[2pt]
xy^6 - y^7 \\[2pt]
\underline{xy^6 - y^7}
\end{array}
$$

$$(x^7 - y^7) \div (x-y) = x^6 + x^5y + x^4y^2 + x^3y^3 + x^2y^4 + xy^5 + y^6$$
$$x^7 - y^7 = (x-y)(x^6 + x^5y + x^4y^2 + x^3y^3 + x^2y^4 + xy^5 + y^6)$$

29. $D^4 - d^3 D = D(D^3 - d^3)$
$$= D(D-d)(D^2 + Dd + d^2)$$

6.5   Equivalent Fractions

1. $\dfrac{2}{3} \cdot \dfrac{7}{7} = \dfrac{14}{21}$

5. $\dfrac{2}{(x+3)} \cdot \dfrac{(x-2)}{(x-2)} = \dfrac{2(x-2)}{(x+3)(x-2)} = \dfrac{2x-4}{x^2 + x - 6}$

9. $\dfrac{28}{44} = \dfrac{\frac{28}{4}}{\frac{44}{4}} = \dfrac{7}{11}$

13. $\dfrac{2(x-1)}{(x-1)(x+1)} = \dfrac{\dfrac{2(x-1)}{(x-1)}}{\dfrac{(x-1)(x+1)}{(x-1)}} = \dfrac{2}{x+1}$

17. $\dfrac{2a}{8a} = \dfrac{2a}{2a \cdot 4} = \dfrac{1}{4}$

21. $\dfrac{a+b}{5a^2+5ab} = \dfrac{(a+b)}{5a(a+b)} = \dfrac{1}{5a}$

25. $\dfrac{4x^2+1}{4x^2-1} = \dfrac{4x^2+1}{(2x-1)(2x+1)}$

29. $\dfrac{2y+3}{4y^3+6y^2} = \dfrac{(2y+3)}{2y^2(2y+3)}$

$= \dfrac{1}{2y^2}$

33. $\dfrac{2w^4+5w^2-3}{w^4+11w^2+24} = \dfrac{(2w^2-1)(w^2+3)}{(w^2+8)(w^2+3)}$

$= \dfrac{2w^2-1}{w^2+8}$

37. $\dfrac{x^4-16}{x+2} = \dfrac{(x^2+4)(x^2-4)}{(x+2)}$

$= \dfrac{(x^2+4)(x+2)(x-2)}{(x+2)}$

$= (x^2+4)(x-2)$

41. $\dfrac{(x-1)(3+x)}{(3-x)(1-x)} = \dfrac{(x-1)(3+x)}{-(3-x)(x-1)}$

$= \dfrac{3+x}{-(3-x)}$

$= \dfrac{x+3}{x-3}$

45. $\dfrac{2x^2-9x+4}{4x-x^2} = \dfrac{(2x-1)(x-4)}{-x(x-4)}$

$= \dfrac{-(2x-1)}{x}$

49. $\dfrac{x^3+y^3}{2x+2y} = \dfrac{(x+y)(x^2-xy+y^2)}{2(x+y)} = \dfrac{x^2-xy+y^2}{2}$

53. (a) $\dfrac{x^2(x+2)}{x^2+4}$ will not reduce further since x^2+4 does not factor.

(b) $\dfrac{x^4+4x^2}{x^4-16} = \dfrac{x^2(x^2+4)}{(x^2+4)(x^2-4)} = \dfrac{x^2}{x^2-4} = \dfrac{x^2}{(x+2)(x-2)}$

57. $\dfrac{mu^2-mv^2}{mu-mv} = \dfrac{m(u^2-v^2)}{m(u-v)} = \dfrac{(u-v)(u+v)}{(u-v)} = u+v$

6.6   Multiplication and Division of Fractions

1. $\dfrac{3}{8} \cdot \dfrac{2}{7} = \dfrac{3}{4} \cdot \dfrac{1}{7} = \dfrac{3}{28}$

5. $\dfrac{2}{9} \div \dfrac{4}{7} = \dfrac{2}{9} \cdot \dfrac{7}{4} = \dfrac{1}{9} \cdot \dfrac{7}{2} = \dfrac{7}{18}$

9. $\dfrac{4x+12}{5} \cdot \dfrac{15t}{3x+9} = \dfrac{4(x+3)}{5} \cdot \dfrac{5(3t)}{3(x+3)}$

$= 4t$

13. $\dfrac{2a+8}{15} \div \dfrac{a^2+8a+16}{25} = \dfrac{2(a+4)}{3 \cdot 5} \cdot \dfrac{5 \cdot 5}{(a+4)(a+4)}$

$= \dfrac{10}{3(a+4)}$

17. $\dfrac{3ax^2 - 9ax}{10x^2 + 5x} \cdot \dfrac{2x^2 + x}{a^2x - 3a^2} = \dfrac{3ax(x-3)}{5x(2x+1)} \cdot \dfrac{x(2x+1)}{a^2(x-3)} = \dfrac{3x}{5a}$

21. $\dfrac{ax + x^2}{2b - cx} \div \dfrac{a^2 + 2ax + x^2}{2bx - cx^2} = \dfrac{x(a+x)}{(2b-cx)} \cdot \dfrac{x(2b-cx)}{(a+x)(a+x)} = \dfrac{x^2}{a+x}$

25. $\dfrac{x^2 - 6x + 5}{4x^2 - 17x - 15} \cdot \dfrac{6x + 21}{2x^2 + 5x - 7} = \dfrac{(x-5)(x-1)}{(4x+3)(x-5)} \cdot \dfrac{3(2x+7)}{(2x+7)(x-1)} = \dfrac{3}{4x+3}$

29. $\dfrac{7x^2}{3a} \div \left(\dfrac{a}{x} \cdot \dfrac{a^2x}{x^2} \right) = \dfrac{7x^2}{3a} \div \dfrac{a^3}{x^2} = \dfrac{7x^2}{3a} \cdot \dfrac{x^2}{a^3} = \dfrac{7x^4}{3a^4}$

33. $\dfrac{x^3 - y^3}{2x^2 - 2y^2} \cdot \dfrac{x^2 + 2xy + y^2}{x^2 + xy + y^2} = \dfrac{(x-y)(x^2 + xy + y^2)}{2(x-y)(x+y)} \cdot \dfrac{(x+y)(x+y)}{(x^2 + xy + y^2)} = \dfrac{x+y}{2}$

37. $\dfrac{d}{2} \div \dfrac{v_1 d + v_2 d}{4v_1v_2} = \dfrac{d}{2} \cdot \dfrac{4v_1v_2}{d(v_1 + v_2)} = \dfrac{2v_1v_2}{v_1 + v_2}$

6.7   Addition and Subtraction of Fractions

1. $\dfrac{3}{5} + \dfrac{6}{5} = \dfrac{3+6}{5} = \dfrac{9}{5}$ **5.** $\dfrac{1}{2} + \dfrac{3}{4} = \dfrac{2}{4} + \dfrac{3}{4} = \dfrac{2+3}{4} = \dfrac{5}{4}$ **9.** $\dfrac{a}{x} - \dfrac{b}{x^2} = \dfrac{ax}{x^2} - \dfrac{b}{x^2} = \dfrac{ax - b}{x^2}$

13. $\dfrac{2}{5a} + \dfrac{1}{a} - \dfrac{a}{10} = \dfrac{4}{10a} + \dfrac{10}{10a} - \dfrac{a^2}{10a} = \dfrac{4 + 10 - a^2}{10a} = \dfrac{14 - a^2}{10a}$

17. $\dfrac{3}{2x - 1} + \dfrac{1}{4x - 2} = \dfrac{3}{(2x-1)} \cdot \dfrac{2}{2} + \dfrac{1}{2(2x-1)} = \dfrac{6+1}{2(2x-1)} = \dfrac{7}{2(2x-1)}$

21. $\dfrac{s}{2s - 6} + \dfrac{1}{4} - \dfrac{3s}{4s - 12} = \dfrac{s}{2(s-3)} \cdot \dfrac{2}{2} + \dfrac{1}{4} \cdot \dfrac{(s-3)}{(s-3)} - \dfrac{3s}{4(s-3)} = \dfrac{2s + (s-3) - 3s}{4(s-3)} = \dfrac{-3}{4(s-3)}$

25. $\dfrac{3}{x^2 - 8x + 16} - \dfrac{2}{4 - x} = \dfrac{3}{(x-4)(x-4)} + \dfrac{2}{(x-4)} \cdot \dfrac{(x-4)}{(x-4)} = \dfrac{3 + 2(x-4)}{(x-4)(x-4)} = \dfrac{3 + 2x - 8}{(x-4)(x-4)}$

$\qquad = \dfrac{2x - 5}{(x-4)(x-4)} = \dfrac{2x - 5}{(x-4)^2}$

29. $\dfrac{x - 1}{3x^2 - 13x + 4} - \dfrac{3x + 1}{4 - x} = \dfrac{(x-1)}{(3x-1)(x-4)} + \dfrac{(3x+1)}{(x-4)} \cdot \dfrac{(3x-1)}{(3x-1)}$

$\qquad = \dfrac{x - 1 + 9x^2 - 1}{(3x-1)(x-4)} = \dfrac{9x^2 + x - 2}{(3x-1)(x-4)}$

33. $\dfrac{1}{w^3 + 1} + \dfrac{1}{w + 1} - 2 = \dfrac{1}{(w+1)(w^2 - w + 1)} + \dfrac{(w^2 - w + 1)}{(w+1)(w^2 - w + 1)} - \dfrac{2(w+1)(w^2 - w + 1)}{(w+1)(w^2 - w + 1)}$

$\qquad = \dfrac{1 + w^2 - w + 1 - 2(w+1)(w^2 - w + 1)}{(w+1)(w^2 - w + 1)}$

$\qquad = \dfrac{w^2 - w + 2 - 2w^3 - 2}{(w+1)(w^2 - w + 1)} = \dfrac{-2w^3 + w^2 - w}{(w+1)(w^2 - w + 1)}$

37. $\dfrac{\dfrac{x}{y} - \dfrac{y}{x}}{1 + \dfrac{y}{x}} \cdot \dfrac{xy}{xy} = \dfrac{x^2 - y^2}{xy + y^2} = \dfrac{(x+y)(x-y)}{y(x+y)} = \dfrac{x-y}{y}$

41. $\dfrac{\dfrac{3}{x} + \dfrac{1}{x^2 + x}}{\dfrac{1}{x+1} - \dfrac{1}{x-1}} = \dfrac{\dfrac{3}{x} + \dfrac{1}{x(x+1)}}{\dfrac{1}{x+1} - \dfrac{1}{x-1}} \cdot \dfrac{x(x+1)(x-1)}{x(x+1)(x-1)} = \dfrac{3(x+1)(x-1) + (x-1)}{x(x-1) - x(x+1)}$

$$= \dfrac{3x^2 - 3 + x - 1}{x^2 - x - x^2 - x} = \dfrac{3x^2 + x - 4}{-2x} = -\dfrac{(3x+4)(x-1)}{2x}$$

45.   For $f(x) = \dfrac{x}{x+1}$, $f(x+h) - f(x) = \dfrac{x+h}{x+h+1} - \dfrac{x}{x+1}$

$$f(x+h) - f(x) = \dfrac{(x+h)}{(x+h+1)} \cdot \dfrac{(x+1)}{(x+1)} - \dfrac{x}{(x+1)} \cdot \dfrac{(x+h+1)}{(x+h+1)} = \dfrac{x^2 + x + hx + h - x^2 - xh - x}{(x+1)(x+h+1)}$$

$$= \dfrac{h}{(x+1)(x+h+1)}$$

49. $\tan\theta \cdot \cot\theta + (\sin\theta)^2 - \cos\theta = \dfrac{y}{x} \cdot \dfrac{x}{y} + \left(\dfrac{y}{r}\right)^2 - \dfrac{x}{r} = 1 + \dfrac{y^2}{r^2} - \dfrac{x}{r} = \dfrac{r^2 + y^2 - rx}{r^2}$

53. $f(x) = x - \dfrac{2}{x}$

$$f(a+1) = a + 1 - \dfrac{2}{a+1} = \dfrac{(a+1)^2 - 2}{a+1} = \dfrac{a^2 + 2a + 1 - 2}{a+1} = \dfrac{a^2 + 2a - 1}{a+1}$$

57. $\dfrac{3}{4\pi} - \dfrac{3H_0}{4\pi H} = \dfrac{3H}{4\pi H} - \dfrac{3H_0}{4\pi H} = \dfrac{3H - 3H_0}{4\pi H} = \dfrac{3(H - H_0)}{4\pi H}$

61. $\left(\dfrac{3Px}{2L^2}\right)^2 + \left(\dfrac{P}{2L}\right)^2 = \dfrac{9P^2 x^2}{4L^4} + \dfrac{P^2}{4L^2} \cdot \dfrac{L^2}{L^2} = \dfrac{9P^2 x^2 + P^2 L^2}{4L^4} = \dfrac{P^2(9x^2 + L^2)}{4L^4}$

6.8   Equations Involving Fractions

1. $\dfrac{x}{2} + 6 = 2x$

$x + 12 = 4x$
$3x = 12$
$x = 4$

5. $\dfrac{1}{2} - \dfrac{t-5}{6} = \dfrac{3}{4}$

$6 - 2(t-5) = 9$
$6 - 2t + 10 = 9$
$2t = 7$
$t = \dfrac{7}{2}$

9. $\dfrac{3}{x} + 2 = \dfrac{5}{3}$

$9 + 6x = 5x$
$x = -9$

13. $\dfrac{2y}{y-1} = 5$

$$2y = 5y - 5$$
$$3y = 5$$
$$y = \frac{5}{3}$$

17. $\dfrac{5}{2x+4} + \dfrac{3}{x+2} = 2$

$$\frac{5}{2(x+2)} + \frac{3}{(x+2)} = 2$$
$$5 + 6 = 2 \cdot 2(x+2)$$
$$11 = 4x + 8$$
$$4x = 3$$
$$x = \frac{3}{4}$$

21. $\dfrac{1}{x} + \dfrac{3}{2x} = \dfrac{2}{x+1}$

$$2(x+1) + 3(x+1) = 2 \cdot 2x$$
$$2x + 2 + 3x + 3 = 4x$$
$$x = -5$$

25. $\dfrac{1}{x^2-x} - \dfrac{1}{x} = \dfrac{1}{x-1}$

$$\frac{1}{x(x-1)} - \frac{1}{x} = \frac{1}{(x-1)}$$
$$1 - (x-1) = x$$
$$1 - x + 1 = x$$
$$2x = 2$$
$$x = 1, \text{ no solution}$$

29. $2 - \dfrac{1}{b} + \dfrac{3}{c} = 0$, for c

$$2bc - c + 3b = 0$$
$$c(2b-1) = -3b$$
$$c = \frac{3b}{1-2b}$$

33. $\dfrac{s-s_0}{t} = \dfrac{v+v_0}{2}$

$$2(s-s_0) = t(v+v_0)$$
$$2(s-s_0) = tv + tv_0$$
$$2(s-s_0) - tv_0 = tv$$
$$v = \frac{2(s-s_0) - t_0 v_0}{t}$$

37. $z = \dfrac{1}{g_m} - \dfrac{jX}{g_m R}$ for R

$$g_m R z = R - jX$$
$$g_m R z - R = -jX$$
$$R(g_m z - 1) = -jX$$
$$R = \frac{jX}{1 - g_m z}$$

41. $\dfrac{1}{R_1} = \dfrac{N_2^2}{N_1^2 R_2} + \dfrac{N_3^2}{N_1^2 R_3}$ for R_1

$$\frac{1}{R_1} = \frac{N_2^2 R_3}{N_1^2 R_2 R_3} + \frac{N_3^2 R_2}{N_1^2 R_2 R_3}$$
$$\frac{1}{R_1} = \frac{N_2^2 R_3 + N_3^2 R_2}{N_1^2 R_2 R_3}$$
$$R_1 = \frac{N_1^2 R_2 R_3}{N_2^2 R_3 + N_3^2 R_2}$$

45. $\dfrac{1}{4} \cdot t + \dfrac{1}{6} \cdot t = 1$

$$\frac{5}{12} \cdot t = 1$$
$$t = 2.4 \text{ h}$$

49. Use $d = rt$: $(450 + w) \cdot t = 2580$ with wind
$(450 - w) \cdot t = 1800$ against wind

$$t = \frac{2580}{450 + w} = \frac{1800}{450 - w}$$

$$2580(450 - w) = 1800(450 + w)$$
$$1{,}161{,}000 - 2580w = 810{,}000 + 1800w$$
$$4380w = 351{,}000$$
$$w = 80 \text{ km/h, wind speed}$$

Chapter 6 Review Exercises

1. $3a(4x + 5a) = 12ax + 15a^2$

5. $(2a + 1)^2 = 4a^2 + 4a + 1$

9. $(2x + 5)(x - 9) = 2x^2 - 13x - 45$

13. $3s + 9t = 3(s + 3t)$

17. $W^2 - 144 = (W + 12)(W - 12)$

21. $9t^2 - 6t + 1 = (3t - 1)(3t - 1) = (3t - 1)^2$

25. $x^2 + x - 56 = (x + 8)(x - 7)$

29. $2x^2 - x - 36 = (2x - 9)(x + 4)$

33. $10b^2 + 23b - 5 = (5b - 1)(2b + 5)$

37. $250 - 16y^6 = 2(125 - 8y^6)$
$$= 2(5^3 - (2y^2)^3)$$
$$= 2(5 - 2y^2)(25 + 10y^2 + 4y^4)$$

41. $ab^2 - 3b^2 + a - 3 = b^2(a - 3) + (a - 3)$
$$= (a - 3)(b^2 + 1)$$

45. $\dfrac{48ax^3y^6}{9a^3xy^6} = \dfrac{16x^2}{3a^2}$

49. $\dfrac{4x + 4y}{35x^2} \cdot \dfrac{28x}{x^2 - y^2} = \dfrac{4(x + y)}{35x^2} \cdot \dfrac{28x}{(x + y)(x - y)}$
$$= \dfrac{4 \cdot 7 \cdot 4}{7 \cdot 5x(x - y)}$$
$$= \dfrac{16}{5x(x - y)}$$

53. $\dfrac{\dfrac{3x}{7x^2 + 13x - 3}}{\dfrac{6x^2}{x^2 + 4x + 4}} = \dfrac{3x}{(7x - 1)(x + 2)} \cdot \dfrac{(x + 2)(x + 2)}{3 \cdot 2x^2}$
$$= \dfrac{x + 2}{2x(7x - 1)}$$

57. $\dfrac{4}{9x} - \dfrac{5}{12x^2} = \dfrac{4}{9x} \cdot \dfrac{4x}{4x} - \dfrac{5}{12x^2} \cdot \dfrac{3}{3}$

$\qquad\qquad = \dfrac{16x - 15}{36x^2}$

61. $\dfrac{a+1}{a+2} - \dfrac{a+3}{a} = \dfrac{(a+1)}{(a+2)} \cdot \dfrac{a}{a} - \dfrac{(a+3)}{a} \cdot \dfrac{(a+2)}{(a+2)} = \dfrac{a(a+1) - (a+3)(a+2)}{a(a+2)}$

$\qquad\qquad = \dfrac{a^2 + a - a^2 - 5a - 6}{a(a+2)} = \dfrac{-4a - 6}{a(a+2)} = \dfrac{-2(a+3)}{a(a+2)}$

65. $\dfrac{3x}{2x^2 - 2} - \dfrac{2}{4x^2 - 5x + 1} = \dfrac{3x}{2(x+1)(x-1)} \cdot \dfrac{(4x-1)}{(4x-1)} - \dfrac{2}{(4x-1)(x-1)} \cdot \dfrac{2(x+1)}{2(x+1)}$

$\qquad\qquad = \dfrac{3x(4x-1) - 4(x+1)}{2(4x-1)(x+1)(x-1)} = \dfrac{12x^2 - 3x - 4x - 4}{2(4x-1)(x+1)(x-1)}$

$\qquad\qquad = \dfrac{12x^2 - 7x - 4}{2(4x-1)(x+1)(x-1)}$

69. $\quad \dfrac{x}{2} - 3 = \dfrac{x-10}{4}$

$\qquad 2x - 12 = x - 10$

$\qquad\qquad\quad x = 2$

73. $\quad \dfrac{2x}{x^2 - 3x} - \dfrac{3}{x} = \dfrac{1}{2x - 6}$

$\qquad \dfrac{2x}{x(x-3)} - \dfrac{3}{x} = \dfrac{1}{2(x-3)}$

$\qquad 4x - 3 \cdot 2(x-3) = x$

$\qquad 4x - 6x + 18 = x$

$\qquad\qquad\quad 3x = 18$

$\qquad\qquad\quad\ x = 6$

77. $\dfrac{1}{4}[(x+y)^2 - (x-y)^2] = \dfrac{1}{4}[x^2 + 2xy + y^2 - (x^2 - 2xy + y^2)] = \dfrac{1}{4}[x^2 + 2xy + y^2 - x^2 + 2xy - y^2]$

$\qquad\qquad = \dfrac{1}{4}[4xy] = xy$

81. $\pi r_1^2 l - \pi r_2^2 l = \pi l(r_1^2 - r_2^2)$

$\qquad\qquad\quad = \pi l(r_1 + r_2)(r_1 - r_2)$

85. $(2R - r)^2 - (r^2 + R^2) = 4R^2 - 4Rr + r^2 - r^2 - R^2$

$\qquad\qquad\qquad\qquad = 3R^2 - 4Rr$

$\qquad\qquad\qquad\qquad = R(3R - 4r)$

89. $10a(T - t) + a(T - t)^2 = 10aT - 10at + a(T^2 - 2Tt + t^2) = 10aT - 10at + aT^2 - 2aTt + at^2$

93. $\dfrac{2wtv^2}{Dg} \cdot \dfrac{b\pi^2 D^2}{n^2} \cdot \dfrac{6}{bt^2} = \dfrac{12wv^2\pi^2 D}{gn^2 t}$

97. $1 - \dfrac{d^2}{2} + \dfrac{d^4}{24} - \dfrac{d^6}{120} = \dfrac{120 - 60d^2 + 5d^4 - d^6}{120}$

101. $1 - \dfrac{3a}{4r} - \dfrac{a^3}{4r^3} = \dfrac{4r^3 - 3ar^2 - a^3}{4r^3}$

105. $W = mgh_2 - mgh_1$ for m

$W = m(gh_2 - gh_1)$

$$m = \dfrac{W}{gh_2 - gh_1}$$

109. $s^2 + \dfrac{cs}{m} + \dfrac{kL^2}{mb^2} = 0$ for c

$s^2mb^2 + csb^2 + kL^2 = 0$

$csb^2 = -kL^2 - s^2mb^2$

$c = \dfrac{-kL^2 - s^2mb^2}{sb^2}$

113. $\dfrac{1}{4} \cdot t + \dfrac{1}{24} \cdot t = 1$

$\dfrac{7}{24} \cdot t = 1$

$t = 3.4$ h

117. $d = \dfrac{w_a}{w_a - w_w} = \dfrac{1.097w_w}{1.097w_w - w_w} = \dfrac{1.097}{1.097 - 1} = 11.3$

121. $\dfrac{2r^2 + 5r - 3}{2r^2 + 7r + 3} = \dfrac{(r+3)(2r-1)}{(r+3)(2r+1)} = \dfrac{2r-1}{2r+1}$

When you "cancel", the basic operation being performed is division.

Chapter 7

QUADRATIC EQUATIONS

7.1 Quadratic Equations; Solution by Factoring

1.
$$x^2 + 5 = 8x$$
$$x^2 - 8x + 5 = 0,$$

quadratic with $a = 1, b = -8, c = 5$.

5. $x^2 = (x + 2)^2$
$$x^2 = x^2 + 4x + 4$$

$4x + 4 = 0$, no x^2 term, not quadratic

9.
$$x^2 - 4 = 0$$
$$(x + 2)(x - 2) = 0$$
$$x + 2 = 0 \quad \text{or} \quad x - 2 = 0$$
$$x = -2 \qquad x = 2$$

13.
$$x^2 - 8x - 9 = 0$$
$$(x - 9)(x + 1) = 0$$
$$x - 9 = 0 \quad \text{or} \quad x + 1 = 0$$
$$x = 9 \qquad x = -1$$

17.
$$x^2 = -2x$$
$$x^2 + 2x = 0$$
$$x(x + 2) = 0$$
$$x = 0 \quad \text{or} \quad x + 2 = 0$$
$$x = -2$$

21.
$$3x^2 - 13x + 4 = 0$$
$$(3x - 1)(x - 4) = 0$$
$$3x - 1 = 0 \quad \text{or} \quad x - 4 = 0$$
$$3x = 1 \qquad x = 4$$
$$x = \frac{1}{3}$$

25.
$$6x^2 = 13x - 6$$
$$6x^2 - 13x + 6 = 0$$
$$(3x - 2)(2x - 3) = 0$$
$$3x - 2 = 0 \quad \text{or} \quad 2x - 3 = 0$$
$$3x = 2 \qquad 2x = 3$$
$$x = \frac{2}{3} \qquad x = \frac{3}{2}$$

29.
$$x^2 - x - 1 = 1$$
$$x^2 - x - 2 = 0$$
$$(x - 2)(x + 1) = 0$$
$$x - 2 = 0 \quad \text{or} \quad x + 1 = 0$$
$$x = 2 \qquad x = -1$$

33. $40x - 16x^2 = 0$
$$2x^2 - 5x = 0$$
$$x(2x - 5) = 0$$
$$x = 0 \quad \text{or} \quad 2x - 5 = 0$$
$$2x = 5$$
$$x = \frac{5}{2}$$

37.
$$(x + 2)^3 = x^3 + 8$$
$$x^3 + 6x^2 + 12x + 8 = x^3 + 8$$
$$6x^2 + 12x = 0$$
$$6x(x + 2) = 0$$
$$6x = 0 \quad \text{or} \quad x + 2 = 0$$
$$x = 0 \qquad x = -2$$

41. $M = \dfrac{1}{2}wLx - \dfrac{1}{2}wx^2 = 0$

$$Lx - x^2 = 0$$
$$x(L - x) = 0$$
$$x = 0 \quad \text{or} \quad L - x = 0$$
$$x = L$$

$M = 0$ for x-values of 0 and L

49. $\dfrac{1}{k_c} = \dfrac{1}{k_1} + \dfrac{1}{k_2} \Rightarrow \dfrac{1}{2} = \dfrac{1}{k} + \dfrac{1}{k + 3}$

$$k(k + 3) = 2(k + 3) + 2k$$
$$k^2 + 3k = 2k + 6 + 2k$$
$$k^2 - k - 6 = 0$$
$$(k - 3)(k + 2) = 0$$

45. $\dfrac{1}{x - 3} + \dfrac{4}{x} = 2$

$$x + 4(x - 3) = 2x(x - 3)$$
$$x + 4x - 12 = 2x^2 - 6x$$
$$2x^2 - 11x + 12 = 0$$
$$(x - 4)(2x - 3) = 0$$
$$x - 4 = 0 \quad \text{or} \quad 2x - 3 = 0$$
$$x = 4 \qquad\qquad 2x = 3$$
$$x = \dfrac{3}{2}$$

$$k - 3 = 0 \quad \text{or} \quad k + 2 = 0$$
$$k = 3 \qquad\qquad k = -2, \text{ reject since } k > 0.$$
$$k + 3 = 6$$

The spring constants are 3 N/cm and 6 N/cm.

7.2   Completing the Square

1. $x^2 = 25$
$$x = \pm\sqrt{25}$$
$$x = \pm 5$$

5. $(x - 2)^2 = 25$
$$x - 2 = \pm\sqrt{25}$$
$$x - 2 = \pm 5$$
$$x = 2 + 5 \quad \text{or} \quad x = 2 - 5$$
$$x = 7 \qquad\qquad x = -3$$

9. $x^2 + 2x - 8 = 0$
$$x^2 + 2x + 1 = 8 + 1$$
$$(x + 1)^2 = 9$$
$$x + 1 = \pm\sqrt{9} = \pm 3$$
$$x + 1 = 3 \quad \text{or} \quad x + 1 = -3$$
$$x = 2 \qquad\qquad x = -4$$

13. $x^2 - 4x + 2 = 0$
$$x^2 - 4x + 4 = -2 + 4$$
$$(x - 2)^2 = 2$$
$$x - 2 = \pm\sqrt{2}$$
$$x - 2 = \sqrt{2} \quad \text{or} \quad x - 2 = -\sqrt{2}$$
$$x = 2 + \sqrt{2} \qquad\qquad x = 2 - \sqrt{2}$$

17. $2s^2 + 5s = 3$

$$s^2 + \dfrac{5}{2}s + \dfrac{25}{16} = \dfrac{3}{2} + \dfrac{25}{16} = \dfrac{49}{16}$$
$$\left(s + \dfrac{5}{4}\right)^2 = \dfrac{49}{16}$$
$$s + \dfrac{5}{4} = \pm\dfrac{7}{4}$$
$$s + \dfrac{5}{4} = \dfrac{7}{4} \quad \text{or} \quad s + \dfrac{5}{4} = -\dfrac{7}{4}$$
$$s = \dfrac{1}{2} \qquad\qquad s = -3$$

21. $2y^2 - y - 2 = 0$

$$y^2 - \dfrac{1}{2}y + \dfrac{1}{16} = 1 + \dfrac{1}{16} = \dfrac{17}{16}$$
$$\left(y - \dfrac{1}{4}\right)^2 = \dfrac{17}{16}$$
$$y - \dfrac{1}{4} = \dfrac{\pm\sqrt{17}}{4}$$
$$y - \dfrac{1}{4} = \dfrac{\sqrt{17}}{4} \quad \text{or} \quad y - \dfrac{1}{4} = \dfrac{-\sqrt{17}}{4}$$
$$y = \dfrac{1 + \sqrt{17}}{4} \qquad\qquad y = \dfrac{1 - \sqrt{17}}{4}$$

25. $9x^2 + 6x + 1 = 0$

$$x^2 + \frac{2}{3}x + \frac{1}{9} = -\frac{1}{9} + \frac{1}{9}$$

$$\left(x + \frac{1}{3}\right)^2 = 0$$

$$x + \frac{1}{3} = 0$$

$$x = -\frac{1}{3} \text{ double root}$$

7.3   The Quadratic Formula

1. $x^2 + 2x - 8 = 0$

$$x = \frac{-2 \pm \sqrt{2^2 - 4(1)(-8)}}{2(1)}$$

$$= \frac{-2 \pm \sqrt{36}}{2}$$

$$= \frac{-2 \pm 6}{2}$$

$$x = 2 \quad \text{or} \quad x = -4$$

5. $x^2 - 4x + 2 = 0$

$$x = \frac{-(-4) \pm \sqrt{(-4)^2 - 4(1)(2)}}{2}$$

$$= \frac{4 \pm \sqrt{8}}{2}$$

$$= \frac{4 \pm 2\sqrt{2}}{2}$$

$$= 2 \pm \sqrt{2}$$

9. $2s^2 + 5s = 3$

$2s^2 + 5s - 3 = 0$

$$s = \frac{-5 \pm \sqrt{5^2 - 4(2)(-3)}}{2(2)}$$

$$= \frac{-5 \pm \sqrt{49}}{4}$$

$$= \frac{-5 \pm 7}{4}$$

$$s = \frac{1}{2} \quad \text{or} \quad s = -3$$

13. $2y^2 - y - 2 = 0$

$2y^2 - y - 2 = 0$

$$y = \frac{-(-1) \pm \sqrt{(-1)^2 - 4(2)(-2)}}{2(2)}$$

$$= \frac{1 \pm \sqrt{17}}{4}$$

17. $2t^2 + 10t = -15$

$2t^2 + 10t + 15 = 0$

$$t = \frac{-10 \pm \sqrt{10^2 - 4(2)(15)}}{2(2)}$$

$$= \frac{-10 \pm \sqrt{-20}}{4}$$

$$= \frac{-10 \pm 2\sqrt{-5}}{4}$$

$$t = \frac{-5 \pm \sqrt{-5}}{2}, \text{ imaginary roots}$$

21. $4x^2 = 9$

$4x^2 + 0x - 9 = 0$

$$x = \frac{-0 \pm \sqrt{0^2 - 4(4)(-9)}}{2(4)}$$

$$= \frac{\pm\sqrt{144}}{8}$$

$$= \frac{\pm 12}{8}$$

$$= \pm\frac{3}{2}$$

25. $x^2 - 0.20x - 0.40 = 0$

$$x = \frac{-(-0.20) \pm \sqrt{(-0.20)^2 - 4(1)(-0.40)}}{2(1)}$$

$$= \frac{0.2 \pm \sqrt{1.64}}{2}$$

$x = -0.54 \quad \text{or} \quad x = 0.74$

29. $x^2 + 2cx - 1 = 0$

$$x = \frac{-2c \pm \sqrt{(2c)^2 - 4(1)(-1)}}{2(1)}$$

$$= \frac{-2c \pm \sqrt{4c^2 + 4}}{2}$$

$$= \frac{-2c \pm 2\sqrt{c^2 + 1}}{2}$$

$$x = -c \pm \sqrt{c^2 + 1}$$

33. For $D = 3.625$, $D_0^2 - DD_0 - 0.25D^2 = 0$ is

$$D_0^2 - 3.625D_0 - 0.25(3.625)^2 = 0$$
$$D_0^2 - 3.625D_0 - 3.28515625 = 0$$

$$D_0 = \frac{-(-3.625) \pm \sqrt{(-3.625)^2 - 4(1)(-3.28515625)}}{2}$$

$D_0 = 4.38$ cm or $D_0 = -.75$, reject since $D_0 > 0$.

37. $2x^2 - 7x = -8$
$2x^2 - 7x + 8 = 0$
$D = \sqrt{(-7)^2 - 4(2)(8)} = \sqrt{-15}$,

unequal imaginary roots

41. $A = l \cdot w = 262$
$(w + 12.8) \cdot w = 262$
$w^2 + 12.8w - 262 = 0$,

using the quadratic formula
$w = 11.0$ m   or $w = -24$, reject since
$w > 0$
$l = w + 12.8 = 23.8$ m. The dimension of
the rectangle are $l = 23.8$ m and $w = 11.0$ m

7.4   The Graph of the Quadratic Function

1. $y = x^2 - 6x + 5$; $a = 1$, $b = -6$, $c = 5$
$c = 5 \Rightarrow y\text{-int} = 5$

$x^2 - 6x + 5 = 0 \Rightarrow x = 1 \quad \text{or} \quad x = 5$, the x-intercepts

x vertex $= \dfrac{-b}{2a} = \dfrac{-(-6)}{2(1)} = 3$

y vertex $= 3^2 - 6(3) + 5 = -4$

 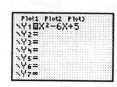 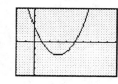 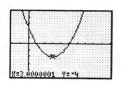

5. $y = x^2 - 4x + 0$; $a = 1$, $b = -4$, $c = 0$

$c = 0 \Rightarrow y\text{-int} = 0$

$x^2 - 4x = 0 \Rightarrow x = 0$, $x = 4$, the x-intercepts

$x \text{ vertex} = \dfrac{-b}{2a} = \dfrac{-(-4)}{2(1)} = 2$

$y \text{ vertex} = 2^2 - 4(2) = -4$

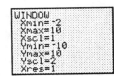 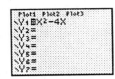 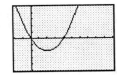 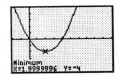

9. $y = x^2 - 4 = x^2 + 0x - 4$; $a = 1$, $b = 0$, $c = -4$

$c = -4 \Rightarrow x = \pm 2$, the x-intercepts

$x \text{ vertex} = \dfrac{-b}{2a} = \dfrac{-0}{2(1)} = 0$

$y \text{ vertex} = 0^2 - 4 = -4$

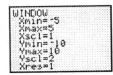 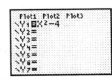 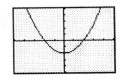 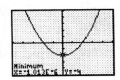

13. $y = 2x^2 + 3 = 2x^2 + 0x + 3$; $a = 2$, $b = 0$, $c = 3$

$c = 3 \Rightarrow y\text{-int} = 3$

$2x^2 + 3 = 0 \Rightarrow D = -24 < 0 \Rightarrow$ no x-intercepts

$x \text{ vertex} = \dfrac{-b}{2a} = \dfrac{-0}{2(2)} = 0$

$y \text{ vertex} = 2(0)^2 + 3 = 3$

 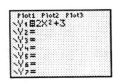 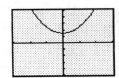

17. $2x^2 - 3 = 0$. Graph $y = 2x - 3$ and find roots.

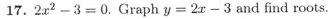

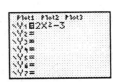 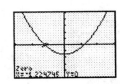 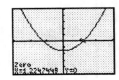

21. $x(2x - 1) = -3$. Graph $y = x(2x - 1) + 3$ and find roots.

 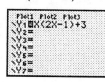 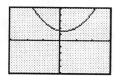

as the graph shows there are no real solutions.

25.

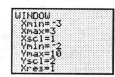  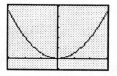 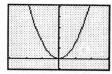 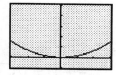

The graph of $y = 3x^2$ is the graph of $y = x^2$ narrowed. The graph of $y = \frac{1}{3}x^2$ is the graph of $y = x^2$ broadened.

29. Graph $y = x(8 - x)$ for $0 > x > 8$.

  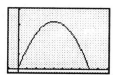

33.

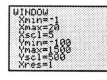 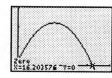 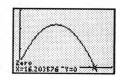 

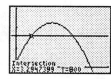 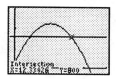

From the graphs:

(a) the missle will hit the gound after 16.2 s
(b) the missle will reach a maximum height of 1130 ft
(c) the missle will have a height of 800 ft at 3.3 s and 12.3 s

---

Chapter 7 Review Exercises

1. $x^2 + 3x - 4 = 0$
$(x + 4)(x - 1) = 0$
$x + 4 = 0 \quad \text{or} \quad x - 1 = 0$
$\quad x = -4 \qquad\qquad x = 1$

5. $3x^2 + 11x = 4$
$3x^2 + 11x - 4 = 0$
$(3x - 1)(x + 4) = 0$
$3x - 1 = 0 \quad \text{or} \quad x + 4 = 0$
$\quad 3x = 1 \qquad\qquad x = -4$
$\quad x = \dfrac{1}{3}$

9.
$$6s^2 = 25s$$
$$6s^2 - 25s = 0$$
$$s(6s - 25) = 0$$
$$s = 0 \quad \text{or} \quad 6s - 25 = 0$$
$$6s = 25$$
$$s = \frac{25}{6}$$

13. $x^2 - x - 110 = 0$
$$x = \frac{-(-1) \pm \sqrt{(-1)^2 - 4(1)(-110)}}{2(1)}$$
$$= \frac{1 \pm 21}{2}$$
$$x = -10 \quad \text{or} \quad x = 11$$

17.
$$2x^2 - x = 36$$
$$2x^2 - x - 36 = 0$$
$$x = \frac{-(-1) \pm \sqrt{(-1)^2 - 4(2)(-36)}}{2(2)}$$
$$= \frac{1 \pm \sqrt{289}}{4}$$
$$= \frac{1 \pm 17}{4}$$
$$x = \frac{9}{2} \quad \text{or} \quad x = -4$$

21. $2.1x^2 + 2.3x + 5.5 = 0$
$$x = \frac{-2.3 \pm \sqrt{2.3^2 - 4(2.1)(5.5)}}{2(2.1)}$$
$$= \frac{-2.3 \pm \sqrt{-40.91}}{4.2} = \frac{-23 \pm \sqrt{-4091}}{42}$$
there are two unequal imaginary roots.

25. $x^2 + 4x - 4 = 0$
$$x = \frac{-4 \pm \sqrt{4^2 - 4(1)(-4)}}{2(1)}$$
$$= \frac{-4 \pm \sqrt{32}}{2}$$
$$= \frac{-4 \pm 4\sqrt{2}}{2}$$
$$x = -2 \pm 2\sqrt{2}$$

29.
$$4v^2 = v + 5$$
$$4v^2 - v - 5 = 0$$
$$(v + 1)(4v - 5) = 0$$
$$v + 1 = 0 \quad \text{or} \quad 4v - 5 = 0$$
$$v = -1 \qquad 4v = 5$$
$$v = \frac{5}{4}$$

33. $a^2x^2 + 2ax + 2 = 0$
$$x = \frac{-2a \pm \sqrt{(2a)^2 - 4(a^2)(2)}}{2(a^2)}$$
$$x = \frac{-2a \pm \sqrt{-4a^2}}{2a^2} \quad \text{and for } a > 0,$$
$$x = \frac{-2a \pm 2a\sqrt{-1}}{2a^2}$$
$$x = \frac{-1 \pm \sqrt{-1}}{a}$$

37. $x^2 - x - 30 = 0$
$$x^2 - x + \frac{1}{4} = 30 + \frac{1}{4} = \frac{121}{4}$$
$$\left(x - \frac{1}{2}\right)^2 = \frac{121}{4}$$
$$x - \frac{1}{2} = \frac{\pm 11}{2}$$
$$x = \frac{1}{2} \pm \frac{11}{2}$$
$$x = -5 \quad \text{or} \quad x = 6$$

41.
$$\frac{x-4}{x-1} = \frac{2}{x}$$

$$x(x-4) = 2(x-1)$$
$$x^2 - 4x = 2x - 2$$
$$x^2 - 6x + 2 = 0$$

$$x = \frac{-(-6) \pm \sqrt{(-6)^2 - 4(1)(2)}}{2(1)} = \frac{6 \pm \sqrt{28}}{2} = \frac{6 \pm \sqrt{4 \cdot 7}}{2} = \frac{6 \pm 2\sqrt{7}}{2}$$

$$x = 3 \pm \sqrt{7}$$

45. $y = 2x^2 - x - 1$; $a = 2$, $b = -1$, $c = -1$

$c = -1 \Rightarrow y\text{-int} = -1$

$2x^2 - x - 1 = 0 \Rightarrow x = -\dfrac{1}{2}$, $x = 1$, the x-intercepts

$x \text{ vertex} = \dfrac{-b}{2a} = \dfrac{-(-1)}{2(2)} = \dfrac{1}{4}$

$y \text{ vertex} = 2\left(\dfrac{1}{4}\right)^2 - \left(\dfrac{1}{4}\right) - 1 = -\dfrac{9}{8}$

 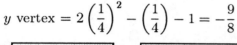 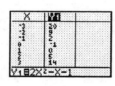 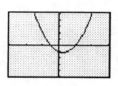 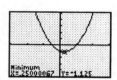

49. Graph $y = 2x^2 + x - 4$ and find roots.

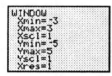 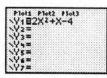 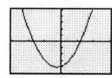 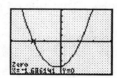 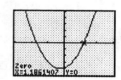

53. $v = 5.2x - x^2$ for $v = 4.8$ is

$x^2 - 5.2x + 4.8 = 0$
$(x - 1.2)(x - 4) = 0$
$x - 1.2 = 0 \qquad \text{or} \qquad x - 5 = 0$
$\qquad x = 1.2 \text{ cm} \qquad\qquad x = 4 \text{ cm}$

57.
$$\frac{n^2}{500,000} = 144 - \frac{n}{500}$$
$$n^2 + 1000n - 72,000,000 = 0$$
$$(n + 9000)(n - 8000) = 0$$
$$n - 8000 = 0 \qquad \text{or} \qquad n + 9000 = 0$$
$$n = 8000 \qquad\qquad n = -9000$$

reject since $n > 0$

61. Graph $y = 0.090x - 0.015x^2$ for $0 < x < 6$ since the percenet of the drug in the blood cannot be negative and $0.090x - 0.015x^2 = 0$ has $x = 0$ and $x = 6$ as solutions.

  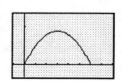

65.
$$V = e^3, \text{ before drying} \qquad (1)$$
$$V - 29 = (e - 0.1)^3, \text{ after drying} \qquad (2)$$
$$e^3 - 29 = (e - 0.1)^3$$
$$e^3 - 29 = e^3 - 0.3e^2 + 0.03e - 0.001$$
$$0.3e^2 - 0.03e - 28.999 = 0$$
$$e = \frac{-(-0.03) \pm \sqrt{(-0.03)^2 - 4(0.3)(-28.999)}}{2(0.3)}$$

$e = 9.88$ or -9.78 which is rejected since $c > 0$ the original edge had a length of 9.88 cm.

69. Suppose n poles are placed along the road for a distance of 1 km and x is the distance, in km, between the poles, then $n \cdot x = 1$. Increasing the distance between the poles to $x + 0.01$ and decreasing the number of poles to $n - 5$ gives $(n - 5)(x + 0.01) = 1$. Substitution gives

$$(n - 5)\left(\frac{1}{n} + 0.01\right) = 1$$
$$1 + 0.01n - \frac{5}{n} - 0.05 = 1$$
$$0.01n^2 - 0.05n - 5 = 0$$
$$n^2 - 5n - 500 = 0$$
$$(n + 20)(n - 25) = 0$$
$$= 0$$
$$n + 20 = 0 \qquad \text{or} \qquad n - 25 = 0$$
$$n = -20, \quad \text{reject} \qquad n = 25$$

There are 25 poles being placed each kilometer.

73.
$$\frac{1}{R_T} = \frac{1}{R} + \frac{1}{R + 1}$$
$$R(R + 1) = R_T(R + 1) + RR_T$$
$$R^2 + R = R_T R + R_T + RR_T$$
$$R^2 - 2R_T R + R - R_T = 0$$
$$R^2 + (1 - 2R_T)R - R_T = 0, \text{ use quadratic formula}$$
$$R = \frac{-(1 - 2R_T) \pm \sqrt{(1 - 2R_T)^2 - 4(1)(-R_T)}}{2(1)}$$
$$R = \frac{2R_T - 1 \pm \sqrt{1 - 4R_T + 4R_T^2 + 4R_T}}{2}$$
$$R = \frac{2R_T - 1 \pm \sqrt{1 + 4R_T^2}}{2}$$

TRIGONOMETRIC FUNCTIONS OF ANY ANGLE

8.1 Signs of the Trigonometric Functions

1. sin 36° is positive since 36° is in QI where sine is positive.
cos 120° is negative since 120° is in QII where cosine is negative.

5. sec 150° is negative since 150° is in QII where secant is negative.
tan 220° is positive since 220° is in QIII where tangent is positive.

9. tan 460° is negative since 460° is QII where tangent is negative.
sin(−110°) is negative since −110° is in QIII where sine is negative.

13. $r = \sqrt{2^2 + 1^2} = \sqrt{4+1} = \sqrt{5}$ for $(2, 1)$.

$$\sin\theta = \frac{y}{r} = \frac{1}{\sqrt{5}} \qquad \csc\theta = \frac{r}{y} = \sqrt{5}$$

$$\cos\theta = \frac{x}{r} = \frac{2}{\sqrt{5}} \qquad \sec\theta = \frac{r}{x} = \frac{\sqrt{5}}{2}$$

$$\tan\theta = \frac{y}{x} = \frac{1}{2} \qquad \cot\theta = \frac{x}{y} = 2$$

17. $r = \sqrt{(-5)^2 + 12^2} = 13$ for $(-5, 12)$

$$\sin\theta = \frac{y}{r} = \frac{12}{13} \qquad \csc\theta = \frac{r}{y} = \frac{13}{12}$$

$$\cos\theta = \frac{x}{r} = \frac{-5}{13} \qquad \sec\theta = \frac{r}{x} = \frac{13}{-5}$$

$$\tan\theta = \frac{y}{x} = \frac{12}{-5} \qquad \cot\theta = \frac{x}{y} = \frac{-5}{12}$$

21. $\sin\theta$ positive in QI and QII
$\cos\theta$ negative in QII and QIII
θ is in QII

25. $\csc\theta$ negative in QIII and QIV
$\tan\theta$ negative in QII and QIV
θ is in QIV

29. $\tan\theta$ is negative in QII and QIV
$\cos\theta$ is positive in QI and QIV
θ is in QIV

8.2 Trigonometric Functions of Any Angle

1.

$$\sin 160° = \sin(180° - 160°) = \sin 20°$$

$$\cos 220° = \cos(180° + 40°)$$
$$= -\cos 40°$$

5.

$$\cos 400° = \cos(360° + 40°)$$
$$= \cos 40°$$

$$\tan(-400°) = \tan(-360° - 40°)$$
$$= \tan(-40°) = -\tan 40°$$

9. The reference angle for $106.3° = 180° - 106.3° = 73.7°$. $\cos 106.3° = -\cos 73.7° = -0.281$

13. $\tan 152.4° = -0.5228$

17. $\csc 194.82° = -3.9096$

```
tan(152.4°)
          -.5227873662
```

```
1/sin(194.82°)
          -3.909560491
```

21. $\cos\theta = 0.4003$
  $\theta = 66.4°$
  $\theta = 293.6°$
  $0° \le \theta < 360°$

25. $\sin\theta = 0.870$
  $\theta = 119.5°$
  $0° \le \theta < 360°$, $\cos\theta < 0$

```
cos-1(.4003)
          66.40306574
360-Ans
          293.5969343
```

```
sin-1(.870)
          60.4586395
180-Ans
          119.5413605
```

29. $\tan\theta = -1.366$
  $\theta = 306.2°$
  $0° \le \theta < 360°$, $\cos\theta > 0$

33. $\sin\theta = -0.5736$, $\cos\theta > 0$
  $\theta = 325° + k \cdot 360°$
  where $k = 0, \pm 1, \pm 2, \cdots$
  $\tan\theta = -0.7003$

```
tan-1(-1.366)
          -53.79346902
Ans+360
          306.206531
```

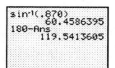

```
sin-1(-.5736)
          -35.00164818
Ans+360
          324.9983518
tan(Ans)
          -.7002504091
```

37. $\sin 90° = 1$, $2 \sin 45° = 2 \cdot \dfrac{\sqrt{2}}{2} = \sqrt{2}$

  $1 < \sqrt{2}$
$\sin 90° < 2 \sin 45°$

41. $i = i_m \sin\theta$
  $i = 0.0259 \cdot \sin 495.2°$
  $i = 0.0183$ A

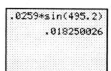

```
.0259*sin(495.2)
          .018250026
```

45.

$$\left.\begin{array}{l} \sin(-\theta) = \dfrac{-y}{r} \\[2mm] -\sin\theta = -\dfrac{y}{r} \end{array}\right\} \Rightarrow \sin(-\theta) = -\sin\theta$$

$$\left.\begin{array}{l} \cot(-\theta) = \dfrac{x}{-y} \\[2mm] \cot\theta = \dfrac{x}{y} \end{array}\right\} \Rightarrow \cot(-\theta) = -\cot\theta$$

$$\left.\begin{array}{l} \cos(-\theta) = \dfrac{x}{r} \\[2mm] \cos\theta = \dfrac{x}{r} \end{array}\right\} \Rightarrow \cos(-\theta) = \cos\theta$$

$$\left.\begin{array}{l} \sec(-\theta) = \dfrac{r}{x} \\[2mm] \sec\theta = \dfrac{r}{x} \end{array}\right\} \Rightarrow \sec(-\theta) = \sec\theta$$

$$\left.\begin{array}{l} \tan(-\theta) = \dfrac{-y}{x} \\[2mm] \tan\theta = \dfrac{y}{x} \end{array}\right\} \Rightarrow \tan(-\theta) = \tan\theta$$

$$\left.\begin{array}{l} \csc(-\theta) = \dfrac{r}{-y} \\[2mm] \csc\theta = \dfrac{r}{y} \end{array}\right\} \Rightarrow \csc(-\theta) = -\csc\theta$$

8.3   Radians

1. $15° \cdot \dfrac{\pi}{180°} = \dfrac{\pi}{12}$

$150° \cdot \dfrac{\pi}{180°} = \dfrac{5\pi}{6}$

5. $210° \cdot \dfrac{\pi}{180°} = \dfrac{7\pi}{6}$

$270° \cdot \dfrac{\pi}{180°} = \dfrac{3\pi}{2}$

9. $\dfrac{2\pi}{5} \cdot \dfrac{180°}{\pi} = 72°$

$\dfrac{3\pi}{2} \cdot \dfrac{180°}{\pi} = 270°$

13. $\dfrac{17\pi}{18} \cdot \dfrac{180°}{\pi} = 170°$

$\dfrac{5\pi}{3} \cdot \dfrac{180°}{\pi} = 300°$

17. $23° \cdot \dfrac{\pi}{180°} = 0.401$

21. $333.5° \cdot \dfrac{\pi}{180°} = 5.821$

25. $0.750 \cdot \dfrac{180°}{\pi} = 43.0°$

29. $2.45 \cdot \dfrac{180°}{\pi} = 140°$

33. $\sin\dfrac{\pi}{4} = \sin 45° = 0.7071$

37. $\cos\dfrac{5\pi}{6} = \cos 150° = -0.8660$

41. $\tan 0.7359 = 0.9056$

45. $\sec 2.07 = \dfrac{1}{\cos 2.07} = -2.09$

49. $\sin\theta = 0.3090,\ 0 \le \theta < 2\pi$
  $\theta = \sin^{-1}(0.3090)$
   $= 0.3141$ from calculator
  $\theta = \pi - \sin^{-1}(0.3090)$
   $= 2.827$ is also a solution

53. $\cos\theta = 0.6742,\ 0 \le \theta < 2\pi$
  $\theta = \cos^{-1}(0.6742)$
   $= 0.8309$ from calculator
  $\theta = 2\pi - \cos^{-1}(0.6742)$
   $= 5.452$ is also a solution

57. $V = \dfrac{1}{2}Wb\theta^2$

$V = \dfrac{1}{2} \cdot (8.75)(0.75)\left(5.5° \cdot \dfrac{\pi}{180°}\right)^2$

$V = 0.030$ ft $\cdot$ lb

8.4   Applications of Radian Measure

1. $s = r \cdot \theta = (3.30)\left(\dfrac{\pi}{3}\right) = 3.46$ in

5. $\theta = \dfrac{s}{r} = \dfrac{0.3913}{0.9449} = 0.4141 = 23.73°$

9. $A = \dfrac{1}{2}r^2\theta \Rightarrow r = \sqrt{\dfrac{2A}{\theta}} = \sqrt{\dfrac{2(0.0119)}{326° \cdot \dfrac{\pi}{180}}}$

$r = 0.0647$ ft

13. $s = r\theta = 3.30 \cdot \left(820° \cdot \dfrac{\pi}{180°}\right) = 47.2$

47.2 cm of tape is played.

17. From $\theta = w \cdot t$,

hour hand: $\theta = \dfrac{2\pi}{12} \cdot t$

minute hand: $\theta + \pi = \dfrac{2\pi}{1} \cdot t$, t in hours

$$\dfrac{\pi}{6} \cdot t + \pi = 2\pi \cdot t \Rightarrow t = \dfrac{6}{11} \text{ hour} = 32.73 \text{ minutes}$$

$$t = 32 \text{ minutes } 44 \text{ seconds}$$

at 32 minutes and 44 seconds after noon the hour and minute hands will be at 180°.

21. $w = \dfrac{\theta}{t} = \dfrac{\pi}{6}$ rad/s $= 0.52$ rad/s

25.

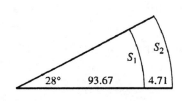

From $s = r\theta$

$s_1 = 93.67 \cdot 28° \cdot \dfrac{\pi}{180°}$

$s_1 = 45.78$

$s_2 = (93.67 + 4.71) \cdot 28° \cdot \dfrac{\pi}{180°}$

$s_2 = 48.08$

$s_2 - s_1 = 2.30$ ft. Outer rail is 2.30 ft longer.

29. $V = A \cdot t = \left[\dfrac{1}{2}r_1^2\theta - \dfrac{1}{2}r_2^2\theta\right] \cdot t = \dfrac{1}{2}\theta\left(r_1^2 - r_2^2\right) \cdot t$

$V = \dfrac{1}{2} \cdot 15.6° \cdot \dfrac{\pi}{180°}((285 + 15.2)^2 - 285^2) \cdot (0.305)$

$V = 369$ m^3

33. $v = r \cdot w \Rightarrow w = \dfrac{v}{r} = \dfrac{3.5 \frac{\text{mi}}{\text{h}}}{\frac{12.0 \text{ ft}}{2}} \cdot \dfrac{5280 \text{ ft}}{\text{mi}} \cdot \dfrac{\text{h}}{60 \text{ min}} \cdot \dfrac{1 \text{ r}}{2\pi \text{ rad}}$

$w = 8.2$ r/min

37.

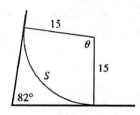

$$82.0° + 2 \cdot 90° + \theta = 360°$$
$$\theta = 98.0°$$

$$s = r\theta$$

$$s = 15.0 \cdot 98.0° \cdot \frac{\pi}{180°}$$

$$s = 25.7 \text{ ft}$$

41. $w = \dfrac{v}{r} = \dfrac{\frac{1}{4} \cdot 6.5}{3.75} = 0.433 \text{ rad/s}$

45. $\theta = w \cdot t = 2400\dfrac{r}{\text{min}} \cdot \dfrac{2\pi \text{ rad}}{r} \cdot \dfrac{\text{min}}{60 \text{ s}} \cdot 1 \text{ s}$

$\theta = 80\pi \text{ rad} = 250 \text{ rad}$

49.

θ	$\dfrac{\sin\theta}{\theta}$	$\dfrac{\tan\theta}{\theta}$
0.0001	0.9999999983	1.000000003
0.001	0.9999998333	1.000000333
0.01	0.9999833334	1.000033335
0.1	0.9983341665	1.003346721

For small θ, in rad, $\theta \approx \sin\theta \approx \tan\theta$

Chapter 8 Review Exercises

1. $r = \sqrt{6^2 + 8^2} = 10$ for $(6, 8)$

$$\sin\theta = \frac{y}{r} = \frac{8}{10} = \frac{4}{5}$$

$$\cos\theta = \frac{x}{r} = \frac{6}{10} = \frac{3}{5}$$

$$\tan\theta = \frac{y}{x} = \frac{8}{6} = \frac{4}{3}$$

$$\csc\theta = \frac{r}{y} = \frac{5}{4}$$

$$\sec\theta = \frac{r}{x} = \frac{5}{3}$$

$$\cot\theta = \frac{x}{y} = \frac{3}{4}$$

5. $\cos 132° = -\cos(180° - 132) = -\cos 48°$

$\tan 194° = \tan(194° - 180°) = \tan 14°$

9. $40° \cdot \dfrac{\pi}{180°} = \dfrac{2\pi}{9}$

$153° \cdot \dfrac{\pi}{180°} = \dfrac{17\pi}{20}$

13. $\dfrac{7\pi}{5} \cdot \dfrac{180°}{\pi} = 252°$; $\dfrac{13\pi}{18} \cdot \dfrac{180°}{\pi} = 130°$

17. $0.560 \cdot \dfrac{180°}{\pi} = 32.1°$

21. $102° \cdot \dfrac{\pi}{180°} = 1.78$

25. $262.05° \cdot \dfrac{\pi}{180°} = 4.5736$

29. $\cos 245.5° = -0.415$

33. $\csc 247.82° = -1.080$

37. $\tan 301.4° = -1.64$

41. $\sin \dfrac{9\pi}{4} = -0.5878$

45. $\sin 0.5906 = 0.5569$

49. $\tan\theta = 0.1817,\ 0 \le \theta < 360°$
$\quad\quad \theta = \tan^{-1}(0.1817) = 10.3°$ in QI
$\quad\quad \theta = 180° + 10.3° = 190.3°$ in QIII

53. $\cos\theta = 0.8387,\ 0 \le \theta < 2\pi$
$\quad\quad \theta = \cos^{-1}(0.8387) = 0.5759$ in QI
$\quad\quad \theta = 2\pi - 0.5759 = 5.707$ in QIV

57. $\cos\theta = -0.7222,\ \sin\theta < 0$ for $0° \le \theta < 360° \Rightarrow \theta$ in QIII
$\quad\quad \theta = \cos^{-1}(-0.7222) = 136.2364165$ from calculator
reference angle $= 180° - \theta = 43.76°$
QIII angle $= 180° +$ reference angle $= 223.76°$.

61. $r = \dfrac{s}{\theta} = \dfrac{20.3 \text{ in}}{107.5° \cdot \dfrac{\pi}{180°}} = 10.8$ in

65. $p = P_m \sin^2 377 \cdot t = .120 \sin^2(377 \cdot 2 \cdot 10^{-3})$
$\quad\quad p = 0.0562\ W$

69. (a) $3960 \cdot 60° \cdot \dfrac{\pi}{180°} = 3960 \cdot \dfrac{\pi}{3}$ over with pole, 4150 mi

$\quad\quad$ (b) $3960 \cdot \sin 30° \cdot \pi = 3960 \cdot \dfrac{\pi}{2}$ along 60°N latitude arc, 6220 mi

The distance over the north pole is shorter.

73.

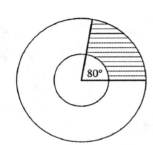

$\begin{aligned}
A_{\text{circle}} &= \pi \cdot 3.25^2 = A_1 \\
A_{\text{hole}} &= \pi \cdot 0.75^2 = A_2
\end{aligned}$

$A_{\text{hatched}} = \dfrac{1}{2} \cdot 80° \cdot \dfrac{\pi}{180°}(3.25^2 - 0.75^2) = A_3$

$A_{\text{hood}} = A - A_2 - A_3 = 24.4 \text{ ft}^2$

77. $A = \underbrace{\dfrac{1}{2} \cdot 15^2 \cdot 60° \cdot \dfrac{\pi}{180°}}_{\substack{\text{area of sector} \\ \text{formed by one arc}}} + (\underbrace{\dfrac{1}{2} \cdot 15^2 \cdot 60° \cdot \dfrac{\pi}{180°} - \dfrac{1}{2} \cdot 15 \cdot 15 \sin 60°}_{\substack{\text{area of equilateral} \\ \text{triangle inside} \\ \text{one sector}}})$

$A = 138 \text{ m}^2$

81. $v = r \cdot w = (1080 \text{ mi} + 70.0 \text{ mi}) \cdot \dfrac{1 \text{ r}}{1.95 \text{ h}} \cdot \dfrac{2\pi \text{ rad}}{\text{r}}$

$\quad\quad v = 3705.5 \text{ mi/h}$

VECTORS AND OBLIQUE TRIANGLES

9.1   Introduction to Vectors

1. (a) 300 km sw is a vector; it has magnitude and direction

   (b) 300 km is a scalar; it has only magnitude

5.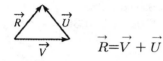

$$\vec{R} = \vec{V} + \vec{U}$$

9.

13.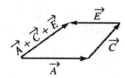

17.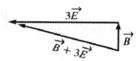

21.

25.

29.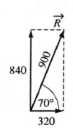

from drawing
$\vec{R}$ is approximately
900 lb at 70°

33.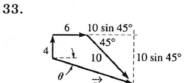

$R = 13$ mi

$\theta = 13°$

9.2   Components of Vectors

1. horizontal component $= 750 \cos 28° = 662$
vertical component $= 750 \sin 28° = 352$

5.

x-component $= 8.6 \cos 68° = 3.22$
y-component $= 8.6 \sin 68° = 7.97$

9.

x-component $= 9.04 \cos 283.3° = 2.08$
y-component $= 9.04 \sin 283.3° = -8.80$

13.

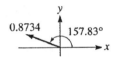

x-component $= 0.8734 \cos 157.83° = -0.8088$
y-component $= 0.8734 \sin 157.83° = 0.3296$

17.

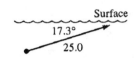

x-component $= 25.0 \cos 17.3° = 23.9$ km/h
y-component $= 25.0 \sin 17.3° = 7.43$ km/h

21.

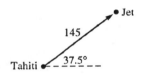

x-component $= 145 \cos 37.5° = 115$ km to the east
y-component $= 145 \sin 37.5° = 88.3$ km to the north

9.3   Vector Addition by Components

1.

$R = \sqrt{14.7^2 + 19.2°} = 24.2$

$\theta = \tan^{-1} \dfrac{19.2}{14.7} = 52.6°$

5.

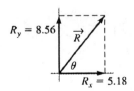

$R = \sqrt{5.18^2 + 8.56^2} = 10.0$

$\theta = \tan^{-1} \dfrac{8.56}{5.18} = 58.8°$

9.

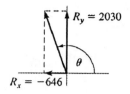

$$R = \sqrt{(-646)^2 + 2030^2} = 2130$$

$$\tan^{-1} \frac{2030}{-646} = -72.3° \text{ from calculator}$$

$$\theta = 180° - 72.3° = 107.7°$$

13.

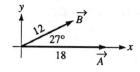

$$R_x = 18 + 12 \cos 27°$$
$$R_y = 0 + 12 \sin 27°$$
$$R = \sqrt{R_x^2 + R_y^2} = 29.2$$

$$\theta = \tan^{-1} \frac{R_y}{R_x} = 10.8°$$

17.

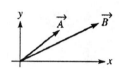

$$R_x = A_x + B_x = 9.821 \cos 34.27° + 17.45 \cos 752.5°$$
$$R_y = 9.821 \sin 34.27° + 17.45 \sin 752.5°$$
$$R = \sqrt{R_x^2 + R_y^2} = 27.27$$

$$\theta = \tan^{-1} \frac{R_y}{R_x} = 33.14°$$

21. $R_x = A_x + B_x + C_x$
$$R_x = 21.9 \cos 236.2° + 96.7 \cos 11.5° + 62.9 \cos 143.4°$$
$$R_y = A_y + B_y + C_y$$
$$R_y = 21.9 \sin 236.2° + 96.7 \sin 11.5° + 62.9 \sin 143.4°$$
$$R = \sqrt{R_x^2 + R_y^2} = 50.2$$

$$\theta = \tan^{-1} \frac{R_y}{R_x} = 50.3°$$

25. $R_x = 302 \cos(180° - 45.4°) + 155 \cos(180° + 53.0°) + 212 \cos 30.8° = -123$
$$R_y = 302 \sin(180° - 45.4°) + 155 \sin(180° + 53°) + 212 \sin 30.8° = 200$$
$$R = \sqrt{R_x^2 + R_y^2} = 235$$

$$\theta = \tan^{-1} \frac{R_y}{R_x} = -58.4° \text{ from calculator}$$

$$\theta_R = 180° - 58.4 = 121.6° \text{ since } \theta_R \text{ is in QII}$$

9.4   Applications of Vectors

1.

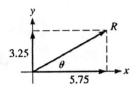

$$R = \sqrt{5.75^2 + 3.25^2} = 6.60 \text{ lb}$$

$$\theta = \tan^{-1} \frac{3.25}{5.75} = 29.5°$$

5.

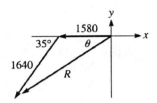

$$R_x = -1580 - 1640 \cos 35.0°$$
$$R_y = -1640 \sin 35.0°$$
$$R = \sqrt{R_x^2 + R_y^2} = 3070 \text{ ft}$$

$$\theta = \tan^{-1} \frac{R_y}{R_x} = 17.8° \text{ S of W}$$

9.

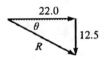

$$R = \sqrt{22.0^2 + 12.5^2} = 25.3 \text{ km/h}$$
$$\theta = \tan^{-1} \frac{12.5}{22.0} = 29.6°$$

13.

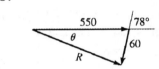

$$R = \sqrt{R_x^2 + R_y^2}$$
$$R = \sqrt{(550 - 60 \cos 78°)^2 + (60 \sin 78°)^2}$$
$$R = 540 \text{ km/h}$$

$$\theta = \tan^{-1} \frac{60 \sin 78°}{550 - 60 \cos 78°}$$

$$\theta = 6°$$

17. $r = \dfrac{d}{2} = \dfrac{8.20}{2} = 4.10$

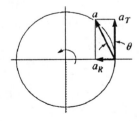

$$a = \sqrt{a_T^2 + a_R^2} = \sqrt{(\alpha r)^2 + (w^2 r)^2}$$
$$a = \sqrt{(318(4.10))^2 + (212^2 \cdot 4.10)^2}$$
$$a = 184,000 \text{ in/min}^2$$

$$\theta = \tan^{-1} \frac{a_R}{a_T} = \tan^{-1} \frac{w_r^2 r}{\alpha r} = \tan^{-1} \frac{212^2}{318}$$

$$\theta = 89.6°$$

21. top view of plane

$$V_H = \sqrt{75.0^2 + 15.0^2} = 76.5$$

$$\theta = \tan^{-1}\frac{15.0}{75.0} = 11.3°$$

$$V_v = 9.80(2.00) = 19.6$$

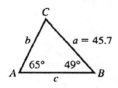

$$V = \sqrt{76.5^2 + 19.6^2}$$
$$V = 79.0 \text{ m/s}$$

$$\alpha = \tan^{-1}\frac{19.6}{76.5}$$

$$\alpha = 14.4°, \; 75.6° \text{ from vertical}$$

9.5   Oblique Triangles, the Law of Sines

1.

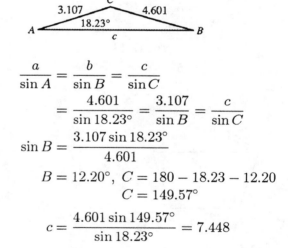

5.

$$C = 180° - 65.0° - 49.0° = 66.0°$$

$$\frac{a}{\sin A} = \frac{b}{\sin B} = \frac{c}{\sin C}$$

$$\frac{45.7}{\sin 65.0°} = \frac{b}{\sin 49.0°} \Rightarrow b = 38.1$$

$$\frac{45.7}{\sin 65.0°} = \frac{c}{\sin 66.0°} \Rightarrow c = 46.1$$

$$\frac{a}{\sin A} = \frac{b}{\sin B} = \frac{c}{\sin C}$$

$$= \frac{4.601}{\sin 18.23°} = \frac{3.107}{\sin B} = \frac{c}{\sin C}$$

$$\sin B = \frac{3.107 \sin 18.23°}{4.601}$$

$$B = 12.20°, \; C = 180 - 18.23 - 12.20$$
$$C = 149.57°$$

$$c = \frac{4.601 \sin 149.57°}{\sin 18.23°} = 7.448$$

9.

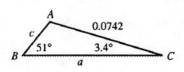

$$A = 180° - 51.0° - 3.4° = 125.6°$$

$$\frac{a}{\sin A} = \frac{b}{\sin B} = \frac{c}{\sin C}$$

$$\frac{a}{\sin 125.6°} = \frac{0.0742}{\sin 51.0°} = \frac{c}{\sin 3.4°}$$

$$a = \frac{0.0742 \sin 125.6°}{\sin 51.0°} = 0.0776$$

$$c = \frac{0.0742 \sin 3.4°}{\sin 51.0°} = 0.00566$$

13.

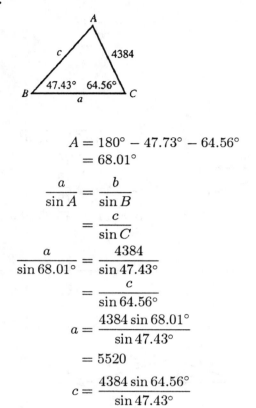

$$A = 180° - 47.73° - 64.56°$$
$$= 68.01°$$

$$\frac{a}{\sin A} = \frac{b}{\sin B}$$
$$= \frac{c}{\sin C}$$

$$\frac{a}{\sin 68.01°} = \frac{4384}{\sin 47.43°}$$
$$= \frac{c}{\sin 64.56°}$$

$$a = \frac{4384 \sin 68.01°}{\sin 47.43°}$$
$$= 5520$$

$$c = \frac{4384 \sin 64.56°}{\sin 47.43°}$$
$$= 5376$$

21.

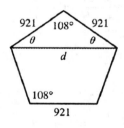

$$2\theta + 108° = 180°$$
$$\theta = 36°$$

$$\frac{921}{\sin \theta} = \frac{d}{\sin 108°}$$
$$d = \frac{921 \sin 108°}{\sin 36°}$$
$$d = 1490 \text{ ft}$$

25. $d =$ distance along Arsenal from Gravois
to Jefferson

17.

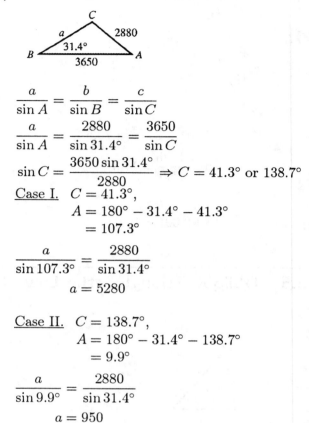

$$\frac{a}{\sin A} = \frac{b}{\sin B} = \frac{c}{\sin C}$$
$$\frac{a}{\sin A} = \frac{2880}{\sin 31.4°} = \frac{3650}{\sin C}$$
$$\sin C = \frac{3650 \sin 31.4°}{2880} \Rightarrow C = 41.3° \text{ or } 138.7°$$
<u>Case I.</u> $C = 41.3°,$
$$A = 180° - 31.4° - 41.3°$$
$$= 107.3°$$

$$\frac{a}{\sin 107.3°} = \frac{2880}{\sin 31.4°}$$
$$a = 5280$$

<u>Case II.</u> $C = 138.7°,$
$$A = 180° - 31.4° - 138.7°$$
$$= 9.9°$$

$$\frac{a}{\sin 9.9°} = \frac{2880}{\sin 31.4°}$$
$$a = 950$$

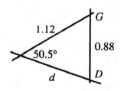

$$\frac{1.12}{\sin D} = \frac{0.88}{\sin 50.5}$$
$$D = 79.1° \quad \text{or} \quad 100.9°$$

from drawing ∢ is acute.

$$C = 180° - 50.5° - 79.1° = 50.4°$$
$$\frac{d}{\sin C} = \frac{0.88}{\sin 50.5°}$$
$$d = \frac{0.88 \sin 50.4°}{\sin 50.5°}$$
$$d = 0.88 \text{ mi}$$

29. $C = 180° - 86.5 - 90.8° = 2.7°$

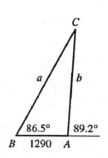

$$\frac{b}{\sin 86.5°} = \frac{1290}{\sin 2.7°}$$

$$b = 27,300 \text{ km}$$

$A = 180° - 89.2° = 90.8°$

9.6   The Law of Cosines

1.

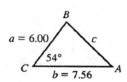

$c^2 = a^2 + b^2 - 2ab \cos C$
$c^2 = 6.00^2 + 7.56^2 - 2(6.00)(7.56) \cos 54.0°$
$\quad c = 6.31$
$b^2 + c^2 - 2bc \cos A = a^2$

$7.56^2 + 6.31^2 - 2(7.56)(6.31) \cos A = 6^2$
$\cos A = 0.639$
$A = 50.3°$

$B = 180° - 54.0° - 50.3°$
$B = 75.7°$

5.

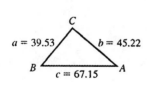

$b^2 = a^2 + c^2 - 2ac \cos B$
$45.22^2 = 39.53^2 + 67.15^2 - 2(39.53)(67.15) \cos B$
$\cos B = 0.7585$
$B = 40.67°$

$b^2 + c^2 - 2bc \cos A = a^2$

$45.22^2 + 67.15^2 - 2(45.22)(67.15) \cos A = 39.53^2$
$\cos A = 0.8219$
$A = 34.73°$

$C = 180 - 40.67° - 34.73° = 104.6°$

9.

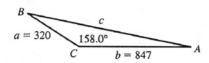

$$c^2 = a^2 + b^2 - 2ab \cos C = 320^2 + 847^2 - 2(320)(847) \cos 158.0$$
$$c = 1150$$
$$b^2 + c^2 - 2bc \cos A = a^2$$
$$847^2 + 1150^2 - 2(847)(1150) \cos A = 320^2$$
$$\cos A = 0.9946$$
$$A = 6.0°$$
$$a^2 + c^2 - 2ac \cos B = b^2$$
$$320^2 + 1150^2 - 2(320)(1150) \cos B = 847^2$$
$$\cos B = 0.9612$$
$$B = 16.0°$$

13.

$$c^2 = a^2 + b^2 - 2ab \cos C$$
$$159.1^2 = a^2 + 103.7^2 - 2a(103.7) \cos 104.67$$
$$a^2 + 52.52a - 14,559.12 = 0$$
$$a = 97.23, \ -149.75 \ (\text{reject, } a > 0)$$
$$\cos B = 0.7761$$
$$B = 39.09°$$

$$a^2 + c^2 - 2ac \cos B = b^2$$
$$97.23^2 + 159.1^2 - 2(97.23)(159.1) \cos B = 103.7^2$$
$$A = 180° - C - B = 180° - 104.67 - 39.09° = 36.24°$$

17.

$$a^2 = b^2 + c^2 - 2bc \cos A$$
$$723^2 = 598^2 + 158^2 - 2(598)(158) \cos A$$
$$\cos A = -0.7417$$
$$A = 137.9°$$
$$b^2 = a^2 + c^2 - 2ac \cos B$$
$$598^2 = 723^2 + 158^2 - 2(723)(158) \cos B$$
$$\cos B = 0.8320$$
$$B = 33.7°$$
$$C = 180° - A - B = 180° - 137.9° - 33.7° = 8.4°$$

21.

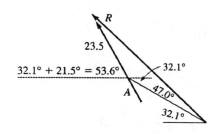

from $d = rt$, $23.5(2.00) = 47.0$ and
$$23.5(1.00) = 23.5$$
$\measuredangle = 32.1° + 90° + (90° - 53.6°)$
$\measuredangle = 158.5°$

$R^2 = 23.5^2 + 47^2 - 2(23.5)(47)\cos 158.5°$
$R = 69.4$ miles from base

25.

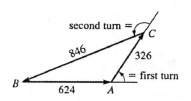

$846^2 = 624^2 + 326^2 - 2(624)(326)\cos A$
$\cos A = -0.5409$
$A = 122.7°$
first turn $= 180° - 122.7° = 57.3°$
$624^2 = 846^2 + 326^2 - 2(846)(326)\cos C$
$\cos C = 0.7843$
$C = 38.3°$
second turn $= 180° - 38.3° = 141.7°$

29.

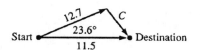

$c^2 = 12.7^2 + 11.5^2 - 2(12.7)(11.5)\cos 23.6°$
$c = 5.09$ km/h

Chapter 9 Review Exercises

1.

y-component $= 65.0\cos 28.0° = 57.4$
x-component $= 65.0\sin 28.0° = 30.5$

5.

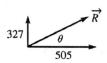

$R = \sqrt{327^2 + 505^2} = 602$

$\theta = \tan^{-1}\dfrac{327}{505} = 32.9°$

9.

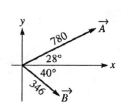

$R_x = 780\cos 28.0° + 346\cos 40.0° = 954$
$R_y = 780\sin 28.0° - 346\sin 40.0° = 144$

$R = \sqrt{R_x^2 + R_y^2} = \sqrt{954^2 + 144^2} = 965$

$\theta_R = \tan^{-1}\dfrac{144}{954} = 8.6°$

13.

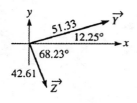

$$Y_x = 51.33 \cos 12.25° = 5016$$
$$Y_y = 51.33 \sin 12.25° = 10.89$$
$$Z_x = 42.61 \cos 68.23° = 15.80$$
$$Z_y = -42.61 \sin 68.23° = -39.57$$
$$R_x = 50.16 + 15.80 = 65.98$$
$$R_y = 10.89 - 39.57 = -28.68$$
$$R = \sqrt{R_x^2 + R_y^2} = \sqrt{65.98^2 + (-28.68)^2} = 71.94$$

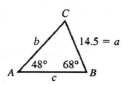

$$\tan\theta = \frac{R_y}{R_x} = \frac{-28.68}{65.98}$$
$$\theta = 336.50°, \ \theta_{\text{ref}} = 23.50°$$

17.

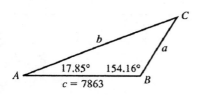

$$C = 180° - 48.0° - 68.0° = 64.0°$$
$$\frac{14.5}{\sin 48.0°} = \frac{b}{\sin 68.0°} = \frac{c}{\sin 64.0°}$$
$$b = \frac{14.5 \sin 68.0°}{\sin 48.0°} = 18.1,$$
$$c = \frac{14.5 \sin 64.0°}{\sin 48.0°} = 17.5$$

21.

$$C = 180° - 17.85° - 154.16° = 7.99°$$
$$\frac{a}{\sin 17.85°} = \frac{b}{\sin 154.16} = \frac{7863}{\sin 7.99°}$$
$$b = \frac{7863 \sin 154.16°}{\sin 7.99°} = 24{,}660$$
$$a = \frac{7863 \sin 17.85°}{\sin 7.99°} = 17{,}340$$

25.

$$\frac{a}{\sin A} = \frac{14.5}{\sin B} = \frac{13.0}{\sin 56.6}$$
$$\sin B = \frac{14.5 \sin 56.6}{13.0}$$
$$B = 68.6° \quad \text{or} \quad 111.4°$$

<u>Case I:</u> $B = 68.6°$, $A = 180° - 68.6° - 56.6° = 54.8°$
$$\frac{a}{\sin 54.8°} = \frac{13.0}{\sin 56.6°} \Rightarrow a = 12.7$$

<u>Case II:</u> $B = 111.4°$, $A = 180° - 111.4° - 56.6° = 12.0°$
$$\frac{a}{\sin 12.0°} = \frac{13.0}{\sin 56.6°} \Rightarrow a = 3.24$$

29.

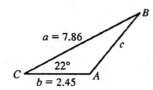

$$c^2 = a^2 + b^2 - 2ab \cos C$$
$$c^2 = 7.86^2 + 2.45^2 - 2(7.86)(2.45) \cos 22.0°$$
$$c = 5.66$$
$$a^2 = b^2 + c^2 - 2bc \cos A$$
$$7.86^2 = 2.45^2 + 5.66^2 - 2(2.45)(5.66) \cos A$$
$$\cos A = -0.8560$$
$$A = 148.9°$$
$$B = 180° = C - A = 180° - 22° - 148.9°$$
$$B = 9.1°$$

33.

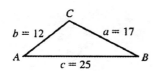

$$a^2 = b^2 + c^2 - 2bc \cos A$$
$$17^2 = 12^2 + 25^2 - 2(12)(25) \cos A$$
$$\cos A = 0.8$$
$$A = 37°$$
$$b^2 = a^2 + c^2 - 2ac \cos B$$
$$12^2 = 17^2 + 25^2 - 2(17)(25) \cos B$$
$$\cos B = 0.9059$$
$$B = 25°$$
$$C = 180° - 37° - 25° = 118°$$

37.

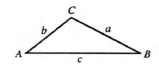

$$a^2 = b^2 + c^2 - 2bc \cos A$$
$$b^2 = a^2 + c^2 - 2ac \cos B$$
$$\underline{c^2 = a^2 + b^2 - 2ab \cos C \quad add}$$

$$a^2 + b^2 + c^2 = 2a^2 + 2b^2 + 2c^2 - 2bc \cos A - 2ac \cos B - 2ab \cos C$$
$$a^2 + b^2 + c^2 = 2bc \cos A + 2ac \cos B + 2ab \cos G$$

$$\frac{a^2 + b^2 + c^2}{2abc} = \frac{\cos A}{a} + \frac{\cos}{b} + \frac{\cos C}{c}$$

41.   horizontal component $= 175.6 \cos 152.48° = -155.7 \text{ lb}$
     vertical component $= 175.6 \sin 152.48° = 81.14 \text{ lb}$

45.

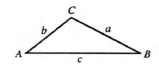

$$d = 480 \text{ km/h} \cdot 3 \text{ min} \cdot \frac{\text{h}}{60 \text{ min}}$$
$$d = 24 \text{ km}$$
$$h = d \sin 24° = 24 \sin 24°$$
$$h = 9.8 \text{ km}$$

49. $x^2 + 2.7^2 - 2 \cdot x \cdot 2.7 \cos 27.5° = 1.25^2$
    $x^2 - 5.4 \cdot \cos 27.5 \cdot x + 5.7275 = 0$

$$x = \frac{5.4 \cos 27.5 \pm \sqrt{(-5.4 \cos 27.5)^2 - 4(1)(5.7275)}}{2(1)}$$

$x = 2.30,\ 2.49$ m

53.

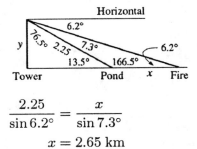

$$\frac{2.25}{\sin 6.2°} = \frac{x}{\sin 7.3°}$$

$$x = 2.65 \text{ km}$$

57.

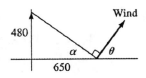

$$\tan \alpha = \frac{480}{650}$$

$$\alpha = 36.4° \text{ N of E}$$
$$F = \sqrt{F_x^2 + F_y^2}$$
$$= \sqrt{650^2 + 480^2}$$
$$= 810 \text{ N}$$

61.

$$\begin{array}{c}
B \\
2.00 \diagup \backslash 3.00 \\
A \underline{\qquad} C \\
4.50
\end{array}$$

Use law of cosines three times.

$$2.00^2 + 4.50^2 - 2(2.00)(4.50) \cos A = 3.00^2$$
$$A = 32.1°$$
$$2.00^2 + 3.00^2 - 2(2.00)(3.00) \cos B = 4.50^2$$
$$B = 127.2°$$
$$3.00^2 + 4.50^2 - 2(3.00)(4.50) \cos C = 2.00^2$$
$$C = 20.7°$$

GRAPHS OF THE TRIGONOMETRIC FUNCTIONS

10.1 Graphs of $y = a \sin x$ and $y = a \cos x$

1. $y = \sin x$

x	y
$-\pi$	0
$-\dfrac{3\pi}{4}$	$-\dfrac{\sqrt{2}}{2}$
$-\dfrac{\pi}{2}$	-1
$-\dfrac{\pi}{4}$	$-\dfrac{\sqrt{2}}{2}$
0	0
$\dfrac{\pi}{4}$	$\dfrac{\sqrt{2}}{2}$
$\dfrac{\pi}{2}$	1
$\dfrac{3\pi}{4}$	$\dfrac{\sqrt{2}}{2}$
π	0

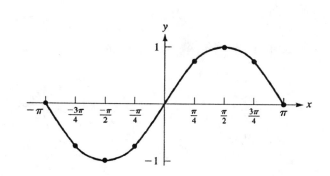

5. $y = 3 \sin x$

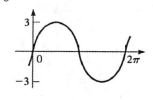

x	y
0	0
$\dfrac{\pi}{2}$	3
π	0
$\dfrac{3\pi}{2}$	-3
2π	0

9. $y = 2 \cos x$

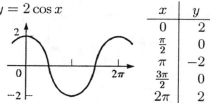

x	y
0	2
$\dfrac{\pi}{2}$	0
π	-2
$\dfrac{3\pi}{2}$	0
2π	2

13. $y = -\sin x$

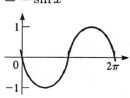

x	y
0	0
$\dfrac{\pi}{2}$	-1
π	0
$\dfrac{3\pi}{2}$	1
2π	0

17. $y = -\cos x$

x	y
0	-1
$\dfrac{\pi}{2}$	0
π	1
$\dfrac{3\pi}{2}$	0
2π	-1

21. $y = \sin x$

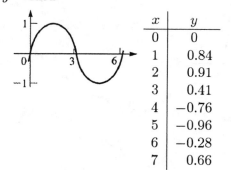

x	y
0	0
1	0.84
2	0.91
3	0.41
4	-0.76
5	-0.96
6	-0.28
7	0.66

25. The graph shown is the graph of $y = 4 \sin x$.

10.2   Graphs of $y = a \sin bx$ and $y = a \cos bx$

1. The period of $y = 2 \sin 6x$ is $\dfrac{2\pi}{6} = \dfrac{\pi}{3}$.

5. The period of $y = -2 \sin 12x$ is $\dfrac{2\pi}{12} = \dfrac{\pi}{6}$.

9. The period of $y = 520 \sin 2\pi x$ is $\dfrac{2\pi}{2\pi} = 1$.

13. The period of $y = 3 \sin \dfrac{1}{3}x$ is $\dfrac{2\pi}{\dfrac{1}{3}} = 6\pi$.

17. The period of $y = 0.4 \sin \dfrac{2\pi x}{3}$ is $\dfrac{2\pi}{\dfrac{2\pi}{3}} = 3$.

21. $y = 2 \sin 6x$

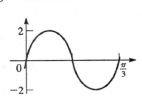

x	y
0	0
$\frac{\pi}{12}$	2
$\frac{\pi}{6}$	0
$\frac{\pi}{4}$	-2
$\frac{\pi}{3}$	0

25. $y = -2 \sin 12x$

x	y
0	0
$\frac{\pi}{24}$	-2
$\frac{\pi}{12}$	0
$\frac{\pi}{8}$	2
$\frac{\pi}{6}$	0

29. $y = 520 \sin 2\pi x$

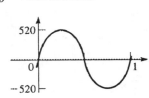

x	y
0	0
0.25	520
0.50	0
0.75	-520
1.0	0

33. $y = 3 \sin \dfrac{1}{3}x$

x	y
0	0
$\frac{3\pi}{2}$	3
3π	0
$\frac{9\pi}{2}$	-3
6π	0

37. $y = 0.4 \sin \dfrac{2\pi x}{3}$

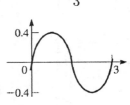

x	y
0	0
0.75	0.4
1.50	0
2.25	-0.4
3.0	0

41. $\text{period} = \dfrac{2\pi}{b} = \dfrac{\pi}{3} \Rightarrow b = 6 \Rightarrow y = \sin 6x$

45. $V = 170 \sin 120\pi t$

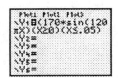

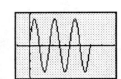

49. period $= \dfrac{2\pi}{b} = \pi \Rightarrow b = 2$. Amplitude $= 0.5 \Rightarrow y = 0.5\cos 2x$

 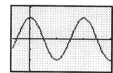

10.3   Graphs of $y = a\sin(bx+c)$ and $y = a\cos(bx+c)$

1.

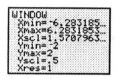 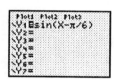 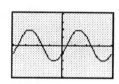

$y = \sin\left(x - \dfrac{\pi}{6}\right)$ has $a = 1$, period $=$

and displacement $= -\dfrac{-\dfrac{\pi}{6}}{1} = \dfrac{\pi}{6}$.

5.

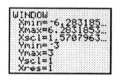 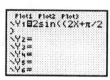 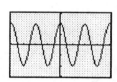

$y = 2\sin\left(2x + \dfrac{\pi}{2}\right)$ has $a = 2$,

period $= \dfrac{2\pi}{2}$ and

displacement $= -\dfrac{\dfrac{\pi}{2}}{2} = -\dfrac{\pi}{4}$.

9.

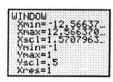 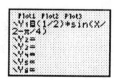 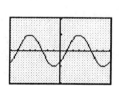

$y = \dfrac{1}{2}\sin\left(\dfrac{x}{2} - \dfrac{\pi}{4}\right)$ has $a = \dfrac{1}{2}$,

period $= \dfrac{2\pi}{\dfrac{1}{2}} = 4\pi$, and

displacement $= -\dfrac{-\dfrac{\pi}{4}}{\dfrac{1}{2}} = \dfrac{\pi}{2}$.

13.

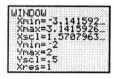 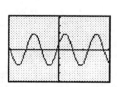

$y = \sin\left(\pi x + \dfrac{\pi}{8}\right)$ has $a = 1$,

period $= \dfrac{2\pi}{\pi} = 2$, and

displacement $= -\dfrac{\dfrac{\pi}{8}}{\pi} = -\dfrac{1}{8}$.

17.

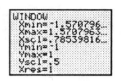   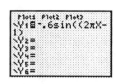   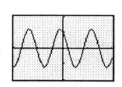

$y = -0.6\sin(2\pi x - 1)$ has $a = |-0.6|$

period $= \dfrac{2\pi}{2\pi} = 1$, and $a = 0.6$

displacement $= -\dfrac{-1}{2\pi} = \dfrac{1}{2\pi}$.

21.

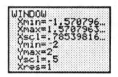      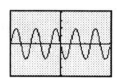

$y = \sin(\pi^2 x - \pi)$ has $a = 1$,

period $= \dfrac{2\pi}{\pi^2} = \dfrac{2}{\pi}$, and

displacement $= -\dfrac{-\pi}{\pi^2} = \dfrac{1}{\pi}$.

25.

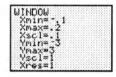   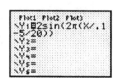   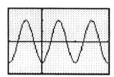

29. From the graph, $a = 5$, period $= 16 = \dfrac{2\pi}{b} \Rightarrow b = \dfrac{\pi}{8}$. Then displacement $= -1 = -\dfrac{c}{b} = -\dfrac{c}{\frac{\pi}{8}} \Rightarrow c = \dfrac{\pi}{8}$.

$y = a\sin(bx + c)$ becomes $y = 5\sin\left(\dfrac{\pi}{8}x + \dfrac{\pi}{8}\right)$.

10.4   Graphs of $y = \tan x$, $y = \cot x$, $y = \sec x$, $y = \csc x$

1.

x	$-\dfrac{\pi}{2}$	$-\dfrac{\pi}{3}$	$-\dfrac{\pi}{4}$	$-\dfrac{\pi}{6}$	0	$\dfrac{\pi}{6}$	$\dfrac{\pi}{4}$	$\dfrac{\pi}{3}$	$\dfrac{\pi}{2}$	$\dfrac{2\pi}{3}$	$\dfrac{3\pi}{4}$	$\dfrac{5\pi}{6}$	π
y	$*$	$-\sqrt{3}$	-1	$\dfrac{-1}{\sqrt{3}}$	0	$\dfrac{1}{\sqrt{3}}$	1	$\sqrt{3}$	$*$	$-\sqrt{3}$	-1	$\dfrac{1}{-\sqrt{3}}$	0

$y = \tan x$

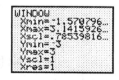   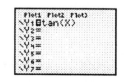   

5. $y = 2\tan x$ is the graph of $y = \tan x$ stretched by a factor of 2.

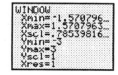   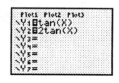   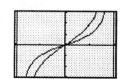

9. The graph of $y = -2\cot x$ is the graph of $y = \tan x$ reflected in the x-axis and stretched by a factor of 2.

  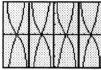

13. $y = \tan 2x$

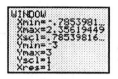  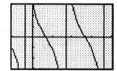

17. $y = 2\cot\left(2x + \dfrac{\pi}{6}\right)$

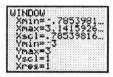 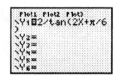 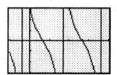

21. $d = 3.00\sec\theta,\ 0 \le \theta \le \dfrac{\pi}{2}$

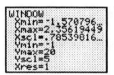 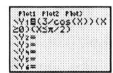 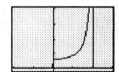

10.5   Applications of the Trigonometric Graphs

1. $d = R\sin wt = 2.40\sin 2t$ has $a = 2.40$, period $= \dfrac{2\pi}{2} = \pi$, and displacement $= 0$.

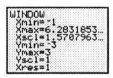 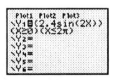 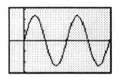

5. $D = A\sin(wt + \partial) = 500\sin(3.6t)$ has $a = 500$, period $= \dfrac{2\pi}{3.6} = \dfrac{5\pi}{9}$, and displacement $= 0$.

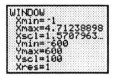  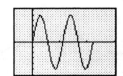

9. $y = A \sin 2\pi \left(\dfrac{t}{T} - \dfrac{x}{\lambda} \right)$

$y = 3.20 \sin 2\pi \left(\dfrac{t}{0.050} - \dfrac{5.00}{40.0} \right)$ has $a = 3.20$, period $= \dfrac{2\pi}{\dfrac{2\pi}{0.050}} = 0.050$, and displacement

$= -\dfrac{\dfrac{-2\pi(5.00)}{40.0}}{\dfrac{2\pi}{0.050}} = 0.00625$

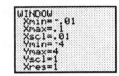 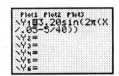 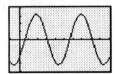

13. $y = 14.0 \sin 40.0\pi t$ has a $a = 14.0$, period $= \dfrac{2\pi}{40\pi} = \dfrac{1}{20}$, and displacement $= 0$.

 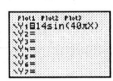 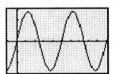

10.6   Composite Trigonometric Curves

1. $y = 1 + \sin x$

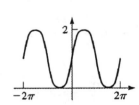

x	-2π	$-\frac{3\pi}{2}$	$-\pi$	$-\frac{\pi}{2}$	0	$\frac{\pi}{2}$	π	$\frac{3\pi}{2}$	2π
y	1	2	1	0	1	2	1	0	1

5. $y = \dfrac{1}{10}x^2 - \sin \pi x$

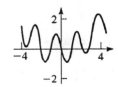

x	-4	-3.43	-2.55	-1.88	-1.47	-1.03	-0.51	0	0.49	0.97	1.53	2.15	2.45	2.73	4
y	1.60	0.20	1.64	0	-0.78	0	1.03	0	-0.98	0	1.23	0	-0.39	0	1.6

9. $y = x^3 + 10\sin 2x$

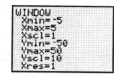 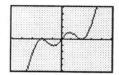

13. $y = 20\cos 2x + 30\sin x$

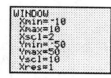 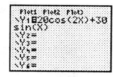 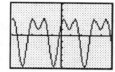

17. $y = \sin \pi x - \cos 2x$

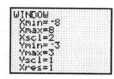 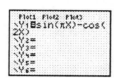 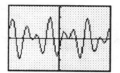

21. $x = \sin t$, $y = \sin t$

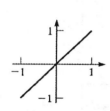

t	x	y
$-\frac{\pi}{2}$	-1	-1
$-\frac{\pi}{4}$	-0.71	-0.71
0	0	0
$\frac{\pi}{4}$	0.71	0.71
$\frac{\pi}{2}$	1	1

25. $x = \cos \pi \left(t + \dfrac{1}{6} \right)$, $y = 2\sin \pi t$

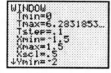 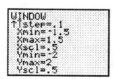 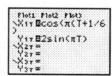 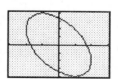

29. $x = \sin t$, $y = \sin 5t$

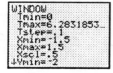 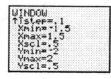 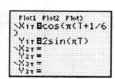 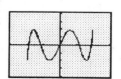

33. $T = 56 - 22 \cos\left[\dfrac{\pi}{6}(x - 0.5)\right]$

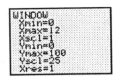   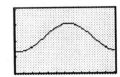

37. $i = 0.32 + 0.50 \sin t - 0.20 \cos 2t$

Chapter 10 Review Exercises

1. $y = \dfrac{2}{3}\sin x$

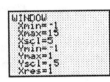   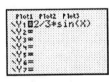   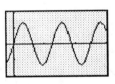

5. $y = 2 \sin 3x$

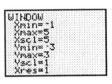   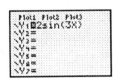   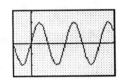

9. $y = 3 \cos \dfrac{1}{3}x$

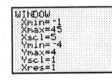   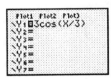   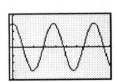

13. $y = 5 \cos 2\pi x$

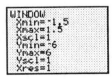   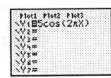   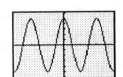

17. $y = 2\sin\left(3x - \dfrac{\pi}{2}\right)$

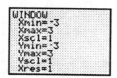 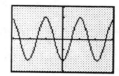

21. $y = -\sin\left(\pi x + \dfrac{\pi}{6}\right)$

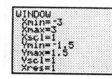 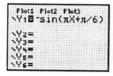 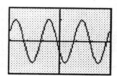

25. $y = 3\tan x$

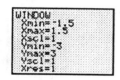 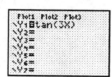 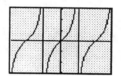

29. $y = 2 + \dfrac{1}{2}\sin 2x$

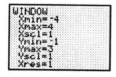 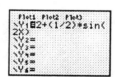 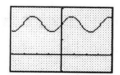

33. $y = 2\sin x - \cos 2x$

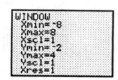  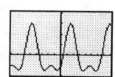

37. $y = \dfrac{\sin x}{x}$

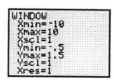 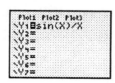 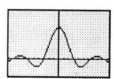

41. From the graph, $a = 2$, period $= \pi = \dfrac{2\pi}{b} \Rightarrow b = 2$, and displacement $= -\dfrac{c}{b} = -\dfrac{\pi}{4} \Rightarrow c = \dfrac{\pi}{2}$.
$y = a\sin(bx + c)$ is $y = 2\sin\left(2x + \dfrac{\pi}{2}\right)$.

45. $x = -\cos 2\pi t$, $y = 2\sin \pi t$

  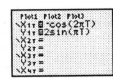 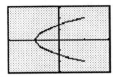

49. $R = \dfrac{v_0^2 \sin 2\theta}{g} = \dfrac{(1000)^2 \sin 2\theta}{9.8}$

 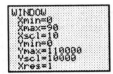 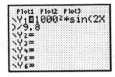 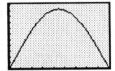

53. $i = 10\,|\sin 120\pi t|$, $0 \le t \le 0.05$, period $= \dfrac{.05}{6} \approx 0.00835$

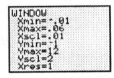  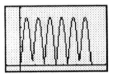

57. $y = 4\sin 2t - 2\cos 2t$

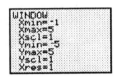 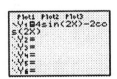 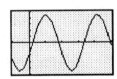

61. $Z = R\sec\theta$, $-\dfrac{\pi}{2} < \theta < \dfrac{\pi}{2}$. Graph shown is for an R value of 1. In general, the y-int would be R.

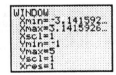 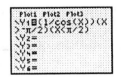 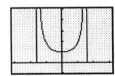

65. (a) If a is doubled in $y = a\sin(bx + c)$ the amplitude will be doubled.

   (b) If b is doubled in $y = a\sin(bx + c)$ the period will be reduced by one half.

   (c) If c is doubled in $y = a\sin(bx + c)$ the displacement will be doubled.

Chapter 11

EXPONENTS AND RADICALS

11.1 Simplifying Expressions with Integral Exponents

1. $x^7 \cdot x^{-4} = x^{7+(-4)} = x^3$

5. $5 \cdot 5^{-3} = 5^{1+(-3)} = 5^{-2} = \dfrac{1}{5^2} = \dfrac{1}{25}$

9. $(5an^{-2})^{-1} = 5^{-1}a^{-1}n^{(-2)(-1)} = \dfrac{n^2}{5a}$

13. $-7x^0 = -7 \cdot 1 = -7$

17. $(7ax)^{-3} = \dfrac{1}{(7ax)^3} = \dfrac{1}{7^3a^3x^3} = \dfrac{1}{343a^3x^3}$

21. $\left(\dfrac{a}{b^{-2}}\right)^{-3} = \dfrac{a^{-3}}{(b^{-2})^{-3}} = \dfrac{\frac{1}{a^3}}{b^{(-2)(-3)}} = \dfrac{\frac{1}{a^3}}{b^6} = \dfrac{1}{a^3b^6}$

25. $3x^{-2} + 2y^{-2} = \dfrac{3}{x^2} + \dfrac{2}{y^2}$

29. $\left(\dfrac{3a^2}{4b}\right)^{-3}\left(\dfrac{4}{a}\right)^{-5} = \dfrac{3^{-3}a^{-6}}{4^{-3}b^{-3}} \cdot \dfrac{4^{-5}}{a^{-5}}$

$$= \dfrac{4^3b^3}{3^3a^6} \cdot \dfrac{a^5}{4^5} = \dfrac{b^3}{432a}$$

33. $2a^{-2} + (2a^{-2})^4 = \dfrac{2}{a^2} + 2^4a^{-8} = \dfrac{2}{a^2} + \dfrac{16}{a^8}$

$$= \dfrac{2a^6 + 16}{a^8}$$

37. $(R_1^{-1} + R_2^{-1})^{-1} = \dfrac{1}{\dfrac{1}{R_1} + \dfrac{1}{R_2}} = \dfrac{R_1R_2}{R_1 + R_2}$

41. $\dfrac{6^{-1}}{4^{-2}+2} = \dfrac{\frac{1}{6}}{\frac{1}{4^2}+2} = \dfrac{\frac{1}{6}}{\frac{1}{16}+2} \cdot \dfrac{48}{48}$

$$= \dfrac{8}{3+96} = \dfrac{8}{99}$$

45. $2t^{-2} + t^{-1}(t+1) = \dfrac{2}{t^2} + t^0 + t^{-1}$

$$= \dfrac{2}{t^2} + 1 + \dfrac{1}{t}$$

$$= \dfrac{2 + t^2 + t}{t^2}$$

49. (a) $4^2 \cdot 64 = 4^2 \cdot 4^3 = 4^5$

(b) $4^2 \cdot 64 = (2^2)^2 \cdot 2^6 = 2^4 \cdot 2^6 = 2^{10}$

53. $kg \cdot s^{-2} \cdot m^2 = \dfrac{kg \cdot m^2}{s^2}$

$$= \dfrac{kg \cdot m}{s^2} \cdot m = N \cdot m$$

57. $v = a^p t^r$

$m \cdot s^{-1} = (m \cdot s^{-2})^p \cdot s^r$

$m^1 \cdot s^{-1} = m^p \cdot s^{-2p+r} \Rightarrow p = 1$ and $-2p + r = -1 \Rightarrow -2(1) + r = -1 \Rightarrow r = 1$

11.2   Fractional Exponents

1. $25^{1/2} = \sqrt{25} = 5$

5. $100^{25/2} = (100^{1/2})^{25} = (\sqrt{100})^{25} = 10^{25}$

9. $64^{-2/3} = \dfrac{1}{(64^{1/3})^2} = \dfrac{1}{(\sqrt[3]{64})^2} = \dfrac{1}{4^2} = \dfrac{1}{16}$

13. $(3^6)^{2/3} = 3^{6 \cdot 2/3} = 3^4 = 81$

17. $\dfrac{15^{2/3}}{5^2 \cdot 15^{-1/3}} = \dfrac{15^{2/3+1/3}}{5^2} = \dfrac{15^1}{25} = \dfrac{3}{5}$

21. $125^{-2/3} - 100^{-3/2} = \dfrac{1}{(125^{1/3})^2} - \dfrac{1}{(100^{1/2})^3} = \dfrac{1}{(\sqrt[3]{125})^2} - \dfrac{1}{(\sqrt{100})^3} = \dfrac{1}{5^2} - \dfrac{1}{10^3} = \dfrac{1}{25} - \dfrac{1}{1000} = \dfrac{39}{1000}$

25. $17.98^{1/4} = 2.059194748$

29. $a^{2/3} \cdot a^{1/2} = a^{2/3+1/2} = a^{7/6}$

33. $\dfrac{x^{3/10}}{x^{-1/5}x^2} = x^{3/10+1/5-2} = x^{-3/2} = \dfrac{1}{x^{3/2}}$

37. $(16a^4b^3)^{-3/4} = 16^{-3/4}a^{4(-3/4)}b^{3(-3/4)}$
$$= \dfrac{1}{16^{3/4}a^3b^{9/4}} = \dfrac{1}{8a^3b^{9/4}}$$

41. $\dfrac{1}{2}(4x^2 + 1)^{-1/2}(8x) = \dfrac{4x}{(4x^2 + 1)^{1/2}}$

45. $(T^{-1} + 2T^{-2})^{-1/2} = \dfrac{1}{\left(\dfrac{1}{T} + \dfrac{2}{T^2}\right)^{1/2}} = \dfrac{1}{\left(\dfrac{T+2}{T^2}\right)^{1/2}} = \dfrac{1}{\dfrac{(T+2)^{1/2}}{(T^2)^{1/2}}} = \dfrac{T}{(T+2)^{1/2}}$

49. $\left[(a^{1/2} - a^{-1/2})^2 + 4\right]^{1/2} = \left[\left(a^{1/2} - \dfrac{1}{a^{1/2}}\right)^2 + 4\right]^{1/2} = \left[\left(\dfrac{a-1}{a^{1/2}}\right)^2 + 4\right]^{1/2} = \left[\dfrac{(a-1)^2}{a} + 4\right]^{1/2}$
$$= \left[\dfrac{a^2 - 2a + 1 + 4a}{a}\right]^{1/2} = \left[\dfrac{a^2 + 2a + 1}{a}\right]^{1/2} = \left[\dfrac{(a+1)^2}{a}\right]^{1/2} = \dfrac{a+1}{a^{1/2}}$$

53. $f(x) = 3x^{1/2}$

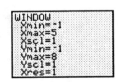 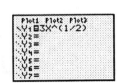 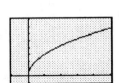

57. $\left(\dfrac{A}{S}\right)^{-1/4} = \left(\dfrac{S}{A}\right)^{1/4} = 0.5$
$$\dfrac{S}{A} = 0.5^4 = \dfrac{1}{16}$$

11.3   Simplest Radical Form

1. $\sqrt{24} = \sqrt{4 \cdot 6} = \sqrt{4} \cdot \sqrt{6} = 2\sqrt{6}$

5. $\sqrt{x^2 y^5} = \sqrt{x^2 y^4 y} = \sqrt{x^2}\sqrt{y^4}\sqrt{y} = xy^2\sqrt{y}$

9. $\sqrt{18a^3bc^4} = \sqrt{9a^2c^4 \cdot 2ab} = \sqrt{9}\sqrt{a^2}\sqrt{c^4}\sqrt{2ab}$
$\qquad = 3ac^2\sqrt{2ab}$

13. $\sqrt[5]{96} = \sqrt[5]{32 \cdot 3} = \sqrt[5]{32}\sqrt[5]{3} = 2\sqrt[5]{3}$

17. $\sqrt[4]{64r^3 s^4 t^5} = \sqrt[4]{16 \cdot 4s^4 t^4 r^3 t}$
$\qquad = \sqrt[4]{16}\sqrt[4]{s^4}\sqrt[4]{t^4}\sqrt[4]{4r^3 t}$
$\qquad = 2st\sqrt[4]{4r^3 t}$

21. $\sqrt[3]{ab^4}\sqrt[3]{a^2 b} = \sqrt[3]{a^3 b^5} = \sqrt[3]{a^3 b^3 b^2} = ab\sqrt[3]{b^2}$

25. $\sqrt[3]{\dfrac{3}{4}} = \sqrt[3]{\dfrac{48}{64}} = \dfrac{\sqrt[3]{48}}{\sqrt[3]{64}} = \dfrac{\sqrt[3]{8 \cdot 6}}{4} = \dfrac{2\sqrt[3]{6}}{4} = \dfrac{\sqrt[3]{6}}{2}$

29. $\sqrt[4]{400} = \sqrt[4]{2^4 \cdot 5^2} = 2 \cdot \sqrt[4]{25} = 2\sqrt{5}$

33. $\sqrt{4 \times 10^4} = \sqrt{4} \cdot \sqrt{10^4} = 2 \times 10^2 = 200$

37. $\sqrt[4]{4a^2} = (4a^2)^{1/4} = (2^2)^{1/4}(a^2)^{1/4}$
$\qquad = 2^{1/2} a^{1/2} = \sqrt{2a}$

41. $\sqrt[4]{\sqrt[3]{16}} = \sqrt[4]{16^{1/3}} = \left(16^{1/3}\right)^{1/4}$
$\qquad = \left(16^{1/4}\right)^{1/3} = 2^{1/3} = \sqrt[3]{2}$

45. $\sqrt{\dfrac{1}{2} - \dfrac{1}{3}} = \sqrt{\dfrac{1}{6}} = \dfrac{1}{\sqrt{6}} \cdot \dfrac{\sqrt{6}}{\sqrt{6}} = \dfrac{\sqrt{6}}{6}$

49. $\sqrt{\dfrac{x}{x^2 + 1}} = \sqrt{\dfrac{x}{(x^2 + 1)} \cdot \dfrac{(x^2 + 1)}{(x^2 + 1)}}$

$\qquad = \dfrac{\sqrt{x(x^2 + 1)}}{\sqrt{(x^2 + 1)^2}} = \dfrac{\sqrt{x(x^2 + 1)}}{x^2 + 1}$

53. $\sqrt{4x^2 - 1}$ is in simplest terms

57. $E = 100\left(1 - \dfrac{1}{\sqrt[5]{R^2}}\right) = 100\left(1 - \dfrac{1}{R^{2/5}}\right) = 100\left(1 - R^{-2/5}\right)$

For $R = 7.35$, $E = 100(1 - 7.35^{-2/5}) = 55\%$

11.4   Addition and Subtraction of Radicals

1. $2\sqrt{3} + 5\sqrt{3} = (2 + 5)\sqrt{3} = 7\sqrt{3}$

5. $\sqrt{5} + \sqrt{20} = \sqrt{5} + \sqrt{4 \cdot 5} = \sqrt{5} + 2\sqrt{5}$
$\qquad = (1 + 2)\sqrt{5} = 3\sqrt{5}$

9. $\sqrt{8a} - \sqrt{32a} = \sqrt{4 \cdot 2a} - \sqrt{16 \cdot 2a}$
$\qquad = 2\sqrt{2a} - 4\sqrt{2a} = -2\sqrt{2a}$

13. $2\sqrt{20} - \sqrt{125} - \sqrt{45} = 2\sqrt{4 \cdot 5} - \sqrt{25 \cdot 5} - \sqrt{9 \cdot 5}$
$\qquad = 4\sqrt{5} - 5\sqrt{5} - 3\sqrt{5} = -4\sqrt{5}$

17. $\sqrt{60} + \sqrt{\dfrac{5}{3}} = \sqrt{4 \cdot 15} + \sqrt{\dfrac{5}{3} \cdot \dfrac{3}{3}}$
$\qquad = 2\sqrt{15} + \dfrac{\sqrt{15}}{3} = \dfrac{7\sqrt{15}}{3}$

21. $\sqrt[3]{81} + \sqrt[3]{3000} = \sqrt[3]{27 \cdot 3} + \sqrt[3]{1000 \cdot 3}$
$\qquad = 3\sqrt[3]{3} + 10\sqrt[3]{3} = 13\sqrt[3]{3}$

25. $\sqrt{a^3b} - \sqrt{4ab^5} = \sqrt{a^2 \cdot ab} - \sqrt{4b^4ab}$
$= a\sqrt{ab} - 2b^2\sqrt{ab}$
$= (a - 2b^2)\sqrt{ab}$

29. $\sqrt[3]{24a^2b^4} - \sqrt[3]{3a^5b} = \sqrt[3]{8b^3 \cdot 3a^2b} - \sqrt[3]{a^3 \cdot 3a^2b}$
$= 2b\sqrt[3]{3a^2b} - a\sqrt[3]{3a^2b}$
$= (2b - a)\sqrt[3]{3a^2b}$

33. $\sqrt[3]{\dfrac{a}{b}} - \sqrt[3]{\dfrac{8b^2}{a^2}} = \sqrt[3]{\dfrac{a}{b} \cdot \dfrac{b^2}{b^2}} - \sqrt[3]{\dfrac{8b^2}{a^2} \cdot \dfrac{a}{a}} = \dfrac{\sqrt[3]{ab^2}}{b} - \dfrac{2\sqrt[3]{ab^2}}{a} = \dfrac{(a - 2b)\sqrt[3]{ab^2}}{ab}$

37. $3\sqrt{45} + 3\sqrt{75} - 2\sqrt{500} = 1.384014361 \cdots$ on calculator
$3\sqrt{45} + 3\sqrt{75} - 2\sqrt{500} = 3\sqrt{9 \cdot 5} + 3\sqrt{25 \cdot 3} - 2\sqrt{100 \cdot 5}$
$= 9\sqrt{5} + 15\sqrt{3} - 20\sqrt{5}$
$= 15\sqrt{3} - 11\sqrt{5} = 1.384014361 \cdots$ on calculator

41. $x^2 - 2x - 2 = 0$ has roots $x = \dfrac{2 \pm \sqrt{(-2)^2 - 4(1)(-2)}}{2} = \dfrac{2 \pm \sqrt{12}}{2}$
$$x = \dfrac{2 \pm 2\sqrt{3}}{2} = 1 \pm \sqrt{3}$$
$$x = 1 + \sqrt{3} \text{ is the positive root}$$

$x^2 + 2x - 11 = 0$ has roots $x = \dfrac{-2 \pm \sqrt{2^2 - 4(1)(-11)}}{2(1)} = \dfrac{-2 \pm \sqrt{48}}{2}$
$$x = \dfrac{-2 \pm 4\sqrt{3}}{2}$$
$$x = -1 + 2\sqrt{3} \text{ is the positive root}$$

sum of positive roots $= 1 + \sqrt{3} - 1 + 2\sqrt{3} = 3\sqrt{3}$

11.5   Multiplication and Division of Radicals

1. $\sqrt{3}\sqrt{10} = \sqrt{3 \cdot 10} = \sqrt{30}$

5. $\sqrt[3]{4} \cdot \sqrt[3]{2} = \sqrt[3]{4 \cdot 2} = \sqrt[3]{8} = 2$

9. $\sqrt{8} \cdot \sqrt{\dfrac{5}{2}} = \sqrt{\dfrac{40}{2}} = \sqrt{20} = \sqrt{4 \cdot 5} = 2\sqrt{5}$

13. $(2 - \sqrt{5})(2 + \sqrt{5}) = 2^2 - \sqrt{5}^2 = 4 - 5 = -1$

17. $(3\sqrt{11} - \sqrt{x})(2\sqrt{11} + 5\sqrt{x}) = 6\sqrt{11}^2 + 15\sqrt{11x} - 2\sqrt{11x} - 5\sqrt{x}^2 = 6 \cdot 11 + 13\sqrt{11x} - 5x$
$= 66 + 13\sqrt{11x} - 5x$

21. $\dfrac{\sqrt{6} - 3}{\sqrt{6}} = \dfrac{\sqrt{6} - 3}{\sqrt{6}} \cdot \dfrac{\sqrt{6}}{\sqrt{6}} = \dfrac{\sqrt{6}^2 - 3\sqrt{6}}{\sqrt{6}^2} = \dfrac{6 - 3\sqrt{6}}{6} = \dfrac{2 - \sqrt{6}}{2}$

25. $(\sqrt{2a} - \sqrt{b})(\sqrt{2a} + 3\sqrt{b}) = \sqrt{2a}^2 + 3\sqrt{2ab} - \sqrt{2ab} - 3\sqrt{b}^2 = 2a + 2\sqrt{2ab} - 3b$

29. $\dfrac{1}{\sqrt{7}+\sqrt{3}} = \dfrac{1}{\sqrt{7}+\sqrt{3}} \cdot \dfrac{\sqrt{7}-\sqrt{3}}{\sqrt{7}-\sqrt{3}}$

$\qquad = \dfrac{\sqrt{7}-\sqrt{3}}{\sqrt{7}^2-\sqrt{3}^2} = \dfrac{\sqrt{7}-\sqrt{3}}{7-3}$

$\qquad = \dfrac{\sqrt{7}-\sqrt{3}}{4}$

33. $\dfrac{\sqrt{2}-1}{\sqrt{7}-3\sqrt{2}} = \dfrac{\sqrt{2}-1}{\sqrt{7}-3\sqrt{2}} \cdot \dfrac{\sqrt{7}+3\sqrt{2}}{\sqrt{7}+3\sqrt{2}}$

$\qquad = \dfrac{\sqrt{14}+3\sqrt{2}^2-\sqrt{7}-3\sqrt{2}}{\sqrt{7}^2-3^2\sqrt{2}^2}$

$\qquad = \dfrac{\sqrt{14}+3\cdot 2-\sqrt{7}-3\sqrt{2}}{7-9\cdot 2}$

$\qquad = \dfrac{\sqrt{14}+6-\sqrt{7}-3\sqrt{2}}{7-18}$

$\qquad = -\dfrac{\sqrt{14}+6-\sqrt{7}-3\sqrt{2}}{11}$

37. $\dfrac{2\sqrt{x}}{\sqrt{x}-\sqrt{y}} = \dfrac{2\sqrt{x}}{\sqrt{x}-\sqrt{y}} \cdot \dfrac{\sqrt{x}+\sqrt{y}}{\sqrt{x}+\sqrt{y}}$

$\qquad = \dfrac{2\sqrt{x}^2+2\sqrt{x}\sqrt{y}}{\sqrt{x}^2-\sqrt{y}^2} = \dfrac{2x+2\sqrt{xy}}{x-y}$

41. $(\sqrt[5]{\sqrt{6}-\sqrt{5}})(\sqrt[5]{\sqrt{6}+\sqrt{5}}) = \sqrt[5]{\sqrt{6}^2-\sqrt{5}^2}$

$\qquad\qquad = \sqrt[5]{6-5} = \sqrt[5]{1} = 1$

45. $\dfrac{\sqrt{x+y}}{\sqrt{x-y}-\sqrt{x}} = \dfrac{\sqrt{x+y}}{\sqrt{x-y}-\sqrt{x}} \cdot \dfrac{\sqrt{x-y}+\sqrt{x}}{\sqrt{x-y}+\sqrt{x}} = \dfrac{\sqrt{x+y}\sqrt{x-y}+\sqrt{x+y}\sqrt{x}}{\sqrt{x-y}^2-\sqrt{x}^2}$

$\qquad = \dfrac{\sqrt{x^2-y^2}+\sqrt{x^2+xy}}{x-y-x} = -\dfrac{\sqrt{x^2-y^2}+\sqrt{x^2+xy}}{y}$

49. $(\sqrt{11}+\sqrt{6})(\sqrt{11}-2\sqrt{6}) = -9.124038405\cdots$ on calculator

$\quad (\sqrt{11}+\sqrt{6})(\sqrt{11}-2\sqrt{6}) = \sqrt{11}^2-2\sqrt{66}+\sqrt{66}-2\sqrt{6}^2 = 11-\sqrt{66}-2\cdot 6$

$\qquad\qquad\qquad\qquad = 11-\sqrt{66}-12 = -1-\sqrt{66}$

$\qquad\qquad\qquad\qquad = -9.124038405\cdots$ on calculator

53. $2\sqrt{x}+\dfrac{1}{\sqrt{x}} = 2\sqrt{x}\cdot\dfrac{\sqrt{x}}{\sqrt{x}}+\dfrac{1}{\sqrt{x}}$

$\qquad = \dfrac{2\sqrt{x}^2}{\sqrt{x}}+\dfrac{1}{\sqrt{x}}$

$\qquad = \dfrac{2x}{\sqrt{x}}+\dfrac{1}{\sqrt{x}} = \dfrac{2x+1}{\sqrt{x}}$

57. $\dfrac{\sqrt{5}+\sqrt{2}}{3\sqrt{6}} = \dfrac{\sqrt{5}+\sqrt{2}}{3\sqrt{6}} \cdot \dfrac{\sqrt{5}-\sqrt{2}}{\sqrt{5}-\sqrt{2}}$

$\qquad = \dfrac{\sqrt{5}^2-\sqrt{2}^2}{3\sqrt{30}-3\sqrt{12}} = \dfrac{5-2}{3\sqrt{30}-3\sqrt{4\cdot 3}}$

$\qquad = \dfrac{3}{3\sqrt{30}-6\sqrt{3}} = \dfrac{1}{\sqrt{30}-2\sqrt{3}}$

61. $x^2-2x-1\big|_{x=1-\sqrt{2}} = (1-\sqrt{2})^2-2(1-\sqrt{2})-1 = 1-2\sqrt{2}+2-2+2\sqrt{2}-1 = 0$

65.

$A = 8\cdot A_{\text{triangle}}$ where each triangle has
base $= b = s$ and altitude $= h$.

Each angle of the octagon is $\dfrac{6\cdot 180°}{8} = 135°$. The base angles, θ, of the triangle are $\dfrac{135°}{2} = 67.5°$.

The altitude may be found from $\tan\theta = \dfrac{h}{\frac{s}{2}} \Rightarrow h = \dfrac{s}{2}\tan\theta$

$$A_{\text{octagon}} = 8 \cdot A_{\text{triangle}} = 8 \cdot \frac{1}{2} \cdot s \cdot \frac{s}{2} \cdot \tan\theta = 2 \cdot s^2 \cdot \tan 67.5°$$

$$A_{\text{octagon}} = 2 \cdot s^2 \cdot \tan\frac{135°}{2} = 2 \cdot s^2 \cdot (1 + \sqrt{2})$$

Chapter 11 Review Exercises

1. $2a^{-2}b^0 = 2a^{-2} \cdot 1 = 2 \cdot \dfrac{1}{a^2} = \dfrac{2}{a^2}$

5. $3(25)^{3/2} = 3\left[(25)^{1/2}\right]^3 = 3\left[\sqrt{25}\right]^3$
$\qquad = 3\,[5]^3 = 3 \cdot 125 = 375$

9. $\left(\dfrac{3}{t^2}\right)^{-2} = \dfrac{1}{\left(\dfrac{3}{t^2}\right)^2} = \dfrac{1}{\dfrac{3^2}{(t^2)^2}} = \dfrac{1}{\dfrac{9}{t^4}} = \dfrac{t^4}{9}$

13. $\left(2a^{1/3}b^{5/6}\right)^6 = 2^6 \cdot \left(a^{1/3}\right)^6 \cdot \left(b^{5/6}\right)^6 = 64 \cdot a^{6/3} \cdot b^{5/6 \cdot 6}$
$\qquad = 64a^2b^5$

17. $2L^{-2} - 4C^{-1} = \dfrac{2}{L^2} - \dfrac{4}{C}$
$\qquad\qquad = \dfrac{2C - 4L^2}{L^2C}$

21. $(a - 3b^{-1})^{-1} = \dfrac{1}{(a - 3b^{-1})^1} = \dfrac{1}{a - \dfrac{3}{b}} \cdot \dfrac{b}{b}$
$\qquad\qquad = \dfrac{b}{ab - 3}$

25. $(8a^3)^{2/3}(4a^{-2} + 1)^{1/2} = 8^{2/3} \cdot (a^3)^{2/3}\sqrt{\dfrac{4}{a^2} + 1} = \sqrt[3]{8^2} \cdot a^{3 \cdot 2/3}\sqrt{\dfrac{4 + a^2}{a^2}} = \sqrt[3]{64} \cdot a^2 \cdot \dfrac{\sqrt{4 + a^2}}{\sqrt{a^2}}$

$\qquad\qquad = 4 \cdot a^2 \cdot \dfrac{\sqrt{4 + a^2}}{a} = 4a\sqrt{4 + a^2}$

29. $\sqrt{68} = \sqrt{4 \cdot 17} = \sqrt{4} \cdot \sqrt{17} = 2\sqrt{17}$

33. $\sqrt{9a^3b^4} = \sqrt{9a^2b^4 \cdot a} = 3ab^2\sqrt{a}$

37. $\dfrac{5}{\sqrt{2s}} = \dfrac{5}{\sqrt{2s}} \cdot \dfrac{\sqrt{2s}}{\sqrt{2s}} = \dfrac{5\sqrt{2s}}{2s}$

41. $\sqrt[4]{8m^6n^9} = \sqrt[4]{8m^4 \cdot m^2 \cdot n^8 \cdot n} = mn^2\sqrt[4]{8m^2n}$

45. $\sqrt{200} + \sqrt{32} = \sqrt{100 \cdot 2} + \sqrt{16 \cdot 2}$
$\qquad\qquad = 10\sqrt{2} + 4\sqrt{2} = 14\sqrt{2}$

49. $a\sqrt{2x^3} + \sqrt{8a^2x^3} = a\sqrt{2x^2 \cdot x} + \sqrt{4 \cdot 2 \cdot a^2x^2 \cdot x}$
$\qquad\qquad = ax\sqrt{2x} + 2ax\sqrt{2x} = 3ax\sqrt{2x}$

53. $\sqrt{5}(2\sqrt{5} - \sqrt{11}) = 2\sqrt{5}^2 - \sqrt{5}\sqrt{11}$
$\qquad\qquad = 2 \cdot 5 - \sqrt{55} = 10 - \sqrt{55}$

57. $(2 - 3\sqrt{17})(3 + \sqrt{17}) = 6 + 2\sqrt{17} - 9\sqrt{17} - 3 \cdot \sqrt{17}^2 = 6 - 7\sqrt{17} - 3 \cdot 17 = 6 - 7\sqrt{17} - 51 = -45 - 7\sqrt{17}$

61. $\dfrac{\sqrt{3x}}{2\sqrt{3x} - \sqrt{y}} = \dfrac{\sqrt{3x}}{(2\sqrt{3x} - \sqrt{y})} \cdot \dfrac{(2\sqrt{3x} + \sqrt{y})}{(2\sqrt{3x} + \sqrt{y})} = \dfrac{2\sqrt{3x}^2 + \sqrt{3x} \cdot \sqrt{y}}{4\sqrt{3x}^2 - \sqrt{y}^2} = \dfrac{2 \cdot 3x + \sqrt{3xy}}{4 \cdot 3x - y} = \dfrac{6x + \sqrt{3xy}}{12x - y}$

65. $\dfrac{\sqrt{7} - \sqrt{5}}{\sqrt{5} + 3\sqrt{7}} = \dfrac{(\sqrt{7} - \sqrt{5})}{(\sqrt{5} + 3\sqrt{7})} \cdot \dfrac{(\sqrt{5} - 3\sqrt{7})}{(\sqrt{5} - 3\sqrt{7})} = \dfrac{\sqrt{35} - 3 \cdot 7 - 5 + 3\sqrt{35}}{5 - 9 \cdot 7} = \dfrac{4\sqrt{35} - 26}{5 - 63} = \dfrac{4\sqrt{35} - 26}{-58}$

$\qquad\qquad = \dfrac{2(2\sqrt{35} - 13)}{2(-29)} = \dfrac{2\sqrt{35} - 13}{-29} = \dfrac{13 - 2\sqrt{35}}{29}$

69. $\sqrt{4b^2 + 1}$ is in simplest form

73. $(\sqrt{7} - 2\sqrt{15})(3\sqrt{7} - \sqrt{15}) = 3 \cdot 7 - \sqrt{105} - 6\sqrt{105} + 2 \cdot 15 = 51 - 7\sqrt{105}$

77. $i = 100 \left[\left(\dfrac{C_2}{C_1} \right)^{1/n} - 1 \right]$

$= 100 \left[\left(\dfrac{172.0}{130.7} \right)^{1/10} - 1 \right] = 2.78\%$

81. $\dfrac{v}{n_2^{-2} - n_1^{-2}} = \dfrac{v}{\dfrac{1}{n_2^2} - \dfrac{1}{n_1^2}} \cdot \dfrac{n_1^2 n_2^2}{n_1^2 n_2^2} = \dfrac{v n_1^2 n_2^2}{n_1^2 - n_2^2}$

85. $\sqrt{3^2 + 3^2} + \sqrt{2^2 + 2^2} + \sqrt{1^2 + 1^2} = \sqrt{18} + \sqrt{8} + \sqrt{2} = \sqrt{9 \cdot 2} + \sqrt{4 \cdot 2} + \sqrt{2} = 3\sqrt{2} + 2\sqrt{2} + \sqrt{2} = 6\sqrt{2}$

89. Answers may vary.

Chapter 12

COMPLEX NUMBERS

12.1 Basic Definitions

1. $\sqrt{-81} = \sqrt{81(-1)} = \sqrt{81}\sqrt{-1} = 9j$

5. $\sqrt{-0.36} = \sqrt{0.36(-1)} = \sqrt{0.36}\sqrt{-1} = 0.6j$

9. $\sqrt{-\dfrac{7}{4}} = \dfrac{\sqrt{-7}}{\sqrt{4}} = \dfrac{\sqrt{7(-1)}}{2}$

$\qquad = \dfrac{\sqrt{7}\sqrt{-1}}{2}$

$\qquad = \dfrac{\sqrt{7}j}{2}$

13. **(a)** $(\sqrt{-7})^2 = (\sqrt{7(-1)})^2 = (\sqrt{7}\cdot\sqrt{-1})$
$\qquad\qquad\qquad\qquad = (\sqrt{7}\cdot j)^2$
$\qquad\qquad\qquad\qquad = \sqrt{7}^2\cdot j^2$
$\qquad\qquad\qquad\qquad = 7(-1) = -7$

(b) $(\sqrt{-7})^2 = \sqrt{49} = 7$

17. **(a)** $j^7 = j^4\cdot j^3 = (1)(-j) = -j$

(b) $j^{49} = j^{48}\cdot j^1 = (1)(j) = j$

21. $j^{15} - j^{13} = j^{12}\cdot j^3 - j^{12}\cdot j = (1)(-j) - (1)(j)$
$\qquad\qquad = -j - j = -2j$

25. $2 + \sqrt{-9} = 2 + \sqrt{9(-1)} = 2 + 3j$

29. $\sqrt{-4j^2} + \sqrt{-4} = \sqrt{(-4)(-1)} + 2j = 2 + 2j$

33. $\sqrt{18} - \sqrt{-8} = \sqrt{9\cdot 2} - \sqrt{4\cdot 2}j$
$\qquad\qquad\quad = 3\sqrt{2} - 2\sqrt{2}j$

37. **(a)** the conjugate of $6 - 7j$ is $6 + 7j$

(b) the conjugate of $8 + j$ is $8 - j$

41. $7x - 2yj = 14 + 4j \Rightarrow 7x = 14 \quad$ and $\quad -2y = 4$
$\qquad\qquad\qquad\qquad\qquad\quad x = 2 \qquad\qquad\quad y = -2$

45. $x - y = 1 - xj - yj - j$
$\quad\; x - y = 1 + (-x - y - 1)j \Rightarrow \quad (1) \qquad x - y = 1 \quad$ and
$\qquad\qquad\qquad\qquad\qquad\qquad\qquad (2) \quad -x - y - 1 = 0$

from (1) $y = x - 1$ which may be substituted into (2) to obtain

$$-x - (x - 1) - 1 = 0$$
$$-x - x + 1 - 1 = 0$$
$$-2x = 0$$
$$x = 0, \;\; y = x - 1 = 0 - 1$$
$$y = -1$$

49. Substituting $8j$ for x gives $(8j)^2 + 64 = 64j^2 + 64$
$$\qquad\qquad\qquad\qquad\qquad\qquad\quad = 64(-1) + 64$$
$$\qquad\qquad\qquad\qquad\qquad\qquad\quad = 0$$

$8j$ is a solution.
Substituting $-8j$ for x gives $(-8j)^2 + 64 = 64j^2 + 64$
$$\qquad\qquad\qquad\qquad\qquad\qquad\qquad\quad = 0 \text{ as before}$$

$-8j$ is a solution.

53. $j + j^2 + j^3 + j^4 + j^5 + j^6 + j^7 + j^8 = j + (-1) + (-j) + 1 + j + (-1) + (-j) + 1$
$$= j - 1 - j + 1 + j - 1 - j + 1$$
$$= 0$$

12.2   Basic Operations with Complex Numbers

1. $(3 - 7j) + (2 - j) = (3 + 2) + (-7 - 1)j = 5 - 8j$

5. $0.23 - (0.46 - 0.9j) + 0.67j = 0.23 - 0.46 + 0.19j + 0.67j = -0.23 + 0.86j$

9. $(7 - j)(7j) = 49j - 7j^2 = 49j - 7(-1) = 7 + 49j$

13. $(20j - 30)(30j + 10) = 600j^2 + 200j - 900j - 300 = 600(-1) - 700j - 300 = -900 - 700j$

17. $7j^3 - 7\sqrt{-9} = 7(-j) - 7 \cdot 3j$
$$= -7j - 21j$$
$$= -28j$$

21. $(3 - 7j)^2 = 3^2 - 2 \cdot 3 \cdot 7j + 7^2 j^2$
$$= 9 - 42j + 49(-1)$$
$$= 9 - 49 - 42j = -40 - 42j$$

25. $\dfrac{6j}{2 - 5j} = \dfrac{6j}{(2 - 5j)} \cdot \dfrac{(2 + 5j)}{(2 + 5j)}$
$$= \dfrac{12j + 30j^2}{2^2 + 5^2}$$
$$= \dfrac{-30 + 12j}{29}$$
$$= \dfrac{-30}{29} + \dfrac{12}{29}j$$

29. $\dfrac{6 + 5j}{3 - 4j} = \dfrac{6 + 5j}{3 - 4j} \cdot \dfrac{3 + 4j}{3 + 4j}$
$$= \dfrac{18 + 39j + 20j^2}{3^2 + 4^2}$$
$$= \dfrac{18 + 39j - 20}{25}$$
$$= \dfrac{-2 + 39j}{25} = \dfrac{-2}{25} + \dfrac{39}{25}j$$

33. $\dfrac{j^2 - j}{2j - j^8} = \dfrac{-1 - j}{2j - 1} \cdot \dfrac{-2j - 1}{-2j - 1}$
$$= \dfrac{1 + 3j + 2j^2}{1^2 + 2^2}$$
$$= \dfrac{1 + 3j - 2}{5}$$
$$= \dfrac{-1 + 3j}{5}$$

37. Substituting $-1 - j$ into $x^2 + 2x + 2$ gives
$$(-1 - j)^2 + 2(-1 - j) + 2 = 1 + 2j + j^2 - 2 - 2j + 2$$
$$= 1 + j^2$$
$$= 1 - 1$$
$$= 0$$

41. $E = I \cdot Z = (0.835 - 0.427j)(250 + 170j) = 208.75 + 141.95j - 106.75j - 72.59j^2$
$$= 208.75 + 35.2j + 72.59$$
$$= 281.34 + 35.2j$$

45. $(a + bi) + (a - bi) = a + bi + a - bi = 2a$, a real number

12.3   Graphical Representation of Complex Numbers

1.

5.

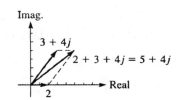

algebra:

$$2 + 3 + 4j = 5 + 4j$$

9.

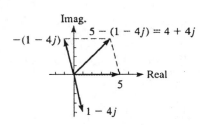

algebra:

$$5 - (1 - 4j) = 5 - 1 + 4j$$
$$= 4 + 4j$$

13.

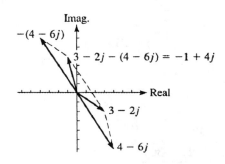

algebra:

$$(3 - 2j) - (4 - 6j) = 3 - 2j - 4 + 6j$$
$$= -1 + 4j$$

17.

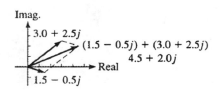

algebra:

$$(1.5 - 0.5j) + (3.0 + 2.5j) = 1.5 - 0.5j + 3.0 + 2.5j$$
$$= 4.5 + 2.0j$$

21.

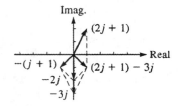

algebra:

$$(2j + 1) - 3j - (j + 1) = (2j + 1) + (-3j) + (-(j + 1))$$
$$= 2j + 1 - 3j - j - 1$$
$$= -2j$$

25.   number: $3 + 2j$
           negative: $-(3 + 2j)$
           conjugate: $3 - 2j$

29. $a + bj = 3 - j$
          $3(a + bj) = 9 - 3j$
         $-3(a + bj) = -9 + 3j$

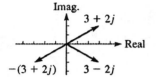

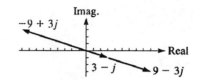

12.4   Polar Form of a Complex Number

1. $8 + 6j \Rightarrow x = 8, y = 6 \Rightarrow r = \sqrt{8^2 + 6^2} = 10.$ $\theta = \tan^{-1} \dfrac{6}{8} = 36.9°$

The polar form is $10(\cos 36.9° + j \sin 36.9°)$

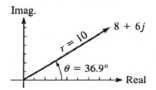

5. $-2.00 + 3.00j \Rightarrow x = -2.00, y = 3.00 \Rightarrow r = \sqrt{(-2.00)^2 + (3.00)^2} = \sqrt{13.0}$

$\theta_{\mathrm{ref}} = \tan^{-1} \dfrac{3}{-2} = -56.3° \Rightarrow \theta = 180° + \theta_{\mathrm{ref}} = 123.7°$

The polar form is

$\sqrt{13.0}(\cos 123.7° + j \sin 123.7°).$

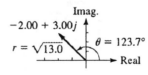

9. $1 + j\sqrt{3} \Rightarrow x = 1, y = \sqrt{3} \Rightarrow$

$r = \sqrt{1^2 + \sqrt{3}^2} = 2.$ $\theta = \tan^{-1} \dfrac{\sqrt{3}}{1} = 60°$

The polar form is $2(\cos 60° + j \sin 60°)$

13. $-3 + 0j \Rightarrow x = -3, y = 0 \Rightarrow$

$r = \sqrt{(-3)^2 + 0^2} = 3.$ $\theta_{\mathrm{ref}} = \tan^{-1} \dfrac{0}{-3} = 0°$

The polar from is $3(\cos 180° + j \sin 180°)$

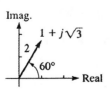

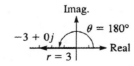

17. $5.00(\cos 54.0° + j\sin 54.0°)$
$= 5.00\cos 54.0° + 5.00\sin 54.0° \cdot j$
$= 2.94 + 4.05j$

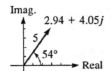

21. $6(\cos 180° + j\sin 180°)$
$= 6(-1) + 6j(0) = -6 + 0j$

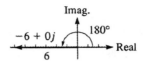

25. $12.36(\cos 345.56° + j\sin 345.56)$
$= 11.97 - 3.082j$

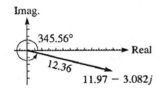

29. $4.75\angle 172.8°$
$= 4.75(\cos 172.8° + j\sin 172.8°)$
$= -4.71 + 0.595j$

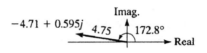

33. $7.32\angle -270° = 7.32(\cos(-270°) + j\sin(-270°))$
$= 0 + 7.32j$

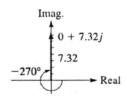

37. $r = \sqrt{2.84^2 + (-1.06)^2} = 3.03$
$\theta = \tan^{-1}\dfrac{-1.06}{2.84} = -20.5° = 339.5°$
$2.84 - 1.06j = 3.03(\cos 339.5° + j\sin 339.5°)$

12.5   Exponential Form of a Complex Number

1. $3.00(\cos 60.0° + j\sin 60.0°)$;
$r = 3.00, \theta = 60.0°$
$3.00e^{1.05j}$

5. $375.5(\cos 95.46° + j\sin 95.46°)$;
$r = 375.5, \theta = 95.46° = 1.666rad$
$375.5e^{1.666j}$

9. $4.06\angle\underline{-61.4°}; r = 4.06, \theta = -61.4° = -1.07$

$\qquad 4.06e^{-1.07j}$

13. $3 - 4j; r = \sqrt{3^2 + (-4)^2} = 5,$

$$\theta_{\text{ref}} = \tan^{-1}\frac{-4}{3} = -0.9273$$

$$\theta = \theta_{\text{ref}} + 2\pi = 5.36$$
$$3 - 4j = 5e^{5.36j}$$

17. $5.90 + 2.40j;$

$\qquad r = \sqrt{5.90^2 + 2.40^2} = 6.37$

$\qquad \theta = \tan^{-1}\dfrac{2.40}{5.90} = 0.386$

$\qquad 5.90 + 2.40j = 6.37e^{0.386j}$

21. $3.00e^{0.500j}; r = 3, \theta = 0.500 \text{ rad} = 28.6°$

$\qquad 3.00e^{0.500j} = 3(\cos 28.6° + j\sin 28.6°)$
$$= 2.63 + 1.44j$$

25. $3.2e^{5.41j}; r = 3.2, \theta = 5.41 \text{ rad} = 310.0°$

$\qquad 3.2e^{5.41j} = 3.2(\cos 310.0° + j\sin 310.0°)$
$$= 2.06 - 2.45j$$

29. $(4.55e^{1.32j})^2 = 20.7e^{2.64j};$

$\qquad r = 20.7, \theta = 2.64 \text{ rad} = 151.3°$

$\qquad (4.55e^{1.32j})^2 = 20.7(\cos 151.3° + j\sin 151.3°)$
$$= -18.2 + 9.94j$$

33. $375 + 110j; r = \sqrt{375^2 + 110^2} = 391, \theta = \tan^{-1}\dfrac{110}{375} = 0.285$

$\qquad 375 + 110j = 391e^{0.285j}$

The magnitude of the impedance is 391 ohms.

12.6 Products, Quotients, Powers, and Roots of Complex Numbers

1. $[4(\cos 60° + j\sin 60°)][2(\cos 20° + j\sin 20°] = 4 \cdot 2[(\cos(60° + 20°) + j\sin(60° + 20°)]$
$$= 8(\cos 80° + j\sin 80°)$$

5. $\dfrac{8(\cos 100° + j\sin 100°)}{4(\cos 65° + j\sin 65°)} = \dfrac{8}{4}(\cos(100° - 65°) + j\sin(100° - 65°))$

$$= 2(\cos 35° + j\sin 35°)$$

9. $[2(\cos 35° + j\sin 35°)]^3 = 2^3(\cos(3 \cdot 35°) + j\sin(3 \cdot 35°)) = 8(\cos 105° + j\sin 105°)$

13. $\dfrac{(50\angle 236°)^2(\angle 1840°)}{25\angle 47°} = \dfrac{100\angle 320°}{25\angle 47°} = 4\angle 273°$

17. $2.78\angle\underline{56.8°} + 1.37\angle\underline{207.3°} = 2.78(\cos 56.8° + j\sin 56.8°) + 1.37(\cos 207.3° + j\sin 207.3°)$
$$= 1.5222 + 2.3262j - 1.2174 - 0.6283j$$
$$= 0.3048 + 1.6979j.$$

$\qquad r = \sqrt{0.3048^2 + 1.6979^2} = 1.73, \theta = \tan^{-1}\dfrac{1.6979}{0.3048} = 79.8°.$

$\qquad 2.78\angle 56.8° + 1.37\angle 207.3° = 0.3048 + 1.6979j = 1.73\angle 79.8°$

21. $3 + 4j = \sqrt{3^2 + 4^2} = \left(\cos\left(\tan^{-1}\frac{4}{3}\right) + j\sin\left(\tan^{-1}\frac{4}{3}\right)\right) = 5(\cos 53.1° + j\sin 53.1°)$

$5 - 12j = \sqrt{5^2 + (-12)^2}\left(\cos\left(\tan^{-1}\frac{-12}{5} + 360°\right) + j\sin\left(\tan^{-1}\frac{-12}{5} + 360°\right)\right)$

$\qquad = 13(\cos 292.6° - j\sin 292.6°)$

polar form:

$(3 + 4j)(5 - 12j) = [5(\cos 53.1° + j\sin 53.1°)][13(\cos 292.6° + j\sin 292.6°)]$

$\qquad = 65(\cos(53.1° + 292.6°) + j\sin(53.1° + 292.6°))$

$\qquad = 65(\cos(345.7°) + j\sin(345.7°))$

$\qquad = 63.0 - 16.1j$

rectangular form:

$(3 + 4j)(5 - 12j) = 15 - 36j + 20j - 48j^2$

$\qquad = 15 - 16j + 48$

$\qquad = 63 - 16j$

25. $\dfrac{7}{1 - 3j} = \dfrac{7(\cos 0° + j\sin 0°)}{\sqrt{1^2 + (-3)^2}\left(\cos\left(\tan^{-1}\frac{-3}{1} + 360°\right) + j\sin\left(\tan^{-1}\frac{-3}{1} + 360°\right)\right)}$

$\qquad = \dfrac{7(\cos 0° + j\sin 0°)}{3.16(\cos 288.4° + j\sin 288.4°)}$

$\qquad = \dfrac{7}{3.16}(\cos(0° - 288.4°) + j\sin(0 - 288.4°))$

$\qquad = 2.22(\cos(-288.4°) + j\sin(-288.4°))$

$\qquad = 2.22(\cos 71.6° + j\sin 71.6°)$

$\qquad = 0.7 + 2.1j$

rectangular form:

$\dfrac{7}{1 - 3j} = \dfrac{7}{(1 - 3j)} \cdot \dfrac{(1 + 3j)}{(1 + 3j)} = \dfrac{7 + 21j}{1^2 + 3^2} = \dfrac{7}{10} + \dfrac{21}{10}j$

29. $(3 + 4j) = \sqrt{3^2 + 4^2}\left(\cos\left(\tan^{-1}\frac{4}{3}\right) + j\sin\left(\tan^{-1}\frac{4}{3}\right)\right) = 5(\cos 53.1° + j\sin 53.1°)$

$(3 + 4j)^4 = [5(\cos 53.1° + j\sin 53.1°)]^4 = 5^4(\cos(4 \cdot 53.1°) + j\sin(4 \cdot 53.1°)$

$\qquad = 625(\cos 212.5° + j\sin 212.5°) = -527 - 336j$

rectangular form:

$(3 + 4j)^4 = [(3 + 4j)^2]^2 = (9 + 24j + (16j^2))^2 = (9 + 24j - 16)^2 = (-7 + 24j)^2 = 49 - 336j + 576j^2$

$\qquad = 49 - 336j - 576 = -527 - 336j$

33. The two square roots of $4(\cos 60° + j\sin 60°)$ are

$$r_1 = \sqrt{4}\left(\cos\frac{60° + 0 \cdot 360°}{2} + j\sin\frac{60° + 0 \cdot 360°}{2}\right)$$

$$r_1 = 2(\cos 30° + j\sin 30°) = \sqrt{3} + j$$

and

$$r_2 = \sqrt{4}\left(\cos\frac{60° + 1 \cdot 360°}{2} + j\sin\frac{60° + 1 \cdot 360°}{2}\right)$$

$$r_2 = 2(\cos 210° + j\sin 210°) = -\sqrt{3} - j$$

37. The two square roots of $1 + j = \sqrt{2}(\cos 45° + j \sin 45°)$ are

$$r_1 = \sqrt{\sqrt{2}} \left(\cos \frac{45° + 0 \cdot 360°}{2} + j \sin \frac{4.5° + 0 \cdot 360°}{2} \right)$$

$$r_1 = 2^{1/4}(\cos 22.5° + j \sin 22.5°) = 1.0987 + 0.4551j$$

and

$$r_2 = 2^{1/4} \left(\cos \frac{45° + 1 \cdot 360°}{2} + j \sin \frac{4.5° + 1 \cdot 360°}{2} \right)$$

$$r_2 = 2^{1/4}(\cos 202.5° + j \sin 202.5°) = -1.0987 - 0.4551j$$

41. The three cube roots of $-27j = 27(\cos 270° + j \sin 270°)$ are

$$r_1 = 27^{1/3} \left(\cos \frac{270° + 0 \cdot 360°)}{3} + j \sin \frac{270° + 0 \cdot 360°}{3} \right) = 3(\cos 90° + j \sin 90°)$$

$$r_1 = 3j$$

$$r_2 = 27^{1/3} \left(\cos \frac{270° + 1 \cdot 360°)}{3} + j \sin \frac{270° + 1 \cdot 360°}{3} \right) = 3(\cos 210° + j \sin 210°)$$

$$r_2 = \frac{-3\sqrt{3}}{2} - \frac{3}{2}j$$

$$r_3 = 27^{1/3} \left(\cos \frac{270° + 2 \cdot 360°)}{3} + j \sin \frac{270° + 2 \cdot 360°}{3} \right) = 3(\cos 330° + j \sin 330°)$$

$$r_3 = \frac{3\sqrt{3}}{2} - \frac{3}{2}j$$

45. $\left[\left(\frac{1}{2} - \frac{\sqrt{3}}{2}j \right) \right]^3 = \left[\frac{1}{2}(1 - \sqrt{3}j) \right]^3 = \frac{1}{8}(1 - \sqrt{3}j)^2(1 - \sqrt{3}j) = \frac{1}{8}(1 - 2\sqrt{3}j + 3j^2)(1 - \sqrt{3}j)$

$$= \frac{1}{8}(1 - 2\sqrt{3}j - 3)(1 - \sqrt{3}j) = \frac{1}{8}(-2 - 2\sqrt{3}j)(1 - \sqrt{3}j)$$

$$= \frac{1}{8}(-2 - 2\sqrt{3}j + 2\sqrt{3}j + 2\sqrt{3}^2j^2)$$

$$= \frac{1}{8}(-2 + 2 \cdot 3(-1)) = \frac{1}{8}(-2 - 6) = \frac{1}{8}(-8) = -1$$

12.7   An Application to Alternating-Current (ac) Circuits

1. $V_{AB} = I_{AB} \cdot R_{AB} = 0.00575(2250) = 12.9$ volts

5. $\overline{X}_L = 2\pi f L = 2\pi(60)(0.0429) = 16.2$

$\overline{X}_C = \dfrac{1}{2\pi f C} = \dfrac{1}{2\pi(60)(86.2 \times 10^{-6})} = 30.8$

$Z = R + (\overline{X}_L - \overline{X}_C)j = -14.6j$

$|Z| = 14.6$

    (a)   The magnitude of the impedance is 14.6 ohms

    (b)   The voltage lags the current by 90°.

9. $V = IZ = (3.90 - 6.04j) \times 10^{-3} \cdot (5.16 + 1.14j) \times 10^3$

$V = 27.0 - 26.7j$

$|V| = \sqrt{27.0^2 + (-26.7)^2} = 38.0$ volts

13. $\overline{X}_L = 2\pi f L = 2\pi(280) \cdot L = 1200$

$L = 0.682$ H

$= 682$ mH

17. $C = \dfrac{1}{(2\pi f)^2 \cdot L} = \dfrac{1}{(2\pi \cdot 680 \times 10^3)^2 \cdot (4.2 \times 10^{-3})}$

$= 1.3 \times 10^{-11}$

$C = 13 \times 10^{-12} = 13$ pF

Chapter 12 Review Exercises

1. $(6 - 2j) + (4 + j) = 6 - 2j + 4 + j$

$= 10 - j$

5. $(2 + j)(4 - j) = 8 - 2j + 4j - j^2$

$= 8 + 2j + 1 = 9 + 2j$

9. $\dfrac{3}{7 - 6j} = \dfrac{3}{(7 - 6j)} \cdot \dfrac{(7 + 6j)}{(7 + 6j)}$

$= \dfrac{21 + 18j}{7^2 + 6^2}$

$= \dfrac{21}{85} + \dfrac{18}{85}j$

13. $\dfrac{5j - (3 - j)}{4 - 2j} = \dfrac{(-3 + 6j)}{(4 - 2j)} \cdot \dfrac{(4 + 2j)}{(4 + 2j)}$

$= \dfrac{-12 - 6j + 24j + 12j^2}{4^2 + 2^2}$

$= \dfrac{-12 + 18j - 12}{16 + 4} = \dfrac{-24 + 18j}{20}$

$= -\dfrac{6}{5} + \dfrac{9}{10}j$

17. $3x - 2j = yj - 2 \Rightarrow 3x = -2, \quad -2y = yj$

$x = -\dfrac{2}{3} \qquad -2 = y$

$x = -\dfrac{2}{3}, \, y = -2$

21.

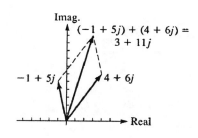

algebraically:

$$(-1 + 5j) + (4 + 6j) = -1 + 5j + 4 + 6j$$
$$= -1 + 4 + 5j + 6j$$
$$= 3 + 11j$$

25. $1 - j \Rightarrow r = \sqrt{1^2 + 1^2} = \sqrt{2}, \; \theta = \tan^{-1}\dfrac{-1}{1} + 360° = 315° = \dfrac{7\pi}{4}$ rad

polar: $1 - j = \sqrt{2}(\cos 315° + j\sin 315°)$

exponential: $1 - j = \sqrt{2}e^{7\pi/4 j}$

29. $1.07 + 4.55j \Rightarrow r = \sqrt{1.07^2 + 4.55^2} = 4.67, \; \theta = \tan^{-1}\dfrac{4.55}{1.07} = 76.7° = 1.34$ rad

polar: $1.07 + 4.55j = 4.67(\cos 76.7° + j\sin 76.7°)$

exponential: $1.07 + 4.55j = 4.67e^{1.34j}$

33. $2(\cos 225° + j\sin 225°) = -\sqrt{2} - \sqrt{2}j$

37. $0.62\angle -72° = 0.62(\cos(-72°) + j\sin(-72°))$
$$= 0.19 - 0.59j$$

41. $2.00e^{0.25j} = 2.00(\cos(0.25) + j\sin(0.25))$
$$= 1.94 + 0.495j$$

45. $[3(\cos 32° + j\sin 32°)] \cdot [5(\cos 52° + j\sin 52°)]$
$$= 3 \cdot 5(\cos(32° + 52°) + j\sin(32° + 52°))$$
$$= 15(\cos 84° + j\sin 84°)$$

49. $\dfrac{24(\cos 165° + j\sin 165°)}{3(\cos 106° + j\sin 106°)} = \dfrac{24}{3}(\cos(165° - 106°) + j\sin(165° - 106°)) = 8(\cos 59° + j\sin 59°)$

53. $0.983\angle 47.2° + 0.366\angle 95.1° = 0.983(\cos 47.2° + j\sin 47.2°) + 0.366(\cos 95.1° + j\sin 95.1°)$
$$= 0.6679 + 0.7213j - 0.03254 + 0.3646j = 0.6354 + 1.0859j$$

in polar form: $r = \sqrt{0.6354^2 + 1.0859^2} = 1.26$

$$\theta = \tan^{-1}\dfrac{1.0859}{0.6354} = 59.7°, \; 1.26\angle 59.7°$$

57. $[2(\cos 16° + j\sin 16°)]^{10} = 2^{10}(\cos(10 \cdot 16°) + j\sin(10 \cdot 16°)] = 1024(\cos 160° + j\sin 160°)$

61. $1 - j = \sqrt{2}(\cos 315° + j\sin 315°)$ from problem 25.

$(1 - j)^{10} = [\sqrt{2}(\cos 315° + j\sin 315°)]^{10} = \sqrt{2}^{10}(\cos(10 \cdot 315°) + j\sin(10 \cdot 315°))$
$$= 32(\cos 3150° + j\sin 3150°), \text{ polar form}$$
$$= 0 - 32j, \text{ rectangular form}$$

$(1 - j)^{10} = ((1 - j)^2)^5 = (1 - 2j + j^2)^5 = (1 - 2j - 1)^5 = (-2j)^5 = (-2)^5 \cdot j^5$
$$= -32 \cdot j^4 \cdot j = -32j$$

65. $-8 = -8 + 0j = 8(\cos 180° + j \sin 180°)$

$$r_1 = \sqrt[3]{8} \left(\cos \frac{180° + 0 \cdot 360°}{3} + j \sin \frac{180° + 0 \cdot 360°}{3} \right) = 2(\cos 60° + j \sin 60°)$$

$$= 2 \left(\frac{1}{2} + j \cdot \frac{\sqrt{3}}{2} \right) = 1 + j\sqrt{3}$$

$$r_2 = \sqrt[3]{8} \left(\cos \frac{180° + 1 \cdot 360°}{3} + j \sin \frac{180° + 1 \cdot 360°}{3} \right) = 2(\cos 180° + j \sin 180°)$$

$$= 2(-1 + j(0)) = -2$$

$$r_3 = \sqrt[3]{8} \left(\cos \frac{180° + 2 \cdot 360°}{3} + j \sin \frac{180° + 2 \cdot 360°}{3} \right) = 2(\cos 300° + j \sin 300°)$$

$$= 2 \left(\frac{1}{2} - \frac{\sqrt{3}}{2} j \right) = 1 - j\sqrt{3}$$

69. Rectangular: $40 + 9j$ from the graph

polar: $40 + 9j \Rightarrow r = \sqrt{40^2 + 9^2} = 41, \theta = \tan^{-1} \frac{9}{40} = 12.7°$

$$40 + 9j = 41(\cos 12.7° + j \sin 12.7°)$$

73. $V_L = 60j, V_c = -60j$
$V = V_R + V_L + V_c$
$60 = V_R + 60j - 60j$
$V_R = 60$ volts

77. $2\pi f L = \dfrac{1}{2\pi f C} \Rightarrow f = \sqrt{\dfrac{1}{4\pi^2 LC}} = \sqrt{\dfrac{1}{4\pi^2 (2.65)(18.3 \times 10^{-6})}}$

$$f = 22.9 Hz$$

81. $\dfrac{1}{u + jwn} = \dfrac{1}{(u + jwn)} \cdot \dfrac{(u - jwn)}{(u - jwn)} = \dfrac{u - jwn}{u^2 + w^2 n^2}$

85. Answers may vary.

Chapter 13

EXPONENTIAL AND LOGARITHMIC FUNCTIONS

13.1   The Exponential and Logarithmic Functions

1. $y = 9^x = 9^{0.5} = 9^{1/2} = \sqrt{9} = 3$

5. $3^3 = 27 \Rightarrow 3 = \log_3 27$,

"3 is the exponent on 3 that gives 27"

9. $4^{-2} = \dfrac{1}{16} \Rightarrow -2 = \log_4 \dfrac{1}{16}$,

"-2 is the exponent on the 4 that gives $\dfrac{1}{16}$"

13. $8^{1/3} = 2 \Rightarrow \dfrac{1}{3} = \log_8 2$,

"$\dfrac{1}{3}$ is the exponent on 8 that gives 2"

17. $\log_3 81 = 4 \Rightarrow 3^4 = 81$,

"4 is the exponent on 3 that gives 81"

21. $\log_{25} 5 = \dfrac{1}{2} \Rightarrow 25^{1/2} = 5$,

"$\dfrac{1}{2}$ is the exponent on 25 that gives 5"

25. $\log_{10} 0.1 = -1 \Rightarrow 10^{-1} = 0.1$,

"-1 is the exponent on 10 that gives 0.1"

29. $\log_4 16 = x \Rightarrow 4^x = 16$
$$4^x = 4^2$$
$$x = 2$$

33. $\log_7 y = 3 \Rightarrow 7^3 = y$
$$y = 343$$

37. $\log_b 81 = 2 \Rightarrow b^2 = 81 = 9^2$
$$b = 9$$

41. $\log_{10} 10^{0.2} = x \Rightarrow 10^x = 10^{0.2}$
$$x = 0.2$$

45. $y = \log_4 x = \log_4 64 = 3$, "3 is the exponent on 4 that gives 64"

49. $V = A(1.05)^t \Rightarrow 1.05^t = \dfrac{V}{A}$
$$t = \log_{1.05} \dfrac{V}{A}$$

53. $\log_e \left(\dfrac{N}{N_0} \right) = -kt \Rightarrow e^{-kt} = \dfrac{N}{N_0}$
$$N = N_0 e^{-kt}$$

13.2   Graphs of $y = b^x$ and $y = \log_b x$

1. $y = 3^x$

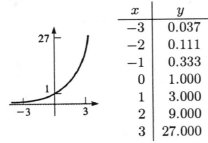

x	y
-3	0.037
-2	0.111
-1	0.333
0	1.000
1	3.000
2	9.000
3	27.000

5. $y = \left(\dfrac{1}{3}\right)^x$

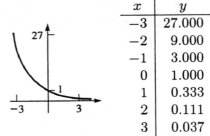

x	y
-3	27.000
-2	9.000
-1	3.000
0	1.000
1	0.333
2	0.111
3	0.037

9. $y = \log_3 x$

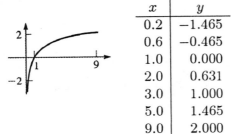

x	y
0.2	-1.465
0.6	-0.465
1.0	0.000
2.0	0.631
3.0	1.000
5.0	1.465
9.0	2.000

13. $N = 2.5 \log_3 v$

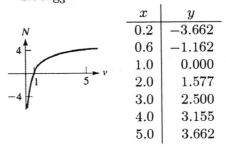

x	y
0.2	-3.662
0.6	-1.162
1.0	0.000
2.0	1.577
3.0	2.500
4.0	3.155
5.0	3.662

17. $y = 0.3(2.55)^x$

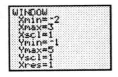   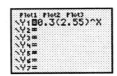   

21. $i = 1.2(2 + 6^{-t}) \Rightarrow y = 1.2(2 + 6^{-x})$

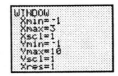   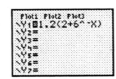   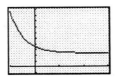

25. $V = P\left(1 + \dfrac{r}{n}\right)^{nt} = 1000\left(1 + \dfrac{0.06}{2}\right)^{2t}, \; 0 \le t \le 8 \Rightarrow y = 1000\left(1 + \dfrac{0.06}{2}\right)^{2x}, \; 0 \le x \le 8$

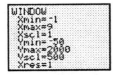      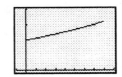

29. $t = N + \log_2 N = N + \dfrac{\log N}{\log 2} \Rightarrow y = x + \dfrac{\log x}{\log 2}$ where the change of base formula, equation

13-2, has been used since the calculator does not have a $\log_2$ key.

 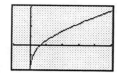

33. For each graph to be the mirror image of the other across $y = x$ the calculator window must be "square." One way to do this is ZOOM 5: ZSquare

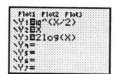 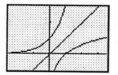

13.3   Properties of Logarithms

1. $\log_5 33 = \log_5(3 \cdot 11) = \log_5 3 + \log_5 11$

5. $\log_2(a^3) = 3\log_2 a$

9. $\log_5 \sqrt[4]{y} = \log_5 y^{1/4} = \dfrac{1}{4}\log_5 y$

13. $\log_b a + \log_b c = \log_b(ac)$

17. $\log_b x^2 - \log_b \sqrt{x} = \log_b \dfrac{x^2}{x^{1/2}}$
$$= \log_b x^{3/2}$$

21. $\log_2\left(\dfrac{1}{32}\right) = \log_2\left(\dfrac{1}{2^5}\right) = \log_2 2^{-5}$
$$= -5\log_2 2 = -5$$

25. $\log_7 \sqrt{7} = \log_7 7^{1/2}$
$$= \dfrac{1}{2}\log_7 7 = \dfrac{1}{2}$$

29. $\log_3 18 = \log_3(9 \cdot 2) = \log_3 9 + \log_3 2$
$$= \log_3 3^2 + \log_3 2 = 2\log_3 3 + \log_3 2$$
$$= 2 + \log_3 2$$

33. $\log_3 \sqrt{6} = \log_3(3 \cdot 2)^{1/2} = \dfrac{1}{2}\log_3(3 \cdot 2)$
$$= \dfrac{1}{2} \cdot [\log_3 3 + \log_3 2]$$
$$= \dfrac{1}{2} \cdot [1 + \log_3 2]$$

37. $\log_b y = \log_b 2 + \log_b x$
$$\log_b y = \log_b(2x)$$
$$y = 2x$$

41. $\log_{10} y = 2\log_{10} 7 - 3\log_{10} x$
$\qquad = \log_{10} 7^2 - \log_{10} x^3$
$\qquad = \log_{10} 49 - \log_{10} x^3$
$\log_{10} y = \log_{10} \dfrac{49}{x^3}$
$\qquad y = \dfrac{49}{x^3}$

45. $\log_2 x + \log_2 y = 1$
$\qquad \log_2(xy) = 1 \Leftrightarrow 2^1 = xy$
$$y = \frac{2}{x}$$

49. $\log_{10}(x+3) = \log_{10} x + \log_{10} 3 = \log_{10}(3x)$
$\qquad x + 3 = 3x$
$\qquad 2x = 3$

$\qquad x = \dfrac{3}{2}$ is the only value for which $\log_{10}(x+3) = \log_{10} x + \log_{10} 3$ is true.

For any other x-value $\log_{10}(x+3) = \log x + \log_{10} 3$ is false and thus not true in general. This can also be seen from the following graphs.

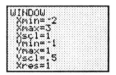 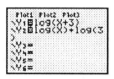 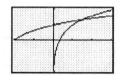 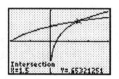

13.4 Logarithms to the Base 10

1. $\log 567 = 2.754$

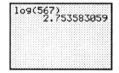

5. $\log 9.24 \times 10^6 = 6.966$

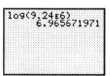

9. $\log(\cos 12.5°) = -0.0104$

13. $10^{4.437} = 27,400$

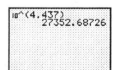

17. $10^{3.30112} = 2000.4$

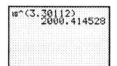

21. $(5.98)(14.3) = 85.5$

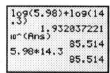

25. $\log 9 \times 10^8 = 8.9542$

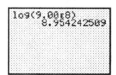

29. $G = 10 \cdot \log\left(\dfrac{P_o}{P_i}\right) = 10 \cdot \log\left(\dfrac{25.0}{0.750}\right)$
$\qquad = 15.2 dB$

13.5   Natural Logarithms

1. $\ln 26.0 = 3.258$

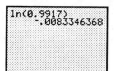

5. $\ln 0.0073267 = -4.91623$

9. $\log_5 245 = \dfrac{\log 245}{\log 5} = 3.418$

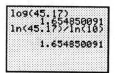

13. $\ln 51.4 = 3.940$

17. $\ln 0.9917 = -0.008335$

21.

$$\log 45.17 = \frac{\ln 45.17}{\ln 10} = 1.6549$$

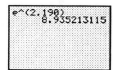

25. $e^{2.190} = 8.935$

29. $e^{-0.7429} = 0.4757$

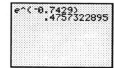

33. $y = \log_5 x = \dfrac{\ln x}{\ln 5}$

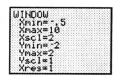

37. $\ln y - \ln x = 1.0986$

$$\ln \frac{y}{x} = 1.0986$$

$$e^{\ln y/x} = e^{1.0986}$$

$$\frac{y}{x} = e^{1.0986}$$

$$y = x \cdot e^{1.0986}$$

$$= 3x$$

41. $t = -\dfrac{L \cdot \ln\left(\dfrac{i}{I}\right)}{R} = -\dfrac{1.25 \ln\left(\dfrac{0.1I}{I}\right)}{7.5}$

$$t = 0.384 \text{ s}$$

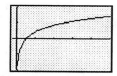

13.6 Exponential and Logarithmic Equations

1. $2^x = 16 = 2^4$
$x = 4$

5. $3^{-x} = 0.525$
$\ln 3^{-x} = \ln 0.525$
$-x \ln 3 = \ln 0.525$
$x = -\dfrac{\ln 0.525}{\ln 3}$
$x = 0.587$

9. $6^{x+1} = 10$
$\ln 6^{x+1} = \ln 10$
$(x+1) \cdot \ln 6 = \ln 10$
$x + 1 = \dfrac{\ln 10}{\ln 6}$
$x = \dfrac{\ln 10}{\ln 6} - 1$
$x = 0.285$

13. $0.8^x = 0.4$
$\ln 0.8^x = \ln 0.4$
$x \cdot \ln 0.8 = \ln 0.4$
$x = \dfrac{\ln 0.4}{\ln 0.8}$
$x = 4.11$

17. $3 \log_8 x = -2$
$\log_8 x = \dfrac{-2}{3}$
$8^{-2/3} = x$
$x = \dfrac{1}{4}$

21. $\log_2 x + \log_2 7 = \log_2 21$
$\log_2(7 \cdot x) = \log_2 21$
$7x = 21$
$x = 3$

25. $\log 4x + \log x = 2$
$\log 4x^2 = 2$
$4x^2 = 10^2$
$x^2 = 25 \Rightarrow x = \pm 5$
$x = 5$, note: -5 must be rejected;
$\ln(-5)$ is not a real number.

29. $3 \ln 2 + \ln(x-1) = \ln 24$
$\ln 2^3 + \ln(x-1) = \ln 24$
$\ln 8 + \ln(x-1) = \ln 24$
$\ln[8(x-1)] = \ln 24$
$8(x-1) = 24$
$x - 1 = 3$
$x = 4$

33. $\log_5(x-3) + \log_5 x = \log_5 4$
$\log_5[x(x-3)] = \log_5 4$
$x(x-3) = 4$
$x^2 - 3x - 4 = 0$
$(x-4)(x+1) = 0$
$x - 4 = 0 \quad \text{or} \quad x + 1 = 0$
$x = 4 \qquad x = -1$, reject since $\log_5 x$ requires $x > 0$.

37. $N = 2^x \Rightarrow 2.6 \times 10^8 = 2^x$

$$\log 2^x = \log 2.6 \times 10^8$$
$$x \cdot \log 2 = \log 2.6 \times 10^8$$
$$x = \frac{\log 2.6 \times 10^8}{\log 2} = 27.95393638\cdots$$
$$x = 28$$

41.
$$pH = -\log(H^+)$$
$$4.764 = -\log(H^+)$$
$$-4.764 = \log(H^+)$$
$$H^+ = 10^{-4.764}$$
$$H^+ = 1.722 \times 10^{-5}$$

45. $\ln n = -0.04t + \ln 20$
$$\ln n - \ln 20 = -0.04t$$
$$\ln \frac{n}{20} = -0.04t$$
$$e^{\ln n/20} = e^{-0.04t}$$
$$\frac{n}{20} = e^{-0.04t}$$
$$n = 20e^{-0.04t}$$

49.

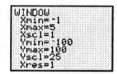 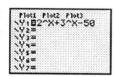 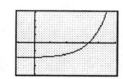 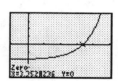

13.7   Graphs on Logarithmic and Semilogarithmic Paper

1. $y = 2^x$

x	y
1	2
2	4
3	8
4	16
5	32
6	64
7	128
8	256
9	512

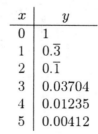

5. $y = 3^{-x}$

x	y
0	1
1	$0.\overline{3}$
2	$0.\overline{1}$
3	0.03704
4	0.01235
5	0.00412

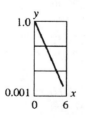

9. $y = 3x^6$

x	y
1.8	102.04
2.0	192
2.2	340.14
2.4	573.31
2.6	926.75
2.8	1445.7
3.0	2187
3.2	3221.2
3.4	4634.4
3.6	6530.3
3.8	9032.8
4.0	12,288
4.2	16,467
4.4	21,769
4.6	28,432
4.8	36,692
5.0	46,875
5.2	59,312
5.4	74,385
5.6	92,523
5.8	114,206
6.0	139,968

13. $y = 0.01x^4$

x	y
4	2.56
5	6.25
6	12.96
7	24.01
8	40.96
9	65.61
10	100

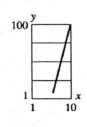

17. $y = x^2 + 2x$

x	y
2	8
3	15
4	24
5	35
6	48
7	63
8	80
9	99

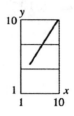

21. $y^3 x = 1 \Rightarrow y = \dfrac{1}{\sqrt[3]{x}}$

x	y
1	1
2	0.7937
3	0.69336
4	0.62996
5	0.5848
6	0.55032
7	0.52276
8	0.5
9	0.48075
10	0.46416

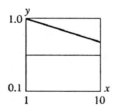

25. (a)

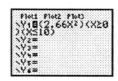 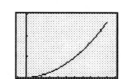

(b) $s = 2.66t^2$

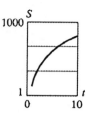

29. $g = 3.99 \times \dfrac{10^{14}}{r^2}$

r	6.37×10^6	1.0×10^7	6.0×10^7
g	9.83	3.98	1.1×10^{-1}

r	9.0×10^7	3.91×10^8
g	4.9×10^{-2}	2.6×10^{-3}

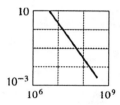

33.

d	0.63	1.3	1.9	2.5	3.8
R	600	190	100	72	46

d	5.0	7.5	10	15
R	29	17	10	6.0

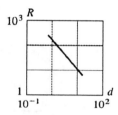

Chapter 13 Review Exercises

1. $\log_{10} x = 4 \Rightarrow x = 10^4 = 10{,}000$

5. $\log_2 64 = x \Rightarrow 2^x = 64 = 2^6$
$$x = 6$$

9. $\log_x 36 = 2 \Rightarrow x^2 = 36 = 6^2$
$$x = 6$$

13. $\log_3 2x = \log_3 2 + \log_3 x$

17. $\log_2 28 = \log_2(2^2 \cdot 7)$
$$= \log_2 2^2 + \log_2 7$$
$$= 2 \log_2 2 + \log_2 7$$
$$= 2 \cdot 1 + \log_2 7$$
$$= 2 + \log_2 7$$

21. $\log_4 \sqrt{48} = \log_4 48^{1/2} = \dfrac{1}{2} \log_4(16 \cdot 3)$
$$= \frac{1}{2}\left[\log_4 16 + \log_4 3\right] = \frac{1}{2}\left[\log_4 4^2 + \log_4 \right.$$
$$= \frac{1}{2}\left[2 \log_4 4 + \log_4 3\right] = \frac{1}{2}\left[2 + \log_4 3\right]$$
$$= 1 + \frac{1}{2} \log_4 3$$

25. $\log_6 y = \log_6 4 - \log_6 x$
$$\log_6 y = \log_6 \frac{4}{x}$$
$$y = \frac{4}{x}$$

29. $\log_5 x + \log_5 y = \log_5 3 + 1 = \log_5 3 + \log_5 5$
$$= \log_5(3 \cdot 5)$$
$$\log_5(x \cdot y) = \log_5(15)$$
$$xy = 15$$
$$y = \frac{15}{x}$$

33. $2(\log_4 y - 3\log_4 x) = 3$

$$\log_4 y - \log_4 x^3 = \frac{3}{2}$$

$$\log_4 \frac{y}{x^3} = \frac{3}{2}$$

$$\frac{y}{x^3} = 4^{3/2} = 8$$

$$y = 8x^3$$

37.

 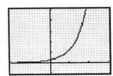

41. $y = \log_{3.15} x$

 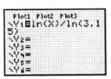 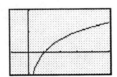

45. $\ln 8.86 = \dfrac{\log_{10} 8.86}{\log_{10} e}$

$\qquad = 2.18$

49. $\log_{10} 65.89 = \dfrac{\ln 65.89}{\ln 10}$

$\qquad = 1.819$

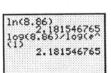

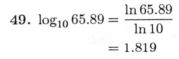

53. $\qquad e^{2x} = 5$

$\qquad \ln e^{2x} = \ln 5$

$\qquad 2x \cdot \ln e = \ln 5$

$\qquad 2x \cdot 1 = \ln 5$

$$x = \frac{\ln 5}{2}$$

57. $\log_4 x + \log_4 6 = \log_4 12$

$\qquad \log(x \cdot 6) = \log_4 12$

$\qquad\qquad 6x = 12$

$\qquad\qquad\ x = 2$

61. $y = 6^x$

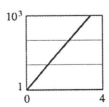

65. $10^{\log 4} = 4$

69.
$$V = Ae^{it}$$
$$\frac{V}{A} = e^{it}$$
$$\ln \frac{V}{A} = \ln e^{it} = it$$
$$t = \frac{1}{i} \ln \frac{V}{A}$$

73. $P = 2.25e^{0.00486t}$

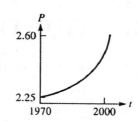

77. $C = B \log_2(1 + R)$
$$\log_2(1 + R) = \frac{C}{B}$$
$$1 + R = 2^{C/B}$$
$$R = 2^{C/B} - 1$$

81.
$$R = 4520(0.750)^{2.50t}$$
$$1950 = 4520(0.750)^{2.50t}$$
$$\frac{1950}{4520} = 0.750^{2.5t}$$
$$\ln \frac{1950}{4520} = \ln(0.750)^{2.5t} = 2.5t \ln(0.750)$$
$$2.5t = \frac{\ln \dfrac{1950}{4520}}{\ln(0.750)}$$
$$t = \frac{1}{2.5} \cdot \frac{\ln \dfrac{1950}{4520}}{\ln(0.750)}$$
$$t = 1.17 \text{ min}$$

85. $\log x^2$ exists for all values of x; $2 \log x$ exists only for positive x-values. The graphs, shown below, are not the same.

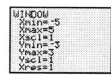 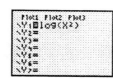 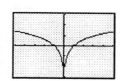

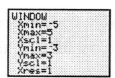 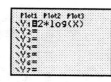 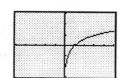

ADDITIONAL TYPES OF EQUATIONS AND SYSTEMS OF EQUATIONS

14.1 Graphical Solution of Systems of Equations

1.

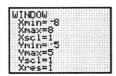 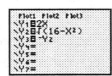 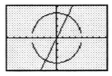 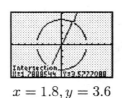 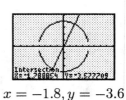

$x = 1.8, y = 3.6$ $x = -1.8, y = -3.6$

5.

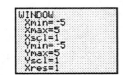 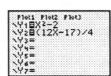 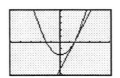 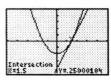

$x = 1.5, y = 0.2$

9.

 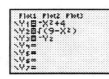 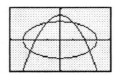 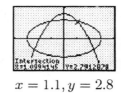 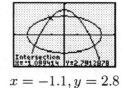

$x = 1.1, y = 2.8$ $x = -1.1, y = 2.8$

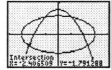 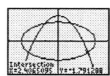

$x = -2.4, y = -1.8$ $x = 2.4, y = -1.8$

13.

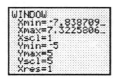 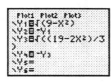 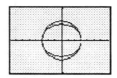 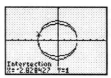 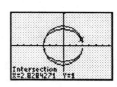

$x = -2.8, y = 1.0$ $x = 2.8, y = 1.0$

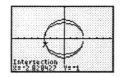 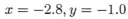 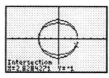

$x = -2.8, y = -1.0$ $x = 2.8, y = -1.0$

17.

 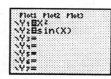 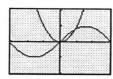 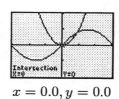 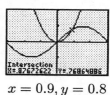

$x = 0.0, y = 0.0 \qquad x = 0.9, y = 0.8$

21.

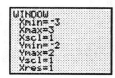 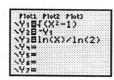 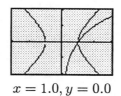

$x = 1.0, y = 0.0$

25. $y = 3x, \; y > 0; \; x^2 + y^2 = 5.2^2 = 27.04, \; x > 0, \; y > 0.$

  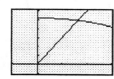 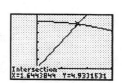

4.9 mi N, 1.6 mi E

14.2   Algebraic Solution of Systems of Equations

1. (1) $y = x + 1$ $\qquad\qquad\qquad y = y$

  (2) $y = x^2 + 1$ $\qquad\qquad x^2 + 1 = x + 1$ with y from (1) and (2)

$$x^2 - x = 0$$
$$x(x - 1) = 0$$
$$x = 0 \quad \text{or} \qquad x - 1 = 0$$
$$y = x + 1 = 1 \qquad\qquad x = 1$$
$$y = x + 1 = 2$$

Solution: $(0, 1), (1, 2)$

5. (1) $x + y = 1 \Rightarrow y = 1 - x$ (2) $\quad x^2 - (1 - x)^2 = 1$ with $y = 1 - x$ from (1)

  (2) $x^2 - y^2 = 1$ $\qquad\qquad\qquad x^2 - 1 + 2x - x^2 = 1$

$$-1 + 2x = 1$$
$$2x = 2$$
$$x = 1 \quad y = 1 - (1) = 0$$

Solution: $(1, 0)$

9. (1) $\quad wh = 1$

  (2) $w + h = 2 \Rightarrow h = 2 - w$ (1) $\quad w(2 - w) = 1$ with h from (2)

$$2w - w^2 = 1$$
$$w^2 - 2w + 1 = 0$$
$$(w - 1)^2 = 0$$
$$w - 1 = 0$$
$$w = 1 \Rightarrow (2) \; 1 + h = 2, \, h = 1$$

Solution: $(1, 1)$

13. (1) $y = x^2$ $y = y$

   (2) $y = 3x^2 - 8$ $x^2 = 3x^2 - 8$ with y from (1) and (2)

$$2x^2 = 8$$
$$x^2 = 4$$
$$x = \pm 2, \ (1) \ y = x^2 = (\pm 2)^2 = 4$$

Solution: $(2, 4), (-2, 4)$

17. (1) $D^2 - 1 = R \Rightarrow D^2 = 1 + R$ $1 + R - 2R^2 = 1$ (2) with D^2 from (1)

   (2) $D^2 - 2R^2 = 1$

$$R - 2R^2 = 0$$
$$R(1 - 2R) = 0$$
$$R = 0 \quad \text{or} \quad 1 - 2R = 0$$
$$2R = 1$$
$$R = \frac{1}{2}$$

(1) with $R = 0$, $D^2 = 1 + 0$

$$D = \pm 1$$

(1) with $R = \frac{1}{2}$, $D^2 = 1 + \frac{1}{2} = \frac{3}{2} = \frac{6}{4}$

$$D = \frac{\pm\sqrt{6}}{2}$$

Solutions: $R = 0, \ D = 1$

$R = 0, \ D = -1$

$R = \frac{1}{2}, \ D = \frac{\sqrt{6}}{2}$

$R = \frac{1}{2}, \ D = \frac{-\sqrt{6}}{2}$

21. (1) $x^2 + 3y^2 = 37$ $3 \cdot (1) \Rightarrow$ $3x^2 + 9y^2 = 111$ (1) $(\pm 5)^2 + 3y^2 = 37$

   (2) $2x^2 - 9y^2 = 14$ (2) $\underline{2x^2 - 9y^2 = \ \ 14}$ $25 + 3y^2 = 37$

$$5x^2 \qquad = 125 \qquad\qquad 3y^2 = 12$$
$$x^2 = 25 \qquad\qquad\qquad y^2 = 4$$
$$x = \pm 5 \qquad\qquad\qquad y = \pm 2$$

Solution: $(5, 2), (5, -2), (-5, 2), (-5, -2)$

25. (1) $h = 3x - 0.05x^2$ $h = h$

   (2) $h = 0.8x - 15 \Rightarrow$ $3x - 0.05x^2 = 0.8x - 15$

$$0.05x^2 - 2.2x - 15 = 0$$
$$5x^2 - 220x - 1500 = 0$$
$$x^2 - 44x - 300 = 0$$
$$(x - 50)(x + 6) = 0$$
$$x - 50 = 0 \quad \text{or} \quad x + 6 = 0$$
$$x = 50 \qquad\qquad x = -6 \text{ reject since } x > 0$$

(2) $h = 0.8(50) - 15$

   $h = 25$

The rocket and missile cross when $h = 25 \ mi$ and $x = 50$ mi.

29.

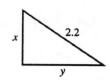

(1) $x + y + 2.2 = 4.6$

$x + y = 2.4 \Rightarrow y = 2.4 - x$

(2) $x^2 + y^2 = 2.2^2$

$x^2 + y^2 = 4.84$

(2) $x^2 + (2.4 - x)^2 = 4.84$, (2) with y from (1).

$x^2 + 5.76 - 4.8x + x^2 = 4.84$

$2x^2 - 4.8x + 0.92 = 0$

$$x = \frac{-(-4.8) \pm \sqrt{(-4.8)^2 - 4(2)(0.92)}}{2(2)} = \frac{4.8 \pm \sqrt{15.68}}{4} = \begin{cases} + \ 2.2 \\ - \ 0.2 \end{cases}$$

$x = 2.2$ in (1), $2.2 + y = 2.4$

$y = 0.2$

$x = 0.2$ in (1), $0.2 + y = 2.4$

$y = 2.2$

The lengths of the sides of the truss are 2.2 m and 0.2 m.

14.3   Equations in Quadratic Form

1. $x^4 - 13x^2 + 36 = 0$.

Let $y = x^2, y^2 = x^4$, then

$y^2 - 13y + 36 = 0$
$(y - 9)(y - 4) = 0$
$y - 9 = 0$ or $y - 4 = 0$
$\quad y = 9 \qquad\qquad y = 4$
$\quad x^2 = 9 \qquad\qquad x^2 = 4$
$\quad x = \pm 3 \qquad\qquad x = \pm 2$

5. $x^{-2} - 2x^{-1} - 8 = 0$.

Let $y = x^{-1}, y^2 = x^{-2}$, then

$y^2 - 2y - 8 = 0$
$(y - 4)(y + 2) = 0$
$y - 4 = 0$ or $y + 2 = 0$
$\quad y = 4 \qquad\qquad y = -2$
$\quad x^{-1} = 4 \qquad\quad x^{-1} = -2$

$\quad x = \dfrac{1}{4} \qquad\qquad x = \dfrac{-1}{2}$

9. $2x - 7\sqrt{x} + 5 = 0$.

Let $y = \sqrt{x}, y^2 = x$, then

$2y^2 - 7y + 5 = 0$
$(2y - 5)(y - 1) = 0$
$2y - 5 = 0$ or $y - 1 = 0$
$\quad 2y = 5 \qquad\qquad y = 1$

$\quad y = \dfrac{5}{2} \qquad\qquad \sqrt{x} = 1$

$\quad \sqrt{x} = \dfrac{2}{5} \qquad\qquad x = 1$

$\quad x = \dfrac{25}{4}$

13. $x^{2/3} - 2x^{1/3} - 15 = 0$.

Let $y = x^{1/3}, y^2 = x^{2/3}$, then

$y^2 - 2y - 15 = 0$
$(y - 5)(y + 3) = 0$
$y - 5 = 0$ or $y + 3 = 0$
$\quad y = 5 \qquad\qquad y = -3$
$\quad x^{1/3} = 5 \qquad\quad x^{1/3} = -3$
$\quad x = 125 \qquad\qquad x = -27$

17. $(x - 1) - \sqrt{x - 1} - 2 = 0$.

Let $y = \sqrt{x - 1}, y^2 = x - 1$, then

$$y^2 - y - 2 = 0$$
$$(y - 2)(y + 1) = 0$$

$y - 2 = 0$ or $y + 1 = 0$
$\quad y = 2 \qquad\qquad y = -1$, reject since
$\sqrt{x - 1} = 2 \qquad\qquad y = \sqrt{x - 1}$
$\quad x - 1 = 4 \qquad\qquad$ requires $y \geq 0$
$\qquad x = 5$

21. $x - 3\sqrt{x - 2} = 6$.

Let $y = \sqrt{x - 2}, y^2 = x - 2 \Rightarrow y^2 + 2 = x$, then

$$y^2 + 2 - 3y = 6$$
$$y^2 - 3y - 4 = 0$$
$$(y - 4)(y + 1) = 0$$

$y - 4 = 0 \quad$ or $\quad y + 1 = 0$
$\quad y = 4 \qquad\qquad\quad y = -1$
$\sqrt{x - 2} = 4 \qquad \sqrt{x - 2} = -1$ has no solution
$\quad x - 2 = 16 \qquad\qquad\qquad$ since $\sqrt{x - 2} \geq 0$
$\qquad x = 18$

25.
$$R_T^{-1} = R_1^{-1} + R_2^{-1}$$
$$1^{-1} = R_1^{-1} + \sqrt{R_1}^{-1}$$
$$1 = R_1^{-1} + R_1^{-1/2}. \text{ Let } y = R_1^{-1/2}, y^2 = R_1^{-1}, \text{ then}$$

$y^2 + y - 1 = 0$. The quadratic formula given

$$y = \frac{-1 \pm \sqrt{1^2 - 4(1)(-1)}}{2(1)}$$

$$= \frac{-1 \pm \sqrt{5}}{2}. \text{ The } - \text{ is rejected since } y > 0$$

$$y = \frac{-1 + \sqrt{5}}{2}$$

$$R_1^{-1/2} = \frac{-1 + \sqrt{5}}{2}$$

$$\sqrt{R_1} = \frac{2}{-1 + \sqrt{5}}$$

$$R_1 = \left[\frac{2}{-1 + \sqrt{5}}\right]^2 \approx 2.62 \ \Omega$$

$$R_2 = R_1^{1/2} = \frac{2}{-1 + \sqrt{5}}$$

$$R_2 \approx 1.62 \ \Omega$$

14.4   Equations with Radicals

1. $\sqrt{x - 8} = 2$
$\quad x - 8 = 4$
$\qquad x = 12$

5.
$$\sqrt{3x+2} = 3x$$
$$3x + 2 = 9x^2$$
$$9x^2 - 3x - 2 = 0$$
$$(3x + 1)(3x - 2) = 0$$
$$3x + 1 = 0 \quad \text{or} \quad 3x - 2 = 0$$
$$3x = -1 \qquad\qquad 3x = 2$$
$$x = \frac{-1}{3} \qquad\qquad x = \frac{2}{3}$$

Check: $\sqrt{3 \cdot \dfrac{1}{3} + 2} \;\bigg|\; 3 \cdot \dfrac{-1}{3}$

$$\sqrt{-1+2} \qquad -1$$
$$\sqrt{1}$$
$$1$$

$x = \dfrac{-1}{3}$ is not a solution.

Check: $\sqrt{3 \cdot \dfrac{2}{3} + 2} \;\bigg|\; 3 \cdot \dfrac{2}{3}$

$$\sqrt{2+2} \qquad 2$$
$$\sqrt{4}$$
$$2$$

$x = \dfrac{2}{3}$ is the solution.

9.
$$2\sqrt{3-x} - x = 5$$
$$2\sqrt{3-x} = x + 5$$
$$4(3 - x) = x^2 + 10x + 25$$
$$12 - 4x = x^2 + 10x + 25$$
$$x^2 + 14x + 13 = 0$$
$$(x + 13)(x + 1) = 0$$
$$x + 13 = 0 \quad \text{or} \quad x + 1 = 0$$
$$x = -13 \qquad\qquad x = -1$$

Check: $2\sqrt{3 - (-13)} - (-13) \;\bigg|\; 5$

$$2\sqrt{16} + 13$$
$$2 \cdot 4 + 13$$
$$8 + 13$$
$$21$$

$x = -13$ is not a solution.

$2\sqrt{3 - (-1)} - (-1) \;\bigg|\; 5$

$$2\sqrt{3 + 1} + 1$$
$$2\sqrt{4} + 1$$
$$2 \cdot 2 + 1$$
$$4 + 1$$
$$5$$

$x = -1$ is a solution.

13.
$$5\sqrt{s} - 6 = s$$
$$5\sqrt{s} = s + 6$$
$$25s = s^2 + 12s + 36$$
$$s^2 - 13s + 36 = 0$$
$$(s - 9)(s - 4) = 0$$
$$s - 9 = 0 \quad \text{or} \quad s - 4 = 0$$
$$s = 9 \qquad\qquad s = 4$$

Check: $5\sqrt{9} - 6 \;\bigg|\; 9$

$$5 \cdot 3 - 6$$
$$15 - 6$$
$$9$$

$s = 9$ is a solution.

$5\sqrt{4} - 6 \;\bigg|\; 4$

$$5 \cdot 2 - 6$$
$$10 - 6$$
$$4$$

$s = 4$ is a solution.

17.
$$\sqrt{x+4} + 8 = x$$
$$\sqrt{x+4} = x - 8$$
$$x + 4 = x^2 - 16x + 64$$
$$x^2 - 17x + 60 = 0$$
$$(x - 12)(x - 5) = 0$$
$$x - 12 = 0 \quad \text{or} \quad x - 5 = 0$$
$$x = 12 \qquad\qquad x = 5$$

Check: $\sqrt{12 + 4} + 8 \;\bigg|\; 12$

$$\sqrt{16} + 8$$
$$4 + 8$$
$$12$$

$x = 12$ is a solution.

$\sqrt{5 + 4} + 8 \;\bigg|\; 5$

$$\sqrt{9} + 8$$
$$3 + 8$$
$$11$$

$x = 5$ is not a solution. This extraneous root was introduced by squaring both sides.

21.
$$3\sqrt{1-2t}+1=2t$$
$$3\sqrt{1-2t}=2t-1$$
$$9(1-2t)=4t^2-4t+1$$
$$9-18t=4t^2-4t+1$$
$$4t^2+14t-8=0$$
$$2t^2+7t-4=0$$
$$(2t-1)(t+4)=0$$

$$2t-1=0 \ \text{ or } \ t+4=0$$
$$2t=1 \qquad\qquad t=-4$$
$$t=\frac{1}{2}$$

Check: $3\sqrt{1-2\cdot\frac{1}{2}}+1 \ \Big| \ 2\cdot\frac{1}{2}$

$$3\sqrt{1-1}+1 \qquad 1$$
$$3\cdot\sqrt{0}+1$$
$$1$$

$t=\dfrac{1}{2}$ is a solution.

Check: $3\sqrt{1-2(-4)}+1 \ \Big| \ 2\cdot(-4)$

$$3\sqrt{9}+1 \qquad\qquad -8$$
$$3\cdot 3+1$$
$$9+1$$
$$10$$

$t=-4$ is not a solution.

25.
$$\sqrt{5x+1}-1=3\sqrt{x}$$
$$\sqrt{5x+1}=3\sqrt{x}+1$$
$$5x+1=9x+6\sqrt{x}+1$$
$$-4x=6\sqrt{x}. \quad \sqrt{x} \text{ requires } x\ge 0 \text{ which implies the } x \text{ on LHS is negative. The RHS}$$
$$\text{is } \ge 0. \text{ Therefore } x=0 \text{ is the only possible solution and it checks.}$$

Check: $\sqrt{5\cdot 0+1}-1 \ \Big| \ 3\sqrt{0}$

$$\sqrt{1}-1 \qquad\quad 3\cdot 0$$
$$0 \qquad\qquad\quad 0$$

29.
$$\sqrt[3]{2x-1}=\sqrt[3]{x+5}$$
$$2x-1=x+5$$
$$x=6$$

Check: $\sqrt[3]{2\cdot 6-1} \ \Big| \ \sqrt[3]{6+5}$

$$\sqrt[3]{12-1} \qquad \sqrt[3]{11}$$
$$\sqrt[3]{11}$$

$x=6$ is a solution.

33.
$$f=\frac{1}{2\pi\sqrt{LC}}$$
$$f^2=\frac{1}{4\pi^2 LC}$$
$$L=\frac{1}{4\pi^2 f^2 C}$$

37.

$$(x+5.2)^2=x^2+8.3^2$$
$$x^2+10.4x+27.04=x^2+68.89$$
$$10.4x=41.85$$
$$x=4.0$$
$$x+5.2=4.0+5.2=9.2$$

The freighter is 9.2 km from the station.

Chapter 14 Review Exercises

1. $x + 2y = 6 \Rightarrow y = \dfrac{6 - x}{2}$

$y = 4x^2$

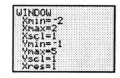 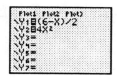 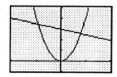 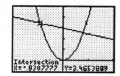 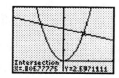

5. $y = x^2 + 1$

$2x^2 + y^2 = 4 \Rightarrow y \pm \sqrt{4 - 2x^2}$

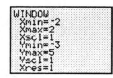 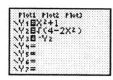 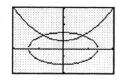 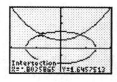 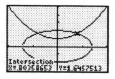

9. $y = x^2 - 2x$

$y = 1 - e^{-x}$

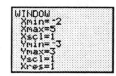 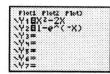 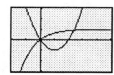 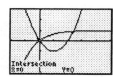 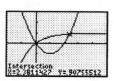

13. (1) $2y = x^2$

(2)

$$
\begin{array}{lll}
x^2 + y^2 = 3 & y + 3 = 0, & y - 1 = 0 \\
2y + y^2 = 3 & y = -3, & y = 1 \\
y^2 + 2y - 3 = 0 & 2(-3) = x^2, & 2(1) = x^2 \\
(y + 3)(y - 1) = 0 & \text{no solution} & x = \pm\sqrt{2}
\end{array}
$$

Solutions: $(\sqrt{2}, 1), (-\sqrt{2}, 1)$

17. (1) $4x^2 - 7y^2 = 21$ (1) $8x^2 - 14y^2 = 42$ (2) $(\pm 7)^2 + 2y^2 = 99$

(2) $x^2 + 2y^2 = 99$ (2) $\dfrac{7x^2 + 14y^2 = 693}{15x^2 \qquad = 735}$ $49 + 2y^2 = 99$

$x^2 = 49$ $2y^2 = 50$

$x = \pm 7$ $y^2 = 25$

$y = \pm 5$

Solution: $(7, 5), (7, -5), (-7, 5), (-7, -5)$

21. $x^4 - 20x^2 + 64 = 0.$
Let $y = x^2$, $y^2 = x^4$

$$y^2 - 20y + 64 = 0$$
$$(y - 16)(y - 4) = 0$$
$$y - 16 = 0 \quad \text{or} \quad y - 4 = 0$$
$$\quad y = 16 \qquad\qquad y = 4$$
$$\quad x^2 = 16 \qquad\qquad x^2 = 4$$
$$\qquad x = \pm 4 \qquad\qquad x = \pm 2$$

25. $D^{-2} + 4D^{-1} - 21 = 0.$

Let $x = D^{-1}$, $x^2 = D^{-2}$

$$x^2 + 4x - 21 = 0$$
$$(x + 7)(x - 3) = 0$$
$$x + 7 = 0 \quad \text{or} \quad x - 3 = 0$$
$$\quad x = -7 \qquad\qquad x = 3$$
$$D^{-1} = -7 \qquad\quad D^{-1} = 3$$
$$\quad D = \frac{-1}{7} \qquad\qquad D = \frac{1}{3}$$

29. $\dfrac{4}{r^2 + 1} + \dfrac{7}{2r^2 + 1} = 2$

$$4(2r^2 + 1) + 7(r^2 + 1) = 2(r^2 + 1)(2r^2 + 1)$$
$$8r^2 + 4 + 7r^2 + 7 = 2(2r^4 + 3r^2 + 1)$$
$$15r^2 + 11 = 4r^4 + 6r^2 + 2$$
$$4r^4 - 9r^2 - 9 = 0. \text{ Let } x = r^2,\ x^2 = r^4$$
$$4x^2 - 9x - 9 = 0$$
$$(4x + 3)(x - 3) = 0$$

$$4x + 3 = 0 \qquad \text{or} \quad x - 3 = 0$$
$$\quad 4x = -3 \qquad\qquad\qquad x = 3$$

$$\quad x = \frac{-3}{4} \qquad\qquad\qquad r^2 = 3$$

$$r^2 = \frac{-3}{4} \qquad\qquad\qquad r = \pm\sqrt{3}$$

$$r = \frac{\pm\sqrt{3}}{2}j$$

33. $\sqrt{5x + 9} + 1 = x$
$$\sqrt{5x + 9} = x - 1$$
$$5x + 9 = x^2 - 2x + 1$$
$$x^2 - 7x - 8 = 0$$
$$(x - 8)(x + 1) = 0$$

$$x - 8 = 0 \quad \text{or} \quad x + 1 \quad 0$$
$$\quad x = 0 \qquad\qquad\quad x \quad -1$$

Check: $\sqrt{5 \cdot 8 + 9} + 1 \ \Big|\ 8$
$$\sqrt{49} + 1$$
$$7 + 1$$
$$8$$

$x = 8$ is a solution.

$\sqrt{5(-1) + 9} + 1 \ \Big|\ -1$
$$\sqrt{-5 + 9} + 1$$
$$\sqrt{4} + 1$$
$$2 + 1$$
$$3$$

$x = -1$ is not a solution.

37.
$$\sqrt{x+4}+2\sqrt{x+2}=3$$
$$\sqrt{x+4}=3-2\sqrt{x+2}$$
$$x+4=9-12\sqrt{x+2}+4(x+2)$$
$$x+4=9-12\sqrt{x+2}+4x+8$$
$$12\sqrt{x+2}=13+3x$$
$$144(x+2)=169+78x+9x^2$$
$$144x+288=169+78x+9x^2$$
$$9x^2-66x-119=0$$

From the quadratic formula,

$$x=\frac{-(-66)\pm\sqrt{(-66)^2-4(9)(-119)}}{2(9)}=\frac{66\pm\sqrt{8640}}{18}$$

Check: $\sqrt{\dfrac{66+\sqrt{8640}}{18}+4}+2\sqrt{\dfrac{66+\sqrt{8640}}{18}+2}=10.1639\cdots$

$\dfrac{66+\sqrt{8640}}{18}$ does not check

$\sqrt{\dfrac{66-\sqrt{8640}}{18}+4}+2\sqrt{\dfrac{66-\sqrt{8640}}{18}+2}=3$

$\dfrac{66-\sqrt{8640}}{18}$ is the solution.

41.
$$L=\frac{h}{2\pi}\sqrt{l(l+1)}$$
$$L^2=\frac{h^2}{4\pi^2}\cdot l(l+1)$$
$$\frac{4\pi^2L^2}{h^2}=l^2+l$$

$l^2+l-\dfrac{4\pi^2L^2}{h^2}=0$; from the quadratic formula,

$$l=\frac{-1\pm\sqrt{1^2-4\cdot1\left(-\dfrac{4\pi^2L^2}{h^2}\right)}}{2(1)}=\frac{-1\pm\sqrt{1+\dfrac{16\pi^2L^2}{h^2}}}{2}$$

$$l=\frac{-1+\sqrt{1+\dfrac{16\pi^2L^2}{h^2}}}{2}.$$ The + was chosen because $l>0$

45. (1) $16t_1^2 + 16t_2^2 = 45$ where $t_1, t_2 > 0$

      (2) $t_2 = 2t_1$

    (1) with t_2 from (2)

$$16t_1^2 + 16(2t_1)^2 = 45$$
$$16t_1^2 + 64t_1^2 = 45$$
$$80t_1^2 = 45$$
$$t_1^2 = \frac{45}{80}$$
$$t_1 = 0.75s$$
$$t_2 = 2 \cdot t_1 = 2(0.75)$$
$$t_2 = 1.5 \text{ s}$$

49. $Z = \sqrt{R^2 + X^2}$, $Z = 2X^2$; $R = 0.800$

      $2X^2 = \sqrt{R^2 + X^2}$ $R^2 = 0.640$

      $4X^4 = R^2 + X^2$

      $4X^4 - X^2 - R^2 = 0$

    $4X^4 - X^2 - 0.640 = 0$. Let $X^2 = y$, $X^4 = y^2$

      $4y^2 - y - 0.640 = 0$

$$y = \frac{-(-1) \pm \sqrt{(-1)^2 - 4(4)(-0.640)}}{2(4)} = \frac{1 \pm \sqrt{11.24}}{8}. \text{ Choose } +, y > 0$$

    $y = 0.544 = X^2$

    $X = 0.738\Omega$. Choose $+$, $X > 0$

    $Z = 2X^2 = 2(0.738)^2$

    $Z = 1.09 \; \Omega$

53. $2x + \sqrt{x + 20} = 12$, $x > 0$

          $\sqrt{x + 20} = 12 - 2x$

             $x + 20 = 144 - 48x + 4x^2$

    $4x^2 - 49x + 124 = 0$

$$x = \frac{-(-49) \pm \sqrt{(-49)^2 - 4(4)(124)}}{2(4)} = \frac{49 \pm \sqrt{417}}{8}$$

$$x = \frac{49 + \sqrt{417}}{8} \quad \text{or} \quad x = \frac{49 - \sqrt{417}}{8}$$

Check: $2 \cdot \dfrac{49 + \sqrt{417}}{8} + \sqrt{\dfrac{49 + \sqrt{417}}{8} + 20} = 22.710\cdots$

    $\dfrac{49 + \sqrt{417}}{8}$ does not check

    $2 \cdot \dfrac{49 - \sqrt{417}}{8} + \sqrt{\dfrac{49 - \sqrt{417}}{8} + 20} = 12$

    $\dfrac{49 + \sqrt{417}}{8}$ checks

  $x = \dfrac{49 + \sqrt{417}}{8} = 3.57$ m is the solution.

57. $x + \sqrt{x-1} + 4.00 = 9.00$ may be solved graphically or algebraically. To solve graphically, find the x-int of $y = x + \sqrt{x-1} - 5$

 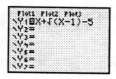 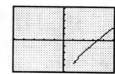 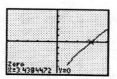

Algebraically, the solution is

$$x + \sqrt{x-1} + 4 = 9$$
$$\sqrt{x-1} = 5 - x$$
$$x - 1 = 25 - 10x + x^2$$
$$x^2 - 11x + 26 = 0$$

$$x = \frac{-(-11) \pm \sqrt{(-11)^2 - 4(1)(26)}}{2(1)} = \frac{11 \pm \sqrt{17}}{2}$$

$$x = \frac{11 + \sqrt{17}}{2} \quad \text{or} \quad x = \frac{11 - \sqrt{17}}{2}$$

Check: $\dfrac{11 + \sqrt{17}}{2} + \sqrt{\dfrac{11 + \sqrt{17}}{2} - 1} + 4 = 14.123\cdots$

$\dfrac{11 + \sqrt{17}}{2}$ does not check

$\dfrac{11 - \sqrt{17}}{2} + \sqrt{\dfrac{11 - \sqrt{17}}{2} - 1} + 4 = 9$

$\dfrac{11 - \sqrt{17}}{2}$ checks.

$\dfrac{11 - \sqrt{17}}{2} = 3.438447\cdots$ in agreement with the graphical solution..

The solution is 3.44 in.

EQUATIONS OF HIGHER DEGREE

15.1 The Remainder Theorem and the Factor Theorem

1.

$$\begin{array}{r} x^2 + 3x + 2 \\ x - 1\overline{)x^3 + 2x^2 - x - 2} \\ \underline{x^3 - x^2} \\ 3x^2 - x \\ \underline{3x^2 - 3x} \\ 2x - 2 \\ \underline{2x - 2} \\ 0 \end{array}$$

$$f(1) = 1^3 + 2 \cdot 1^2 - 1 - 2$$
$$f(1) = 0$$

5.

$$\begin{array}{r} 2x^4 - 4x^3 + 8x^2 - 17x + 42 \\ x + 2\overline{)2x^5 + 0x^4 + 0x^3 - x^2 + 8x + 44} \\ \underline{2x^5 + 4x^4} \\ -4x^4 + 0x^3 \\ \underline{-4x^4 - 8x^3} \\ 8x^3 - x^2 \\ \underline{8x^3 + 16x^2} \\ -17x^2 + 8x \\ \underline{-17x^2 - 34x} \\ 42x + 44 \\ \underline{42x + 84} \\ -40 \end{array}$$

$$f(-2) = 2(-2)^5 - (-2)^2 + 8(-2) + 44$$
$$f(-2) = -40$$

9. $f(-1) = (-1)^3 + 2(-1)^2 - 3(-1) + 4 = 8$
   $R = 8$

13. $f(3) = 2 \cdot 3^4 - 7 \cdot 3^3 - 3^2 + 8 = -28$
   $R = -28$

17. Evaluate $y^3 - 8y - 3$ for $y = 3$

$$3^3 - 8 \cdot 3 - 3 = 27 - 24 - 3 = 0 \Rightarrow y - 3 \text{ is a factor.}$$

21. Evaluate $3V^4 - 7V^3 + V + 8$ for $V = 2$

$$3 \cdot 2^4 - 7 \cdot 2^3 + 2 + 8 = 2 \Rightarrow V - 2 \text{ is not a factor.}$$

25. $f(x) = x^3 - 2x^2 - 9x + 18$
   $f(2) = 2^3 - 2 \cdot 2^2 - 9 \cdot 2 + 18 = 0$, 2 is a zero.

29.

$$
\begin{array}{r}
2x^2 + 5x + 2 \\
2x - 1\overline{)\,4x^3 + 8x^2 - x - 2} \\
\underline{4x^3 - 2x^2} \\
10x^2 - x \\
\underline{10x^2 - 5x} \\
4x - 2 \\
\underline{4x - 2} \\
0
\end{array}
$$

$2x - 1$ is a factor of $4x^3 + 8x^2 - x - 2$.

We may not conclude $f(-1) = 0$ because $2x - 1$ is not in the form $x - r$.

15.2   Synthetic Division

1. $(x^3 + 2x^2 - x - 2) \div (x - 1)$
$ = x^2 + 3x + 2$

$$
\begin{array}{rrrr|l}
1 & 2 & -1 & -2 & \underline{1} \\
  & 1 & 3 & 2 & \\
\hline
1 & 3 & 2 & 0 &
\end{array}
$$

5. $(2x^5 - x^2 + 8x + 44) \div (x + 2)$

$ = 2x^4 - 4x^3 + 8x^2 - 17x + 42 + \dfrac{-40}{x + 2}$

$$
\begin{array}{rrrrrr|l}
2 & 0 & 0 & -1 & 8 & 44 & \underline{-2} \\
  & -4 & 8 & -16 & 34 & -84 & \\
\hline
2 & -4 & 8 & -17 & 42 & -40 &
\end{array}
$$

9. $(x^3 + 2x^2 - 3x + 4) \div (x + 1)$

$ = x^2 + x - 4 + \dfrac{8}{x + 1}$

$$
\begin{array}{rrrr|l}
1 & 2 & -3 & 4 & \underline{-1} \\
  & -1 & -1 & 4 & \\
\hline
1 & 1 & -4 & 8 &
\end{array}
$$

13. $(2x^4 - 7x^3 - x^2 + 8) \div (x - 3)$

$ = 2x^3 - x^2 - 4x - 12 + \dfrac{-28}{x - 3}$

$$
\begin{array}{rrrrr|l}
2 & -7 & -1 & 0 & 8 & \underline{3} \\
  & 6 & -3 & -12 & -36 & \\
\hline
2 & -1 & -4 & -12 & -28 &
\end{array}
$$

17. $(p^6 - 6p^3 - 2p^2 - 6) \div (p - 2)$

$ = p^5 + 2p^4 + 4p^3 + 2p^2 + 2p + 4 + \dfrac{2}{p - 2}$

$$
\begin{array}{rrrrrrr|l}
1 & 0 & 0 & -6 & -2 & 0 & -6 & \underline{2} \\
  & 2 & 4 & 8 & 4 & 4 & 8 & \\
\hline
1 & 2 & 4 & 2 & 2 & 4 & 2 &
\end{array}
$$

21.
$$
\begin{array}{rrrr|l}
1 & 1 & -1 & 2 & \underline{-2} \\
  & -2 & 2 & -2 & \\
\hline
1 & -1 & 1 & 0 &
\end{array}
$$
$, R = 0 \Rightarrow x + 2$ is a factor.

25.
$$
\begin{array}{rrrrrr|l}
2 & 0 & -1 & 3 & 0 & -4 & \underline{-1} \\
  & -2 & 2 & -1 & -2 & 2 & \\
\hline
2 & -2 & 1 & 2 & -2 & -2 &
\end{array}
$$
$, R = -2, \; x + 1$ is not a factor.

29.
$$
\begin{array}{rrrrr|l}
2 & -1 & -4 & 0 & 1 & \underline{\frac{1}{2}} \\
  & 1 & 0 & -2 & -1 & \\
\hline
2 & 0 & -4 & -2 & 0 &
\end{array}
$$
$, R = 0, \; Z - \frac{1}{2}$ is a factor $\Rightarrow 2z - 1$ is a factor.

33.
$$
\begin{array}{rrrrr|l}
1 & -5 & -15 & 5 & 14 & \underline{7} \\
  & 7 & 14 & -7 & -14 & \\
\hline
1 & 2 & -1 & -2 & 0 &
\end{array}
$$
$, R = 0, \; 7$ is a zero.

15.3   The Roots of an Equation

1.
$$\begin{array}{rrrr} 1 & 2 & -1 & -2 \\ & 1 & 3 & 2 \\ \hline 1 & 3 & 2 & 0 \end{array} \quad \underline{|\,1}$$

$$\begin{aligned} x^3 + 2x^2 - x - 2 &= (x-1)(x^2 + 3x + 2) \\ &= (x-1)(x+2)(x+1) \end{aligned}$$

$$r_1 = 1,\ r_2 = -2,\ r_3 = -1$$

5.
$$\begin{array}{rrrr} 2 & 11 & 20 & 12 \\ & -3 & -12 & -12 \\ \hline 2 & 8 & 8 & 0 \end{array} \quad \underline{\left|-\tfrac{3}{2}\right.}$$

$$\begin{aligned} 2x^3 + 11x^2 + 20x + 12 &= \left(x + \frac{3}{2}\right)(2x^2 + 8x + 8) \\ &= 2\left(x + \frac{3}{2}\right)(x^2 + 4x + 4) \\ &= 2\left(x + \frac{3}{2}\right)(x+2)(x+2) \end{aligned}$$

$$r_1 = -\frac{3}{2},\ r_2 = r_3 = -2$$

9.
$$\begin{array}{rrrrr} 1 & 1 & -2 & 4 & -24 \\ & 2 & 6 & 8 & 24 \\ \hline 1 & 3 & 4 & 12 & 0 \end{array} \quad \underline{|\,2}$$

$$t^4 + t^3 - 2t^2 + 4t - 24 = (t-2)(t^3 + 3t^2 + 4t + 12)$$

$$\begin{array}{rrrr} 1 & 3 & 4 & 12 \\ & -3 & 0 & -12 \\ \hline 1 & 0 & 4 & 0 \end{array} \quad \underline{|-3}$$

$$t^4 + t^3 - 2t^2 + 4t - 24 = (t-2)(t+3)(t^2+4) = (t-2)(t+3)(t-2j)(t+2j)$$

$$r_1 = 2,\ r_2 = -3,\ r_3 = -2j,\ r_4 = 2j$$

13.
$$\begin{array}{rrrrr} 6 & 5 & -15 & 0 & 4 \\ & -3 & -1 & 8 & -4 \\ \hline 6 & 2 & -16 & 8 & 0 \end{array} \quad \underline{\left|-\tfrac{1}{2}\right.}$$

$$6x^4 + 5x^3 - 15x^2 + 4 = \left(x + \frac{1}{2}\right)(6x^3 + 2x^2 - 16x + 8) = 2\left(x + \frac{1}{2}\right)(3x^3 + x^2 - 8x + 4)$$

$$\begin{array}{rrrr} 3 & 1 & -8 & 4 \\ & 2 & 2 & -4 \\ \hline 3 & 3 & -6 & 0 \end{array} \quad \underline{\left|\tfrac{2}{3}\right.}$$

$$\begin{aligned} 6x^4 + 5x^3 - 15x^2 + 4 &= 2\left(x + \frac{1}{2}\right)\left(x - \frac{2}{3}\right)(3x^2 + 3x - 6) \\ &= 6\left(x + \frac{1}{2}\right)\left(x - \frac{2}{3}\right)(x^2 + x - 2) \\ &= 6\left(x + \frac{1}{2}\right)\left(x - \frac{2}{3}\right)(x+2)(x-1) \end{aligned}$$

$$r_1 = -\frac{1}{2},\ r_2 = \frac{2}{3},\ r_3 = -2,\ r_4 = 1$$

17.

2	11	16	−8	−32	−16		−2
	−4	−14	−4	24	16		
2	7	2	−12	−8	0		

$$2x^5 + 11x^4 + 16x^3 - 8x^2 - 32x - 16 = (x+2)(2x^4 + 7x^3 + 2x^2 - 12x - 8)$$

2	7	2	−12	−8		−2
	−4	−6	8	8		
2	3	−4	−4	0		

$$2x^5 + 11x^4 + 16x^3 - 8x^2 - 32x - 16 = (x+2)(x+2)(2x^3 + 3x^2 - 4x - 4)$$

2	3	−4	−4		−2
	−4	2	4		
2	−1	−2	0		

$$2x^5 + 11x^4 + 16x^3 - 8x^2 - 32x - 16 = (x+2)(x+2)(x+2)(2x^2 - x - 2)$$

$$r_1 = r_2 = r_3 = -2, \ r_4 = \frac{-(-1) + \sqrt{(-1)^2 - 4(2)(-2)}}{2(2)}, \ r_5 = \frac{-(-1) - \sqrt{(-1)^2 - 4(2)(-2)}}{2(2)}$$

$$r_1 = r_2 = r_3 = -2, \ r_4 = \frac{1 + \sqrt{17}}{4}, \ r_5 = \frac{1 - \sqrt{17}}{4}$$

21.

1	−3	0	0	−1	3		3
	3	0	0	0	−3		
1	0	0	0	−1	0		

$$\begin{aligned}
P^5 - 3P^4 - P + 3 &= (P-3)(P^4 - 1) \\
&= (P-3)(P^2 - 1)(P^2 + 1) \\
&= (P-3)(P-1)(P+1)(P+j)(P-j)
\end{aligned}$$

$$r_1 = 3, \ r_2 = 1, \ r_3 = -1, \ r_4 = -j, \ r_5 = j$$

15.4   Rational and Irrational Roots

1. $f(x) = x^3 + 2x^2 - x - 2 = 0$ has $n = 3$ and therefore three roots. $f(x)$ has one sign change; therefore, there is one positive root. $f(-x) = -x^3 + 2x^2 + x - 2$ has two sign changes; therefore, there are at most two negative roots. Possible rational roots are $\frac{\pm 2}{\pm 1}$.

1	2	−1	−2		1
	1	3	2		
1	3	2	0		

$$\begin{aligned}
x^3 + 2x^2 - x - 2 &= (x-1)(x^2 + 3x + 2) \\
&= (x-1)(x+2)(x+1)
\end{aligned}$$

$$r_1 = 1, \ r_2 = -2, \ r_3 = -1$$

5. $f(x) = 2x^3 - 5x^2 - 28x + 15 = 0$ has $n = 3$ and therefore three roots. $f(x)$ has two sign changes; therefore, there are at most two positive roots. $f(-x) = -2x^3 - 5x^2 + 28x + 15 = 0$ has one sign change and therefore one negative root.

$$\text{Possible rational roots} = \frac{\text{factors of } 15}{\text{factors of } 12} = \frac{\pm 1, \ \pm 3, \ \pm 5, \ \pm 15}{\pm 1, \ \pm 2}$$

$$
\begin{array}{rrrr|l}
2 & -5 & -28 & 15 & \underline{-3} \\
  & -6 & 33 & -15 & \\
\hline
2 & -11 & 5 & 0, & \ -3 \text{ is the one negative root.}
\end{array}
$$

$$
\begin{aligned}
2x^3 - 5x^2 - 28x + 15 &= (x+3)(2x^2 - 11x + 5) \\
&= (x+3)(2x-1)(x-5)
\end{aligned}
$$

The two positive roots are $\frac{1}{2}$ and 5.

9. $f(x) = x^4 - 11x^2 - 12x + 4 = 0$ has $n = 4$ and therefore four roots. $f(x)$ has two sign changes and therefore at most two positive roots. $f(-x) = x^4 - 11x^2 + 12x + 4$ has two sign changes and therefore at most two negative roots.

$$\text{Possible rational roots} = \frac{\text{factors of } 4}{\text{factors of } 1} = \frac{\pm 1, \ \pm 2, \ \pm 4}{\pm 1}$$

$$
\begin{array}{rrrrr|l}
1 & 0 & -11 & -12 & 4 & \underline{-2} \\
  & -2 & 4 & 14 & -4 & \\
\hline
1 & -2 & -7 & 2 & 0, & \ -2 \text{ is a root.}
\end{array}
$$

$$x^4 - 11x^2 - 12x + 4 = (x+2)(x^3 - 2x^2 - 7x + 2)$$

$$\text{Possible rational roots of } x^3 - 2x^2 - 7x + 2 = \frac{\text{factors of } 2}{\text{factors of } 1} = \frac{\pm 1, \ \pm 2}{\pm 1}$$

$$
\begin{array}{rrrr|l}
1 & -2 & -7 & 2 & \underline{-2} \\
  & -2 & 8 & -2 & \\
\hline
1 & -4 & 1 & 0, & \ -2 \text{ is a repeated root.}
\end{array}
$$

$$x^4 - 11x^2 - 12x + 4 = (x+2)(x+2)(x^2 - 4x + 1)$$

$$x^2 - 4x + 1 = 0 \text{ has roots} = \frac{-(-4) \pm \sqrt{(-4)^2 - 4(1)(1)}}{2(1)} = \frac{4 \pm 2\sqrt{3}}{2} = 2 \pm \sqrt{3}$$

roots: $-2, \ -2, \ 2 \pm \sqrt{3}$

13. $f(x) = 2x^4 - 5x^3 - 3x^2 + 4x + 2 = 0$ has $n = 4$ and therefore four roots. $f(x)$ has two sign changes and therefore at most two positive roots. $f(-x) = 2x^4 + 5x^3 - 3x^2 - 4x + 2$ has two sign changes and therefore at most two negative roots.

$$\text{Positive rational roots} = \frac{\text{factors of } 2}{\text{factors of } 2} = \frac{\pm 1, \ \pm 2}{\pm 1, \ \pm 2}.$$

$$
\begin{array}{rrrrr|l}
2 & -5 & -3 & 4 & 2 & \underline{1} \\
  & 2 & -3 & -6 & -2 & \\
\hline
2 & -3 & -6 & -2 & 0, & \ 1 \text{ is a root.}
\end{array}
$$

$$2x^4 - 5x^3 - 3x^2 + 4x + 2 = (x-1)(2x^3 - 3x^2 - 6x - 2)$$

$$\begin{array}{rrrr|l}
2 & -3 & -6 & -2 & \underline{-\frac{1}{2}} \\
  & -1 & 2 & 2 & \\
\hline
2 & -4 & -4 & 0, & -\frac{1}{2}\text{ is a root}
\end{array}$$

$$2x^4 - 5x^3 - 3x^2 + 4x + 2 = (x - 1)\left(x + \frac{1}{2}\right)(2x^2 - 4x - 4)$$

$$2x^2 - 4x - 4 \text{ has roots} = \frac{-(-4) \pm \sqrt{(-4)^2 - 4(2)(-4)}}{2(2)} = \frac{4 \pm \sqrt{48}}{4} = \frac{4 \pm 4\sqrt{3}}{4} = 1 \pm \sqrt{3}$$

roots: $1, -\dfrac{1}{2}, 1 \pm \sqrt{3}$

17. $f(D) = D^5 + D^4 - 9D^3 - 5D^2 + 16D + 12 = 0$ has $n = 5$ and therefore five roots. $f(D)$ has two sign changes and therefore at most two positive roots. $f(-D) = -D^5 + D^4 + 9D^3 - 5D^2 - 16D + 12$ has three sign changes and therefore at most three negative roots.

$$\text{Possible rational roots} = \frac{\text{factors of } 12}{\text{factors of } 1} = \frac{\pm 1, \ \pm 2, \ \pm 3, \ \pm 4, \ \pm 6, \ \pm 12}{\pm 1}$$

$$\begin{array}{rrrrrr|l}
1 & 1 & -9 & -5 & 16 & 12 & \underline{2} \\
  & 2 & 6 & -6 & -22 & -12 & \\
\hline
1 & 3 & -3 & -11 & -6 & 0, & 2 \text{ is a root}
\end{array}$$

$$D^5 + D^4 - 9D^3 - 5D^2 + 16D + 12 = (D - 2)(D^4 + 3D^3 - 3D^2 - 11D - 6)$$

$$\begin{array}{rrrrr|l}
1 & 3 & -3 & -11 & -6 & \underline{2} \\
  & 2 & 10 & 14 & 6 & \\
\hline
1 & 5 & 7 & 3 & 0, & 2 \text{ is a root of } D^4 + 3D^3 - 3D^2 - 11D - 6
\end{array}$$

$$D^5 + D^4 - 9D^3 - 5D^2 + 16D + 12 = (D - 2)(D - 2)(D^3 + 5D^2 + 7D + 3)$$

$$\begin{array}{rrrr|l}
1 & 5 & 7 & 3 & \underline{-1} \\
  & -1 & -4 & -3 & \\
\hline
1 & 4 & 3 & 0, & -1 \text{ is a root of } D^3 + 5D^2 + 7D + 3
\end{array}$$

$$\begin{aligned}
D^5 + D^4 - 9D^3 - 5D^2 + 16D + 12 &= (D - 2)(D - 2)(D + 1)(D^2 + 4D + 3) \\
&= (D - 2)(D - 2)(D + 1)(D + 1)(D + 3)
\end{aligned}$$

roots: $2, 2, -1, -1, -3$

21. $x^3 - 2x^2 - 5x + 4 = 0$

 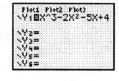 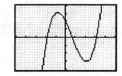 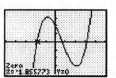

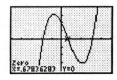 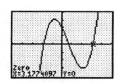

25. $x^3 - 6x^2 + 10x - 4 = 0$ (0 and 1)

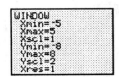 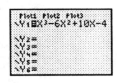 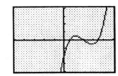 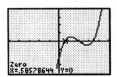

29. $f(x) = 4x^3 + 3x^2 - 20x - 15$.

Possible rational roots $= \dfrac{\text{factors of } 15}{\text{factors of } 4} = \dfrac{\pm 1, \ \pm 3, \ \pm 5, \ \pm 15}{\pm 1, \ \pm 2, \ \pm 4}$

$$
\begin{array}{rrrr|l}
4 & 3 & -20 & -15 & \ -\frac{3}{4} \\
  & -3 & 0 & 15 & \\
\hline
4 & 0 & -20 & 0, & -\frac{3}{4} \text{ is a root}
\end{array}
$$

$4x^3 + 3x^2 - 20x - 15 = \left(x + \dfrac{3}{4}\right)(4x^2 - 20) = 0$

$$4x^2 - 20 = 0$$
$$4x^2 = 20$$
$$x^2 = 5$$
$$x = \pm\sqrt{5}$$

roots: $-\dfrac{3}{4}, \pm\sqrt{5}$

33.

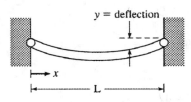

$y = deflection$

$y = k(x^4 - 2Lx^3 + L^3x) = 0$ for a deflection of zero.
$$x^4 - 2Lx^3 + L^3x = 0$$
$$x(x^3 - 2Lx^2 + L^3) = 0, \ x = 0 \text{ is a root}$$
$$x^3 - 2Lx^2 + L^3 = 0$$

$$
\begin{array}{rrrr|l}
1 & -2L & 0 & L^3 & \ L \\
  & L & -L^2 & -L^3 & \\
\hline
1 & -L & -L^2 & 0, & L \text{ is a root of } x^3 - 2Lx^2 + L^3 = 0
\end{array}
$$

$x^3 - 2Lx^2 + L^3 = (x - L)(x^2 - Lx - L^2)$

$x^2 - Lx - L^2 = 0$ has roots $= \dfrac{-(-L) \pm \sqrt{(-L)^2 - 4(1)(-L^2)}}{2} = \dfrac{L \pm L\sqrt{5}}{2} = \dfrac{L(1 \pm \sqrt{5})}{2}$

$\dfrac{L}{2}(1 - \sqrt{5}) < 0$ and $\dfrac{L}{23}(1 + \sqrt{5}) > L$; reject both

Beam has a deflection of 0 for $x = 0$ and $x = L$.

37. Let $r =$ smallest radius, the radii are $r, r+1, r+2, r+3$.

$$\frac{4}{3}\pi(r+3)^3 = \frac{4}{3}\pi r^3 + \frac{4}{3}\pi(r+1)^3 + \frac{4}{3}\pi(r+2)^3$$

$$r^3 + 3 \cdot r^2 \cdot 3 + 3 \cdot r \cdot 3^2 + 3^3 = r^3 + r^3 + 3r^2 + 3r + 1 + r^3 + 3 \cdot r^2 \cdot 2 + 3 \cdot r \cdot 2^2 + 2^3$$

$$r^3 + 9r^2 + 27r + 27 = 3r^3 + 9r^2 + 15r + 9$$

$$2r^3 - 12r - 18 = 0 \text{ has one sign change and therefore one positive root.}$$

Possible rational roots $= \dfrac{\text{factors of } 18}{\text{factors of } 2} = \dfrac{\pm 1, \ \pm 2, \ \pm 3, \ \pm 6, \ \pm 9, \ \pm 18}{\pm 1, \ \pm 2}$

$$\begin{array}{rrrr|l}
2 & 0 & -12 & -18 & \underline{3} \\
  & 6 & 18 & 18 & \\
\hline
2 & 6 & 6 & 0, & \text{3 is a root of } 2r^3 - 12r - 18
\end{array}$$

$$2r^3 - 12r - 18 = (x-3)(2x^2 + 6x + 6)$$
$$2x^2 + 6x + 6 = 0$$
$$x^2 + 3x + 3 = 0, \text{ has no real solutions}$$

The radii are 3.0 mm, 4.0 mm, 5.0 mm, and 6.0 mm.

41.

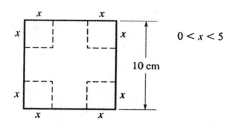

$$V = (10 - 2x)^2 \cdot x = 70 \Rightarrow (100 - 40x + 4x^2) \cdot x = 70$$

$$4x^3 - 40x^2 + 100x - 70 = 0$$

  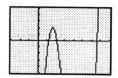 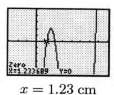

$x = 1.23$ cm

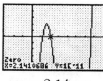 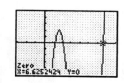

$x = 2.14$ cm

6.62 > 5 and must be rejected.

Chapter 15 Review Exercises

1. $2(1)^3 - 4(1)^2 - (1) + 4 = 1$

$(2x^3 - 4x^2 - x + 4) \div (x - 1)$ has remainder 1.

5.

$$
\begin{array}{rrrrr|r}
1 & 1 & 1 & -2 & -3 & \underline{-1} \\
  & -1 & 0 & -1 & 3 & \\
\hline
1 & 0 & 1 & -3 & 0, &
\end{array}
$$
remainder $= 0$, therefore $x + 1$ is a factor of $x^4 + x^3 + x^2 - 2x - 3$

9.

$$
\begin{array}{rrrr|r}
1 & 3 & 6 & 1 & \underline{1} \\
  & 1 & 4 & 10 & \\
\hline
1 & 4 & 10 & 11 &
\end{array}
$$

$(x^3 + 3x^2 + 6x + 1) \div (x - 1)$

$= (x^2 + 4x + 10) + \dfrac{11}{x - 1}$

13.

$$
\begin{array}{rrrrr|r}
1 & -2 & -3 & -4 & -8 & \underline{-1} \\
  & -1 & 3 & 0 & 4 & \\
\hline
1 & -3 & 0 & -4 & -4 &
\end{array}
$$

$(x^4 - 2x^3 - 3x^2 - 4x - 8) \div (x + 1)$

$= x^3 - 3x^2 - 4 + \dfrac{-4}{x + 1}$

17.

$$
\begin{array}{rrrr|r}
1 & 5 & 0 & -6 & \underline{-3} \\
  & -3 & -6 & 18 & \\
\hline
1 & 2 & -6 & 12, &
\end{array}
$$
remainder $= 12$, therefore

-3 is not a root of $y^3 + 5y^2 - 6 = 0$

21.

$$
\begin{array}{rrrr|r}
1 & 8 & 17 & 6 & \underline{-3} \\
  & -3 & -15 & -6 & \\
\hline
1 & 5 & 2 & 0 &
\end{array}
$$

$x^3 + 8x^2 + 17x + 6 = (x + 3)(x^2 + 5x + 2)$

$x^2 + 5x + 2$ has roots $= \dfrac{-5 \pm \sqrt{5^2 - 4(1)(2)}}{2(1)} = \dfrac{-5 \pm \sqrt{17}}{2}$

roots: $-3, \dfrac{-5 \pm \sqrt{17}}{2}$

25.

$$
\begin{array}{rrrrr|r}
4 & 0 & -1 & -18 & 9 & \underline{\frac{1}{2}} \\
  & 2 & 1 & 0 & -9 & \\
\hline
4 & 2 & 0 & -18 & 0 &
\end{array}
$$

$4p^4 - p^2 - 18p + 9 = \left(p - \dfrac{1}{2}\right)(4p^3 + 2p^2 - 18)$

$$
\begin{array}{rrrr|r}
4 & 2 & 0 & -18 & \underline{\frac{3}{2}} \\
  & 6 & 12 & 18 & \\
\hline
4 & 8 & 12 & 0 &
\end{array}
$$

$4p^4 - p^2 - 18p + 9 = \left(p - \dfrac{1}{2}\right)\left(p - \dfrac{3}{2}\right)(4p^2 + 8p + 12) = 4\left(p - \dfrac{1}{2}\right)\left(p - \dfrac{3}{2}\right)(p^2 + 2p + 3)$

$p^2 + 2p + 3$ has roots $= -1 \pm \sqrt{2}j$

roots $\dfrac{1}{2}, \dfrac{3}{2}, -1 \pm \sqrt{2}j$

29.

$$
\begin{array}{rrrrrr|l}
1 & 3 & -1 & -11 & -12 & -4 & \underline{-1} \\
  & -1 & -2 & 3 & 8 & 4 & \\
\hline
1 & 2 & -3 & -8 & -4 & 0 &
\end{array}
$$

$x^5 + 3x^4 - x^3 - 11x^2 - 12x - 4 = (x + 1)(x^4 + 2x^3 - 3x^2 - 8x - 4)$

$$
\begin{array}{rrrrr|l}
1 & 2 & -3 & -8 & -4 & \underline{-1} \\
  & -1 & -1 & 4 & 4 & \\
\hline
1 & 1 & -4 & -4 & 0 &
\end{array}
$$

$x^5 + 3x^4 - x^3 - 11x^2 - 12x - 4 = (x + 1)(x + 1)(x^3 + x^2 - 4x - 4)$

$$
\begin{array}{rrrr|l}
1 & 1 & -4 & -4 & \underline{-1} \\
  & -1 & 0 & 4 & \\
\hline
1 & 0 & -4 & 0 &
\end{array}
$$

$x^5 + 3x^4 - 11x^3 - 11x^2 - 4 = (x + 1)(x + 1)(x + 1)(x^2 - 4)$
$$= (x + 1)(x + 1)(x + 1)(x + 2)(x - 2)$$

roots: $-1, -1, -1, 2, -2$

33. $x^3 + x^2 - 10x + 8 = 0$ has three roots.

Possible rational roots $= \dfrac{\pm 1, \ \pm 2, \ \pm 4, \ \pm 8}{\pm 1}$

$$
\begin{array}{rrrr|l}
1 & 1 & -10 & 8 & \underline{1} \\
  & 1 & 2 & -8 & \\
\hline
1 & 2 & -8 & 0, & \text{1 is a root}
\end{array}
$$

$x^3 + x^2 - 10x + 8 = (x - 1)(x^2 + 2x - 8)$
$$= (x - 1)(x + 4)(x - 2)$$

roots: $1, -4, 2$

37. $6x^3 - x^2 - 12x - 5 = 0$ has three roots.

Possible rational roots $= \dfrac{\pm 1, \ \pm 5}{\pm 1, \ \pm 2, \ \pm 3, \ \pm 6}$

$$
\begin{array}{rrrr|l}
6 & -1 & -12 & -5 & \underline{\frac{5}{3}} \\
  & 10 & 15 & 5 & \\
\hline
6 & 9 & 3 & 0, & \frac{5}{3} \text{ is a root}
\end{array}
$$

$6x^3 - x^2 - 12x - 5 = \left(x - \dfrac{5}{3}\right)(6x^2 + 9x + 3)$

$$= 3\left(x - \dfrac{5}{3}\right)(2x^2 + 3x + 1)$$

$$= 3\left(x - \dfrac{5}{3}\right)(2x + 1)(x + 1)$$

roots: $\dfrac{5}{3}, \dfrac{-1}{2}, -1$

41.

$$
\begin{array}{ccccc}
3 & k & -8 & -8 & \quad\underline{|-2} \\
 & -6 & 12-2k & -8+4k & \\
\hline
3 & -6+k & 4-2k & -16+4k &
\end{array}
$$

remainder $= 4k - 16 = 0$

$$4k = 16$$

$$k = 4$$

45. From the calculator, $x = 1.91$ is the irrational root between 1 and 2.

None of the possible rational roots $= \dfrac{\text{factors of } 2}{\text{factors of } 3} = \dfrac{\pm 1, \ \pm 2}{\pm 1, \ \pm 3}$ is a root.

 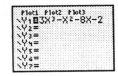 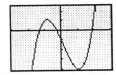 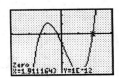

49. $f(d) = 64d^3 - 144d^2 + 108d - 27$ has $n = 3$ and therefore three roots at most. $f(d)$ has three sign changes and therefore at most three positive roots.

$f(-d) = -64d^3 - 144d^2 + 108d - 27$ has sign changes and therefore no negative roots.

Possible rational roots $= \dfrac{\text{factors of } 27}{\text{factors of } 64} = \dfrac{\pm 1, \ \pm 3, \ \pm 9, \ \pm 27}{\pm 1, \ \pm 2, \ \pm 4, \ \pm 8, \ \pm 16, \ \pm 32, \ \pm 64}$

From the graph, the root is between 0 and 1.

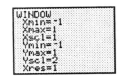 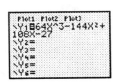 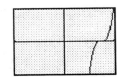 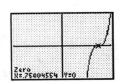

$$
\begin{array}{ccccc}
64 & -144 & 108 & -27 & \quad\underline{|\frac{3}{4}} \\
 & 48 & -72 & 27 & \\
\hline
64 & -96 & 36 & 0, & \frac{3}{4} \text{ is a repeated (multiplicity } = 3) \text{ root.}
\end{array}
$$

$d = 0.75$ cm

53. $h = r + 3.2$, h and $r > 0$.
$V = \pi r^2 h = \pi r^2 (r + 3.2) = \pi r^3 + 3.2\pi r^2 = 680$

$\pi r^3 + 3.2\pi r^2 - 680 = 0$ has one sign change and therefore one positive root. From the graph $r = 5.1$ m and $h = 5.1 + 3.2 = 8.3$ m.

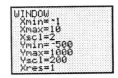 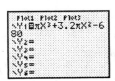 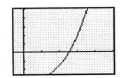 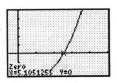

57. Perform synthetic division of $x^4 + r^4$ by $x + r$ for $n = 1, 2, 3, 4, 5$

$n = 1$ $\quad$ $1 \quad r \quad \lfloor\underline{-r}$

$\quad\quad\quad\quad\quad\quad \underline{-r}$

$\quad\quad\quad\quad 1 \quad 0, \quad x + r$ is a factor of $x + r$

$n = 2$ $\quad$ $1 \quad 0 \quad r^2 \quad \lfloor\underline{-r}$

$\quad\quad\quad\quad\quad \underline{-r \quad r^2}$

$\quad\quad\quad\quad 1 \quad -r \quad 2r^2, \quad x + r$ is not a factor of $x^2 + r^2$

$n - 3$ $\quad$ $1 \quad 0 \quad 0 \quad r^3 \quad \lfloor\underline{-r} \quad\quad\quad (x^3 + r^3) \div (x + r)$

$\quad\quad\quad\quad\quad \underline{-r \quad r^2 \quad -r^3}$

$\quad\quad\quad\quad 1 \quad -r \quad r^2 \quad 0 \quad x + r$ is a factor of $x^3 + r^3$

$n = 4$ $\quad$ $1 \quad 0 \quad 0 \quad 0 \quad r^4 \quad \lfloor\underline{-r} \quad\quad\quad (x^4 + r^4) \div (x + r)$

$\quad\quad\quad\quad\quad \underline{-r \quad r^2 \quad -r^3 \quad r^4}$

$\quad\quad\quad\quad 1 \quad -r \quad r^2 \quad -r^3 \quad 2r^4, \quad x + r$ is not a factor of $x^4 + r^4$

$n = 5$ $\quad$ $1 \quad 0 \quad 0 \quad 0 \quad 0 \quad r^5 \quad \lfloor\underline{-r} \quad\quad\quad (x^5 + r^5) \div (x + r)$

$\quad\quad\quad\quad\quad \underline{-r \quad r^2 \quad -r^3 \quad r^4 \quad -r^5}$

$\quad\quad\quad\quad 1 \quad -r \quad r^2 \quad -r^3 \quad r^4 \quad 0, \quad x + r$ is a factor of $x^5 + r^5$

Apparently, $x + r$ is a factor of $x^n + r^n$ for $n = 1, 3, 5, 7, \cdots$.

DETERMINANTS AND MATRICES

16.1   Determinants: Expansion by Minors

1. Expand along the first row since it has two zeros.

$$
\begin{vmatrix} 3 & 0 & 0 \\ -2 & 1 & 4 \\ 4 & -2 & 5 \end{vmatrix} = +(3)\begin{vmatrix} 1 & 4 \\ -2 & 5 \end{vmatrix} - (0)\begin{vmatrix} -2 & 4 \\ 4 & 5 \end{vmatrix} + (0)\begin{vmatrix} -2 & 1 \\ 4 & -2 \end{vmatrix}
$$

$$
= 3\left[(1)(5) - (-2)(4)\right]
$$

$$
= 39
$$

5.
$$
\begin{vmatrix} -6 & -1 & 3 \\ 2 & -2 & -3 \\ 10 & 1 & -2 \end{vmatrix} = +(-6)\begin{vmatrix} -2 & -3 \\ 1 & -2 \end{vmatrix} - (-1)\begin{vmatrix} 2 & -3 \\ 10 & -2 \end{vmatrix} + (3)\begin{vmatrix} 2 & -2 \\ 10 & 1 \end{vmatrix}
$$

$$
= -6\left[(-2)(-2) - (1)(-3)\right] + \left[2(-2) - 10(-3)\right] + 3\left[2(1) - 10(-2)\right]
$$

$$
= 50
$$

9.
$$
\begin{vmatrix} 1 & 0 & 1 & 0 \\ 2 & 4 & -3 & 1 \\ 1 & 1 & 1 & 1 \\ 3 & 5 & 0 & 2 \end{vmatrix} = +(1)\begin{vmatrix} 4 & -3 & 1 \\ 1 & 1 & 1 \\ 5 & 0 & 2 \end{vmatrix} - (0)\begin{vmatrix} 2 & -3 & 1 \\ 1 & 1 & 1 \\ 3 & 0 & 2 \end{vmatrix} + (1)\begin{vmatrix} 2 & 4 & 1 \\ 1 & 1 & 1 \\ 3 & 5 & 2 \end{vmatrix} - (0)\begin{vmatrix} 2 & 4 & -3 \\ 1 & 1 & 1 \\ 3 & 5 & 0 \end{vmatrix}
$$

$$
= \begin{vmatrix} 4 & -3 & 1 \\ 1 & 1 & 1 \\ 5 & 0 & 2 \end{vmatrix} + \begin{vmatrix} 2 & 4 & 1 \\ 1 & 1 & 1 \\ 3 & 5 & 2 \end{vmatrix}
$$

$$
= +(4)\begin{vmatrix} 1 & 1 \\ 0 & 2 \end{vmatrix} - (-3)\begin{vmatrix} 1 & 1 \\ 5 & 2 \end{vmatrix} + (1)\begin{vmatrix} 1 & 1 \\ 5 & 0 \end{vmatrix}
$$

$$
+ (2)\begin{vmatrix} 1 & 1 \\ 5 & 2 \end{vmatrix} - (4)\begin{vmatrix} 1 & 1 \\ 3 & 2 \end{vmatrix} + (1)\begin{vmatrix} 1 & 1 \\ 3 & 5 \end{vmatrix}
$$

$$
= 4\left[(1)(2) - (0)(1)\right] + 3\left[(1)(2) - 5(1)\right] + \left[(1)(0) - (5)(1)\right]
$$

$$
+ 2\left[(1)(2) - (5)(1)\right] - 4\left[(1)(2) - (3)(1)\right] + \left[(1)(5) - (3)(1)\right]
$$

$$
= -6
$$

13. Expand along the second row since it has three zeros.

$$\begin{vmatrix} 1 & 2 & 1 & 2 & 1 \\ 1 & 0 & 0 & 1 & 0 \\ 0 & 1 & 1 & 0 & 1 \\ 1 & 1 & 2 & 2 & 1 \\ 0 & 1 & 1 & 0 & 2 \end{vmatrix} = -(1)\begin{vmatrix} 2 & 1 & 2 & 1 \\ 1 & 1 & 0 & 1 \\ 1 & 2 & 2 & 1 \\ 1 & 1 & 0 & 2 \end{vmatrix} + (1)\begin{vmatrix} 1 & 2 & 1 & 1 \\ 0 & 1 & 1 & 1 \\ 1 & 1 & 2 & 1 \\ 0 & 1 & 1 & 2 \end{vmatrix}$$

$$= -1\left[+(2)\begin{vmatrix} 1 & 1 & 1 \\ 1 & 2 & 1 \\ 1 & 1 & 2 \end{vmatrix} + (2)\begin{vmatrix} 2 & 1 & 1 \\ 1 & 1 & 1 \\ 1 & 1 & 2 \end{vmatrix}\right] + \begin{vmatrix} 1 & 1 & 1 \\ 1 & 2 & 1 \\ 1 & 1 & 2 \end{vmatrix} + \begin{vmatrix} 2 & 1 & 1 \\ 1 & 1 & 1 \\ 1 & 1 & 2 \end{vmatrix}$$

$$= -2\left[(1)\left[(2)(2) - (1)(1)\right] - (1)\left[(1)(2) - (1)(1)\right] + (1)\left[(1)(1) - (1)(2)\right]\right.$$
$$+(2)\left[(1)(2) - (1)(1)\right] - (1)\left[(1)(2) - (1)(1)\right] + (1)\left[(1)(1) - (1)(1)\right]$$
$$+(1)\left[(2)(2) - (1)(1)\right] - (1)\left[(1)(2) - (1)(1)\right] + (1)\left[(1)(1) - (1)(2)\right]$$
$$\left.+(2)\left[(1)(2) - (1)(1)\right] - (1)\left[(1)(2) - (1)(1)\right] + (1)\left[(1)(1) - (1)(1)\right]\right]$$
$$= -2\left[3 - 1 - 1 + 2 - 1\right] + 3 - 1 - 1 + 2 - 1 = -2\left[2\right] + 2 = -2$$

17. $D = \begin{vmatrix} 2 & 1 & 1 \\ 1 & -2 & 2 \\ 3 & -1 & -1 \end{vmatrix} = 2\left[(-2)(-1) - (-1)(2)\right] - \left[(1)(-1) - (3)(2)\right] + \left[(1)(-1) - (3)(-2)\right]$

$D = 20$

$N_x = \begin{vmatrix} 6 & 1 & 1 \\ 10 & -2 & 2 \\ 4 & -1 & -1 \end{vmatrix} = 6\left[(-2)(-1) - (-1)(2)\right] - \left[(10)(-1) - (4)(2)\right] + \left[(10)(-1) - (4)(-2)\right]$

$N_x = 40$

$N_y = \begin{vmatrix} 2 & 6 & 1 \\ 1 & 10 & 2 \\ 3 & 4 & -1 \end{vmatrix} = 2\left[(10)(-1) - (4)(2)\right] - 6\left[(1)(-1) - (3)(2)\right] + \left[(1)(4) - (3)(10)\right]$

$N_y = -20$

$N_z = \begin{vmatrix} 2 & 1 & 6 \\ 1 & -2 & 10 \\ 3 & -1 & 4 \end{vmatrix} = 2\left[(-2)(4) - (-1)(10)\right] - \left[(1)(4) - (3)(10)\right] + 6\left[(1)(-1) - (3)(-2)\right]$

$N_z = 60$

$$x = \frac{N_x}{D} = \frac{40}{20} = 2, \; y = \frac{N_y}{D} = \frac{-20}{20} = -1, \; z = \frac{D_z}{D} = \frac{60}{20} = 3$$

Solution: $(2, -1, 3)$

21. Rewrite the system as

$$1 \cdot x + 0 \cdot y + 0 \cdot z + 1 \cdot t = 0$$
$$3 \cdot x + 1 \cdot y + 1 \cdot z + 0 \cdot t = -1$$
$$0 \cdot x + 2 \cdot y - 1 \cdot z + 3 \cdot t = 1$$
$$0 \cdot x + 0 \cdot y + 2z - 3t = 1$$

$$x = \frac{\begin{vmatrix} 0 & 0 & 0 & 1 \\ -1 & 1 & 1 & 0 \\ 1 & 2 & -1 & 3 \\ 1 & 0 & 2 & 3 \end{vmatrix}}{\begin{vmatrix} 1 & 0 & 0 & 1 \\ 3 & 1 & 1 & 0 \\ 0 & 2 & -1 & 3 \\ 0 & 0 & 2 & -3 \end{vmatrix}} = \frac{9}{-9} = -1$$

$$y = \frac{\begin{vmatrix} 1 & 0 & 0 & 1 \\ 3 & -1 & 1 & 0 \\ 0 & 1 & -1 & 3 \\ 0 & 1 & 2 & -3 \end{vmatrix}}{-9} = \frac{0}{-9} = 0$$

$$z = \frac{\begin{vmatrix} 1 & 0 & 0 & 1 \\ 3 & 1 & -1 & 0 \\ 0 & 2 & 1 & 3 \\ 0 & 0 & 1 & -3 \end{vmatrix}}{-9} = \frac{-18}{-9} = 2$$

$$t = \frac{\begin{vmatrix} 1 & 0 & 0 & 0 \\ 3 & 1 & 1 & -1 \\ 0 & 2 & -1 & 1 \\ 0 & 0 & 2 & 1 \end{vmatrix}}{-9} = \frac{-9}{-9} = 1$$

Solution: $(-1, 0, 2, 1)$

25. $I_A + I_B + I_C + \ I_D = 0$
$2I_A - I_B \qquad\qquad = -2$
$\qquad\qquad 3I_C - 2I_D = 0$
$\qquad I_B - 3I_C \qquad = 6$

$$I_A = \frac{\begin{vmatrix} 0 & 1 & 1 & 1 \\ -2 & -1 & 0 & 0 \\ 0 & 0 & 3 & -2 \\ 6 & 1 & -3 & 0 \end{vmatrix}}{\begin{vmatrix} 1 & 1 & 1 & 1 \\ 2 & -1 & 0 & 0 \\ 0 & 0 & 3 & -2 \\ 0 & 1 & -3 & 0 \end{vmatrix}} = \frac{8}{28} = \frac{2}{7}A$$

$$I_B = \frac{\begin{vmatrix} 1 & 0 & 1 & 1 \\ 2 & -2 & 0 & 0 \\ 0 & 0 & 3 & -2 \\ 0 & 6 & -3 & 0 \end{vmatrix}}{28} = \frac{72}{28} = \frac{18}{7}A$$

$$I_C = \frac{\begin{vmatrix} 1 & 1 & 0 & 1 \\ 2 & -1 & -2 & 0 \\ 0 & 0 & 0 & -2 \\ 0 & 1 & 6 & 0 \end{vmatrix}}{28} = \frac{-32}{28} = \frac{-8}{7}A$$

$$I_D = \frac{\begin{vmatrix} 1 & 1 & 1 & 0 \\ 2 & -1 & 0 & -2 \\ 0 & 0 & 3 & 0 \\ 0 & 1 & -3 & 6 \end{vmatrix}}{28} = \frac{-48}{28} = \frac{-12}{7}A$$

16.2 Some Properties of Determinants

1. $\begin{vmatrix} 4 & -5 & 8 \\ 0 & 3 & -8 \\ 0 & 0 & -5 \end{vmatrix} = (4)(3)(-5) = -60$

The determinant of a triangular matrix, one with zeros in an off diagonal triangle, is the product of the diagonal elements.

5. $\begin{vmatrix} 2 & 1 & -3 \\ -4 & 3 & 1 \\ 1 & -2 & -3 \end{vmatrix} = 40$

The value of the determinant is unchanged when two columns are switched.

9. $\begin{vmatrix} 3 & 1 & 0 \\ -2 & 3 & -1 \\ 4 & 2 & 5 \end{vmatrix}$ $\xrightarrow[\text{added to first column}]{-3 \text{ times second column}}$ $\begin{vmatrix} 0 & 1 & 0 \\ -11 & 3 & -1 \\ -2 & 2 & 5 \end{vmatrix}$, then

$\begin{vmatrix} 0 & 1 & 0 \\ -11 & 3 & -1 \\ -2 & 2 & 5 \end{vmatrix} = -(1) \cdot \begin{vmatrix} -11 & -1 \\ -2 & 5 \end{vmatrix} = -1 \cdot [(-11)(5) - (-2)(-1)] = 57$

13. $\begin{vmatrix} 1 & 3 & -3 & 5 \\ 4 & 2 & 1 & 2 \\ 3 & 2 & -2 & 2 \\ 0 & 1 & 2 & -1 \end{vmatrix}$ $\xrightarrow[\substack{-3 * R1 + R3 \to R3}]{-4 * R1 + R2 \to R2}$ $\begin{vmatrix} 1 & 3 & -3 & 5 \\ 0 & -10 & 13 & -18 \\ 0 & -7 & 7 & -13 \\ 0 & 1 & 2 & -1 \end{vmatrix}$ $\xrightarrow[\substack{\frac{1}{10} * R2 + R4 \to R4}]{\frac{-7}{10} * R2 + R3 \to R3}$

$\begin{vmatrix} 1 & 3 & -3 & 5 \\ 0 & -10 & 13 & -18 \\ 0 & 0 & -2.1 & -0.4 \\ 0 & 0 & 3.3 & -2.8 \end{vmatrix}$ $\xrightarrow{\frac{3.3}{2.1} * R3 + R4 \to R4}$ $\begin{vmatrix} 1 & 3 & -3 & 5 \\ 0 & -10 & 13 & -18 \\ 0 & 0 & -2.1 & -0.4 \\ 0 & 0 & 0 & -\frac{24}{7} \end{vmatrix}$

which has value $= (1)(-10)(-2.1)(-\frac{24}{7}) = -72$

17. $D = \begin{vmatrix} 2 & -1 & 1 \\ 1 & 2 & 3 \\ 3 & 3 & 2 \end{vmatrix}$ $\xrightarrow[\substack{-\frac{3}{2} * R1 + R3 \to R3}]{-\frac{1}{2} * R1 + R2 \to R2}$ $\begin{vmatrix} 2 & -1 & 1 \\ 0 & 2.5 & 2.5 \\ 0 & 4.5 & 0.5 \end{vmatrix}$ $\xrightarrow{\frac{-4.5}{2.5} * R2 + R3 \to R3}$

$\begin{vmatrix} 2 & -1 & 1 \\ 0 & 2.5 & 2.5 \\ 0 & 0 & -4 \end{vmatrix} = 2(2.5)(-4) = -20$

$N_x = \begin{vmatrix} 5 & -1 & 1 \\ 10 & 2 & 3 \\ 5 & 3 & 2 \end{vmatrix}$ $\xrightarrow[\substack{-1 * R1 + R3 \to R3}]{-2 * R1 + R2 \to R2}$ $\begin{vmatrix} 5 & -1 & 1 \\ 0 & 4 & 1 \\ 0 & 4 & 1 \end{vmatrix} = 0$, two rows are equal.

$N_y = \begin{vmatrix} 2 & 5 & 1 \\ 1 & 10 & 3 \\ 3 & 5 & 2 \end{vmatrix}$ $\xrightarrow[\substack{-\frac{3}{2} * R1 + R3 \to R3}]{-\frac{1}{2} * R1 + R2 \to R2}$ $\begin{vmatrix} 2 & 5 & 1 \\ 0 & 7.5 & 2.5 \\ 0 & -2.5 & 0.5 \end{vmatrix}$ $\xrightarrow{\frac{2.5}{7.5} * R2 + R3 \to R3}$

$\begin{vmatrix} 2 & 5 & 1 \\ 0 & \frac{15}{2} & \frac{5}{2} \\ 0 & 0 & \frac{4}{3} \end{vmatrix} = 2\left(\frac{15}{2}\right)\left(\frac{4}{3}\right) = 20$

$$N_z = \begin{vmatrix} 2 & -1 & 5 \\ 1 & 2 & 10 \\ 3 & 3 & 5 \end{vmatrix} \xrightarrow[\substack{-\frac{3}{2}*R1+R3 \to R3}]{-\frac{1}{2}*R1+R2 \to R2} \begin{vmatrix} 2 & -1 & 5 \\ 0 & 2.5 & 7.5 \\ 0 & 4.5 & -2.5 \end{vmatrix} \xrightarrow{\frac{-4.5}{2.5}*R2+R3 \to R3}$$

$$\begin{vmatrix} 2 & -1 & 5 \\ 0 & 2.5 & 7.5 \\ 0 & 0 & -16 \end{vmatrix} = 2(2.5)(-16) = -80$$

$$x = \frac{N_x}{D} = \frac{0}{-20} = 0, \quad y = \frac{N_y}{D} = \frac{20}{-20} = -1, \quad z = \frac{N_z}{D} = \frac{-80}{-20} = 4$$

Solution: $(0, -1, 4)$

21. $D = \begin{vmatrix} 2 & 1 & 1 & 0 \\ 0 & 3 & -1 & 2 \\ 0 & 1 & 2 & 1 \\ 3 & 0 & 2 & 0 \end{vmatrix} \xrightarrow{\frac{-3}{2}*R1+R4 \to R4} \begin{vmatrix} 2 & 1 & 1 & 0 \\ 0 & 3 & -1 & 2 \\ 0 & 1 & 2 & 1 \\ 0 & -\frac{3}{2} & \frac{1}{2} & 0 \end{vmatrix} \xrightarrow[\substack{\frac{1}{2}*R2+R4 \to R4}]{-\frac{1}{3}*R2+R3 \to R3}$

$$\begin{vmatrix} 2 & 1 & 1 & 0 \\ 0 & 3 & -1 & 2 \\ 0 & 0 & \frac{7}{3} & \frac{1}{3} \\ 0 & 0 & 0 & 1 \end{vmatrix} = (2)(3)\left(\frac{7}{3}\right)(1) = 14$$

$$N_x = \begin{vmatrix} 2 & 1 & 1 & 0 \\ 4 & 3 & -1 & 2 \\ 0 & 1 & 2 & 1 \\ 4 & 0 & 2 & 0 \end{vmatrix} \xrightarrow[\substack{-2*R1+R4 \to R4}]{-2*R1+R2 \to R2} \begin{vmatrix} 2 & 1 & 1 & 0 \\ 0 & 1 & -3 & 2 \\ 0 & 1 & 2 & 1 \\ 0 & -2 & 0 & 0 \end{vmatrix} \xrightarrow[\substack{2*R2+R4 \to R4}]{-1*R2+R3 \to R3}$$

$$\begin{vmatrix} 2 & 1 & 1 & 0 \\ 0 & 1 & -3 & 2 \\ 0 & 0 & 5 & -1 \\ 0 & 0 & -6 & 4 \end{vmatrix} \xrightarrow{\frac{6}{5}*R3+R4 \to R4} \begin{vmatrix} 2 & 1 & 1 & 0 \\ 0 & 1 & -3 & 2 \\ 0 & 0 & 5 & -1 \\ 0 & 0 & 0 & 2.8 \end{vmatrix} = (2)(1)(5)(2.8) = 28$$

$$N_y = \begin{vmatrix} 2 & 2 & 1 & 0 \\ 0 & 4 & -1 & 2 \\ 0 & 0 & 2 & 1 \\ 3 & 4 & 2 & 0 \end{vmatrix} \xrightarrow{\frac{-3}{2}*R1+R4 \to R4} \begin{vmatrix} 2 & 2 & 1 & 0 \\ 0 & 4 & -1 & 2 \\ 0 & 0 & 2 & 1 \\ 0 & 1 & \frac{1}{2} & 0 \end{vmatrix} \xrightarrow{\frac{-1}{4}*R2+R4 \to R4}$$

$$\begin{vmatrix} 2 & 2 & 1 & 0 \\ 0 & 4 & -1 & 2 \\ 0 & 0 & 2 & 1 \\ 0 & 0 & \frac{3}{4} & -\frac{1}{2} \end{vmatrix} \xrightarrow{\frac{-3}{8}*R3+R4 \to R4} \begin{vmatrix} 2 & 2 & 1 & 0 \\ 0 & 4 & -1 & 2 \\ 0 & 0 & 2 & 1 \\ 0 & 0 & 0 & -\frac{7}{8} \end{vmatrix} = (2)(4)(2)\left(-\frac{7}{8}\right) = -14$$

$$N_z = \begin{vmatrix} 2 & 1 & 2 & 0 \\ 0 & 3 & 4 & 2 \\ 0 & 1 & 0 & 1 \\ 3 & 0 & 4 & 0 \end{vmatrix} \xrightarrow{\frac{-3}{2}*R1+R4 \to R4} \begin{vmatrix} 2 & 1 & 2 & 0 \\ 0 & 3 & 4 & 2 \\ 0 & 1 & 0 & 1 \\ 0 & -\frac{3}{2} & 1 & 0 \end{vmatrix} \xrightarrow[\frac{1}{2}*R2+R4 \to R4]{\frac{-1}{3}*R2+R3 \to R3}$$

$$\begin{vmatrix} 2 & 1 & 2 & 0 \\ 0 & 3 & 4 & 2 \\ 0 & 0 & -\frac{4}{3} & \frac{1}{3} \\ 0 & 0 & 3 & 1 \end{vmatrix} \xrightarrow{\frac{9}{4}*R3+R4 \to R4} \begin{vmatrix} 2 & 1 & 2 & 0 \\ 0 & 3 & 4 & 2 \\ 0 & 0 & \frac{-4}{3} & \frac{1}{3} \\ 0 & 0 & 0 & \frac{7}{4} \end{vmatrix} = (2)(3)\left(\frac{-4}{3}\right)\left(\frac{7}{4}\right) = -14$$

$$N_t = \begin{vmatrix} 2 & 1 & 1 & 2 \\ 0 & 3 & -1 & 4 \\ 0 & 1 & 2 & 0 \\ 3 & 0 & 2 & 4 \end{vmatrix} \xrightarrow{\frac{-3}{2}*R1+R4 \to R4} \begin{vmatrix} 2 & 1 & 1 & 2 \\ 0 & 3 & -1 & 4 \\ 0 & 1 & 2 & 0 \\ 0 & \frac{-3}{2} & \frac{1}{2} & 1 \end{vmatrix} \xrightarrow[\frac{1}{2}*R2+R4 \to R4]{\frac{-1}{3}*R2+R3 \to R3}$$

$$\begin{vmatrix} 2 & 1 & 1 & 2 \\ 0 & 3 & -1 & 4 \\ 0 & 0 & \frac{7}{3} & \frac{-4}{3} \\ 0 & 0 & 0 & 3 \end{vmatrix} = (2)(3)\left(\frac{7}{3}\right)(3) = 42$$

$$x = \frac{N_x}{D} = \frac{28}{14} = 2, \quad y = \frac{N_y}{D} = \frac{-14}{14} = -1, \quad z = \frac{N_z}{D} = \frac{-14}{14} = -1$$

$$t = \frac{N_t}{D} = \frac{42}{14} = 3.$$

Solution: $(2, -1, -1, 3)$

25.
$$\begin{aligned} I_A + I_B + I_C + I_D + I_E &= 0 \\ -2I_A + 3I_B &= 0 \\ 3I_B - 3I_C &= 6 \\ -3I_C + I_D &= 0 \\ -I_D + 2I_E &= 0 \end{aligned}$$

$$I_A = \frac{\begin{vmatrix} 0 & 1 & 1 & 1 & 1 \\ 0 & 3 & 0 & 0 & 0 \\ 6 & 3 & -3 & 0 & 0 \\ 0 & 0 & -3 & 1 & 0 \\ 0 & 0 & 0 & -1 & 2 \end{vmatrix}}{\begin{vmatrix} 1 & 1 & 1 & 1 & 1 \\ -2 & 3 & 0 & 0 & 0 \\ 0 & 3 & -3 & 0 & 0 \\ 0 & 0 & -3 & 1 & 0 \\ 0 & 0 & 0 & -1 & 2 \end{vmatrix}} = \frac{-198}{-96} = \frac{33}{16}A$$

$$I_B = \frac{\begin{vmatrix} 1 & 0 & 1 & 1 & 1 \\ -2 & 0 & 0 & 0 & 0 \\ 0 & 6 & -3 & 0 & 0 \\ 0 & 0 & -3 & 1 & 0 \\ 0 & 0 & 0 & -1 & 2 \end{vmatrix}}{-96} = \frac{-132}{-96} = \frac{11}{8}A$$

$$I_C = \dfrac{\begin{vmatrix} 1 & 1 & 0 & 1 & 1 \\ -2 & 3 & 0 & 0 & 0 \\ 0 & 3 & 6 & 0 & 0 \\ 0 & 0 & 0 & 1 & 0 \\ 0 & 0 & 0 & -1 & 2 \end{vmatrix}}{-96} = \dfrac{60}{-96} = \dfrac{-5}{8} A$$

$$I_D = \dfrac{\begin{vmatrix} 1 & 1 & 1 & 0 & 1 \\ -2 & 3 & 0 & 0 & 0 \\ 0 & 3 & -3 & 6 & 0 \\ 0 & 0 & -3 & 0 & 0 \\ 0 & 0 & 0 & 0 & 2 \end{vmatrix}}{-96} = \dfrac{180}{-96} = \dfrac{-15}{8} A$$

$$I_E = \dfrac{\begin{vmatrix} 1 & 1 & 1 & 1 & 0 \\ -2 & 3 & 0 & 0 & 0 \\ 0 & 3 & -3 & 0 & 6 \\ 0 & 0 & -3 & 1 & 0 \\ 0 & 0 & 0 & -1 & 0 \end{vmatrix}}{-96} = \dfrac{90}{-96} = \dfrac{-15}{16} A$$

16.3   Matrices: Definitions and Basic Operations

1. $\begin{bmatrix} a & b \\ c & d \end{bmatrix} = \begin{bmatrix} 1 & -3 \\ 4 & 7 \end{bmatrix}$ Corresponding elements are equal. $a = 1, b = -3, c = 4, d = 7$

5. $\begin{bmatrix} x & x+y \\ x-z & y+z \\ x+t & y-t \end{bmatrix} = \begin{bmatrix} 2 & 3 \\ 4 & -1 \end{bmatrix}$

$\qquad\quad 3 \times 2 \qquad\qquad 2 \times 2$

Matrices have different dimensions and therefore cannot be equal.

9. $\begin{bmatrix} 2 & 3 \\ -5 & 4 \end{bmatrix} + \begin{bmatrix} -1 & 7 \\ 5 & -2 \end{bmatrix} = \begin{bmatrix} 2-1 & 3+7 \\ -5+5 & 4-2 \end{bmatrix} = \begin{bmatrix} 1 & 10 \\ 0 & 2 \end{bmatrix}$

13. $A + B = \begin{bmatrix} -1 & 4 & -7 & 0 \\ 2 & -6 & -1 & 2 \end{bmatrix} + \begin{bmatrix} 1 & 5 & -6 & 3 \\ 4 & -1 & 8 & -2 \end{bmatrix}$, adding corresponding entries

$\qquad = \begin{bmatrix} 0 & 9 & -13 & 3 \\ 6 & -7 & 7 & 0 \end{bmatrix}$

17. $2A + B = 2\begin{bmatrix} -1 & 4 & -7 & 0 \\ 2 & -6 & -1 & 2 \end{bmatrix} + \begin{bmatrix} 1 & 5 & -6 & 3 \\ 4 & -1 & 8 & -2 \end{bmatrix} = \begin{bmatrix} -2 & 8 & -14 & 0 \\ 4 & -12 & -2 & 4 \end{bmatrix} + \begin{bmatrix} 1 & 5 & -6 & 3 \\ 4 & -1 & 8 & -2 \end{bmatrix}$

$\qquad = \begin{bmatrix} -1 & 13 & -20 & 3 \\ 8 & -13 & 6 & 2 \end{bmatrix}$

21. $A + B = \begin{bmatrix} -1 & 2 & 3 & 7 \\ 0 & -3 & -1 & 4 \\ 9 & -1 & 0 & -2 \end{bmatrix} + \begin{bmatrix} 4 & -1 & -3 & 0 \\ 5 & 0 & -1 & 1 \\ 1 & 11 & 8 & 2 \end{bmatrix} = \begin{bmatrix} -1+4 & 2-1 & 3-3 & 7+0 \\ 0+5 & -3+0 & -1-1 & 4+1 \\ 9+1 & -1+11 & 0+8 & -2+2 \end{bmatrix}$

$= \begin{bmatrix} 4-1 & -1+2 & -3+3 & 0+7 \\ 5+0 & 0-3 & -1-1 & 1+4 \\ 1+9 & 11-1 & 8+0 & 2-2 \end{bmatrix} = B + A = \begin{bmatrix} 3 & 1 & 0 & 7 \\ 5 & -3 & -2 & 5 \\ 10 & 10 & 8 & 0 \end{bmatrix}$

25. Equating corresponding entries

$v_p \cos 14.5° - v_w \cos 45.0° = 235 \cos 21.0°$

$v_p \sin 14.5° + v_w \sin 45.0° = 235 \sin 21.0°$, adding $v_p(\cos 14.5° + \sin 14.5°) = (235 \cos 21.0° + 235 \sin 21.0°)$

$v_p = \dfrac{235(\cos 21.0° + \sin 21.0°)}{\cos 14.5° + \sin 14.5} = 249$ km/h; $v_w = \dfrac{235 \sin 21.0° - v_p \sin 14.5°}{\sin 45.0°} = 31.0$ km/h

---

16.4   Multiplication of Matrices

1. $[4 \quad -2] \begin{bmatrix} -1 & 0 \\ 2 & 6 \end{bmatrix} = [-8 \quad -12]$

5. $\begin{bmatrix} 2 & -3 & 1 \\ 0 & 7 & -3 \end{bmatrix} \begin{bmatrix} 9 \\ -2 \\ 5 \end{bmatrix} = \begin{bmatrix} 29 \\ -29 \end{bmatrix}$

9. $\begin{bmatrix} -1 & 7 \\ 3 & 5 \\ 10 & -1 \\ -5 & 12 \end{bmatrix} \begin{bmatrix} 2 & 1 \\ 5 & -3 \end{bmatrix} = \begin{bmatrix} 33 & -22 \\ 31 & -12 \\ 15 & 13 \\ 50 & -41 \end{bmatrix}$

13. $AB = [1 \quad -3 \quad 8] \begin{bmatrix} -1 \\ 5 \\ 7 \end{bmatrix} = [40]$

$BA = \begin{bmatrix} -1 \\ 5 \\ 7 \end{bmatrix} [1 \quad -3 \quad 8] = \begin{bmatrix} -1 & 3 & -8 \\ 5 & -15 & 40 \\ 7 & -21 & 56 \end{bmatrix}$

17. $AI = \begin{bmatrix} 1 & 8 \\ -2 & 2 \end{bmatrix} \begin{bmatrix} 1 & 0 \\ 0 & 1 \end{bmatrix} = \begin{bmatrix} 1 & 8 \\ -2 & 2 \end{bmatrix} = A$

$IA = \begin{bmatrix} 1 & 0 \\ 0 & 1 \end{bmatrix} \begin{bmatrix} 1 & 8 \\ -2 & 2 \end{bmatrix} = \begin{bmatrix} 1 & 8 \\ -2 & 2 \end{bmatrix} = A$

21. $AB = \begin{bmatrix} 5 & -2 \\ -2 & 1 \end{bmatrix} \begin{bmatrix} 1 & 2 \\ 2 & 5 \end{bmatrix} = \begin{bmatrix} 1 & 0 \\ 0 & 1 \end{bmatrix}$

$BA = \begin{bmatrix} 1 & 2 \\ 2 & 5 \end{bmatrix} \begin{bmatrix} 5 & -2 \\ -2 & 1 \end{bmatrix} = \begin{bmatrix} 1 & 0 \\ 0 & 1 \end{bmatrix}$

Therefore $B = A^{-1}$.

25. $\begin{bmatrix} 3 & -2 \\ 4 & 1 \end{bmatrix} \begin{bmatrix} 1 \\ 2 \end{bmatrix} = \begin{bmatrix} -1 \\ 6 \end{bmatrix}$

$A = \begin{bmatrix} 1 \\ 2 \end{bmatrix}$ is the proper matrix of solution values.

29. $I = \begin{bmatrix} 1 & 0 \\ 0 & 1 \end{bmatrix}, -I = \begin{bmatrix} -1 & 0 \\ 0 & -1 \end{bmatrix}$

$(-I)^2 = \begin{bmatrix} -1 & 0 \\ 0 & -1 \end{bmatrix} \begin{bmatrix} -1 & 0 \\ 0 & -1 \end{bmatrix}$

$= \begin{bmatrix} 1 & 0 \\ 0 & 1 \end{bmatrix} = I$

33. $S_y^2 = \begin{bmatrix} 0 & -j \\ j & 0 \end{bmatrix} \begin{bmatrix} 0 & -j \\ j & 0 \end{bmatrix} = \begin{bmatrix} 0(0) - j^2 & 0(-j) - j(0) \\ j(0) + 0(j) & j(-j) - 0(0) \end{bmatrix} = \begin{bmatrix} -j^2 & 0 \\ 0 & -j^2 \end{bmatrix} = \begin{bmatrix} -(-1) & 0 \\ 0 & -(-1) \end{bmatrix}$

$S_y^2 = \begin{bmatrix} 1 & 0 \\ 0 & 1 \end{bmatrix} = I$

16.5   Finding the Inverse of a Matrix

1. $\det \begin{bmatrix} 2 & -5 \\ -2 & 4 \end{bmatrix} = 2(4) - (-2)(-5) = -2$

5. $\det \begin{bmatrix} 0 & -4 \\ 2 & 6 \end{bmatrix} = 0(6) - 2(-4) = 8$

$\begin{bmatrix} 2 & -5 \\ -2 & 4 \end{bmatrix} \longrightarrow \begin{bmatrix} 4 & -5 \\ -2 & 2 \end{bmatrix} \longrightarrow \begin{bmatrix} 4 & 5 \\ 2 & 2 \end{bmatrix}$

$\begin{bmatrix} 0 & -4 \\ 2 & 6 \end{bmatrix} \longrightarrow \begin{bmatrix} 6 & -4 \\ 2 & 0 \end{bmatrix} \longrightarrow \begin{bmatrix} 6 & 4 \\ -2 & 0 \end{bmatrix}$

$\begin{bmatrix} 2 & -5 \\ -2 & 4 \end{bmatrix}^{-1} = \begin{bmatrix} \frac{4}{-2} & \frac{5}{-2} \\ \frac{2}{-2} & \frac{2}{-2} \end{bmatrix} = \begin{bmatrix} -2 & -\frac{5}{2} \\ -1 & -1 \end{bmatrix}$

$\begin{bmatrix} 0 & -4 \\ 2 & 6 \end{bmatrix}^{-1} = \begin{bmatrix} \frac{6}{8} & \frac{4}{8} \\ \frac{-2}{8} & \frac{0}{8} \end{bmatrix} = \begin{bmatrix} \frac{3}{4} & \frac{1}{2} \\ -\frac{1}{4} & 0 \end{bmatrix}$

9.

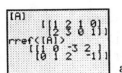 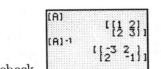 as a check,

13.

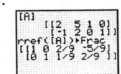 as a check,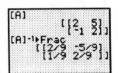

17.

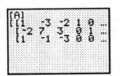 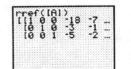 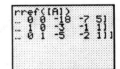

as a check,

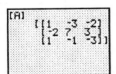 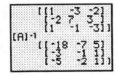

21.

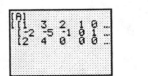 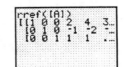 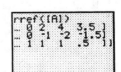

as a check,

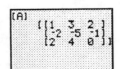

25.

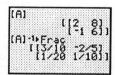

29.

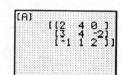

33. $\dfrac{1}{ad-bc} \cdot \begin{bmatrix} a & b \\ c & d \end{bmatrix} \cdot \begin{bmatrix} d & -b \\ -c & a \end{bmatrix} = \dfrac{1}{ad-bc} \cdot \begin{bmatrix} ad-bc & -ab+ab \\ cd-cd & -bc+ad \end{bmatrix}$

$\dfrac{1}{ad-bc} \cdot \begin{bmatrix} ad-bc & 0 \\ 0 & ad-bc \end{bmatrix} = \begin{bmatrix} \frac{ad-bc}{ad-bc} & \frac{0}{ad-bc} \\ \frac{0}{ad-bc} & \frac{ad-bc}{ad-bc} \end{bmatrix} = \begin{bmatrix} 1 & 0 \\ 0 & 1 \end{bmatrix}$

16.6   Matrices and Linear Equations

1. $\quad A = \begin{bmatrix} 2 & -5 \\ -2 & 4 \end{bmatrix}; A^{-1} = \begin{bmatrix} -2 & -2.5 \\ -1 & -1 \end{bmatrix}$

$A^{-1}B = \begin{bmatrix} -2 & -2.5 \\ -1 & -1 \end{bmatrix} \begin{bmatrix} -14 \\ 11 \end{bmatrix} = \begin{bmatrix} \frac{1}{2} \\ 3 \end{bmatrix} = \begin{bmatrix} x \\ y \end{bmatrix}$

5. $\quad A = \begin{bmatrix} 1 & -3 & -2 \\ -2 & 7 & 3 \\ 1 & -1 & -3 \end{bmatrix}; A^{-1} = \begin{bmatrix} -18 & -7 & 5 \\ -3 & -1 & 1 \\ -5 & -2 & 1 \end{bmatrix}$

$A^{-1}B = \begin{bmatrix} -18 & -7 & 5 \\ -3 & -1 & 1 \\ -5 & -2 & 1 \end{bmatrix} \begin{bmatrix} -8 \\ 19 \\ -3 \end{bmatrix} = \begin{bmatrix} -4 \\ 2 \\ -1 \end{bmatrix} = \begin{bmatrix} x \\ y \\ z \end{bmatrix}$

9. $\quad A = \begin{bmatrix} 2 & -3 \\ 4 & -5 \end{bmatrix}; A^{-1} \begin{bmatrix} -\frac{5}{2} & \frac{3}{2} \\ -2 & 1 \end{bmatrix}$

$A^{-1}B = \begin{bmatrix} -\frac{5}{2} & \frac{3}{2} \\ -2 & 1 \end{bmatrix} \begin{bmatrix} 3 \\ 4 \end{bmatrix} = \begin{bmatrix} -1.5 \\ -2 \end{bmatrix} = \begin{bmatrix} x \\ y \end{bmatrix}$

13. $\quad A = \begin{bmatrix} 1 & 2 & 2 \\ 4 & 9 & 10 \\ -1 & 3 & 7 \end{bmatrix}; A^{-1} \begin{bmatrix} -33 & 8 & -2 \\ 38 & 9 & 2 \\ -21 & 5 & 1 \end{bmatrix}$

$A^{-1}B = \begin{bmatrix} -33 & -8 & -2 \\ 38 & 9 & 2 \\ -21 & 5 & -1 \end{bmatrix} \begin{bmatrix} -4 \\ -18 \\ -7 \end{bmatrix} = \begin{bmatrix} 2 \\ -4 \\ 1 \end{bmatrix} = \begin{bmatrix} x \\ y \\ z \end{bmatrix}$

17. $A = \begin{bmatrix} 2 & -1 & -1 \\ 4 & -3 & 2 \\ 3 & 5 & 1 \end{bmatrix}; A^{-1} \begin{bmatrix} \frac{13}{57} & \frac{4}{57} & \frac{5}{57} \\ \frac{-2}{57} & \frac{-5}{57} & \frac{8}{57} \\ \frac{-29}{57} & \frac{13}{57} & \frac{2}{57} \end{bmatrix}$

$$A^{-1}B = \begin{bmatrix} \frac{13}{57} & \frac{4}{57} & \frac{5}{57} \\ \frac{-2}{57} & \frac{-5}{57} & \frac{8}{57} \\ \frac{-29}{57} & \frac{13}{57} & \frac{2}{57} \end{bmatrix} \begin{bmatrix} 7 \\ 4 \\ -10 \end{bmatrix} = \begin{bmatrix} 1 \\ -2 \\ -3 \end{bmatrix} = \begin{bmatrix} x \\ y \\ z \end{bmatrix}$$

21. $A^{-1}B = \begin{bmatrix} 1 & -5 & 2 & -1 \\ 3 & 1 & -3 & 2 \\ 4 & -2 & 1 & -1 \\ -2 & 3 & -1 & 4 \end{bmatrix}^{-1} \begin{bmatrix} -18 \\ 17 \\ -1 \\ 11 \end{bmatrix} = \begin{bmatrix} \frac{-7}{85} & \frac{2}{85} & \frac{23}{85} & \frac{3}{85} \\ \frac{-47}{170} & \frac{-23}{170} & \frac{33}{170} & \frac{4}{85} \\ \frac{-13}{170} & \frac{-57}{170} & \frac{67}{170} & \frac{21}{85} \\ \frac{5}{34} & \frac{1}{34} & \frac{3}{34} & \frac{5}{17} \end{bmatrix} \begin{bmatrix} -18 \\ 17 \\ -1 \\ 11 \end{bmatrix} = \begin{bmatrix} 2 \\ 3 \\ -2 \\ 1 \end{bmatrix} = \begin{bmatrix} x \\ y \\ z \\ t \end{bmatrix}$

25. $\begin{bmatrix} A \\ B \end{bmatrix} = \begin{bmatrix} \sin 47.2° & \sin 64.4° \\ \cos 47.2° & -\cos 64.4° \end{bmatrix}^{-1} \begin{bmatrix} 254 \\ 0 \end{bmatrix} = \begin{bmatrix} 118 \\ 186 \end{bmatrix}$

$A = 118$ N, $B = 186$ N

Chapter 16 Review Exercises

1. $\begin{vmatrix} 1 & 2 & -1 \\ 4 & 1 & -3 \\ -3 & -5 & 2 \end{vmatrix} = (1) \begin{vmatrix} 1 & -3 \\ -5 & 2 \end{vmatrix} - (2) \begin{vmatrix} 4 & -3 \\ -3 & 2 \end{vmatrix} + (-1) \begin{vmatrix} 4 & 1 \\ -3 & -5 \end{vmatrix}$

$$= 1(-13) - 2(-1) - 1(-17) = -13 + 2 + 17 = 6$$

The determinant was expanded by the first row.

5. $\begin{vmatrix} 2 & 6 & 2 & 5 \\ 2 & 0 & 4 & -1 \\ 4 & -3 & 6 & 1 \\ 3 & -1 & 0 & -2 \end{vmatrix}$

$$= -2 \begin{vmatrix} 6 & 2 & 5 \\ -3 & 6 & 1 \\ -1 & 0 & -2 \end{vmatrix} - 4 \begin{vmatrix} 2 & 6 & 5 \\ 4 & -3 & 1 \\ 3 & -1 & -2 \end{vmatrix} - 1 \begin{vmatrix} 2 & 6 & 2 \\ 4 & -3 & 6 \\ 3 & -1 & 0 \end{vmatrix}$$

$$= -2 \left[-1 \begin{vmatrix} 2 & 5 \\ 6 & 1 \end{vmatrix} + (-2) \begin{vmatrix} 6 & 2 \\ -3 & 6 \end{vmatrix} \right] - 4 \left[-4 \begin{vmatrix} 6 & 5 \\ -1 & -2 \end{vmatrix} + (-3) \begin{vmatrix} 2 & 5 \\ 3 & -2 \end{vmatrix} - 1 \begin{vmatrix} 2 & 6 \\ 3 & -1 \end{vmatrix} \right]$$

$$- 1 \left[3 \begin{vmatrix} 6 & 2 \\ -3 & 6 \end{vmatrix} - (-1) \begin{vmatrix} 2 & 2 \\ 4 & 6 \end{vmatrix} \right]$$

$$= -2[-1(-28) - 2(42)] - 4[-4(-7) - 3(-19) - 1(-20)] - 1[3(42) + 1(4)]$$

$$= -2(-56) - 4(105) - 1(130) = -438$$

9.
$$\begin{vmatrix} 1 & 2 & -1 \\ 4 & 1 & -3 \\ -3 & -5 & 2 \end{vmatrix} = \begin{vmatrix} 1 & 2 & 0 \\ 4 & 1 & 1 \\ -3 & -5 & -1 \end{vmatrix} = \begin{vmatrix} 1 & 0 & 0 \\ 4 & -7 & 1 \\ -3 & 1 & -1 \end{vmatrix} = 1 \begin{vmatrix} -7 & 1 \\ 1 & -1 \end{vmatrix} = 1(7-1) = 6$$

13.
$$\begin{vmatrix} 2 & 6 & 2 & 5 \\ 2 & 0 & 4 & -1 \\ 4 & -3 & 6 & 1 \\ 3 & -1 & 0 & -2 \end{vmatrix} = \begin{vmatrix} 2 & 6 & -2 & 5 \\ 2 & 0 & 0 & -1 \\ 4 & -3 & -2 & 1 \\ 3 & -1 & -6 & -2 \end{vmatrix} = \begin{vmatrix} 12 & 6 & -2 & 5 \\ 0 & 0 & 0 & -1 \\ 6 & -3 & -2 & 1 \\ -1 & -1 & -6 & -2 \end{vmatrix} = (-1) \begin{vmatrix} 12 & 6 & -2 \\ 6 & -3 & -2 \\ -1 & -1 & -6 \end{vmatrix}$$

$$= (-1) \begin{vmatrix} 6 & 6 & -2 \\ 9 & -3 & -2 \\ 0 & -1 & -6 \end{vmatrix} = (-1) \begin{vmatrix} 6 & 6 & -38 \\ 9 & -3 & 16 \\ 0 & -1 & 0 \end{vmatrix} = -(-1)(-1) \begin{vmatrix} 6 & -38 \\ 9 & 16 \end{vmatrix} = -438$$

17.

21. $\begin{pmatrix} 2a \\ a - b \end{pmatrix} = \begin{pmatrix} 8 \\ 5 \end{pmatrix}$; $2a = 8$; $a = 4$; $a - b = 5$; $4 - b = 5$; $b = -1$

25. $A + B = \begin{pmatrix} 2 & -3 \\ 4 & 1 \\ -5 & 0 \\ 2 & -3 \end{pmatrix} + \begin{pmatrix} -1 & 0 \\ 4 & -6 \\ -3 & -2 \\ 1 & -7 \end{pmatrix} = \begin{pmatrix} 1 & -3 \\ 8 & -5 \\ -8 & -2 \\ 3 & -10 \end{pmatrix}$

29. $A - C = \begin{pmatrix} 2 & -3 \\ 4 & 1 \\ -5 & 0 \\ 2 & -3 \end{pmatrix} - \begin{pmatrix} 5 & -6 \\ 2 & 8 \\ 0 & -2 \end{pmatrix}$

Cannot be subtracted since A and C do not have the same number of rows.

33.

37. $\begin{pmatrix} 2 & -5 \\ 2 & -4 \end{pmatrix}$

Interchange the elements of the principal diagonal and change the signs of the off-diagonal elements.

$\begin{pmatrix} -4 & 5 \\ -2 & 2 \end{pmatrix}$

Find the determinant of the original matrix.

$\begin{vmatrix} 2 & -5 \\ 2 & -4 \end{vmatrix} = 2$

Divide each element of the second matrix by 2.

$$\frac{1}{2} \begin{pmatrix} -4 & 5 \\ -2 & 2 \end{pmatrix} = \begin{pmatrix} -2 & \frac{5}{2} \\ -1 & 1 \end{pmatrix}$$

41. ① $\left(\begin{array}{ccc|ccc} 1 & 1 & -2 & 1 & 0 & 0 \\ -1 & -2 & 1 & 0 & 1 & 0 \\ 0 & 3 & 4 & 0 & 0 & 1 \end{array} \right)$; ② $\left(\begin{array}{ccc|ccc} 1 & 1 & -2 & 1 & 0 & 0 \\ 0 & -1 & -1 & 1 & 1 & 0 \\ 0 & 3 & 4 & 0 & 0 & 1 \end{array} \right)$; ③ $\left(\begin{array}{ccc|ccc} 1 & 1 & -2 & 1 & 0 & 0 \\ 0 & -1 & -1 & 1 & 1 & 0 \\ 0 & 0 & 1 & 3 & 3 & 1 \end{array} \right)$

④ $\left(\begin{array}{ccc|ccc} 1 & 0 & -3 & 2 & 1 & 0 \\ 0 & -1 & -1 & 1 & 1 & 0 \\ 0 & 0 & 1 & 3 & 3 & 1 \end{array} \right)$; ⑤ $\left(\begin{array}{ccc|ccc} 1 & 0 & -3 & 2 & 1 & 0 \\ 0 & -1 & 0 & 4 & 4 & 1 \\ 0 & 0 & 1 & 3 & 3 & 1 \end{array} \right)$; ⑥ $\left(\begin{array}{ccc|ccc} 1 & 0 & 0 & 11 & 10 & 3 \\ 0 & -1 & 0 & 4 & 4 & 1 \\ 0 & 0 & 1 & 3 & 3 & 1 \end{array} \right)$;

⑦ $\left(\begin{array}{ccc|ccc} 1 & 0 & 0 & 11 & 10 & 3 \\ 0 & 1 & 0 & -4 & -4 & -1 \\ 0 & 0 & 1 & 3 & 3 & 1 \end{array} \right)$

$$A^{-1} = \begin{pmatrix} 11 & 10 & 3 \\ -4 & -4 & -1 \\ 3 & 3 & 1 \end{pmatrix}$$

① Original setup.   ④ Row two added to row one.

② Row one added to row two.   ⑤ Row three added to row two.

③ 3 times row two added to row three.   ⑥ 3 times row three added to row one.

                                              ⑦ -1 times row two.

45. $A = \begin{pmatrix} 2 & -3 \\ 4 & -1 \end{pmatrix}$; $C = \begin{pmatrix} -9 \\ -13 \end{pmatrix}$; $\begin{vmatrix} 2 & -3 \\ 4 & -1 \end{vmatrix} = 10$; $A^{-1} = \frac{1}{10} \begin{pmatrix} -1 & 3 \\ -4 & 2 \end{pmatrix} = \begin{pmatrix} -\frac{1}{10} & \frac{3}{10} \\ -\frac{4}{10} & \frac{2}{10} \end{pmatrix}$

$$A^{-1}C = \begin{pmatrix} -\frac{1}{10} & \frac{3}{10} \\ -\frac{4}{10} & \frac{2}{10} \end{pmatrix} \begin{pmatrix} -9 \\ -13 \end{pmatrix} = \begin{pmatrix} \frac{9}{10} & -\frac{39}{10} \\ \frac{36}{10} & -\frac{26}{10} \end{pmatrix} \begin{pmatrix} -3 \\ 1 \end{pmatrix}$$

$x = -3, y = 1$

49. $A = \begin{pmatrix} 2 & -3 & 2 \\ 3 & 1 & -3 \\ 1 & 4 & 1 \end{pmatrix}; C = \begin{pmatrix} 7 \\ -6 \\ -13 \end{pmatrix}$

① $\left(\begin{array}{ccc|ccc} 2 & -3 & 2 & 1 & 0 & 0 \\ 3 & 1 & -3 & 0 & 1 & 0 \\ 1 & 4 & 1 & 0 & 0 & 1 \end{array} \right)$; ② $\left(\begin{array}{ccc|ccc} 2 & -3 & -2 & 1 & 0 & 0 \\ 3 & 1 & -3 & 0 & 1 & 0 \\ 0 & -11 & -6 & 0 & 1 & -3 \end{array} \right)$; ③ $\left(\begin{array}{ccc|ccc} 2 & -3 & 2 & 1 & 0 & 0 \\ 0 & 11 & -12 & -3 & 2 & 0 \\ 0 & -11 & -6 & 1 & 0 & -3 \end{array} \right)$

④ $\left(\begin{array}{ccc|ccc} 1 & -\frac{3}{2} & 1 & \frac{1}{2} & 0 & 0 \\ 0 & -11 & -12 & -3 & 2 & 0 \\ 0 & -11 & -6 & 0 & 1 & -3 \end{array} \right)$; ⑤ $\left(\begin{array}{ccc|ccc} 1 & -\frac{3}{2} & 1 & \frac{1}{2} & 0 & 0 \\ 0 & 11 & -12 & -3 & 2 & 0 \\ 0 & 0 & -18 & -3 & 3 & -3 \end{array} \right)$;

⑥ $\left(\begin{array}{ccc|ccc} 1 & -\frac{3}{2} & 1 & \frac{1}{2} & 0 & 0 \\ 0 & 1 & -\frac{12}{11} & -\frac{3}{11} & \frac{2}{11} & 0 \\ 0 & 0 & -18 & -3 & 3 & -3 \end{array} \right)$; ⑦ $\left(\begin{array}{ccc|ccc} 1 & 0 & -\frac{7}{11} & \frac{1}{11} & \frac{3}{11} & 0 \\ 0 & 1 & -\frac{12}{11} & -\frac{3}{11} & \frac{2}{11} & 0 \\ 0 & 0 & -18 & -3 & 3 & -3 \end{array} \right)$;

⑧ $\left(\begin{array}{ccc|ccc} 1 & 0 & -\frac{7}{11} & \frac{1}{11} & \frac{3}{11} & 0 \\ 0 & 1 & -\frac{12}{11} & -\frac{3}{11} & \frac{2}{11} & 0 \\ 0 & 0 & 1 & \frac{1}{6} & -\frac{1}{6} & \frac{1}{6} \end{array} \right)$; ⑨ $\left(\begin{array}{ccc|ccc} 1 & 0 & -\frac{7}{11} & \frac{1}{11} & \frac{3}{11} & 0 \\ 0 & 1 & 0 & -\frac{1}{11} & 0 & \frac{2}{11} \\ 0 & 0 & 1 & \frac{1}{6} & -\frac{1}{6} & \frac{1}{6} \end{array} \right)$;

⑩ $\left(\begin{array}{ccc|ccc} 1 & 0 & 0 & \frac{13}{66} & \frac{11}{66} & \frac{7}{66} \\ 0 & 1 & 0 & -\frac{1}{11} & 0 & \frac{2}{11} \\ 0 & 0 & 1 & \frac{1}{6} & -\frac{1}{6} & \frac{1}{6} \end{array} \right)$;

$A^{-1}C = \begin{pmatrix} \frac{13}{66} & \frac{11}{66} & \frac{7}{66} \\ -\frac{1}{11} & 0 & \frac{2}{11} \\ \frac{1}{6} & -\frac{1}{6} & \frac{1}{6} \end{pmatrix} \begin{pmatrix} 7 \\ -6 \\ -13 \end{pmatrix} = \begin{pmatrix} \frac{91}{66} - \frac{66}{66} - \frac{91}{66} \\ -\frac{7}{11} + 0 - \frac{26}{11} \\ \frac{7}{6} + 1 - \frac{13}{6} \end{pmatrix} = \begin{pmatrix} -1 \\ -3 \\ 0 \end{pmatrix}$

① Original.

② Row three multiplied by -3 and added to row two.

③ Row two multiplied by 2. Row one multiplied by -3 and added to row two.

④ Row one divided by 2.

⑤ Row two added to row three.

⑥ Row two divided by 11.

⑦ Row two multiplied by $\frac{3}{2}$ and added to row one.

⑧ Row three divided by -18.

⑨ Row three multiplied by $\frac{12}{11}$ and added to row two.

⑩ Row three multiplied by $\frac{7}{11}$ and added to row one.

53. $3x - 2y + z = 6$
$2x + 0y + 3z = 3$
$4x - y + 5z = 6$

The denominator will be found first.

$$\begin{vmatrix} 3 & -2 & 1 \\ 2 & 0 & 3 \\ 4 & -1 & 5 \end{vmatrix} = -\begin{vmatrix} 3 & -2 & 1 \\ 2 & 0 & 3 \\ -4 & 1 & -5 \end{vmatrix} = (-1)\begin{vmatrix} -5 & 0 & -9 \\ 2 & 0 & 3 \\ -4 & 1 & -5 \end{vmatrix}$$

$$= -(-1)\begin{vmatrix} -5 & -9 \\ 2 & 3 \end{vmatrix} = (1)(-15 + 18) = 3$$

$$x = \frac{\begin{vmatrix} 6 & -2 & 1 \\ 3 & 0 & 3 \\ 6 & -1 & 5 \end{vmatrix}}{3} = (-1)\frac{\begin{vmatrix} 6 & -2 & 1 \\ 3 & 0 & 3 \\ -6 & 1 & -5 \end{vmatrix}}{3} = (-1)\frac{\begin{vmatrix} -6 & 0 & -9 \\ 3 & 0 & 3 \\ -6 & 1 & -5 \end{vmatrix}}{3}$$

$$= (-1)(-1)\frac{\begin{vmatrix} -6 & -9 \\ 3 & 3 \end{vmatrix}}{3} = \frac{-18 + 27}{3} = 3$$

Substitute the value for x into the second equation.

$2(3) + 3z = 3;\ 6 + 3z = 3;\ 3z = -3;\ z = -1$

Substitute the values of x and z into the first equation.

$3(3) - 2y + (-1) = 6;\ 9 - 2y - 1 = 6;\ -2y + 8 = 6;\ -2y = -2;\ y = 1$

57.

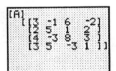 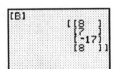 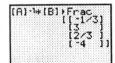

Check

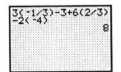 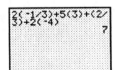 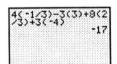 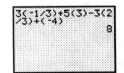

61. $\begin{vmatrix} 1+\sqrt{2} & 2-\sqrt{3} & 0 \\ 3+\sqrt{5} & 7+\sqrt{6} & \sqrt{2} \\ 2+\sqrt{3} & 1-\sqrt{2} & 0 \end{vmatrix} = 0\begin{vmatrix} 3+\sqrt{5} & 7+\sqrt{6} \\ 2+\sqrt{3} & 1-\sqrt{2} \end{vmatrix} - 2\begin{vmatrix} 1+\sqrt{2} & 2-\sqrt{3} \\ 2+\sqrt{3} & 1-\sqrt{2} \end{vmatrix} + 0\begin{vmatrix} 1+\sqrt{2} & 2-\sqrt{3} \\ 3+\sqrt{5} & 7+\sqrt{6} \end{vmatrix}$

$= -\sqrt{2}(1+\sqrt{2})(1-\sqrt{2}) - (2+\sqrt{3})(2-\sqrt{3})$
$= -\sqrt{2}[(1-2) - (4-3)] = -\sqrt{2}[-1-1] = -\sqrt{2}(-2)$
$= 2\sqrt{2}$

65.

69. $N = \begin{pmatrix} 0 & -1 \\ 1 & 0 \end{pmatrix}$; $N^{-1} \dfrac{1}{0-(-1)} \begin{pmatrix} 0 & 1 \\ -1 & 0 \end{pmatrix} = 1 \begin{pmatrix} 0 & 1 \\ -1 & 0 \end{pmatrix} = -1 \begin{pmatrix} 0 & -1 \\ 1 & 0 \end{pmatrix} = -N$

73. $A = \begin{pmatrix} 1 & -2 \\ 0 & 3 \end{pmatrix}$; $2a = \begin{pmatrix} 2 & -4 \\ 0 & 6 \end{pmatrix}$; $(2A)^{-1} = \dfrac{1}{12+0} \begin{pmatrix} 6 & 4 \\ 0 & 2 \end{pmatrix} = \begin{pmatrix} \frac{1}{2} & \frac{1}{3} \\ 0 & \frac{1}{6} \end{pmatrix}$

$\dfrac{A^{-1}}{2} = \dfrac{1}{2} \left(\dfrac{1}{3-(0)} \right) \begin{pmatrix} 3 & 2 \\ 0 & 1 \end{pmatrix} = \dfrac{1}{6} \begin{pmatrix} 3 & 2 \\ 0 & 1 \end{pmatrix} = \begin{pmatrix} \frac{1}{2} & \frac{1}{3} \\ 0 & \frac{1}{6} \end{pmatrix} = (2A)^{-1}$

77. $0.500F = 0.866T$; $0.500F - 0.866T = 0$

$$0.866F + 0.500T = 350;\ C = \begin{pmatrix} 0 \\ 350 \end{pmatrix}$$

$$A^{-1} = \dfrac{1}{1} \begin{pmatrix} 0.500 & 0.866 \\ -0.866 & 0.500 \end{pmatrix} = \begin{pmatrix} 0.500 & 0.866 \\ -0.866 & 0.500 \end{pmatrix}$$

$$A^{-1}C = \begin{pmatrix} 0.500 & 0.866 \\ -0.866 & 0.500 \end{pmatrix} \begin{pmatrix} 0 \\ 350 \end{pmatrix} = \begin{pmatrix} 0 + 0.866(350) \\ 0 + 0.500(350) \end{pmatrix} = \begin{pmatrix} 303 \\ 175 \end{pmatrix}$$

$F = 303$ lb, $T = 175$ lb

81. Let A be the number of grams of alloy A and B be the number of grams of alloy B. Let C be the number of grams of alloy C.

Lead: $0.60A + 0.40B + 0.30C = 0.44(100)$; $6A + 4B + 3C = 440$
Zinc: $0.30A + 0.30B + 0.70C = 0.38(100)$; $3A + 3B + 7C = 280$
Copper: $0.10a + 0.30B = 0.18(100)$; $A + 3B\quad\quad = 180$

Using determinants with expansion by minors.

$$A = \dfrac{\begin{vmatrix} 440 & 4 & 3 \\ 380 & 3 & 7 \\ 180 & 3 & 0 \end{vmatrix}}{\begin{vmatrix} 6 & 4 & 3 \\ 3 & 3 & 7 \\ 1 & 3 & 0 \end{vmatrix}} = \dfrac{440 \begin{vmatrix} 37 \\ 30 \end{vmatrix} - 380 \begin{vmatrix} 4 & 3 \\ 3 & 0 \end{vmatrix} + 180 \begin{vmatrix} 4 & 3 \\ 3 & 7 \end{vmatrix}}{6 \begin{vmatrix} 37 \\ 30 \end{vmatrix} - 3 \begin{vmatrix} 4 & 3 \\ 3 & 0 \end{vmatrix} + 1 \begin{vmatrix} 4 & 3 \\ 3 & 7 \end{vmatrix}} = \dfrac{440(-21) - 380(-9) + 180(19)}{6(-21) - 3(-9) + 1(19)}$$

$$= \dfrac{-2400}{-80} = 30 \text{ g}$$

Substituting into the third equation, $30 + 3B = 180$; $B = 50$ g; substituting into the first equation $6(30) + 4(50) + 3C = 440$; $C = 20$ g.

85.

	Standard Transmission	Automatic Transmission

$$A = \begin{array}{l} \text{4 cylinders} \\ \text{6 cylinders} \\ \text{8 cylinders} \end{array} \begin{pmatrix} 12\ 000 & 15\ 000 \\ 24\ 000 & 8\ 000 \\ 4\ 000 & 30\ 000 \end{pmatrix} \quad B = \begin{pmatrix} 15\ 000 & 20\ 000 \\ 12\ 000 & 3\ 000 \\ 2\ 000 & 22\ 000 \end{pmatrix}$$

$$A + B = \begin{pmatrix} 27\ 000 & 35\ 000 \\ 36\ 000 & 11\ 000 \\ 6\ 000 & 52\ 000 \end{pmatrix}$$

89. Answers may vary, but the basic idea is that the matrix entries show the inventory of a particular product in a particular store in a compact way.

Chapter 17

INEQUALITIES

17.1 Properties of Inequalities

1. $4 + 3 < 9 + 3$; $7 < 12$; property 1

5. $\dfrac{4}{-1} > \dfrac{9}{-1}$; $-4 > -9$; property 3

9. $x > -2$

13. $1 < x < 7$

17. $x < 1$ or $3 < x \le 5$

21. x is greater than 0 and less than or equal to 2

25. $x < 3$

29. $0 \le x < 5$

33. $-3 < x < -1$ or $1 < x \le 3$

37. $d > 3 \times 10^{12}$ mi

41. $0 < n \le 2565$ steps

17.2 Solving Linear Inequalities

1.
$$x - 3 > -4$$
$$x > -4 + 3$$
$$x > -1$$

5.
$$3x - 5 \le -11$$
$$3x \le -11 + 5$$
$$3x \le -6$$
$$x \le -2$$

9.
$$4x - 5 \le 2x$$
$$4x - 2x \le 5$$
$$2x \le 5$$
$$x \le \frac{5}{2}$$

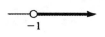

13.
$$2.50(1.50 - 3.40x) < 3.84 - 8.45x$$
$$3.75 - 8.50x < 3.84 - 8.45x$$
$$-0.09 < 0.05x$$
$$x > -1.80$$

17.
$$-1 < 2x + 1 < 3$$
$$-2 < 2x < 2$$
$$-1 < x < 1$$

21.
$$2x < x - 1 \le 3x + 5$$
$$0 < -x - 1 \le x + 5$$
$$0 < -x - 1 \text{ and } -x - 1 \le x + 5$$
$$x < -1 \text{ and } -6 \le 2x$$
$$x < -1 \text{ and } x \ge -3$$
$$-3 \le x < -1$$

25. The solution is $y < 2.5$.

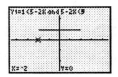

29. The solution is $-2 < t < 2$.

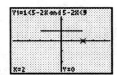

33. $f(x) = \sqrt{2x - 10}$

$2x - 10 \geq 0$

$2x \geq 10$

$x \geq 5$

37. $V = 40\,000 + 4000t; V \leq 64\,000$ and $t \geq 0$

$0 \leq 4000t + 40\,000 \leq 64\,000$

$-40\,000 \leq 4000t \leq 24\,000$

$-10 \leq t \leq 6$

$0 \leq t \leq 6$ years

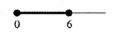

41. $0 \leq x \leq 800 - 300$

$0 \leq x \leq 500$

$y = x + 400 - 200 = x + 200$

$x = y - 200$

$0 \leq y - 200 \leq 500$

$200 \leq y \leq 700$

17.3   Solving Nonlinear Inequalities

1. $x^2 - 1 < 0$

$(x + 1)(x - 1) < 0$

The critical values are
$x = -1$ and $x = 1$.

	$(x+1)(x-1)$		Sign
$x < -1$	$-$	$-$	$+$
$-1 < x < 1$	$+$	$-$	$-$
$x > 1$	$+$	$+$	$+$

$(x + 1)(x - 1) < 0$ for $-1 < x < 1$

5. $2x^2 - 12 \leq -5x$

$2x^2 + 5x - 12 \leq 0$

$(2x - 3)(x + 4) \leq 0$

The critical values are $x = \dfrac{3}{2}, x = -4$.

	$(2x-3)(x+4)$		Sign
$x < -4$	$-$	$-$	$+$
$-4 < x < 3/2$	$-$	$+$	$-$
$0 < x < 3/2$	$+$	$+$	$+$

$(2x - 3)(x + 4) \leq 0$ for $-4 \leq x \leq \dfrac{3}{2}$

9. $x^2 + 4 > 0$

$x^2 + 4$ is never less than 4.

so all values of x are solutions.

13.
$$s^3 + 2s^2 - s \geq 2$$
$$s^2(s + 2) - 1(s + 2) \geq 0$$
$$(s^2 - 1)(s + 2) \geq 0$$
$$(s + 1)(s - 1)(s + 2) \geq 0$$

The critical values are $s = -1$, $s = 1$, $s = -2$.

	$(s - 1)(s + 1)(s + 2)$			Sign
$s < -2$	$-$	$-$	$-$	$-$
$-2 < s < -1$	$-$	$-$	$+$	$+$
$-1 < s < 1$	$-$	$+$	$+$	$-$
$s > 1$	$+$	$+$	$+$	$+$

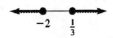

$(s + 1)(s - 1)(s + 2) \geq 0$
for $-2 \leq s \leq -1$ or $s \geq 1$

21.
$$3x^2 + 5x \geq 2$$
$$3x^2 + 5x - 2 \geq 0$$
$$(3x - 1)(x + 2) \geq 0$$

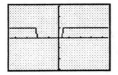

The critical values are $x = \dfrac{1}{3}$ and $x = -2$.

	$(3x - 1)(x + 2)$		Sign
$x < -2$	$-$	$-$	$+$
$-2 < x < 1/3$	$-$	$+$	$-$
$x > 1/3$	$+$	$+$	$+$

$(3x - 1)(x + 2) \geq 0$ for $x \leq -2$ or $x \geq \dfrac{1}{3}$

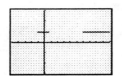

17.
$$\frac{x^2 - 6x - 7}{x + 5} > 0$$
$$\frac{(x - 7)(x + 1)}{x + 5} > 0$$

The critical values are $x = 7$, $x = -1, x = -5$.

	$(x - 7)(x + 1)(x + 5)$			Sign
$x < -5$	$-$	$-$	$-$	$-$
$-5 < x < -1$	$-$	$-$	$+$	$+$
$-1 < x < 7$	$-$	$+$	$+$	$-$
$x > 7$	$+$	$+$	$+$	$+$

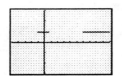

$$\frac{(x - 7)(x + 1)}{x + 5} > 0 \text{ for } -5 < x < -1 \text{ or } x > 7$$

25.
$$\frac{6 - x}{3 - x - 4x^2} \geq 0$$
$$\frac{6 - x}{(1 + x)(3 - 4x)} \geq 0; \left(x \neq -1, x \neq \frac{3}{4} \right)$$

The critical values are $x = 6, x = -1$, and $x = \dfrac{3}{4}$

	$(6 - x)/(1 + x)(3 - 4x)$			Sign
$x < -1$	$+$	$-$	$+$	$-$
$-1 < x < 3/4$	$+$	$+$	$+$	$+$
$3/4 < x < 6$	$+$	$+$	$-$	$-$
$x > 6$	$-$	$+$	$-$	$+$

$$\frac{6 - x}{(1 + x)(3 - 4x)} \geq 0 \text{ for } -1 < x < \frac{3}{4} \text{ or } x \geq 6$$

29. $\sqrt{(x-1)(x+2)}$ is real if $(x-1)(x+2) \geq 0$

The critical values are $x = 1$ and $x = -2$.

	$(x-1)(x+2)$		Sign
$x < -2$	$-$	$-$	$+$
$-2 < x < 1$	$-$	$+$	$-$
$x > 1$	$+$	$+$	$+$

$(x-1)(x+2) > 0$ for $x \leq -2$ or $x \geq 1$

33. To solve $x^3 - x > 2$ using a graphing calculator, let $y_1 = x^3 - x - 2$.

$y > 0$ for $x > 1.52$

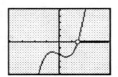

37. To solve $2^x > x + 2$ using a graphing calculator, let $y_1 = 2^x - x - 2$.

$y > 0$ for $x < -1.69, x > 2.00$

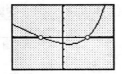

41. $p = 6i - 4i^2, 6i - 4i^2 > 2$ and $i = 1$.

$$4i^2 - 6i + 2 < 0$$
$$2i^2 - 3i + 1 < 0$$
$$(2i - 1)(i - 1) < 0$$

	$(2i-1)(i-1)$		Sign
$i < 0.5$	$-$	$-$	$+$
$0.5 < i < 1$	$+$	$-$	$-$
$i > 1$	$+$	$+$	$+$

$(2i - 1)(i - 1) < 0$ for $0.5 < i < 1$ A

45. $l = w + 2.0; w(w + 2.0) < 35; w \geq 3.0$ mm
$$w^2 + 2.0w - 35 < 0$$
$$(w + 7.0)(w - 5.0) < 0$$

The critical values are $w = -7.0$ and $w = 5.0$.

	$(w+7)(w-5)$		Sign
$w < -7$	$-$	$-$	$+$
$-7 < w < 5$	$+$	$-$	$-$
$w > 5$	$+$	$+$	$+$

$(w + 7.0)(w - 5.0) < 0$ for $-7.0 < w < 5.0$;
$w \geq 3.0$, so $3.0 \leq w < 5.0$ mm

17.4   Inequalities Involving Absolute Values

1. $|x - 4| < 1$
    $-1 < x - 4 < 1$
        $3 < x < 5$

5. $|6 - 5| \leq 4$
    $-4 \leq 6x - 5 \leq 4$
        $1 \leq 6x \leq 9$

$$\frac{1}{6} \leq x \leq \frac{3}{2}$$

9.
$$\left|\frac{t+1}{5}\right| < 5$$

$$-5 < \frac{t+1}{5} < 5$$

$$-25 < t+1 < 25$$
$$-26 < t < 24$$

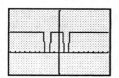

13. $2|x-4| > 8$
$$|x-4| > 4$$
$$x-4 > 4 \text{ or } x-4 < -4$$
$$x > 8 \text{ or } x < 0$$

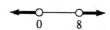

17.
$$\left|\frac{3R}{5}+1\right| < 8$$

$$-8 < \frac{3R}{5}+1 < 8$$

$$-15 < R < \frac{35}{3}$$

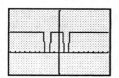

21. $|2x-5| < 3 \Leftrightarrow 1 < x < 4$

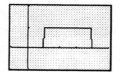

25. $|x^2 + x - 4| > 2$
$$x^2 + x - 4 > 2 \text{ or } x^2 + x - 4 < -2$$
$$x^2 + x - 6 > 0 \text{ or } x^2 + x - 2 < 0$$
(A) $(x+3)(x-2) > 0$
(B) $(x-1)(x+2) < 0$

(A) Critical values are $x = -3, x = 2$

	$(x+3)(x-2)$		Sign
$x < -3$	$-$	$-$	$+$
$-3 < x < 2$	$+$	$-$	$-$
$x > 2$	$+$	$+$	$+$

$(x+3)(x-2) > 0$ for $x < -3$ or for $x > 2$

(B) Critical values are $x = 1, x = -2$.

	$(x-1)(x+2)$		Sign
$x < -2$	$-$	$-$	$+$
$-2 < x < 1$	$-$	$+$	$-$
$x > 1$	$+$	$+$	$+$

$(x-1)(x+2) < 0$ for $-2 < x < 1$

The solution consists of values of x that are
in (A) or (B): $x < -3, -2 < x < 1, x > 2$

29. $|p - 2\,000\,000| < 200\,000$
        $-200\,000 < p - 2\,000\,000 < 200\,000$
                $1\,800\,000 < p < 2\,200\,000$ barrels

The production will be between 1 800 000 barrels and 2 200 000 barrels.

17.5   Graphical Solution of Inequalities with Two Variables

1. $y > x - 1$; graph $y = x - 1$. Use a dashed line to indicate that points on it do not satisfy the inequality. Shade the region above the line.

5. $3x + 2y + 6 > 0$; $2y > -3x - 6$; $y > -\dfrac{3}{2}x - 3$; graph $y = -\dfrac{3}{2}x - 3$. Use a dashed line to indicate that points on it do not satisfy the inequality. Shade the region above the line.

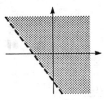

9. $2x^2 - 4x - y > 0$; $-y > 4x - 2x^2$; $y < 2x^2 - 4x$; graph $y = 2x^2 - 4x$. Use a dashed curve to indicate that points on it do not satisfy the inequality. Shade the region below the curve.

13. $y > \dfrac{1}{x^2 + 1}$; graph $y = \dfrac{1}{x^2 + 1}$. Use a dashed curve to indicate that the points on it do not satisfy the inequality. Shade the region above the curve.

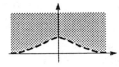

17. $y > x$ and $y > 1 - x$. Graph $y = x$ using a dashed line. Shade the region above the line. Graph $y = 1 - x$ using a dashed line. Shade the region above the curve. The region where the shadings overlap satisfies both inequalities.

21. $y > \dfrac{1}{2}x^2$ and $y \le 4x - x^2$

Graph $y = \dfrac{1}{2}x^2$ with a dashed curve, and shade the region above the curve. Graph $y = 4x - x^2$ with a solid curve, and shade below it. The region where the shaded regions overlap satisfies both inequalities.

25. $2x + y < 5$; $y < -2x + 5$; graph $y = -2x + 5$. Use a dashed line to indicate that points on it do not satisfy the inequality. Shade the region below the line.

29. $y > 2x - 1$
$y < x^4 - 8$

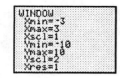 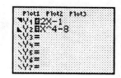 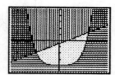

33. $A \le 300$ m; 200 m $\le B \le 400$ m

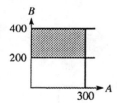

37. Let x = the number of business models produced. Let y = the number of scientific models produced. Let p = the profit. $p = 8x + 10y$. The maximum time worked = 8 hours or 480 minutes. For the first operation, $3x + 6y \leq 480$. For the second operation, $6x + 4y \leq 480$. $3x + 6y \leq 480$; $y \leq -\frac{1}{2}x + 80$. Graph $y = -\frac{1}{2}x + 80$ and shade the region below the graph. $6x + 4y \leq 480$; $y \leq -\frac{3}{2}x + 120$; graph $y = -\frac{3}{2}x + 120$ and shade the region below the graph. The overlapping shaded regions satisfy both inequalities. Vertices of the region are $(80, 0), (0, 80)$ and the point p. p is the intersection of $y = -\frac{1}{2}x + 80$ and $y = -\frac{3}{2}x + 120$. Solve these equations simultaneously by substitution.

$$-\frac{1}{2}x + 80 = -\frac{3}{2}x + 120; x = 40, y = 60.$$

OR

The profit is calculated using the coordinates of the vertices. $p = 8(80) + 10(0) = 640$; $p = 8(0) + 10(80) = 800$; $p = 8(40) + 10(60) = 920$. The maximum profit occurs if 40 business models and 60 scientific models are produced.

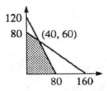

Chapter 17 Review Exercises

1. $2x - 12 > 0$; $2(x - 6) > 0$. The critical value is 6. If $x > 6$, $2x - 12 > 0$. Thus the values which satisfy the inequality are $x > 6$.

5.

$$5x^2 + 9x < 2$$
$$(x + 2)(5x - 1) < 0$$
$$-2 < x < \frac{1}{5}$$

interval	$(x + 2)(5x - 1)$
$x < -2$	$(-)(-) = (+)$
$-2 < x < \frac{1}{5}$	$(+)(-) = (-)$
$x > \frac{1}{5}$	$(+)(+) = (+)$

9. $\dfrac{(2x - 1)(3 - x)}{(x + 4)} > 0$

		-4		$\frac{1}{2}$		3	
$2x - 1$		$-$		$-$		$+$	$+$
$3 - x$		$+$		$+$		$+$	$-$
$x + 4$		$-$		$+$		$+$	$+$
$\dfrac{(2x - 1)(3 - x)}{(x + 4)}$		$+$		$-$		$+$	$-$

$$x < -4 \qquad\qquad \frac{1}{2} < x < 3$$

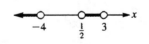

13.
$$|3x + 2| \leq 4$$
$$-4 \leq 3x + 2 \leq 4$$
$$-6 \leq 3x \leq 2$$
$$-2 \leq x \leq \frac{2}{3}$$

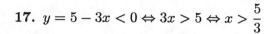

17. $y = 5 - 3x < 0 \Leftrightarrow 3x > 5 \Leftrightarrow x > \dfrac{5}{3}$

 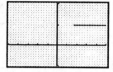

21. $\dfrac{8 - R}{2R + 1} \leq 0$

		$\frac{1}{2}$	8	
$8 - R$	$+$	$+$	$-$	R
$2R + 1$	$-$	$+$	$+$	
$\dfrac{8 - R}{2R + 1}$	$-$	$+$	$-$	

$$R < -\frac{1}{2} \qquad\qquad R \geq 8$$

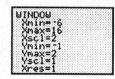 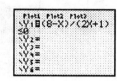 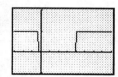

25. $x^3 + x + 1 < 0$; graph $x^3 + x + 1 = y$.

x	-1	0	1	$\cdots$
y	-1	1	3	

$x < -0.68$

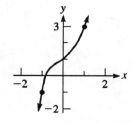

29. $y > 4 - x$. Graph $y = 4 - x$. Use a dashed line to indicate that points on it do not satisfy the inequality. Shade the region above the line.

33. $y > x^2 + 1$. Graph $y = x^2 + 1$. Use a dashed line to indicate that points on it do not satisfy the inequality. Shade the region above the line.

37. $y > x + 1$
     $y < 4 - x^2$

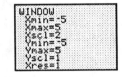  

41. $y < 3x + 5$

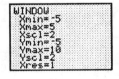 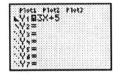 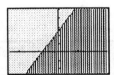

45. $y < 32x - x^4$

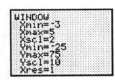 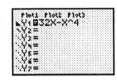 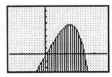

49. $\sqrt{3 - x}$ will be a real number if $3 - x \geq 0$. The critical values are is ± 3. If $x \leq 3$, $3 - x \geq 0$. Thus the values which satisfy the inequalities are $x \leq 3$.

53. $101 + 10.1d > 500$
         $10.1d > 399$
         $d > 39.5 \text{ m}$

57. $p = Ri^2 = 12.0i^2$; $2.50 < 12.0i^2 < 8.00$

$$\frac{2.50}{12.0} < i^2 < \frac{8.00}{12.0}$$

$$\sqrt{\frac{2.50}{12.0}} < i < \sqrt{\frac{8.00}{12.0}}$$

$$0.46 < i < 0.82 \text{ A}$$

61. Let $R =$ minutes research time
$D =$ minutes development time

$R \leq 1200$
$D \leq 1000$

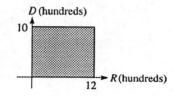

65.

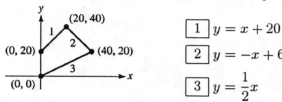

Write the equation of each line:

$\boxed{1}$ $y = x + 20$

$\boxed{2}$ $y = -x + 60$

$\boxed{3}$ $y = \frac{1}{2}x$

The park region is described by the inequalities.

$$x \geq 0$$
$$y \geq 0$$
$$y \leq x + 20$$
$$y \leq -x + 60$$
$$y \geq \frac{1}{2}x$$

VARIATION

18.1 Ratio and Proportion

1. $\dfrac{18 \text{ V}}{3 \text{ V}} = 6$

5. $\dfrac{20 \text{ qt}}{2.5 \text{ gal}} = \dfrac{20 \text{ qt}}{10 \text{ qt}} = 2$

9. $\dfrac{2.6 \text{ W}}{9.6 \text{ W}} = 0.27$

13. $R = \dfrac{s^2}{A_w} = \dfrac{32.0^2}{195} = 5.25$

17. $\dfrac{2540 - 2450}{2540} = \dfrac{90}{2540}$
$= 0.035$
$= 3.5\%$

21.
$\dfrac{6.00}{R_2} = \dfrac{62.5}{15.0}$
$62.5 R_2 = 90.0; \; R_2 = 1.44 \; \Omega$

25. $\dfrac{1.00 \text{ in}^2}{6.45 \text{ cm}^2} = \dfrac{x}{36.3 \text{ cm}^2}; \; x = \dfrac{1.00(36.3)}{6.45} = 5.63 \text{ in}^2$

29. $\dfrac{2.00 \text{ km}}{1.24 \text{ mi}} = \dfrac{x}{5.00 \text{ mi}}; \; 1.24x = 10.0; \; x = 8.06 \text{ km}$

33. $\dfrac{62{,}500}{2.00} = \dfrac{x}{0.75}; \; x = \dfrac{62{,}500(0.75)}{2.00} = 23{,}400 \text{ cm}^3$

37. $\dfrac{17}{595} = \dfrac{500}{x}; \; 17x = 297{,}500; \; x = 17{,}500 \text{ chips}$

18.2 Variation

1. $y = kz$

5. $f = k\sqrt{x}$

9. The area varies directly as the square of the radius.

13. $V = kH^2; \; 2 = k \cdot 64^2$
$k = \dfrac{2}{64^2}; \; V = \dfrac{2H^2}{64^2} = \dfrac{H^2}{2048}$

17. $y = kx; \; 20 = k(8); \; k = 2.5; \; y = 2.5x$
$y = 2.5(10) = 25$

21. $y = \dfrac{kx}{z}; \; 60 = \dfrac{k(4)}{10}; \; k = 150; \; y = \dfrac{150x}{z}$
$y = \dfrac{150(6)}{5} = 180$

25. $V = kV_0; \; 75 = k(160); \; k = \dfrac{15}{32}; \; V = \dfrac{15}{32}V_0$
$V = \dfrac{15}{32}(130) = 61 \text{ ft}^3$

29. $H = kP; \; 1800 = k \cdot 720; \; k = 2.5$

33. (a) a varies inversely with mass.

(b) $a = \dfrac{k}{m}; \; 30 = \dfrac{k}{2}; \; k = 60 \text{ g} \cdot \text{cm/s}^2$
$a = \dfrac{60}{m}$

37. $F = kAv^2$; $19.2 = k(3.72)(31.4)^2$; $k = 5.23 \times 10^{-3}$ lb $\cdot$ s/ft^3
$F = 5.23 \times 10^{-3} Av^2$

41. $R = \dfrac{kl}{A}$; $0.200 = \dfrac{k(225)}{0.0500}$

$k = 4.44 \times 10^{-5}$ $\Omega \cdot$ in^2/ft

$R = \dfrac{4.44 \times 10^{-5} l}{A}$

45. $G = \dfrac{kd^2}{\lambda^2}$; $5.5 \times 10^4 = \dfrac{k(2.9)^2}{(0.030)^2}$; $k = \dfrac{5.5 \times 10^4 (0.030)^2}{(2.9)^2} = 5.9$

$G = \dfrac{5.9d^2}{\lambda^2}$

Chapter 18 Review Exercises

1. $\dfrac{4 \text{ Mg}}{20 \text{ kg}} = \dfrac{4000 \text{ kg}}{20 \text{ kg}} = 200$

5. $\dfrac{28 \text{ kN}}{5000 \text{ N}} = \dfrac{28 \text{ kN}}{5 \text{ kN}} = 5.6$

9. $\dfrac{1.3 \text{ in.}}{20.0 \text{ mi}} = \dfrac{6.0 \text{ in.}}{x}$
$x = 92$ mi

13. $\dfrac{3600}{30 \text{ s}} = \dfrac{x}{300 \text{ s}}$ since 5 min = 300 s
$30x = (360)(300)$
$x = 36{,}000$ characters in 5 min

17. $\dfrac{25.0 \text{ ft}}{2.00 \text{ in.}} = \dfrac{x}{5.75 \text{ in.}}$; $x = 71.9$ ft

21. $y = kx^2$; $27 = k(3^2)$; $k = 3$; $y = 3x^2$

25. $\dfrac{F_1}{F_2} = \dfrac{L_2}{L_1}$; $F_1 = 4.50$ lb, $F_2 = 6.75$ lb, $L_1 = 17.5$ in

$\dfrac{4.50}{6.75} = \dfrac{L_2}{17.5}$

$(6.75)L_2 = (4.50)(17.5)$
$L_2 = 11.7$ in.

29. $p = kA$; $30.0 = k(8.00)$; $k = 3.75 \dfrac{\text{hp}}{\text{in}^2}$.

$p = 3.75A$; $p = 3.75(6.00) = 22.5$ hp

33. $\dfrac{(\log 8000)^2}{(\log 2000)^2} = 1.4$ times longer to sort 8000 numbers.

37. $f = \dfrac{v}{\lambda}$; $v = f\lambda = 90.0 \times 10^6 (3.29) = 299 \times 10^6 = 2.99 \times 10^8$ m/s

41. $d = kv^2$; $52 = k \cdot 32^2$; $d = \dfrac{52}{32^2} \cdot 55^2 = 150$ ft

45. $V = \dfrac{kr^4}{d}$; $V_1 = \dfrac{k(1.25r)^4}{0.98d} = \dfrac{2.44kr^4}{0.98d} = 2.49 \left(\dfrac{kr^4}{d} \right) = 2.49V$

An increase of $V_1 - V = 2.49V - V = 1.49V$ or 149% increase.

49. $L = \dfrac{kt}{d}$; $1200 = \dfrac{k \cdot 30}{8.0}$; $k = 320$ Btu $\cdot$ in/min

$L = \dfrac{320t}{d} = \dfrac{320(90)}{6.0} = 4800$ Btu

53. Let $V_1 = \pi r_1^2 h_1$ be the original volume then the new volume is
$V_2 = \pi r_2^2 h_2 = 0.9\pi r_1^2 h_1$ from which

$\left(\dfrac{r_2}{r_1}\right)^2 \cdot \dfrac{h_2}{h_1} = 0.9$ and since $\dfrac{r_2}{r_1} = \dfrac{h_2}{h_1}$

$\dfrac{r_2^3}{r_1^3} = 0.9$

$r_2 = \sqrt[3]{0.9}\, r_1 = 0.97 r_1$

Reducing the radius and height by 3% will reduce the volume by 10%.

Chapter 19

SEQUENCES AND THE BINOMIAL THEOREM

19.1   Arithmetic Sequences

1. 4, 6, 8, 10, 12

9. $a_{80} = -7 + (80 - 1)4 = 309$

5. $a_8 = 1 + (8 - 1)(3) = 22$

13. $S_{20} = \dfrac{20}{2}(4 + 40) = 440$

17. $45 = 5 + (n - 1)8$
$45 = 8n - 3$
$n = 6$

$S_6 = \dfrac{6}{2}(5 + 45) = 150$

21. $a_{30} = a_1 + (29)(3) = a_1 + 87$

$1875 = \dfrac{30}{2}(a_1 + a_1 + 87)$

$125 = 2a_1 + 87$
$a_1 = 19;\ a_{30} = 106$

25. $a_n = -5k + (n - 1)\left(\dfrac{1}{2}k\right)$

$S_n = \dfrac{n}{2}\left[-5k + \left(-5k + (n - 1)\left(\dfrac{1}{2}k\right)\right)\right]$

$\dfrac{23}{2}k = \dfrac{n}{2}\left(-5k - 5k + \dfrac{1}{2}kn - \dfrac{1}{2}k\right)$

$23k = n\left(-\dfrac{21}{2}k + \dfrac{1}{2}kn\right)$

$46k = n(-21k + kn)$
$46k = -21kn + kn^2$
$kn^2 - 21kn - 46k = 0$
$n^2 - 21n - 46 = 0$
$(n + 2)(n - 23) = 0$
$n = -2$ (not valid)

$n = 23;\ a_{23} = -5k + (22)\left(\dfrac{1}{2}k\right) = -5k + 11k = 6k$

29. $d = \dfrac{72 - 56}{10 - 6} = 4;\ a_6 = 56 = a_1 + (5)(4);$

$a_1 = 56 - 20 = 36$
$S_{10} = 5(36 + 72) = 540$

33. $a_1 = 1,\ a_n = 100,\ n = 100;\ S_{10} = \dfrac{100}{2}(1 + 100) = 50(101) = 5050$

37. Area losses for next 8 years are: 500 m², 600 m², 700 m², ..., $a_8 = 500 + 7(100) = 1200$ m²

$S_8 = \dfrac{8}{2}(500 + 1200) = 6800$ m²

Area in 8 years = $9500 - 6800 = 2700$ m²

41. $a_1 = 1800$, $d = -150$, $a_n = 0$

$0 = 1800 + (n-1)(-150) = 1800 - 150n + 150$

$150n = 1800 + 150 = 1950$; $n = 13$ (12 more years)

$S_{13} = \dfrac{13}{2}(1800 + 0) = \$11,700$, the sum of all depreciations, which is the cost of the car

45. $S_n = \dfrac{n}{2}(a_1 + a_n) = \dfrac{n}{2}[a_1 + (a_1 + (n-1))d] = \dfrac{n}{2}[2a_1 + (n-1)d]$

19.2   Geometric Sequences

1. $45\left(\dfrac{1}{3}\right)^{1-1}$, $45\left(\dfrac{1}{3}\right)^{2-1}$, $45\left(\dfrac{1}{3}\right)^{3-1}$, $45\left(\dfrac{1}{3}\right)^{4-1}$, $45\left(\dfrac{1}{3}\right)^{5-1}$ is $45, 15, 5, \dfrac{5}{3}, \dfrac{5}{9}$

5. $r = 1 \div \dfrac{1}{2} = 2$, $a = \dfrac{1}{2}$, $n = 6$

$a_6 = \dfrac{1}{2}(2)^{6-1} = \dfrac{1}{2}(32) = 16$

9. $a_{10} = -27\left(-\dfrac{1}{3}\right)^{10-1} = -3^3\left(-\dfrac{1}{3^9}\right) = \dfrac{1}{729}$

13. $S_n = \dfrac{\frac{1}{8}(1 - 4^5)}{1 - 4} = \dfrac{-1023}{8(-3)} = \dfrac{341}{8}$

17. $a_6 = \left(\dfrac{1}{16}\right)(4)^{6-1} = \left(\dfrac{1}{16}\right)(4)^5 = 64$

$S_6 = \dfrac{\frac{1}{16}(1 - 4^6)}{1 - 4} = \dfrac{\frac{1}{16}(1 - 4096)}{-3} = \dfrac{4095}{48} = \dfrac{1365}{16}$

21. $27 = a_1 r^{4-1}$; $a_1 = \dfrac{27}{r^3}$

$40 = a_1\dfrac{(1 - r^4)}{1 - r}$

$= a_1\dfrac{(1 + r^2)(1 + r)(1 - r)}{1 - r}$

$= a_1(1 + r^2)(1 + r)$

Substitute a from first equation in second equation:

$40 = \dfrac{27}{r^3}(1 + r^2)(1 + r)$; $40r^3 = 27 + 27r + 27r^2 + 27r^3$; $13r^3 - 27r^2 - 27r - 27 = 0$

Using synthetic division, 3 gives a remainder of zero. Therefore, $r = 3$. $27 = a_1(3^{4-1})$; $27a_1 = 27$; $a_1 = 1$

25. $3, 3^{x+1}, 3^{2x+1}, \cdots$ is a GS, since

$\dfrac{3^{x+1}}{3} = 3^x$

$\dfrac{3^{2x+1}}{3^{x+1}} = 3^x$.

$a_1 = 3$, $r = 3^x$

$a_{20} = 3 \cdot (3^x)^{20-1} = 3^{19x+1}$

29. $r = 1 - 0.125 = 0.875$, $a_1 = 3.27$ mA, $n = 9.2$

$a_{9.2} = 3.27(0.875)^{8.2} = 1.09$ mA

33. $a_1 = 80.8$ cm, $n = 12$, $r = 0.85$

$$S_{12} = \frac{80.8(1 - 0.85^{12})}{1 - 0.85} = \frac{80.8(1 - 0.142)}{0.15}$$

$$= \frac{80.8(0.858)}{0.15} = 462 \text{ cm}$$

37. $a_1 = 0.01$ (dollars); $r = 2$; $n = 27$ ($a_1 + 26$ more)

$$a_{27} = 0.01(2^{27-1}) = 0.01(2^{26}) = \$671,088.64$$

19.3   Infinite Geometric Series

1. $a_1 = 4$, $r = \dfrac{1}{2}$, $S = \dfrac{4}{1 - \frac{1}{2}} = 8$

5. $a_1 = 20$, $r = -\dfrac{1}{20}$

$$S = \frac{20}{1 + \frac{1}{20}} = \frac{20}{\frac{21}{20}} = \frac{400}{21}$$

9. $a_1 = 1$, $r = \dfrac{1}{10,000}$

$$S = \frac{1}{1 - \frac{1}{10,000}} = \frac{1}{\frac{9999}{10,000}} = \frac{10,000}{9999}$$

13. $0.333\ 333\ldots = 0.3 + 0.03 + 0.003 + \cdots$

$a_1 = 0.3$, $r = 0.1$

$$S = \frac{0.3}{1 - 0.1} = \frac{0.3}{0.9} = \frac{1}{3}$$

17. $0.181\ 818\ldots = 0.18 + 0.0018 + 0.000\ 018 + \cdots$

$a_1 = 0.18$, $r = 0.01$

$$S = \frac{0.18}{1 - 0.01} = \frac{2}{11}$$

21. $0.366\ 66\ldots = 0.3 + 0.066\ 66\ldots$

For the GS $0.066\ldots$, $a = 0.06$, $r = 0.1$

$$S = \frac{0.06}{1 - 0.1} = \frac{0.06}{0.9} = \frac{1}{15}$$

Therefore, $0.366\ 66\ldots = \dfrac{3}{10} + \dfrac{1}{15} = \dfrac{11}{30}$

25. $r = 0.92$, $a_1 = 28.0$ gal

$$S = \frac{28.0}{1 - 0.92} = \frac{28.0}{0.08} = 350 \text{ gal}$$

19.4   The Binomial Theorem

1. $(t + 1)^3 = t^3 + 3t^2(1) + \dfrac{3(2)}{2}t(1)^2 + \dfrac{3(2)(1)(1)}{6} = t^3 + 3t^2 + 3t + 1$

5. $(2 + 0.1)^5 = 2.1^5 = 40.84101$ or

$(2 + 0.1)^5 = 2^5 + 5(2)^4(0.1) + 10(2)^3(0.1)^2 + 10(2)^2(0.1)^3 + 5(2)^1(0.1)^4 + 0.1^5$

$\qquad = 40.84101$

9. From Pascal's triangle, the coefficients for $n = 4$ are 1, 4, 6, 4, 1.

$(5x - 3)^4 = [5x + (-3)]^4$

$\qquad = 1(5x)^4 + 4(5x)^3(-3) + 6(5x)^2(-3)^2 + 4(5x)(-3)^3 + (-3)^4$

$\qquad = 625x^4 - 1500x^3 + 1350x^2 - 540x + 81$

13. $(x + 2)^{10} = x^{10} + 10x^9(2) + \dfrac{(10)(9)}{2}x^8(2^2) + \dfrac{(10)(9)(8)}{6}x^7(2)^3 + \cdots = x^{10} + 20x^9 + 180x^8 + 960x^7 + \cdots$

17. $(x^{1/2} - y)^{12} = (x^{1/2})^{12} - 12(x^{1/2})^{11}y + \dfrac{12 \cdot 11}{2!}(x^{1/2})^{10}y^2 - \dfrac{12 \cdot 11 \cdot 10}{3!}(x^{1/2})^9 y^3 + \cdots$

$\qquad = x^6 - 12x^{11/2}y + 66x^5y^2 - 220x^{9/2}y^3 + \cdots$

21. $(1+x)^8 = 1 + 8x + \dfrac{8(7)}{2}x^2 + \dfrac{8(7)(6)}{6}x^3 + \cdots = 1 + 8x + 28x^2 + 56x^3 + \cdots$

25. $\sqrt{1+x} = (1+x)^{1/2} = 1 + \dfrac{1}{2}x + \dfrac{\frac{1}{2}\left(-\frac{1}{2}\right)}{2}x^2 + \dfrac{\frac{1}{2}\left(-\frac{1}{2}\right)\left(-\frac{3}{2}\right)}{6}x^3 + \cdots$

$\qquad = 1 + \dfrac{1}{2}x - \dfrac{1}{8}x^2 + \dfrac{1}{16}x^3 + \cdots$

29. (a) $17! + 4! = 3.557 \times 10^{14}$ (b) $21! = 5.109 \times 10^{19}$

  (c) $17! \times 4! = 8.536 \times 10^{15}$ (d) $68! = 2.480 \times 10^{96}$

33. The term involving b^5 will be the sixth term. $r = 5$, $n = 8$

The sixth term is $\dfrac{8(7)(6)(5)(4)}{5(4)(3)(2)}a^3b^5 = 56a^3b^5$.

37. $V = A(1-r)^5$; expand $(1-r)^5$

$1^5 + 5(1)^4(-r) + \dfrac{5(4)}{2}(1)^3(-r)^2 + \dfrac{5(4)(3)}{6}(1)^2(-r)^3 + \dfrac{5(4)(3)(2)}{24}(1)(-r)^4 + (-r)^5$

$= 1 - 5r + 10r^2 - 10r^3 + 5r^4 - r^5$

$A(1-r)^5 = A(1 - 5r + 10r^2 - 10r^3 + 5r^4 - r^5)$

Chapter 19 Review Exercises

1. $d = 5$; $a_n = 1 + (17-1)5 = 1 + 80 = 81$

5. $d = \dfrac{7}{2} - 8 = -\dfrac{9}{2}$; $a_n = 8 + (16-1)\left(-\dfrac{9}{2}\right) = 8 - \dfrac{135}{2} = -\dfrac{119}{2}$

9. $s_n = \dfrac{15}{2}(-4 + 17) = \dfrac{15}{2}(13) = \dfrac{195}{2}$

13. $a_n = 17 + (9-1)(-2) = 17 - 16 = 1$

$s_n = \dfrac{9}{2}(1 + 17) = 81$

17. $s_n = \dfrac{n}{2}(a + a_n)$; $a_1 = 8$, $a_n = -2.5$, $s_n = 22$

$22 = \dfrac{n}{2}(8 - 2.5)$ $\qquad\qquad$ $a_n = a_1 + (n-1)d$

$44 = n(8 - 2.5) = 5.5n$ $\qquad$ $-2.5 = 8 + (8-1)d = 8 + 7d$

$n = \dfrac{44}{5.5} = 8$ $\qquad\qquad\qquad$ $d = \dfrac{-10.5}{7} = -1.5$

21. $s_{12} = \dfrac{12}{2}(-1 + 32) = 6(31) = 186$

25. $r = \dfrac{6}{9} = \dfrac{2}{3}$; $s = \dfrac{a_1}{1 - r} = \dfrac{9}{1 - \frac{2}{3}} = \dfrac{9}{\frac{1}{3}} = 27$

29. $0.030303\ldots = 0.03 + 0.0003 + 0.000003\ldots$

$a_a = 0.03$; $r = 0.01$

$s = \dfrac{0.03}{1 - 0.01} = \dfrac{0.03}{0.99} = \dfrac{3}{99} = \dfrac{1}{33}$

33. $(x - 2)^4 = [x + (-2)]^4$

$$= x^4 + 4x^3(-2) + \frac{4(3)x^2(-2)^2}{2} + \frac{4(3)(2)}{3(2)}x(-2)^3 + (-2)^4$$

$$= x^4 - 8x^3 + 24x^2 - 32x + 16$$

37. $(a + 2b^2)^{10} = a^{10} + 10a^{10-1}(2b^2) + \dfrac{10(10-1)}{2!}a^{10-2}(2b^2)^2 + \dfrac{10(10-1)(10-2)}{3!}a^{10-3}(2b^2)^3 + \cdots$

$$= a^{10} + 10a^9(2b^2) + 45a^8(4b^4) + 120a^7(8b^6) + \cdots$$

$$= a^{10} + 20a^9b^2 + 180a^8b^4 + 960a^7b^6 + \cdots$$

41. $(1 + x)^{12} = 1 + 12x + \dfrac{12(12-1)}{2!}x^2 + \dfrac{12(12-1)(12-2)}{3!}x^3 + \cdots$

$$= 1 + 12x + 66x^2 + 220x^3 + \cdots$$

45. $[1 + (-a^2)]^{1/2} = 1 + \dfrac{1}{2}(-a^2) + \dfrac{\left(\frac{1}{2}\right)\left(-\frac{1}{2}\right)(-a^2)^2}{2} + \dfrac{\frac{1}{2}\left(-\frac{1}{2}\right)\left(-\frac{3}{2}\right)(-a^2)}{3(2)}$

$$= 1 - \frac{1}{2}a^2 - \frac{1}{8}a^4 - \frac{1}{16}a^6 - \cdots$$

49. $a = 2$, $d = 2$, $n = 1000$

$a_n = 2 + (1000 - 1)2 = 2 + 999(2) = 2000$

$s = \dfrac{1000}{2}(2 + 2000) = 1{,}001{,}000$

53. n = lens thickness in mm; $a_n = a_1(r^{n-1})$

$r = (1 - 0.12) = 0.88$ $0.20 = 1.00(0.88^{n-1}) = 0.88^{n-1}$

$a_1 = 100\% = 1.00$ $\log 0.20 = (-1)\ \log 0.88$

$a_n = 20\% = 0.20$ $n - 1 = \dfrac{\log 0.20}{\log 0.88}$

$n - 1 = 12.6$; $n = 13.6$

57. $V = 8600(0.99)^{60} = 4705.547125$

$V = \$4700$, value after 60 months.

61. The value of each investment is $a_n = 1000(1.0375)^n$; (n = twice the number of invested years.) The total value is the sum of each investment value. Investment values (20 terms) are as follows:

$$1000(1.0375)^{40},\ 1000(1.0375)^{38},\ 1000(1.0375)^{36},\ \ldots\ 1000(1.0375)^2$$

These terms are changing by a factor of 1.0375^{-2}

$$s = \frac{\left[1000(1.0375)^{40}\right]\left[1 - (1.0375^{-2})^{20}\right]}{1 - (1.0375)^{-2}} = \frac{\left[1000(1.0375)^{40}\right]\left[1 - 1.0375^{-40}\right]}{\left[1 - 1.0375^{-2}\right]}$$

$$\approx \frac{[4360.38]\,[0.7706621247]}{0.0709827261} = \$47{,}340.81$$

Value increases as additional decimal places are used.

65. Let $x = \dfrac{a-1}{2}m^2$ and $y = \dfrac{a}{a-1}$.

$$(1+x)^y = 1 + yx + \frac{y(y-1)}{2}x^2 \ldots \text{(3 terms)}$$

$$= 1 + \left(\frac{a}{a-1}\right)\left(\frac{a-1}{2}m^2\right) + \frac{\left(\frac{a}{a-1}\right)\left(\frac{a}{a-1}-1\right)}{2}\left(\frac{a-1}{2}m^2\right)^2$$

$$= 1 + \frac{a}{2}m^2 + \frac{\left(\frac{a}{a-1}\right)\left(\frac{1}{a-1}\right)}{2}\left(\frac{(a-1)^2}{2^2}m^4\right)$$

$$= 1 + \frac{a}{2}m^2 + \frac{a}{2(a-1)^2}\left(\frac{(a-1)^2}{2^2}m^4\right)$$

$$= 1 + \frac{a}{2}m^2 + \frac{a}{2^3}m^4 = 1 + \frac{1}{2}am^2 + \frac{1}{8}am^4$$

69. Let $a_1 = 1000$ units. If 75% are killed, 25% remain after the first application.

$$r = \frac{a_2}{a_1} = \frac{250}{1000} = 0.25$$

If 99.9% are destroyed, 0.1% remain. $0.001 \times 1000 = 1$ insect remains.

$$1 = 1000(0.25)^n$$
$$0.001 = 0.25^n$$
$$\log 0.001 = \log 0.25^n$$
$$\log 0.001 = n \log 0.25$$

$$n = \frac{\log 0.001}{\log 0.25} = 5 \text{ applications}$$

73. Let $A =$ initial deposit, $t =$ time of a compounding period.
$V = A + Art = A(1 + rt)$, after one compounding period
$V = A(1 + rt) + A(1 + rt)rt = A(1 + rt)^2$, after two compounding periods

$$\vdots$$

$V = A(1 + rt)^n$, at the end of one year. For $A = \$1000$ and $r = 0.1 = 10\%$
$V = 1000(1 + 0.1t)^n$.

As n, the number of compounding periods, increases t, the length of a compounding period decreases. The product $nt = 1$. For example, if the compounding is done monthly $n = 12$ and $t = \frac{1}{12}$, so that $nt = 12 \cdot \frac{1}{2} = 1$. Thus, $V = 1000\left(1 + 0.1 \cdot \frac{1}{n}\right)^n$ which will increase as n increases-the more compounding periods the more interest. Write V as

$$V = 1000\left(1 + \frac{0.1}{n}\right)^{\frac{0.1}{0.1}\cdot\frac{n}{1}} = 1000\left[\left(1 + \frac{0.1}{n}\right)^{\frac{1}{0.1}\cdot\frac{n}{1}}\right]^{0.1}.$$

As n increases $\left(1 + \dfrac{0.1}{n}\right)^{\frac{1}{\frac{0.1}{n}}}$ approaches e. Hence, the maximum value is $V_{\text{max}} = 1000 \cdot e^{0.1} = \$1105.17$ as compared with $V = 1000\left(1 + 0.1 \cdot \dfrac{1}{12}\right)^{12} = \$1104.71$ for monthly compounding.

ADDITIONAL TOPICS IN TRIGONOMETRY

20.1 Fundamental Trigonometric Identities

1. Verify $\tan\theta = \dfrac{1}{\cot\theta}$ for $\theta = 56°$

$\tan 56° = 1.483; \cot 56° = 0.6745$

$\dfrac{1}{0.6745} = 1.483 = \tan 56°$

5. $\dfrac{\cot\theta}{\cos\theta} = \cot\theta \times \dfrac{1}{\cos\theta}$

$\quad = \dfrac{\cos\theta}{\sin\theta} \times \dfrac{1}{\cos\theta}$

$\quad = \dfrac{1}{\sin\theta} = \csc\theta$

9. $\sin y \cot y = \dfrac{\sin y}{1} \times \dfrac{\cos y}{\sin y}$

$\quad = \cos y$

13. $\csc^2 x (1 - \cos^2 x) = \dfrac{1}{\sin^2 x} \times \dfrac{\sin^2 x}{1}$

$\quad = \dfrac{\sin^2 x}{\sin^2 x} = 1$

17. $\tan y (\cot y + \tan y) = \tan y \cot y + \tan^2 y = \tan y \times \left(\dfrac{1}{\tan y}\right) + \tan^2 y = 1 + \tan^2 y = \sec^2 y$

21. $\cos\theta \cot\theta + \sin\theta = \cos\theta \times \dfrac{\cos\theta}{\sin\theta} + \sin\theta = \dfrac{\cos^2\theta}{\sin\theta} + \sin\theta = \dfrac{\cos^2\theta + \sin^2\theta}{\sin\theta} = \dfrac{1}{\sin\theta} = \csc\theta$

25. $\tan x + \cot x = \dfrac{\sin x}{\cos x} + \dfrac{\cos x}{\sin x}$

$\quad = \dfrac{\sin^2 x + \cos^2 x}{\cos x \sin x}$

$\quad = \dfrac{1}{\cos x \sin x}$

$\quad = \sec x \csc x$

29. $\dfrac{\sin x}{1 - \cos x} = \dfrac{\sin x (1 + \cos x)}{(1 - \cos x)(1 + \cos x)}$

$\quad = \dfrac{\sin x (1 + \cos x)}{1 - \cos^2 x} = \dfrac{\sin x (1 + \cos x)}{\sin^2 x}$

$\quad = \dfrac{1 + \cos x}{\sin x} = \dfrac{1}{\sin x} + \dfrac{\cos x}{\sin x}$

$\quad = \csc x + \cot x$

33. $\dfrac{\sec\theta}{\cos\theta} - \dfrac{\tan\theta}{\cot\theta}$

$\quad = \dfrac{1}{\cos\theta}\left(\dfrac{1}{\cos\theta}\right) - \dfrac{\sin\theta}{\cos\theta}\left(\dfrac{\sin\theta}{\cos\theta}\right)$

$\quad = \dfrac{1}{\cos^2\theta} - \dfrac{\sin^2\theta}{\cos^2\theta}$

$\quad = \dfrac{\cos^2\theta}{\cos^2\theta}$

$\quad = 1$

37. $\dfrac{1}{2}\sin\pi t \left(\dfrac{\sin\pi t}{1 - \cos\pi t} + \dfrac{1 - \cos\pi t}{\sin\pi t}\right)$

$\quad = \dfrac{1}{2}\sin\pi t \left(\dfrac{\sin^2\pi t + (1 - \cos\pi t)^2}{\sin\pi t(1 - \cos\pi t)}\right)$

$\quad = \dfrac{1}{2}\left(\dfrac{\sin^2\pi t + 1 - 2\cos\pi t + \cos^2\pi t}{1 - \cos\pi t}\right)$

$\quad = \dfrac{1}{2}\left(\dfrac{(\sin^2\pi t + \cos^2\pi t) + 1 - 2\cos\pi t}{1 - \cos\pi t}\right)$

$\quad = \dfrac{1}{2}\left(\dfrac{1 + 1 - 2\cos\pi t}{1 - \cos\pi t}\right)$

$\quad = \dfrac{2 - 2\cos\pi t}{2 - 2\cos\pi t} = 1$

41. $\dfrac{\tan x \csc^2 x}{1 + \tan^2 x} = \dfrac{\dfrac{\sin x}{\cos x} \cdot \dfrac{1}{\sin^2 x}}{1 + \dfrac{\sin^2 x}{\cos^2 x}}$

$= \dfrac{\dfrac{1}{\sin x \cos x}}{\dfrac{\cos^2 x + \sin^2 x}{\cos^2 x}}$

$= \dfrac{1}{\sin x \cos x} \cdot \dfrac{\cos^2 x}{1}$

$= \dfrac{\cos x}{\sin x} = \cot x$

45. $\dfrac{\tan x + \cot x}{\csc x}$

$= \dfrac{\dfrac{\sin x}{\cos x} + \dfrac{\cos x}{\sin x}}{\dfrac{1}{\sin x}} \cdot \dfrac{\sin x \cos x}{\sin x \cos x}$

$= \dfrac{\sin^2 x + \cos^2 x}{\cos x}$

$= \dfrac{1}{\cos x} = \sec x$

49.

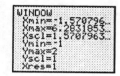 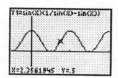 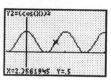

53.

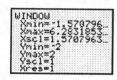 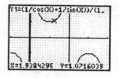 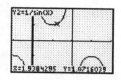

57. $\cos \theta = \cos A \cos B \cos C + \sin A \sin B$;
$\theta = 90°, \ \cos \theta = 0$

$0 = \cos A \cos B \cos C + \sin A \sin B$

$\dfrac{\cos A \cos B}{\cos A \cos B} \cos C = \dfrac{-\sin A \sin B}{\cos A \cos B}$

$\cos C = -\tan A \tan B$

61. $\sin^2 x(1 - \sec^2 x) + \cos^2 x(1 + \sec^4 x)$
$= \sin^2 x - \sin^2 x \sec^2 x + \cos^2 x + \cos^2 x \sec^4 x$

$= \sin^2 x - \dfrac{\sin^2 x}{\cos^2 x} + \cos^2 x + \dfrac{\cos^2 x}{\cos^4 x}$

$= \sin^2 x - \tan^2 x + \cos^2 x + \sec^2 x$
$= 1 - \tan^2 x + \sec^2 x$
$= 1 - (\sec^2 x - 1) + \sec^2 x$
$= 1 - \sec^2 x + 1 + \sec^2 x = 2$

65. $x = \cos \theta; \ \sqrt{1 - x^2} = \sqrt{1 - \cos^2 \theta} = \sqrt{\sin^2 \theta} = \sin \theta$

20.2   The Sum and Difference Formulas

1. Given: $105° = 60° + 45°$

$\sin(\alpha + \beta) = \sin \alpha \cos \beta + \cos \alpha \sin \beta$
$\sin 105° = \sin(60° + 45°)$
$\quad = \sin 60° \cos 45° + \cos 60° \sin 45°$

$= \dfrac{\sqrt{3}}{2} \times \dfrac{\sqrt{2}}{2} + \dfrac{1}{2} \times \dfrac{\sqrt{2}}{2}$

$= \dfrac{\sqrt{6}}{4} + \dfrac{\sqrt{2}}{4} = \dfrac{\sqrt{6} + \sqrt{2}}{4} = 0.9659$

5. $\sin \alpha = \dfrac{4}{5}(Q\ I);\ \cos \alpha = 1 - \sin^2 \alpha;\ \cos^2 \alpha = 1 - \dfrac{16}{25}$

$\cos^2 \alpha = \dfrac{9}{25}$

$\cos \alpha = \dfrac{3}{5}$

$\cos \beta = -\dfrac{12}{13}\ (Q\ II);\ \sin^2 \beta = 1 - \cos^2 \beta$

$\sin^2 \beta = 1 - \dfrac{144}{169};\ \sin^2 \beta = \dfrac{25}{169}$

$\sin \beta = \dfrac{5}{13}(\sin \beta$ is $+$ in $Q\ II)$

$\sin(\alpha + \beta) = \sin \alpha \cos \beta + \cos \alpha \sin \beta$

$\qquad = \dfrac{4}{5}\left(-\dfrac{12}{13}\right) + \dfrac{3}{5}\left(\dfrac{5}{13}\right)$

$\qquad = -\dfrac{48}{65} + \dfrac{15}{65} = -\dfrac{33}{65}$

9. $\sin \alpha \cos \beta + \sin \beta \cos \alpha = \sin(\alpha + \beta)$
$\sin x \cos 2x + \sin 2x \cos x = \sin(x + 2x)$
$\qquad\qquad\qquad\qquad = \sin 3x$

13. $\dfrac{\tan(x - y) + \tan y}{1 - \tan(x - y)\tan y} = \tan(x - y + y)$

$\qquad\qquad\qquad\qquad\ = \tan x$

17. $\sin 122° \cos 32° - \cos 122° \sin 32°$ is of the form
$\sin \alpha \cos \beta - \cos \alpha \sin \beta$, where $\alpha = 122°$ and $\beta = 32°$
$\sin \alpha \cos \beta - \cos \alpha \sin \beta = \sin(\alpha - \beta)$ so
$\sin 122° \cos 32° - \cos 122° \sin 32° = \sin(122° - 32°) = \sin 90° = 1$

21. $\sin(180° - x) = \sin 180° \cos(-x) + \sin(-x)\cos 180°$
$\qquad\qquad\ = (0)\cos x + (-\sin x)(-1) = \sin x$

25. $\tan(180° + x) = \dfrac{\tan 180° + \tan x}{1 - \tan 180° \tan x} = \tan x$

29. $\sin(x + y)\sin(x - y)$
$\qquad = (\sin x \cos y + \cos x \sin y)(\sin x \cos y - \cos x \sin y)$
$\qquad = \sin^2 x \cos^2 y - \cos^2 x \sin^2 y$
$\qquad = \sin^2 x (1 - \sin^2 y) - (1 - \sin^2 x)(\sin^2 y)$
$\qquad = \sin^2 x - \sin^2 x \sin^2 y - \sin^2 y + \sin^2 x \sin^2 y$
$\qquad = \sin^2 x - \sin^2 y$

33.

 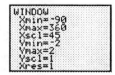 

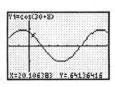 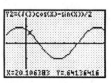

37. $\tan(\alpha \pm \beta)$

$$= \frac{\sin(\alpha \pm \beta)}{\cos(\alpha \pm \beta)} = \frac{\sin\alpha\cos\beta \pm \cos\alpha\sin\beta}{\cos\alpha\cos\beta \mp \sin\alpha\sin\beta}$$

(divide numerator and denominator by $\cos\alpha\cos\beta$)

$$= \frac{\dfrac{\sin\alpha\cos\beta}{\cos\alpha\cos\beta} \pm \dfrac{\cos\alpha\sin\beta}{\cos\alpha\cos\beta}}{\dfrac{\cos\alpha\cos\beta}{\cos\alpha\cos\beta} \mp \dfrac{\sin\alpha\sin\beta}{\cos\alpha\cos\beta}} = \frac{\dfrac{\sin\alpha}{\cos\alpha} \pm \dfrac{\sin\beta}{\cos\beta}}{1 \mp \dfrac{\sin\alpha}{\cos\alpha} \times \dfrac{\sin\beta}{\cos\beta}}$$

$$= \frac{\tan\alpha \pm \tan\beta}{1 \mp \tan\alpha\tan\beta}$$

41. $\alpha + \beta = x; \ \alpha - \beta = y; \ \alpha = \dfrac{1}{2}(x+y); \ \beta = \dfrac{1}{2}(x-y)$

$\sin x + \sin y$
$$= \sin(\alpha + \beta) + \sin(\alpha - \beta)$$
$$= 2\left[\frac{1}{2}\sin(\alpha + \beta) + \frac{1}{2}\sin(\alpha - \beta)\right]$$
$$= 2\sin\alpha\cos\beta = 2\sin\frac{1}{2}(x+y)\cos\frac{1}{2}(x-y)$$

45. $i_0\sin(\omega t + \alpha) = i_0[\sin\omega t\cos\alpha + \sin\alpha\cos\omega t]$
$$= i_0\cos\alpha\sin\omega t + i_0\sin\alpha\cos\omega t$$
$$= i_1\sin\omega t + i_2\cos\omega t$$

20.3   Double-Angle Formulas

1. $60° = 2(30°); \ \sin 2\alpha = 2\sin\alpha\cos\alpha$

$\sin 2(30°) = 2\sin 30°\cos 30°$
$$= 2\left(\frac{1}{2}\right)\left(\frac{\sqrt{3}}{2}\right)$$
$$= \frac{\sqrt{3}}{2} = \frac{1}{2}\sqrt{3}$$

5. $\sin 258° = -0.9781476$

$\sin 258° = \sin 2(129°)$
$$= 2\sin 129°\cos 129°$$
$$= -0.9781476$$

9. Given: $\cos x = \dfrac{4}{5}$ (Q I)

$$\sin^2 x = 1 - \cos^2 x = 1 - \left(\frac{4}{5}\right)^2$$
$$= 1 - \frac{16}{25} = \frac{9}{25}$$
$$\sin x = \frac{3}{5}$$
$$\sin 2x = 2\sin x\cos x = 2\left(\frac{3}{5}\right)\left(\frac{4}{5}\right) = \frac{24}{25}$$

13. $4\sin 4x\cos 4x = 2(2\sin 4x\cos 4x)$
$$= 2\sin 2(4x)$$
$$= 2\sin 8x$$

17. $2\cos^2\dfrac{1}{2}x - 1 = \cos 2\left(\dfrac{1}{2}x\right) = \cos x$

21. $\cos^2 \alpha - \sin^2 \alpha = \cos^2 \alpha - (1 - \cos^2 \alpha)$
$$= \cos^2 \alpha - 1 + \cos^2 \alpha$$
$$= 2\cos^2 \alpha - 1$$

25. $\dfrac{\sin 4\theta}{\sin 2\theta} = \dfrac{2\sin 2\theta \cos 2\theta}{\sin 2\theta} = 2\cos 2\theta$

29. $1 - \cos 2\theta = 1 - (1 - 2\sin^2 \theta)$
$$= 2\sin^2 \theta = \frac{2}{\csc^2 \theta}$$
$$= \frac{2}{1 + \cot^2 \theta}$$

33. $\tan 2\theta = \tan(\theta + \theta) = \dfrac{\tan \theta + \tan \theta}{1 - \tan \theta \cdot \tan \theta}$
$$= \frac{2\tan \theta}{1 - \tan^2 \theta} \cdot \frac{\dfrac{1}{\tan \theta}}{\dfrac{1}{\tan \theta}}$$
$$= \frac{2}{\cot \theta - \tan \theta}$$

37. $\sin 3x = \sin(2x + x) = \sin 2x \cos x + \cos 2x \sin x$
$$= (2\sin x \cos x)(\cos x) + (\cos^2 x - \sin^2 x)(\sin x)$$
$$= 2\sin x \cos^2 x + \sin x \cos^2 x - \sin^3 x$$
$$= 3\sin x \cos^2 x - \sin^3 x$$

41. $R = vt \cos \alpha; \quad t = \dfrac{(2v \sin \alpha)}{g}$
$$R = v\left(\frac{2v \sin \alpha}{g}\right)\cos \alpha = \frac{v^2(2\sin \alpha \cos \alpha)}{g}$$
$$= \frac{v^2 \sin 2\alpha}{g}$$

20.4   Half-Angle Formulas

1. $\cos 15° = \cos \dfrac{30°}{2} = \sqrt{\dfrac{1 + \cos 30°}{2}}$
$$= \sqrt{\frac{1 + 0.8660}{2}} = 0.9659$$

5. $\sqrt{\dfrac{1 - \cos 236°}{2}} = \sin \dfrac{1}{2}(236°) = \sin 118°$
$$= 0.8829476$$

9. $\sin \dfrac{\alpha}{2} = \sqrt{\dfrac{1 - \cos \alpha}{2}}$
$$\sqrt{\frac{1 - \cos 6x}{2}} = \sin \frac{6x}{2}$$
$$= \sin 3x$$

13. $\sin \dfrac{\alpha}{2} = \sqrt{\dfrac{1 - \cos \alpha}{2}} = \sqrt{\dfrac{1 - \frac{12}{13}}{2}}$
$$= \sqrt{\frac{1}{13} \times \frac{1}{2}} = \sqrt{\frac{1}{26}}$$
$$= \sqrt{\frac{1}{26} \times \frac{26}{26}} = \frac{1}{26}\sqrt{26}$$

17. $\csc \dfrac{\alpha}{2} = \dfrac{1}{\sin \dfrac{\alpha}{2}} = \dfrac{1}{\pm\sqrt{\dfrac{1 - \cos \alpha}{2}}}$

$= \dfrac{\sqrt{2}}{\pm\sqrt{1 - \cos \alpha}}$

$= \pm\sqrt{\dfrac{2}{1 - \cos \alpha}}$

21. $\dfrac{1 - \cos \alpha}{2 \sin \dfrac{\alpha}{2}} = \dfrac{1 - \cos \alpha}{2\sqrt{\dfrac{1 - \cos \alpha}{2}}} \times \dfrac{\sqrt{\dfrac{1 - \cos \alpha}{2}}}{\sqrt{\dfrac{1 - \cos \alpha}{2}}}$

$= \dfrac{(1 - \cos \alpha)\sqrt{\dfrac{1 - \cos \alpha}{2}}}{2\left(\dfrac{1 - \cos \alpha}{2}\right)}$

$= \sqrt{\dfrac{1 - \cos \alpha}{2}} = \sin \dfrac{\alpha}{2}$

25. $\cos \dfrac{\theta}{2} = \sqrt{\dfrac{(1 + \cos \theta)(1 - \cos \theta)}{2(1 - \cos \theta)}} = \sqrt{\dfrac{1 - \cos^2 \theta}{2(1 - \cos \theta)}} = \sqrt{\dfrac{\sin^2 \theta}{2(1 - \cos \theta)}} = \dfrac{\sin \theta}{\sqrt{\dfrac{4(1 - \cos \theta)}{2}}}$

$= \dfrac{\sin \theta}{2\sqrt{\dfrac{1 - \cos \theta}{2}}} = \dfrac{\sin \theta}{2 \sin \dfrac{\theta}{2}}$

29. $\sin^2 \omega t = \sin^2 \left[\left(\dfrac{1}{2}\right)(2\omega t)\right] = \left(\sqrt{\dfrac{1 - \cos 2\omega t}{2}}\right)^2 = \dfrac{1 - \cos 2\omega t}{2}$

20.5   Solving Trigonometric Equations

1. $\sin x - 1 = 0,\ 0 \le x < 2\pi;\ \sin x = 1;\ x = \dfrac{\pi}{2}$

5. $4\cos^2 x - 1 = 0;\ 0 \le x < 2\pi$

$4\cos^2 x = 1;\ \cos^2 x = \dfrac{1}{4};\ \cos x = \pm\dfrac{1}{2}$

$x = \dfrac{\pi}{3},\ \dfrac{2\pi}{3},\ \dfrac{4\pi}{3},\ \dfrac{5\pi}{3}$

9. $\sin 2x \sin x + \cos x = 0,\ 0 \le x < 2\pi$

$(2\ \sin x \cos x)(\sin x) + \cos x = 0$

$2\sin^2 x \cos x + \cos x = 0;\ \cos x(2\sin^2 x + 1) = 0$

$\cos x = 0;\ x = \dfrac{\pi}{2},\ \dfrac{3\pi}{2};\ 2\sin x + 1 = 0;\ 2\sin^2 x = -1$

$\sin^2 x = -\dfrac{1}{2},$ which has no real solution; thus, $x = \dfrac{\pi}{2},\ \dfrac{3\pi}{2}$

13. $4\tan x - \sec^2 x = 0;\ 4\tan x - (1 + \tan^2 x) = 0$

$4\tan x - 1 - \tan^2 x = 0;\ \tan^2 x - 4\tan x + 1 = 0$

$\tan^2 x - 4\tan x = -1;\ \tan^2 x - 4\tan x + 4 = -1 + 4$ (completing the square)

$(\tan x - 2)^2 = 3;\ \tan x - 2 = \pm\sqrt{3}$

$\tan x = 2 \pm \sqrt{3} = 3.732,\ 0.2679$

$x = \tan^{-1} 0.2679 = 0.2618,\ \pi + 0.2618 = 3.403$

$x = \tan^{-1} 3.732 = 1.309,\ \pi + 1.309 = 4.451$

$x = 0.2618,\ 1.309,\ 3.403,\ 4.451$

17. $\tan x + 1 = 0$; $\tan x = -1$; $x_{ref} = \dfrac{\pi}{4}$

(tan negative QII, QIV)

$x = \pi - \dfrac{\pi}{4} = \dfrac{3\pi}{4} \approx 2.36$ or $x = 2\pi - \dfrac{\pi}{4} = \dfrac{7\pi}{4} = 5.50$

On a graphing a calculator, enter $y_1 = \tan x + 1$. Be sure mode is set for radian measure. Use trace to check x-intercepts, which should be approximately 2.36 and 5.50.

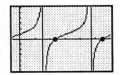

21. $4\sin^2 x - 3 = 0$; $4\sin^2 x = 3$; $\sin^2 x = \dfrac{3}{4}$; $\sin x = \pm\sqrt{\dfrac{3}{4}} = \pm\dfrac{\sqrt{3}}{2}$; $x_{ref} = \dfrac{\pi}{3}$

(sin positive or negative—all quadrants)

$x = \dfrac{\pi}{3} = 1.05$; $\pi - \dfrac{\pi}{3} = \dfrac{2\pi}{3} = 2.09$

$x = \pi + \dfrac{\pi}{3} = \dfrac{4\pi}{3} = 4.19$; $x = 2\pi - \dfrac{\pi}{3} = \dfrac{5\pi}{3} = 5.24$

On graphing calculator, enter $y_1 = 4\sin^2 x - 3$. Use trace to check x-intercepts, which are approximately 1.05, 2.09, 4.19, and 5.24.

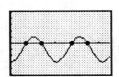

25. $2\sin x - \tan x = 0$; $2\sin x - \dfrac{\sin x}{\cos x} = 0$;

$\sin x \left(2 - \dfrac{1}{\cos x}\right) = 0$; $\sin x = 0$; $x = 0.00$; $x = \pi = 3.14$

$2 - \dfrac{1}{\cos x} = 0$; $\dfrac{1}{\cos x} = 2$; $\cos x = \dfrac{1}{2}$; $x_{ref} = \dfrac{\pi}{3}$;

$x = \dfrac{\pi}{3} = 1.05$; $x = 2\pi - \dfrac{\pi}{3} = \dfrac{5\pi}{3} = 5.24$

On graphing calculator, enter $y_1 = 2\sin x - \tan x$. Use trace to check x-intercepts, which are approximately 0.00, 1.05, 3.14, and 5.24.

29. $\tan x + 3 \cot x = 4$; $\tan x + \dfrac{3}{\tan x} = 4$

$\tan^2 x + 3 = 4 \tan x$; $\tan^2 x - 4 \tan x + 3 = 0$;

$(\tan x - 1)(\tan x - 3) = 0$

$\tan x = 1$; $x = \dfrac{\pi}{4} = 0.7854$ or $x = \pi + \dfrac{\pi}{4} = \dfrac{5\pi}{4} = 3.927$;

$\tan x - 3 = 0$; $\tan x = 3$; $x = 1.249$ or $x = \pi + 1.249 = 4.391$

On graphing calculator, enter $y_1 = \tan x + 3/\tan x - 4$. Use trace to check x-intercepts, which are approximately 0.79, 1.25, 3.93, and 4.39.

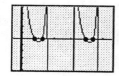

33. $\dfrac{p^2 \tan \theta}{0.0063 + p \tan \theta} = 1.6$; $p = 4.8$

$\dfrac{4.8^2 \tan \theta}{0.0063 + 4.8 \tan \theta} = 1.6$

$4.8^2 \tan \theta = 1.6(0.0063 + 4.8 \tan \theta) = 0.01008 + 7.68 \tan \theta$

$4.8^2 \tan \theta - 7.68 \tan \theta = 0.01008$

$15.36 \tan \theta = 0.01008$

$\tan \theta = \dfrac{0.01008}{15.36} = 6.5625 \times 10^{-4}$

$\theta = 6.56 \times 10^{-4}$

37. $3 \sin x - x = 0$. To solve graphically with a calculator, enter $y_1 = 3 \sin x - x$.
Use the trace to find x-intercepts -2.28, 0.00, 2.28. Using the zoom feature, more accurate values can be found.

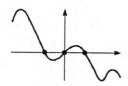

41. $2 \ln x = 1 - \cos 2x$; $2 \ln x - 1 + \cos 2x = 0$

To solve graphically with a calculator, enter $y_1 = 2 \ln x - 1 + \cos 2x$. Use trace to find x-intercept of 2.10. Use the zoom box to find a more accurate value.

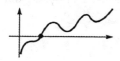

20.6 The Inverse Trigonometric Functions

1. y is an angle whose tangent is x.

5. y is twice the angle whose sine is x.

9. $\cos^{-1} 0.5 = \dfrac{\pi}{3}$ since $\cos \dfrac{\pi}{3} = 0.5$ and $0 \leq \dfrac{\pi}{3} \leq \pi$

13. $\tan^{-1}(-\sqrt{3}) = -\dfrac{\pi}{3}$ since $\tan\left(-\dfrac{\pi}{3}\right) = -\sqrt{3}$ and $-\dfrac{\pi}{2} < -\dfrac{\pi}{3} < \dfrac{\pi}{2}$

17. $\sin^{-1}\left(-\dfrac{\sqrt{2}}{2}\right) = -\dfrac{\pi}{4}$ since $\sin\left(-\dfrac{\pi}{4}\right) = -\dfrac{\sqrt{2}}{2}$ and $-\dfrac{\pi}{2} \leq -\dfrac{\pi}{4} \leq \dfrac{\pi}{2}$

21. $\tan(\sin^{-1} 0) = \tan 0 = 0$

25. $\cos[\tan^{-1}(-1)] = \cos\left(-\dfrac{\pi}{4}\right) = \dfrac{1}{2}\sqrt{2}$

29. $\tan^{-1} x = \sin^{-1}\dfrac{2}{5}$

$x = \tan\left(\sin^{-1}\dfrac{2}{5}\right)$

$x = \dfrac{2}{\sqrt{21}}$

33. $\tan^{-1}(-3.7321) = -1.3090$

37. $\cos^{-1} 0.1291 = 1.4413$

41. $\tan[\cos^{-1}(-0.6281)] = \tan 2.250 = -1.2389$

45. $y = \sin 3x$; $3x = \sin^{-1} y$; $x = \dfrac{1}{3}\sin^{-1} y$

49. $y = 1 + 3\sec 3x$; $\sec 3x = \dfrac{y - 1}{3}$

$3x = \sec^{-1}\left(\dfrac{y - 1}{3}\right)$; $x = \dfrac{1}{3}\sec^{-1}\left(\dfrac{y - 1}{3}\right)$

53. $\tan(\sin^{-1} x) = \tan\theta = \dfrac{x}{\sqrt{1 - x^2}}$

In a triangle, θ is set up such that its sine is x. This gives an opposite side x, hypotenuse 1, and adjacent side $\sqrt{1 - x^2}$.

57. $\sec(\csc^{-1} 3x) = \sec\theta = \dfrac{3x}{\sqrt{9x^2 - 1}}$

In a triangle, θ is set up such that is cosecant is $3x$. This gives an opposite side 1, hypotenuse $3x$, and adjacent side $\sqrt{9x^2 - 1}$.

61. $y = A \cos 2(\omega t + \phi)$

$\dfrac{y}{A} = \cos 2(\omega t + \phi)$

$\cos^{-1} \dfrac{y}{A} = 2(\omega t + \phi) = 2\omega t + 2\phi$

$\cos^{-1} \dfrac{y}{A} - 2\phi = 2\omega t$

$\dfrac{\cos^{-1} \dfrac{y}{A} - 2\phi}{2\omega} = t$

$t = \dfrac{1}{2\omega} \cos^{-1} \dfrac{y}{A} - \dfrac{\phi}{\omega}$

65. Let $\alpha = \sin^{-1} \dfrac{3}{5}$ and $\beta = \sin^{-1} \dfrac{5}{13}$; $\sin \alpha = \dfrac{3}{5}$

$\cos \alpha = \sqrt{1 - \dfrac{9}{25}} = \sqrt{\dfrac{16}{25}} = \dfrac{4}{5}$; $\sin \beta = \dfrac{5}{13}$

$\cos \beta = \sqrt{1 - \dfrac{25}{169}} = \sqrt{\dfrac{144}{169}} = \dfrac{12}{13}$

$\sin^{-1} \dfrac{3}{5} + \sin^{-1} \dfrac{5}{13} = \alpha + \beta$

$\sin(\alpha + \beta) = \sin \alpha \cos \beta + \cos \alpha \sin \beta$

$\qquad = \dfrac{3}{5}\left(\dfrac{12}{13}\right) + \dfrac{4}{5}\left(\dfrac{5}{13}\right)$

$\qquad = \dfrac{36}{65} + \dfrac{20}{65} = \dfrac{56}{65}$

69. Since $\sin A = \dfrac{a}{c}$, $A = \sin^{-1}\left(\dfrac{a}{c}\right)$

Chapter 20 Review Exercises

1. $\sin 120° = \sin(90° + 30°)$
$= \sin 90° \cos 30° + \cos 90° \sin 30°)$
$= 1\left(\dfrac{\sqrt{3}}{2}\right) + 0\left(\dfrac{1}{2}\right) = \dfrac{\sqrt{3}}{2} = \dfrac{1}{2}\sqrt{3}$

5. $\cos 180° = \cos 2(90°)$
$= \cos^2 90° - \sin^2 90°$
$= 0 - (1)^2 = -1$

9. $\sin 14° \cos 38° + \cos 14° \sin 38°$
$= \sin(14° + 38°)$
$= \sin 52°$
$= 0.7880108$

```
sin(14)*cos(38)+
cos(14)*sin(38)
        .7880107536
sin(52)
        .7880107536
```

13. $\cos 73° \cos 142° + \sin 73° \sin 142°$
$= \cos(73° - 142°)$
$= \cos(-69°)$
$= 0.3583679495$

```
cos(73)*cos(142)
+sin(73)sin(142)
        .3583679495
cos(-69)
        .3583679495
```

17. $\sin 2x \cos 3x + \cos 2x \sin 3x$
$= \sin \alpha \cos \beta + \cos \alpha \sin \beta$
$= \sin(\alpha + \beta)$ where $\alpha = 2x$, $\beta = 3x$
$\sin(\alpha + \beta) = \sin(2x + 3x) = \sin 5x$

21. $2 - 4\sin^2 6x = 2(1 - 2\sin^2 6x) = 2(1 - 2\sin^2 \alpha) = 2(\cos 2\alpha) = 2\cos 12x$ where $\alpha = 6x$

25. $\sin^{-1}(-1) = -\dfrac{\pi}{2}$ since $\sin\left(-\dfrac{\pi}{2}\right) = -1$ and $-\dfrac{\pi}{2} \le -\dfrac{\pi}{2} \le \dfrac{\pi}{2}$

29. $\tan\left[\sin^{-1}(-0.5)\right] = \tan\left(-\dfrac{\pi}{6}\right) = -\dfrac{\sqrt{3}}{3} = -\dfrac{1}{3}\sqrt{3}$

33. $\dfrac{\sec y}{\csc y} = \sec y \times \dfrac{1}{\csc y} = \dfrac{1}{\cos y} \times \dfrac{\sin y}{1} = \dfrac{\sin y}{\cos y} = \tan y$

37. $\dfrac{1}{\sin \theta} - \dfrac{\sin \theta}{1} = \dfrac{1 - \sin^2 \theta}{\sin \theta} = \dfrac{\cos^2 \theta}{\sin \theta} = \dfrac{\cos \theta}{\sin \theta} \times \dfrac{\cos \theta}{1} = \cot \theta \cos \theta$

41. $\dfrac{\sec^4 x - 1}{\tan^2 x} = \dfrac{(\sec^2 x - 1)(\sec^2 x + 1)}{\tan^2 x}$

$\dfrac{(\sec^2 x + 1)(\tan^2 x)}{\tan^2 x} = \sec^2 x + 1$

$1 + \tan^2 x + 1 = 2 + \tan^2 x$

45. $\dfrac{1 - \sin^2 \theta}{1 - \cos^2 \theta} = \dfrac{\cos^2 \theta}{\sin^2 \theta}$ since $\sin^2 \theta + \cos^2 \theta = 1$

$\qquad = \left(\dfrac{\cos \theta}{\sin \theta}\right)^2 = (\cot \theta)^2 = \cot^2 \theta$

49. $\sin \dfrac{\theta}{2} \cos \dfrac{\theta}{2} = \dfrac{1}{2}\left(2 \sin \dfrac{\theta}{2} \cos \dfrac{\theta}{2}\right) = \dfrac{1}{2}\left[\sin 2\left(\dfrac{\theta}{2}\right)\right]$

$\qquad \dfrac{1}{2} \sin \theta = \dfrac{\sin \theta}{2}$

53.

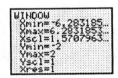  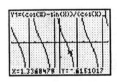 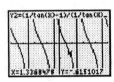

57.

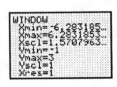  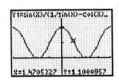 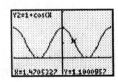

61. $y = 2 \cos 2x;\ \dfrac{y}{2} = \cos 2x$

$\cos^{-1} \dfrac{y}{2} = 2x;\ \dfrac{1}{2} \cos^{-1} \dfrac{1}{2}y = x$

65. $3(\tan x - 2) = 1 + \tan x$

$3 \tan x - 6 = 1 + \tan x$

$2 \tan x = 7$

$\tan x = \dfrac{7}{2}$

$x = \tan^{-1} \dfrac{7}{2} = 1.2925$

Since $\tan x$ is positive in quadrant III also, $\pi + 1.2925 = 4.4341$ is also a value for x that is within the specified range of values for x.

69. $\cos^2 2x - 1 = 0$; $\cos^2 2x = 1$
$\cos 2x = \pm\sqrt{1} = \pm 1$; $2x = 0, \pi, 2\pi, 3\pi$

$x = 0, \dfrac{\pi}{2}, \pi, \dfrac{3\pi}{2}$

73. $\sin 2x = \cos 3x$

$\sin 2x = \cos(2x + x)$

$2\sin x \cos x = \cos 2x \cos x - \sin 2x \sin x$

$2\sin x \cos x = \cos 2x \cos x - 2\sin x \cos x \sin x$

$2\sin x \cos x - \cos 2x \cos x + 2\sin^2 x \cos x = 0$

$\cos x(2\sin x - \cos 2x + \sin^2 x) = 0$

$\cos x = 0$ or $2\sin x - (1 - 2\sin^2 x) + 2\sin^2 x = 0$

$x = \dfrac{\pi}{2}, \dfrac{3\pi}{2}$ $2\sin x - 1 + 2\sin^2 x + 2\sin^2 x = 0$

$4\sin^2 x + 2\sin x - 1 = 0$

$$\sin x = \frac{-2 \pm \sqrt{2^2 - 4(4)(-1)}}{2(4)} = \frac{-2 \pm 2\sqrt{5}}{8} = \frac{-1 \pm \sqrt{5}}{4}$$

$\sin x = \dfrac{-1 + \sqrt{5}}{4}$ or $\sin x = \dfrac{-1 - \sqrt{5}}{4}$

$= \dfrac{\pi}{10}, \dfrac{9\pi}{10}$ $x = \dfrac{13\pi}{10}, \dfrac{17\pi}{10}$

$\sin 2x = \cos 3x$ has solutions $\left\{ \dfrac{\pi}{10}, \dfrac{\pi}{2}, \dfrac{9\pi}{10}, \dfrac{13\pi}{10}, \dfrac{3\pi}{2}, \dfrac{17\pi}{10} \right\}$

77. $x + \ln x - 3\cos^2 x = 2$

$x + \ln x - 3\cos^2 x - 2 = 0$

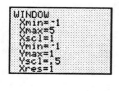 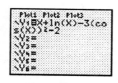 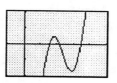

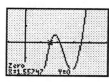 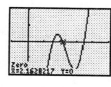 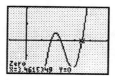

81. $\tan(\cot^{-1} x) = \tan\theta = \dfrac{1}{x}$

$\theta = \cot^{-1} x$

85. $x = 2\cos\theta$; $\sqrt{4 - x^2} = \sqrt{4 - 4\cos^2\theta} = \sqrt{4(1 - \cos^2\theta)} = 2\sqrt{1 - \cos^2\theta}$
$= 2\sqrt{\sin^2\theta} = 2\sin\theta$

89.
$$\sin 2x > 2\sin x$$
$$2\sin x \cos x - 2\sin x > 0$$
$$\sin x(\cos x - 1) > 0$$

I. $\sin x > 0$ and $\cos x - 1 > 0$
     $0 \le x > \pi$ $\cos x > 1$, no solution

II. $\sin x < 0$ and $\cos x - 1 < 0$
     $\pi < x < 2\pi$ $\cos x < 1$
                     $0 < x < 2\pi$

$\sin 2x > 2\sin x$ for $\pi < x < 2\pi$

93. $R = \sqrt{Rx^2 + Ry^2} = \sqrt{(A\cos\theta - B\sin\theta)^2 + (A\sin\theta + B\cos\theta)^2}$
$$= \sqrt{(A^2\cos^2\theta - 2AB\cos\theta\sin\theta + B^2\sin^2\theta) + (A^2\sin^2\theta + 2AB\cos\theta\sin\theta + B^2\cos^2\theta)}$$
$$= \sqrt{A^2\cos^2\theta + A^2\sin^2\theta + B^2\cos^2\theta + B^2\sin^2\theta}$$
$$= \sqrt{A^2(\cos^2\theta + \sin^2\theta) + B^2(\cos^2\theta + \sin^2\theta)}$$
$$= \sqrt{(\cos^2\theta + \sin^2\theta)(A^2 + B^2)} = \sqrt{1(A^2 + B^2)}$$

97.
$$\omega t = \sin^{-1}\frac{\theta - \alpha}{R}$$
$$\sin(\omega t) = \frac{\theta - \alpha}{R}$$
$$R\sin(\omega t) = \theta - \alpha$$
$$\theta = R\sin(\omega t) + \alpha$$

101.

$$\tan\theta = \frac{6.4}{y} \Rightarrow y = \frac{6.4}{\tan\theta}$$
$$\sin\theta = \frac{6.4}{x} \Rightarrow x = \frac{6.4}{\sin\theta}$$
$$2x + 2y = 51.2$$
$$2\cdot\frac{6.4}{\sin\theta} + 2\cdot\frac{6.4}{\tan\theta} = 51.2$$
$$\frac{1}{\sin\theta} + \frac{1}{\tan\theta} = 4.00 \Rightarrow 1 + \cos\theta = 4\sin\theta \Rightarrow \theta = 28.1°$$
$$x = \frac{6.4}{\sin 28.1°} = 13.6 \text{ ft}$$

105. The equation $\tan\theta = 0.4250$ has an infinite number of solutions while $\tan^{-1} 0.4250 = 0.4019$.

Chapter 21

PLANE ANALYTIC GEOMETRY

21.1   Basic Definitions

1. Given: $(x_1, y_1) = (3, 8)$; $(x_2, y_2) = (-1, -2)$

$$d = \sqrt{(x_2 - x_1)^2 + (y_2 - y_1)^2}$$
$$= \sqrt{(-1 - 3)^2 + (-2 - 8)^2}$$
$$= \sqrt{(-4)^2 + (-10)^2}$$
$$= \sqrt{16 + 100} = \sqrt{116}$$
$$= \sqrt{4 \times 29} = 2\sqrt{29}$$

5. Given: $(x_1, y_1) = (-12, 20)$; $(x_2, y_2) = (32, -13)$

$$d = \sqrt{(x_2 - x_1)^2 + (y_2 - y_1)^2}$$
$$= \sqrt{(32 + 12)^2 + (-13 - 20)^2}$$
$$= \sqrt{(44)^2 + (-33)^2}$$
$$= \sqrt{1936 + 1089}$$
$$= \sqrt{3025} = 55$$

9. Given: $(x_1, y_1) = (1.22, -3.45)$;
$(x_2, y_2) = (-1.07, -5.16)$

$$d = \sqrt{(x_2 - x_1)^2 + (y_2 - y_1)^2}$$
$$= \sqrt{(-1.07 - 1.22)^2 + (-5.16 - (-3.45))^2}$$
$$= \sqrt{(-2.29)^2 + (-5.16 + 3.45)^2}$$
$$= \sqrt{(-2.29)^2 + (-1.71)^2} = \sqrt{8.1682} = 2.86$$

13. Given: $(x_1, y_1) = (4, -5)$; $(x_2, y_2) = (4. - 8)$

$$m = \frac{y_2 - y_1}{x_2 - x_1} = \frac{-8 - (-5)}{4 - 4}$$

Since $x_2 - x_1 = 4 - 4 = 0$, the slope is undefined.

17. Given: $(x_1, y_1) = (\sqrt{32}, -\sqrt{18})$;
$(x_2, y_2) = (-\sqrt{50}, \sqrt{8})$

$$m = \frac{y_2 - y_1}{x_2 - x_1} = \frac{\sqrt{8} - (-\sqrt{18})}{-\sqrt{50} - \sqrt{32}} = \frac{-5}{9}$$

21. Given: $\alpha = 30°$; $m = \tan\alpha, 0° < \alpha < 180°$

$$\tan 30° = \frac{\sqrt{3}}{3} \text{ or } \frac{1}{3}\sqrt{3}$$

25. Given: $m = 0.364$; $m = \tan\alpha$; $0.364 = \tan\alpha$; $\alpha = 20.0°$

29. Given: $(x, y) = (6, -1)$; $(x_1, y_1) = (4, 3)$
$(x_2, y_2) = (-5, 2)$; $(x_3, y_3) = (-7, 6)$

$$m_1 = \frac{y - y_1}{x - x_1} = \frac{-1 - 3}{6 - 4} = \frac{-4}{2} = -2$$

$$m_2 = \frac{y_2 - y_3}{x_2 - x_3} = \frac{2 - 6}{-5 - (-7)} = \frac{-4}{-5 + 7}$$
$$= \frac{-4}{2} = -2$$

$m_3 = m_2$ for all parallel lines.

33. Given: distance between $(-1, 3)$ and $(11, k)$ is 13.

$$d = \sqrt{(x_1 - x_2)^2 + (y_1 - y_2)^2}$$
$$13 = \sqrt{(-1 - 11)^2 + (3 - k)^2}$$
$$= \sqrt{(-12)^2 + (3 - k)^2}$$
$$= \sqrt{144 + (3 - k)^2}$$
$$169 = 144 + (3 - k)^2;$$
$$(3 - k)^2 = 25; 3 - k = \pm 5$$
$$-k = -3 \pm 5; k = -2, 8 \qquad \text{Eq. (20-1)}$$

37. $d_1 = \sqrt{(9 - 7)^2 + [4 - (-2)]^2} = \sqrt{2^2 + 6^2} = \sqrt{40} = 2\sqrt{10}$
$d_2 = \sqrt{(9 - 3)^2 + (4 - 2)^2} = \sqrt{6^2 + 2^2} = \sqrt{40} = 2\sqrt{10}$
$d_1 = d_2$ so the triangle is isosceles.

41. $d_1 = \sqrt{(3-5)^2 + (-1-3)^2} = \sqrt{(-2)^2 + (-4)^2} = \sqrt{4+16} = \sqrt{20}$

$$m_1 = \frac{y - y_1}{x - x_1} = \frac{5-3}{3-(-1)} = \frac{5-3}{3+1} = \frac{2}{4} = \frac{1}{2}$$

$$d_2 = \sqrt{(5-1)^2 + (3-5)^2} = \sqrt{(4)^2 + (-2)^2} = \sqrt{16+4} = \sqrt{20}$$

$$m_2 = \frac{y - y_1}{x - x_1} = \frac{5-1}{3-5} = \frac{4}{-2} = -2$$

$$m_1 = \frac{-1}{m_2}, \ m_1 \perp m_2$$

$$A = \frac{1}{2}d_1 d_2 = \frac{1}{2}\sqrt{20}\sqrt{20} = \frac{1}{2}(20) = 10$$

45. $\left(\dfrac{-4+6}{2}, \dfrac{9+1}{2}\right) = \left(\dfrac{2}{2}, \dfrac{10}{2}\right) = (1,5)$

21.2   The Straight Line

1. Given: $m = 4; (x_1, y_1) = (-3, 8)$

$y - y_1 = m(x - x_1)$
$y - 8 = 4[x - (-3)] = 4(x + 3) = 4x + 12$
$y = 4x + 20$
or $4x - y + 20 = 0$

5. Given: $(x_1, y_1) = (1, 3); \alpha = 45°$

$m = \tan \alpha = \tan 45° = 1$
$y - y_1 = m(x - x_1)$
$y - 3 = 1(x - 1) = x - 1; y = x + 2$
or $x - y + 2 = 0$

9. Parallel to y-axis and
3 units left of y-axis.

$x = -3$

13. Perpendicular to line with slope 3;
$(x_1, y_1) = (1, -2)$

$y - y_1 =, (x - x_1); y - (-2) = -\frac{1}{3}(x - 1)$
$y + 2 = -\frac{1}{3}x + \frac{1}{3}; y = -\frac{1}{3}x - \frac{5}{3}$
or $-\frac{1}{3}x - y - \frac{5}{3} = 0; x + 3y + 5 = 0$

17. Given: $6.0x - 2.4y - 3.9 = 0$ or $y = 2.5x - 1.625$
Perpendicular to this line
Therefore, slope of desired line $= -\frac{1}{2.5} = -0.4 = m$

$(x_1, y_1) = (7.5, -4.7)$
$y - y_1 = m(x - x_1)$
$y - (-4.7) = -0.4(x - 7.5)$
$y + 4.7 = -0.4x + 3$
$y = -0.4x - 1.7$ or $0.4x + y + 1.7 = 0$

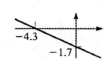

21. Given: $4x - y = 8$
When $x = 0, y = -8$
$y = 0, x = 2$

25. Given: $3x - 2y - 1 = 0$

$3x - 2y - 1 = 0; -2y = -3x + 1$

$y = \frac{-3}{-2}x + \frac{1}{-2}; y = \frac{3}{2}x - \frac{1}{2}$

Slope $= \frac{3}{2} = m$;

y-intercept $= -\frac{1}{2} = b$

29. Given: $4x - ky = 6 \parallel 6x + 3y + 2 = 0$

$6x + 3y + 2 = 0; 3y = -6x - 2$

$y = \frac{-6}{3}x - \frac{2}{3}; y = -2x - \frac{2}{3}$; slope is -2

$4x - ky = 6; -ky = -4x + 6$

$y = \frac{-4}{-k}x + \frac{6}{-k}; y = \frac{4}{k}x - \frac{6}{k}$; slope is $\frac{4}{k}$

Since the lines are parallel, the slopes are equal.

$\frac{4}{k} = -2; 4 = -2k; k = -2$

33. $3x - 2y + 5 = 0; -2y = -3x - 5;$

$y = \frac{-3}{-2}x + \frac{-5}{-2}; y = \frac{3}{2}x + \frac{5}{2};$

slope $= \frac{3}{2} = m_1$

$4y = 6x - 1; y = \frac{6}{4}x - \frac{1}{4};$

$y = \frac{3}{2}x - \frac{1}{4}$; slope $= \frac{3}{2} = m_2$

$m_1 = m_2$ for all parallel lines.

37. $5x + 2y = 3 \Rightarrow y = \frac{-5}{2} \cdot x + \frac{3}{2}$

$10y = 7 - 4x \Rightarrow y = \frac{-4}{10}x + \frac{7}{10}$

$m_1 \cdot m_2 = \frac{-5}{2} \cdot \frac{-4}{10} = 1 \neq -1$

$m_1 \neq m_2$

Lines are neither perpendicular nor parallel.

41. $v = v_0 + at$
$35.4 = 12.2 + a(4.50)$
$4.50a = 35.4 - 12.2$
$a = 5.16 \text{ ft/s}^2$
$v = 12.2 + 5.16t$

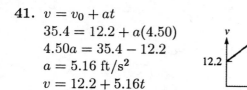

45. $50x + 60y = 12\ 200; 5x + 6y = 1220$

49. $m = \tan(180° - 0.0032°)$
$b = 24\mu\text{m} = 24 \times 10^{-6} \text{ m} = 2.4 \times 10^{-5} \text{ m}$
$m = -5.6 \times 10^{-5}$
$y = mx + b = -5.6 \times 10^{-5}x + 2.4 \times 10^{-5}$
$\quad = (-5.6x + 2.4)10^{-5}$

53. $n = 1200\sqrt{t} + 0$
$m = 1200$
$b = 0$

57. Slope is found by measuring between points. The vertical displacement and the horizontal displacement between the extreme points is in a 1 to 2 ratio; $m = \frac{1}{2}$.

Since the graph is linear, the log equation is of the form $\log y = m \log x + \log a$, where a is the intercept $(1, a)$.

$y = ax^n$; $y = 3x^4$
$a = 3, n = 4$

x	y
1.0	3.0
1.1	4.4
1.2	6.2
1.3	8.6

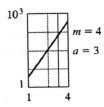

$\log y = \log a + n \log x$
$\log y = \log 3 + 4 \log x$
Verify

(1) Slope is $\dfrac{\log y - \log a}{\log x} = 4$.

Vertical and horizontal measures in millimeters between points are shown. Each slope is 4.

(2) The intercept is $a = 3$.
The line crosses the vertical axis at $x = 1.0, y = 3.0$.

21.3 The Circle

1. $(x - 2)^2 + (y - 1)^2 = 25$
Center at $(2, 1)$, radius is 5.

5. $(x - h)^2 + (y - k)^2 = r^2$; $C(0, 0), r = 3$
$(x - 0)^2 + (y - 0)^2 = 3^2$; $x^2 + y^2 = 9$

9. $(x - h)^2 + (y - k)^2 = r^2$; $C(-2, 5), r = \sqrt{5}$
$[x - (-2)]^2 + (y - 5)^2 = (\sqrt{5})^2$;
$(x + 2)^2 + (y - 5)^2 = \sqrt{25}$
$x^2 + 4x + 4 + y^2 - 10y + 25 = 5$
$x^2 + y^2 + 4x - 10y + 4 + 25 - 5 = 0$;
$x^2 + y^2 + 4x - 10y + 24 = 0$

13. $C(2, 1)$, passes through $(4, -1)$
$r^2 = (2 - 4)^2 + (1 + 1)^2 = (-2)^2 + (2)^2 = 8$
$(x - h)^2 + (y - k)^2 = r^2$;
$(x - 2)^2 + (y - 1)^2 = 8$
$x^2 - 4x + 4 + y^2 - 2y + 1 = 8$;
$x^2 + y^2 - 4x - 2y - 3 = 0$

17. The center is $(2,2)$ and radius is 2.
$(x - h)^2 + (y - k)^2 = r^2$
$(x - 2)^2 + (y - 2)^2 = 2^2$
$x^2 - 4x + 4 + y^2 - 4y + 4 = 4$
$x^2 + y^2 - 4x - 4y + 4 = 0$

21. $x^2 + (y - 3)^2 = 4$
is the same as

$(x - 0)^2 + (y - 3)^2 = 2^2$, so
Therefore,
$h = 0, k = 3, r = 2$
$C(0, 3)$

25. $x^2 + y^2 - 2x - 8 = 0$
$x^2 - 2x + 1 + y^2 = 9$
$(x-1)^2 + (y-0)^2 = 9$
$h = 1, k = 0, r = 3$
$C(1, 0)$

29. $4x^2 + 4y^2 - 16y = 9$
$4(x-0)^2 + 4(y^2 - 4y + 4) = 9 + 16$
$(x-0)^2 + (y-2) = \dfrac{25}{4}$
$h = 0, k = 2, r = \dfrac{5}{2}$
$C(0, 2)$

33. $(-x)^2 + y^2 = 100; x^2 + y^2 = 100$
Symmetrical to y-axis
$x^2 + (-y)^2 = 100; x^2 + y^2 = 100$
Symmetrical to x-axis
$(-x)^2 + (-y)^2 = 100; x^2 + y^2 = 100$
Symmetrical to origin

37. Find all points for which $y = 0$.
$x^2 - 6x + (0)^2 - 7 = 0; x^2 - 6x - 7 = 0 \Rightarrow$
$(x+1)(x-7) = 0; x = -1$ or $x = 7$;
$(-1, 0)$ and $(7, 0)$

41. $x^2 + y^2 + 5y - 4 = 0$
$y^2 + 5y + (x^2 - 4) = 0$; solve for y

$$y = \frac{-5 \pm \sqrt{5^2 - 4(x^2 - 4)}}{2} = \frac{-5 \pm \sqrt{41 - 4x^2}}{2} = -2.5 \pm \sqrt{10.25 - x^2}$$

Set the range for $x_{\min} = -6, x_{\max} = 6, y_{\min} = -6, y_{\max} = 2$

$y_1 = -2.5 + \sqrt{10.25 - x^2}$
$y_2 = -2.5 - \sqrt{10.25 - x^2}$

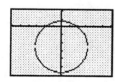 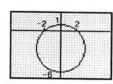

45. 60 Hz = 60 cycles/s = 37.7 m/s; $(h, k) = (0, 0)$
60 cycles = 37.7 m; 1 cycle = 0.628 m
$r = 0.628$ m $\div 2\pi; r = 0.10$
$x^2 + y^2 = (0.10)^2; x^2 + y^2 = 0.0100$

21.4   The Parabola

1. $y^2 = 4x$
$y^2 = 4px$
$y^2 = 4x = 4(1)x; p = 1$
$F(1, 0)$; directrix $x = -1$

5. $x^2 = 8y$

$x^2 = 4py$

$x^2 = 8y = 4(2)y; p = 2$

$F(0, 2)$; directrix $y = -2$

9. $2y^2 = 5x$

$y^2 = \dfrac{5}{2}x = 4px$

$p = \dfrac{5}{8}$

$F(\dfrac{5}{8}, 0)$; directrix: $x = -\dfrac{5}{8}$

13. $F(3, 0)$; directrix $x = -3$; $p = 3$

$y^2 = 4px$

$y^2 = 4(3)x$;

$y^2 = 12x$

17. $V(0, 0)$, directrix $y = -0.16$

$F(0, 0.16)$, $p = 0.16$

$x^2 = 4py = 4(0.16)y$

$x^2 = 0.64y$

21. $F(6, 1)$; directrix $x = 0$; $V(3, 1)$

$d_1 = d_2$

$d_1 = x$

$d_2 = \sqrt{(x-6)^2 + (y-1)^2}$

$x = \sqrt{(x-6)^2 + (y-1)^2}$

$x^2 = (x-6)^2 + (y-1)^2$

$x^2 = x^2 - 12x + 36 + y^2 - 2y + 1$

$0 = -12x + 36 + y^2 - 2y + 1$

$y^2 - 2y - 12x + 37 = 0$

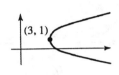

25. Vertex at $(2, -3)$

Focus is 2 units from

vertex at $(4, -3)$.

$(y + 3)^2 = 8(x - 2)$

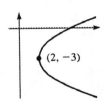

29. Let the vertex of the parabola at the origin. $x^2 = 4py$. A point on the parabola will be $(2100, 300)$. Substitute this into the equation and solve for p.

$2100^2 = 4p(300)$; $1200p = 4\,410\,000$; $p = 3675$

$x^2 = 4(3675)y = 14\,700y$

33. $y^2 = 4px$

$(1.20)^2 = 4p(0.00625)$

$p = 57.6$ m

37. The graph is parabolic since it can be transformed into the form $f^2 = 4pA$.

$$f = 0.065\sqrt{A}$$
$$= 0.065 = \sqrt{200} = 0.92$$

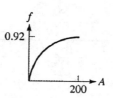

21.5   The Ellipse

1. $\dfrac{x^2}{4} + \dfrac{y^2}{1} = 1$

$a^2 = 4, b^2 = 1$

$\dfrac{x^2}{a^2} + \dfrac{y^2}{b^2} = 1$

$c^2 = a^2 - b^2$

$c^2 = 4 - 1 = 3, c = \sqrt{3}$

$V(\pm 2, 0), F(\pm\sqrt{3}, 0),$

y-intercepts $(0, \pm 1)$

5. $4x^2 + 9y^2 = 36$

$\dfrac{4x^2}{36} + \dfrac{9y^2}{36} = 1$

$\dfrac{x^2}{9} + \dfrac{y^2}{4} = 1$

$a^2 = 9, b^2 = 4$

$c^2 = 9 - 4 = 5; c = \sqrt{5}$

$V(\pm 3, 0), F(\pm\sqrt{5}, 0),$

y-intercepts $(0, \pm 2)$

9. $8x^2 + y^2 = -16$

$\dfrac{8x^2}{16} + \dfrac{y^2}{16} = 1$

$\dfrac{x^2}{2} + \dfrac{y^2}{16} = 1$

$\dfrac{y^2}{16} + \dfrac{x^2}{2} = 1$

$a^2 = 16, b^2 = 2, c^2 = 16 - 2 = 14$

$V(0, \pm 4), F(0, \pm\sqrt{14}), x$-intercepts $(\pm\sqrt{2}, 0)$

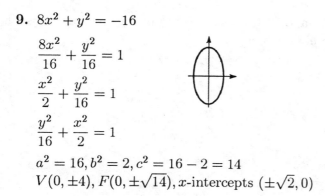

13. $V(15, 0); F(9, 0)$

$a = 15, a^2 = 225;$

$c = 9, c^2 = 81; a^2 - c^2 = b^2$

$b^2 = 144; \dfrac{x^2}{a^2} + \dfrac{y^2}{b^2} = 1;$

$\dfrac{x^2}{225} + \dfrac{y^2}{144} = 1$

$144x^2 + 225y^2 = 32\,400$

17. Vertex $(8, 0); (x, y)$ is $(2, 3); a^2 = 8^2 = 64$

$\dfrac{x^2}{a^2} + \dfrac{y^2}{b^2} = 1; \dfrac{x^2}{64} + \dfrac{y^2}{b^2} = 1$

$\dfrac{(2)^2}{64} + \dfrac{(3)^2}{b^2} = 1; \dfrac{9}{b^2} = \dfrac{16}{16} - \dfrac{1}{16}; 15b^2 = 144; b^2 = \dfrac{144}{15}$

$\dfrac{x^2}{64} + \dfrac{15y^2}{144} = 1; 144x^2 + 960y^2 = 9216$

$3x^2 + 20y^2 = 192$

21. $F(-2, 1)$ and $(4, 1)$, major axis 10

$$\sqrt{[x - (-2)]^2 + (y-1)^2} + \sqrt{(x-4)^2 + (y-1)^2} = 10$$
$$\sqrt{(x+2)^2 + (y-1)^2} = 10 - \sqrt{(x-4)^2 + (y-1)^2}$$
$$(x+2)^2 + (y-1)^2 = 100 - 20\sqrt{(x-4)^2 + (y-1)^2} + (x-4)^2 + (y-1)^2$$
$$x^2 + 4x + 4 + y^2 - 2y + 1 = 100 - 20\sqrt{(x-4)^2 + (y-1)^2} + x^2 - 8x + 16 + y^2 - 2y + 1$$
$$x^2 + 4x + 4 + y^2 - 2y + 1 - 100 - x^2 + 8x - 16 - y^2 + 2y - 1 = -20\sqrt{(x-4)^2 + (y-1)^2}$$

$$12x - 112 = -20\sqrt{(x-4)^2 + (y-1)^2}$$
$$3x - 28 = -5\sqrt{(x-4)^2 + (y-1)^2}$$
$$(3x - 28)^2 = 25[(x-4)^2 + (y-1)^2]$$
$$9x^2 - 168x + 784 = 25(x^2 - 8x + 16 + y^2 - 2y + 1)$$
$$9x^2 - 168x + 784 = 25x^2 - 200x + 400 + 25y^2 - 50y + 25$$
$$-16x^2 - 25y^2 + 32x + 50y + 359 = 0$$
$$16x^2 + 25y^2 - 32x - 50y - 359 = 0$$

25. $2a = 6; a = 3; 2b = 4; b = 2; (h, k) = (2, -1)$

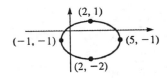

29. Given: $2x^2 + 3y^2 - 8x - 4 = 0$
$$2x^2 + 3(-y)^2 - 8x - 4 = 2x^2 + 3y^2 - 8x - 4$$

33. If the two vertices of each base are fixed at $(-3, 0)$ and $(3, 0)$, and the sum of the two leg lengths is also fixed, the third vertex lies on an ellipse. The base is 6 cm, so

$d_1 + d_2 = 14$ cm -6 cm $= 8$ cm
$(-3, 0)$ and $(3, 0)$ are foci $(-c, 0)$ and $(c, 0)$
$d_1 + d_2 = 2a = 8; a = 4$
$a^2 - c^2 = b^2$
$4^2 - 3^2 = b^2$
$b^2 = 7, a^2 = 16$

The equation is $\dfrac{x^2}{16} + \dfrac{y^2}{7} = 1$, or $7x^2 + 16y^2 = 112$

37. $a = \dfrac{64}{2} = 32, b = 18$

Let the center be at the origin. The equation of the ellipse is $\dfrac{x^2}{32^2} + \dfrac{y^2}{18^2} = 1$; $\dfrac{x^2}{1024} + \dfrac{y^2}{324} = 1$

If $x = 22, \dfrac{22^2}{1024} + \dfrac{y^2}{324} = 1$

$y = 13$ ft

21.6   The Hyperbola

1. $\dfrac{x^2}{25} - \dfrac{y^2}{144} = 1$

$a^2 = 25; a = 5$
$b^2 = 144; b = 12$
$c^2 = a^2 + b^2$
$c^2 = 169; c = 13$
$V(\pm 5, 0), F(\pm 13, 0)$

5. $4x^2 - y^2 = 4$

$\dfrac{4x^2}{4} - \dfrac{y^2}{4} = 1;$

$\dfrac{x^2}{1} - \dfrac{y^2}{4} = 1$

$a^2 = 1; b^2 = 4; c^2 = 5$
$V(\pm 1, 0); F(\pm\sqrt{5}, 0)$

9. $4x^2 - y^2 + 4 = 0$
$4x^2 - y^2 = -4$

$\dfrac{4x^2}{-4} - \dfrac{y^2}{-4} = \dfrac{-4}{-4}$

$-x^2 + \dfrac{y^2}{4} = 1$

$\dfrac{y^2}{4} - \dfrac{x^2}{1} = 1$

$a^2 = 4; b^2 = 1; c^2 = 5$
$V(0, \pm 2), F(0, \pm\sqrt{5})$

13. $V(3, 0); F(5, 0)$

$a = 3; c = 5; a^2 = 9; c^2 = 25$
$b^2 = c^2 - a^2 = 25 - 9 = 16$
$\dfrac{x^2}{a^2} - \dfrac{y^2}{b^2} = 1; \dfrac{x^2}{9} - \dfrac{y^2}{16} = 1;$
$16x^2 - 9y^2 = 144$

17. (x, y) is $(2, 3); F(2, 0), (-2, 0); c = \pm 2, c^2 = 4$

$d_1 = \sqrt{(2 - [-2])^2 + (3 - 0)^2} = \sqrt{4^2 + 3^2} = \sqrt{16 + 9} = \sqrt{25} = 5$
$d_2 = \sqrt{(2 - 2)^2 + (3 - 0)^2} = \sqrt{0 + 9} = \sqrt{9} = 3$
$d_1 - d_2 = 2a; 5 - 3 = 2a; 2 = 2a; 1 = a; a^2 = 1$
$c^2 = 4; b^2 = c^2 - a^2 = 3$

$\dfrac{x^2}{a^2} - \dfrac{y^2}{b^2} = 1; \dfrac{x^2}{1} - \dfrac{y^2}{3} = 1$

$3x^2 - y^2 = 3$

21. $xy = 2; y = \dfrac{2}{x}$

x	y
$\pm\frac{1}{2}$	± 4
± 1	± 2
± 2	± 1
± 4	$\pm\frac{1}{2}$
± 8	$\pm\frac{1}{4}$

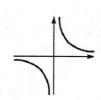

25. $x^2 - 4y^2 + 4x + 32y - 64 = 0$; solve for y
$4y^2 - 34y + (-x^2 - 4x + 64) = 0$

$y = \dfrac{32 \pm \sqrt{(-32)^2 - 4(4)(-x^2 - 4x + 64)}}{2(4)}$

$= \dfrac{32 \pm \sqrt{16x^2 + 64x}}{8}$

$y_1 = 4 + 0.5\sqrt{x^2 + 4x}, y_2 = 4 - 0.5\sqrt{x^2 + 4x}$

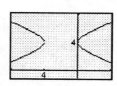

29. $V(0,1), F(0,\sqrt{3}); c^2 = a^2 + b^2$ where $c = \sqrt{3}$ and $a = 1; b^2 = \sqrt{3}^2 - 1^2 = 2$

$$\frac{y^2}{1^2} - \frac{x^2}{\sqrt{2}^2} = 1$$

The transverse axis of the first equation is length $2a = 2\sqrt{1}$ along the y-axis. Its conjugate axis is length $2b = 2\sqrt{2}$ along the x-axis.

The transverse axis of the conjugate hyperbola is length $2\sqrt{2}$ along the x-axis, and its conjugate axis is length $2\sqrt{1}$ along the y-axis.

The equation, then, is $\dfrac{x^2}{\sqrt{2}^2} - \dfrac{y^2}{\sqrt{1}^2} = 1$

$$\frac{x^2}{2} - \frac{y^2}{1} = 1 \text{ or } x^2 - 2y^2 = 2$$

33. $V = iR$ (Ohm's law)

$6.00 = iR$

Therefore, $i = \dfrac{6.00}{R}$

R	i
0.5	12
1	6
2	3
3	2
4	1.5
6	1
9	0.7
12	0.5

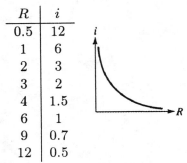

21.7   Translation of Axes

1. $(y-2)^2 = 4(x+1)$; parabola

$y - 2 = y'; x + 1 = x'; y'^2 = 4x'$

Origin O' at $(h,k) = (-1, 2)$

$y'^2 = 4(1)x'; p = 1;$

Focus $(-1 + p, 2)$ or $(0, 2)$

Directrix $x' = -p; x + 1 = -1; x = -2$

Vertex $(-1, 2)$

5. $\dfrac{(x+1)^2}{1} + \dfrac{y^2}{9} = 1$; ellipse

$x' = x + 1; x - h = x + 1; h = -1$

$y' = y - 0; y - k = y - 0; k = 0$

$\dfrac{x'^2}{1} - \dfrac{y'^2}{9} = 1; \dfrac{y'^2}{9} + \dfrac{x'^2}{1} = 1$

Center (h, k) at $(-1, 0)$

$a^2 = 9; a = 3; b^2 = 1; b = 1;$

$c^2 = 8; c = 2\sqrt{2}$

9. Parabola: $V(-1, 3); p = 4;$ parallel to x-axis; vertex (h, k) at $(-1, 3)$

$y'^2 = 4px'; (y - k)^2 = 4p(x - h)$

$(y - 3)^2 = 4(4)[x - (-1)]; (y - 3)^2 = 16(x + 1)$

$y^2 - 6y + 9 = 16x + 16; y^2 - 6y - 16x + 9 - 16 = 0$

$y^2 - 6y - 16x - 7 = 0$

13. Ellipse: center $(-2, 2)$; foci $(-5, 2), (1, 2)$; vertices $(-7, 2), (3, 2)$
(h, k) at $(-2, 2)$; $c = 3$; $c^2 = 9$; $a = 5$; $a^2 = 25$; $b^2 = a^2 - c^2 = 16$

$$\frac{x'^2}{a^2} + \frac{y'^2}{b^2} = 1; \frac{[x - (-2)]^2}{25} + \frac{(y - 2)^2}{16} = 1$$

$$\frac{(x + 2)^2}{25} + \frac{(y - 2)^2}{16} = 1; 16(x + 2)^2 + 25(y - 2)^2 = 400$$

$16(x^2 + 4x + 4) + 25(y^2 - 4y + 4) = 400$
$16x^2 + 64x + 64 + 25y^2 - 100y + 100 - 400 = 0$
$16x^2 + 25y^2 + 64x - 100y - 236 = 0$

17. Hyperbola: center $(h, k) = (-1, 2)$; $F_1 = (-1, 4)$ and $V_1 = (-1, 1)$
$c^2 = a^2 + b^2$ where $c = \sqrt{(-1 + 1)^2 + (4 - 2)^2} = 2$ is the distance between F_1 and (h, k).
Substituting known values:

$$\frac{(y - 2)^2}{a^2} - \frac{(x + 1)^2}{b^2} = 1 \quad (h, k) \text{ substituted}$$

$$\frac{(1 - 2)^2}{a^2} - \frac{(-1 + 1)^2}{b^2} = 1 \quad V_1 \text{ substituted}$$

$$\frac{1}{a^2} - \frac{0}{b^2} = 1; a^2 = 1$$

Since $c^2 = a^2 + b^2$; $2^2 = 1^2 + b^2$; $b^2 = 3$

$$\frac{(y - 2)^2}{1} - \frac{(x + 1)^2}{3} = 1; 3(y - 2)^2 - (x + 1)^2 = 3 \text{ or } x^2 - 3y^2 + 2x + 12y - 8 = 0$$

21. $x^2 + 2x - 4y - 3 = 0$; $x^2 + 2x - 3 = 4y$
$x^2 + 2x = 4y + 3$; $x^2 + 2x + 1 = 4y + 3 + 1$
$(x + 1)^2 = 4y + 4$; $(x + 1)^2 = 4(y + 1)$
Parabola, $p = 1$; $x - h = x + 1$; $h = -1$; $y - k = y + 1$; $k = -1$
Vertex is $(-1, -1)$

25. $9x^2 - y^2 + 8y - 7 = 0$
$9x^2 - (y^2 - 8y + 7 + 9) = -9$
$9x^2 - (y - 4)^2 = -9$

$$\frac{9x^2}{-9} + \frac{(y - 4)^2}{-9} = 1$$

$$-x^2 + \frac{(y - 4)^2}{9} = 1$$

$$\frac{(y - 4)^2}{9} - \frac{x^2}{1} = 1$$

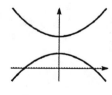

Hyperbola, (h, k) is $(0, 4)$; $a = 3$; $b = 1$

29. $4x^2 - y^2 + 32x + 10y + 35 = 0$

$4(x^2 + 8x) - (y^2 - 10y) = -35$

$4(x^2 + 8x + 16) - (y^2 - 10y + 25) = -35 + 64 - 25$

$\dfrac{(x+4)^2}{1^2} - \dfrac{(y-5)^2}{2^2} = 1$, hyperbola

$C(-4, 5)$

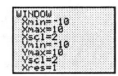   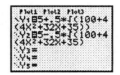   

33. $7x^2 - y^2 - 14x - 16y - 64 = 0$

$7x^2 - 14x - y^2 - 16y - 64 = 0$

$7(x^2 - 2x + 1) - (y^2 + 16y + 64) = 64 + 7 - 64$

$\dfrac{(x-1)^2}{1^2} - \dfrac{(y+8)^2}{\sqrt{7}^2} = 1$

hyperbola, $C(1, -8)$

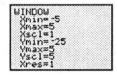      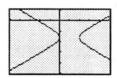

37. Hyperbola: asymptotes: $x - y = -1$ or $x + 1 = y$, and $x + y = -3$ or $y = -x - 3$; vertices $(3, -1)$ and $(-7, -1)$. The center is at the point of interaction of the asymptotes. The equations for the asymptotes are solved simultaneously by adding, $2y = -2; y = -1; -1 = x + 1; x = -2$. Therefore, the coordinates of the center are $(-2, -1)$. Since the slopes are 1 and $-1, a = b$, where a is the distance from the center $(-2, -1)$ to the vertex $(3, -1); a = 5, b = 5$.

$\dfrac{(x-h)^2}{a^2} - \dfrac{(y-k)^2}{b^2} = 1$; $\dfrac{[x-(-2)]^2}{25} - \dfrac{[y-(-1)]^2}{25} = 1$

$\dfrac{(x+2)^2}{25} - \dfrac{(y+1)^2}{25} = 1$

$x^2 + 4x + 4 - (y^2 + 2y + 1) = 25$; $x^2 + 4x + 4 - 2y - 1 = 25$

$x^2 - y^2 + 4x - 2y - 22 = 0$

41. $(x-h)^2 = 4p(y-k)$

$(x-95)^2 = 4p(y-60)$

Solve for $4p$ using $(x, y) = (0, 0)$.

$(-95)^2 = 4p(-60)$

$4p = \dfrac{95^2}{-60}$

$(x-95)^2 = \dfrac{95^2}{-60}(y-60)$

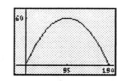

21.8   The Second Degree Equation

1. $x^2 + 2y^2 - 2 = 0$
$A \neq C$, they have the same sign,
and $B = 0$; ellipse

5. $2x^2 + 2y^2 - 3y - 1 = 0$
$A = C$; $B = 0$; circle

9. $x^2 = y^2 - 1$; $x^2 - y^2 + 1 = 0$
A and C have different signs;
$B = 0$; hyperbola

13. $y(3 - 2x) = x(5 - 2y)$
$3y - 2xy = 5x - 2xy$
$$y = \frac{5}{3}x, \text{ line}$$

17. $2x(x - y) = y(3 - y - 2x)$;
$2x^2 - 2xy = 3y - y^2 - 2xy$
$2x^2 - 2xy + 2xy + y^2 - 3y = 0$;
$2x^2 + y^2 - 3y = 0$
$A \neq 0$, same sign; $B = 0$; ellipse

21. $x^2 = 8(y - x - 2)$
$x^2 = 8y - 8x - 16$
$x^2 + 8x - 8y + 16 = 0$
$A \neq 0$; $B = 0$;
$C = 0$; parabola
$x^2 + 8x - 8y + 16 = 0$
$x^2 + 8x + 16 = 8y$
$(x + 4)^2 = 4(2)y$; $p = 2$
Vertex $(-4, 0)$, focus $(-4, 2)$

25. $y^2 + 42 = 2x(10 - x)$
$y^2 + 42 = 20x - 2x^2$
$y^2 + 2x^2 - 20x + 42 = 0$; ellipse
$$\frac{y^2}{2} + x^2 - 10x = -21$$
$$\frac{y^2}{2} + x^2 - 10x + 25 = -21 + 25$$
$$\frac{y^2}{2} + (x - 5)^2 = 4$$
$$\frac{y^2}{8} + \frac{(x - 5)^2}{4} = 1$$
(h, k) at $(5, 0)$
$a = \sqrt{8} = 2\sqrt{2}$; $b = 2$

29. $x^2 + 2y^2 - 4x + 12y + 14 = 0$; solve for y
$x^2 + 0xy + 2y^2 - 4x + 12y + 14 = 0$
$A \neq C$, they have the same sign, and $B = 0$;
ellipse
Solve for y.
$2y^2 + 12 + (x^2 - 4x + 14) = 0$
$$y = \frac{-12 \pm \sqrt{12^2 - 4(2)(x^2 - 4x + 14)}}{2(2)}$$
$$= \frac{-12 \pm \sqrt{-8x^2 + 32x + 32}}{4}$$
$y_1 = -3 + 0.5\sqrt{-2x^2 + 8x + 8}$
$y_2 = -3 - 0.5\sqrt{-2x^2 + 8x + 8}$

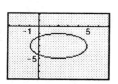

33. (a)   If $k = 1$, $x^2 + ky^2 = a^2$; $x^2 + (1)y^2 = a^2$
$x^2 + y^2 = a^2$ (circle)

(b)   If $k < 0$, $x^2 + ky^2 = a^2$; $x^2 - |k|\,y^2 = a^2$

$$\frac{x^2}{a^2} - \frac{y^2}{a^2/|k|} = 1 \text{ (hyperbola)}$$

(c)   If $k > 0$ $(k \neq 1)$, $x^2 + ky^2 = a^2$

$$\frac{x^2}{a^2} + \frac{y^2}{a^2/k} = 1 \text{ (ellipse)}$$

37. $x^2 + y^2 = (x + 3)^2$
$x^2 + y^2 = x^2 + 6x + 9$

$$y^2 = 6x + 9 = 6\left(x + \frac{3}{2}\right)$$

Therefore a parabola.

21.9   Polar Coordinates

1. $\left(3, \dfrac{\pi}{6}\right)$; $r = 3$, $\theta = \dfrac{\pi}{6}$

5. $\left(-2, \dfrac{7\pi}{6}\right)$; negative r is reversed in direction from positive r.

9. $\left(0.5, -\dfrac{8\pi}{3}\right)$

13. $(\sqrt{3}, 1)$ is (x, y), quadrant I

$\tan \theta = \dfrac{y}{x}$

$\theta = \tan^{-1} \dfrac{y}{x} = \tan^{-1} \dfrac{1}{\sqrt{3}} = \tan^{-1} \dfrac{\sqrt{3}}{3}$;

$\theta = 30° = \dfrac{\pi}{6}$

$r = \sqrt{x^2 + y^2} = \sqrt{(\sqrt{3})^2 + 1^2}$
$ = \sqrt{3 + 1} = \sqrt{4} = 2$

(r, θ) is $\left(2, \dfrac{\pi}{6}\right)$

17. (r, θ) is $\left(8, \dfrac{4\pi}{3}\right)$, quadrant III

$x = r\cos\theta = 8\cos\dfrac{4\pi}{3} = 8\left(-\dfrac{1}{2}\right) = -4$

$y = r\sin\theta = 8\left(-\dfrac{\sqrt{3}}{2}\right) = -4\sqrt{3}$

(x, y) is $\left(-4, -4\sqrt{3}\right)$

21. $x = 3$

$r\cos\theta = x = 3$; $r = \dfrac{3}{\cos\theta} = 3\sec\theta$

25. $x^2 + (y - 2)^2 = 4$
$x^2 + y^2 - 4y + 4 = 4$
$r^2 - 4 \cdot r\sin\theta = 0$
$r = 4\sin\theta$

29. $r = \sin\theta$; $r^2 = r\sin\theta$; $r^2 = x^2 + y^2$
$x^2 + y^2 = r^2 = r\sin\theta = y$; $x^2 + y^2 - y = 0$,
circle

33. $r = \dfrac{2}{\cos\theta - 3\sin\theta}$

$r\cos\theta - 3r\sin\theta = 2$
$x - 3y = 2$, line

37. $r = 2(1 + \cos\theta)$; $x = r\cos\theta$; $\dfrac{x}{r} = \cos\theta$

$r^2 = x^2 + y^2$; $r = \sqrt{x^2 + y^2}$

$r = 2(1 + \cos\theta) = 2\left(1 + \dfrac{x}{r}\right) = 2 + \dfrac{2x}{r}$; $r^2 = 2r + 2x$

Multiply through by r.

$x^2 + y^2 = 2\sqrt{x^2 + y^2} + 2x$; $x^2 + y^2 - 2x = 2\sqrt{x^2 + y^2}$
$(x^2 + y^2 - 2x)^2 = 4(x^2 + y^2)$
$x^4 + y^4 - 4x^3 + 2x^2y^2 - 4xy^2 + 4x^2 = 4x^2 + 4y^2$
$x^4 + y^4 - 4x^3 + 2x^2y^2 - 4xy^2 + 4x^2 - 4x^2 - 4y^2 = 0$
$x^4 + y^4 - 4x^3 + 2x^2y^2 - 4xy^2 - 4y^2 = 0$

41. $B_x = \dfrac{-ky}{x^2 + y^2}$

$\quad = -\dfrac{-ky}{r^2}$

$\quad = \dfrac{-kr\sin\theta}{r^2}$

$\quad = -\dfrac{k\sin\theta}{r}$

$B_y = \dfrac{kx}{x^2 + y^2}$

$\quad = \dfrac{kx}{r^2}$

$\quad = \dfrac{kr\cos\theta}{r^2}$

$\quad = \dfrac{k\cos\theta}{r}$

21.10   Curves in Polar Coordinates

1. $r = 4$ for all θ. Graph is a circle with radius 4.

5. $r = 4\sec\theta = \dfrac{4}{\cos\theta}$

θ	r
0	4
$\frac{\pi}{6}$	4.6
$\frac{\pi}{4}$	5.7
$\frac{\pi}{3}$	8
$\frac{\pi}{2}$	*
$\frac{2\pi}{3}$	-8
$\frac{3\pi}{4}$	-5.7
$\frac{5\pi}{6}$	4.6
π	-4
$\frac{5\pi}{4}$	-5.7
$\frac{3\pi}{2}$	*
$\frac{7\pi}{4}$	5.7
2π	4

*denotes undefined

9. $r = 1 - \cos\theta$; cardioid

θ	r
0	0
$\frac{\pi}{4}$	0.3
$\frac{\pi}{2}$	1
$\frac{3\pi}{4}$	1.7
π	2
$\frac{5\pi}{4}$	1.7
$\frac{3\pi}{2}$	1
$\frac{7\pi}{4}$	0.3

13. $r = 4\sin 2\theta$; rose (4 petals)

θ	r
0	0
$\frac{\pi}{8}$	2.8
$\frac{\pi}{4}$	4
$\frac{3\pi}{8}$	-2.8
$\frac{\pi}{2}$	0
$\frac{5\pi}{8}$	2.8
$\frac{3\pi}{4}$	-4
$\frac{7\pi}{8}$	-2.8
π	0
$\frac{9\pi}{8}$	2.8
$\frac{5\pi}{4}$	4
$\frac{11\pi}{8}$	2.8
$\frac{3\pi}{2}$	0
$\frac{13\pi}{8}$	-2.8
$\frac{7\pi}{4}$	-4
2π	-2.8

17. $r = 2^{\theta}$; spiral

θ	r
0	1
$\frac{\pi}{4}$	1.7
$\frac{\pi}{2}$	3.0
$\frac{3\pi}{4}$	5.1
π	8.8
$\frac{5\pi}{4}$	15.2
$\frac{3\pi}{2}$	26.2
$\frac{7\pi}{4}$	45.2
2π	77.9

21. $r = \dfrac{1}{2 - \cos\theta}$; ellipse

θ	r
0	1
$\frac{\pi}{4}$	0.77
$\frac{\pi}{2}$	0.50
$\frac{3\pi}{4}$	0.37
π	0.33
$\frac{5\pi}{4}$	0.37
$\frac{3\pi}{2}$	0.50
$\frac{7\pi}{4}$	0.77
2π	1

25. $r = 4\cos\dfrac{1}{2}\theta$

θ	r	θ	r
0	4.0	$\frac{13\pi}{6}$	-3.9
$\frac{\pi}{6}$	3.9	$\frac{9\pi}{4}$	-3.7
$\frac{\pi}{4}$	3.7	$\frac{7\pi}{3}$	-3.5
$\frac{\pi}{3}$	3.5	$\frac{5\pi}{2}$	-2.8
$\frac{\pi}{2}$	2.8	$\frac{8\pi}{3}$	-2.0
$\frac{2\pi}{3}$	2.0	$\frac{11\pi}{4}$	-1.5
$\frac{3\pi}{4}$	1.5	$\frac{17\pi}{6}$	-1.0
$\frac{5\pi}{6}$	1.0	3π	0
π	0	$\frac{19\pi}{6}$	1.0
$\frac{7\pi}{6}$	-1.0	$\frac{13\pi}{4}$	1.5
$\frac{5\pi}{4}$	-1.5	$\frac{10\pi}{3}$	2.0
$\frac{4\pi}{3}$	-2.0	$\frac{7\pi}{2}$	2.8
$\frac{3\pi}{2}$	-2.8	$\frac{11\pi}{3}$	3.5
$\frac{5\pi}{3}$	-3.5	$\frac{15\pi}{4}$	3.7
$\frac{7\pi}{4}$	-3.7	$\frac{23\pi}{6}$	3.9
$\frac{11\pi}{6}$	-3.9	4π	4.0
2π	-4.0		

29. $r = \theta$ $(-20 \le \theta \le 20)$

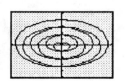

33. $r = 4.0 - \sin\theta$

θ	r
0	4.0
$\frac{\pi}{4}$	3.3
$\frac{\pi}{2}$	3.0
$\frac{3\pi}{4}$	3.3
π	4.0
$\frac{5\pi}{4}$	4.7
$\frac{3\pi}{2}$	5.0
$\frac{7\pi}{4}$	4.7

Chapter 21 Review Exercises

1. Given: straight line; (x_1, y_1) is $(1, -7)$; $m = 4$

$y - y_1 = m(x - x_1)$; $y - (-7) = 4(x - 1)$

$y + 7 = 4x - 4$; $y = 4x - 4 - 7$

$y = 4x - 11$ or $4x - y - 11 = 0$

5. Given: circle; (h, k) is $(1, -2)$; (x, y) is $(4, -3)$

$r = \sqrt{(1-4)^2 + [-2 - (-3)]^2} = \sqrt{(-3)^2 + (-2+3)^2}$
$= \sqrt{9+1} = \sqrt{10}$

$(x - h)^2 + (y - k)^2 = r^2$; $(x - 1)^2 + (y + 2)^2 = \sqrt{10}^2$

$x^2 - 2x + 1 + y^2 + 4y + 4 = 10$

$x^2 + y^2 - 2x + 4y + 1 + 4 - 10 = 0$

$x^2 + y^2 - 2x + 4y + 1 + 4 - 10 = 0$; $x^2 + y^2 - 2x + 4y - 5 = 0$

9. Given: ellipse; vertex $(10, 0)$; focus $(8, 0)$; (h, k) is $(0, 0)$

$\dfrac{(x - h)^2}{a^2} + \dfrac{(y - k)^2}{b^2} = 1$

$\dfrac{x^2}{100} + \dfrac{y^2}{36} = 1$

$9x^2 + 25y^2 = 900$

$a = 10$

$b^2 = a^2 - c^2$

$b^2 = 100 - 64$

$b^2 = 36$

$c = 8$

13. Given: $x^2 + y^2 + 6x - 7 = 0$

$(x^2 + 6x) + (y^2) = 7$; $(x^2 + 6x + 9) + y^2 = 7 + 9$

$(x + 3)^2 + (y + 0)^2 = 16$

$[x - (-3)]^2 + (y - 0)^2 = 4^2$

center $(h, k) = (-3, 0)$; radius $r = 4$

17. Given: $16x^2 + y^2 = 16$

$\dfrac{16x^2}{16} + \dfrac{y^2}{16} = 1$; $\dfrac{x^2}{1^2} + \dfrac{y^2}{4^2} = 1$; $\dfrac{y^2}{4^2} + \dfrac{x^2}{1^2} = 1$

$a = 4, b = 1, c = \sqrt{16-1} = \sqrt{15}$

vertices $(0, a)$, $(0, -a)$ or $(0, 4)$, $(0, -4)$

foci $(0, c)$, $(0, -c)$ or $(0, \sqrt{15})$, $(0, -\sqrt{15})$

21. Given: $x^2 - 8x - 4y - 16 = 0$

$x^2 - 8x = 4y + 16$; $x^2 - 8x + 16 = 4y + 16 + 16$

$(x-4)^2 = 4y + 32$; $(x-4)^2 = 4(y+8)$

$(x-4)^2 = 4(1)(y+8)$; $p = 1$

vertex (h, k) is $(4, -8)$; focus is $(4, -7)$

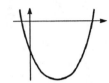

25. Given: $r = 4(1 + \sin\theta)$

θ	r
0	4
$\frac{\pi}{4}$	6.8
$\frac{\pi}{2}$	8
$\frac{3\pi}{4}$	6.8
π	4
$\frac{5\pi}{4}$	1.2
$\frac{3\pi}{2}$	0
$\frac{7\pi}{4}$	1.2

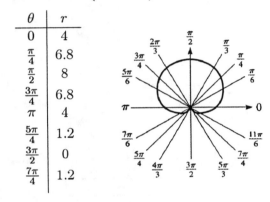

29. Given: $r = \dfrac{3}{\sin\theta + 2\cos\theta}$

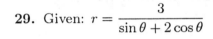

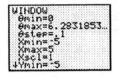

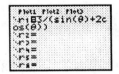
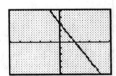

33. Given: $y = 2x$

$\dfrac{y}{x} = 2$; $\tan\theta = 2$; $\theta = \tan^{-1} 2 = 1.11$

37. Given: $r = 2\sin 2\theta$; $r = \sqrt{x^2 + y^2}$

$\sin\theta = \dfrac{y}{r} = \dfrac{y}{\sqrt{x^2+y^2}}$; $\cos = \dfrac{x}{r} = \dfrac{x}{\sqrt{x^2+y^2}}$

$r = 2(2\sin\theta\cos\theta)$

$r = 4\sin\theta\cos\theta$

$\sqrt{x^2+y^2} = 4\dfrac{y}{\sqrt{x^2+y^2}} \times \dfrac{x}{\sqrt{x^2+y^2}}$; $\sqrt{x^2+y^2} = \dfrac{4xy}{x^2+y^2}$

$x^2 + y^2 = \dfrac{(4xy)^2}{(x^2+y^2)^2} = \dfrac{16x^2y^2}{(x^2+y^2)^2}$

$\left(x^2+y^2\right)^3 = 16x^2y^2$

41. $x^2 + y^2 = 9$ circle; center $(0,0)$; radius 3

$x^2 + y^2 = 3^2$

$4x^2 + y^2 = 16$ ellipse; centered at $(0,0)$

                       $(\pm 2, 0)$ and $(0, \pm 4)$ vertices

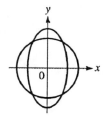

$\dfrac{x^2}{4} + \dfrac{y^2}{16} = 1$

$\dfrac{x^2}{2^2} + \dfrac{y^2}{4^2} = 1$

four real solutions

45. $x^2 - 4y^2 + 4x + 24y - 48 = 0$. Solve for y by completing the square.

$y^2 - 6y = 0.25x^2 + x - 12$

$y^2 - 6y + 9 = 0.25x^2 + x - 3$

$(y - 3)^2 = 0.25x^2 + x - 3$

$y = \pm\sqrt{0.25x^2 + x - 3} + 3$

 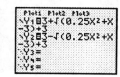 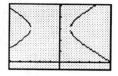

49. (a) The slope of the line between $(-3, 11)$ and $(2, -1)$ is $-\frac{12}{5}$. The slope of the line between $(14, 4)$ and $(2, -1)$ is $\frac{5}{12}$. The product of the slopes is negative one which shows that the line segments form a right triangle.

(b) The distance between $(-3, 11)$ and $(2, -1)$ is 13. The distance between $(14, 4)$ and $(2, -1)$ is 13. The distance between $(-3, 11)$ and $(14, 4)$ is $\sqrt{338}$. Since $13^2 + 13^2 = 338$ the line segments form a right triangle by the Pythagorean Theorem.

53. Given: focus $(3, 1)$; directrix $y = -3$; vertex $(3, -1)$ is (h, k); $p = 2$

By definition, $d_1 = d_2$ where d_1 is from (x, y) to $(x, -3)$, a point on the directrix, and d_2 is from (x, y) to $(3, 1)$, the focus.

$d_1 = \sqrt{(x - x)^2 + [y - (-3)]^2}$; $d_2 = \sqrt{(x - 3)^2 + (y - 1)^2}$

$\sqrt{0^2 + (y + 3)^2} = \sqrt{(x - 3)^2 + (y - 1)^2}$; $0 + (y + 3)^2 = (x - 3)^2 + (y - 1)^2$

$y^2 + 6y + 9 = x^2 - 6x + 9 + y^2 - 2y + 1$

$0 = x^2 + y^2 - y^2 - 6x - 2y - 6y + 9 + 1 - 9$; $0 = x^2 - 6x - 8y + 1$

By translation, $(x - h)^2 = 4(p)(y - k)$

$(x - 3)^2 = 4(2)[y - (-1)]$; $(x - 3)^2 = 8(y + 1)$

$x^2 - 6x + 9 = 8y + 8$; $x^2 - 6x - 8y + 9 - 8 = 0$

$x^2 - 6x - 8y + 1 = 0$

57. $2500x + 1500y = 37,500$

           $1500y = -2500x + 37\,500$

                 $y = -\dfrac{5}{3}x + 25$

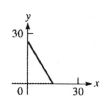

61. $A = \pi \cdot r^2 = \pi(490 \cdot \tan 7°)^2$

     $A = 11,400 \text{ ft}^2$

65.

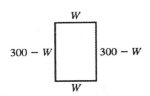

$$A = w(300 - w) = 300w - w^2$$
$$-A = w^2 - 300w$$
$$-A + 150^2 = w^2 - 300w + 150^2 = (w - 150)^2$$
$$-1(A - 150^2) = (w - 150)^2$$
$$(w - 150)^2 = -1(A - 150^2)$$
Parabola with $(h, k) = (150, 150^2)$
$$= (150, 22\ 500)$$

69. $a = \dfrac{120}{2} = 60; \quad c = 60 - 15 = 45; \quad b = \sqrt{a^2 - b^2} = \sqrt{60^2 - 45^2} = \sqrt{1575}$

$A = \pi ab = \pi \cdot 60 \cdot \sqrt{1575} = 7500 \text{ ft}^2$

73.

$$\frac{x^2}{a^2} - \frac{y^2}{b^2} = 1$$
$y = 40$ when $x = 40$
$y = 100$ when $x = 50$

$$\left\{ \begin{array}{l} \dfrac{40^2}{a^2} - \dfrac{40^2}{b^2} = 1 \text{ or } 40^2 b^2 - 40^2 a^2 = a^2 b^2 \\[2mm] \dfrac{50^2}{a^2} - \dfrac{100^2}{b^2} = 1 \text{ or } 50^2 b^2 - 100^2 a^2 = a^2 b^2 \end{array} \right\}$$

M by 100^2
M by -40^2 $\quad \left\{ \begin{array}{l} 40^2 b^2 - 40^2 a^2 = a^2 b^2 \\ 50^2 b^2 - 100^2 a^2 = a^2 b^2 \end{array} \right\}$

$$\left\{ \begin{array}{l} 100^2 40^2 b^2 - 100^2 40^2 a^2 = 100^2 a^2 b^2 \\ -40^2 50^2 b^2 + 40^2 100^2 a^2 = -40^2 a^2 b^2 \end{array} \right\}$$

Add $\quad \left\{ \begin{array}{l} 16 \times 10^6 b^2 - 1.6 \times 10^7 a^2 = 10 \times 10^3 a^2 b^2 \\ 4 \times 10^6 b^2 + 1.6 \times 10^7 a^2 = -1.6 \times 10^3 a^2 b^2 \end{array} \right\}$

$$12 \times 10^6 b^2 = 8.4 \times 10^3 a^2 b^2$$
$$12 \times 10^6 = 8.4 \times 10^3 a^2$$

$$a^2 = \frac{12 \times 10^6}{8.4 \times 10^3} = 1.42 \times 10^3$$

$$a = 37.8 \text{ ft}$$

77. $r^2 = R^2 \cos 2 \left(\theta + \dfrac{\pi}{2} \right)$

$r = R \sqrt{\cos 2 \left(\theta + \dfrac{\pi}{2} \right)} = R \sqrt{\cos(2\theta + \pi)}$

Since the square root is real only when $\cos 2 \left(\theta + \frac{\pi}{2} \right)$ is not negative, the range is $\dfrac{\pi}{4} \leq \dfrac{3\pi}{4}$ or $\dfrac{5\pi}{4} \leq r \leq \dfrac{7\pi}{4}$. (See Chapter 10.)

θ	r
$\frac{\pi}{4}$	0
$\frac{\pi}{3}$	$0.5R$
$\frac{\pi}{2}$	$1R$
$\frac{2\pi}{3}$	$0.5R$
$\frac{3\pi}{4}$	0
$\frac{5\pi}{4}$	0
$\frac{4\pi}{3}$	$0.5R$
$\frac{3\pi}{2}$	$1R$
$\frac{5\pi}{3}$	$0.5R$
$\frac{7\pi}{4}$	0

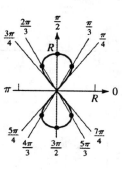

81. $\dfrac{1}{r} = a + b \cos \theta$

$1 = ar + br \cos \theta$

$1 = a\sqrt{x^2 + y^2} + bx$

$a\sqrt{x^2 + y^2} = 1 - bx$

$a^2(x^2 + y^2) = 1 - 2bx + b^2x^2$

$a^2x^2 + a^2y^2 + 2bx - b^2x^2 = 1$

$(a^2 - b^2)x^2 + 2bx + a^2y^2 = 1$

$x^2 + \dfrac{2b}{a^2 - b^2} \cdot x + \dfrac{a^2}{a^2 - b^2} \cdot y^2 = \dfrac{1}{a^2 - b^2}$

$x^2 + \dfrac{2b}{a^2 - b^2} \cdot x + \dfrac{b^2}{(a^2 - b^2)^2} + \dfrac{a^2}{a^2 - b^2} \cdot y^2 = \dfrac{1}{a^2 - b^2} + \dfrac{b^2}{(a^2 - b^2)^2}$

$\left(x + \dfrac{b}{a^2 - b^2} \right)^2 + \dfrac{a^2}{a^2 - b^2} \cdot y^2 = \dfrac{a^2 - b^2 + b^2}{(a^2 - b^2)^2} = \dfrac{a^2}{(a^2 - b^2)^2}$

$\dfrac{\left(x + \frac{b}{a^2-b^2} \right)^2}{\dfrac{a^2}{(a^2 - b^2)^2}} + \dfrac{y^2}{\dfrac{1}{a^2 - b^2}} = 1$ which is the equation of a conic.

INTRODUCTION TO STATISTICS

22.1 Frequency Distributions

1.

Number	103	104	105	106	107	108	109	110	111	112	113
Frequency	1	3	1	3	2	4	3	1	1	0	1

5.

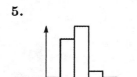

9.

Number	18	19	20	21	22	23	24	25
Frequency	1	3	2	4	3	1	0	1

13.

Time(s)	2.21	2.22	2.23	2.24	2.25	2.26	2.27	2.28	2.29
Frequency	2	7	18	41	56	32	8	3	3

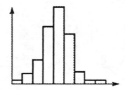

17.

Dist. (ft)	155 − 159	160 − 164	165 − 169	170 − 174	175 − 179	180 − 184	185 − 189
f, (%)	1.7	12.5	26.7	30.0	20.0	8.3	0.8

21.

Dosage (mR)	3.73-3.87	3.88-4.02	4.03-4.17	4.18-4.32	4.33-4.47	4.48-4.62
Frequency	1	2	2	7	7	1

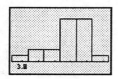

25.

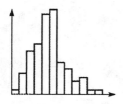

22.2   Measures of Central Tendency

1. Arrange the numbers in numerical order:

2, 3, 3, 3, 4, 4, 4, 4, 5, 5, 6, 6, 6, 7, 7

There are 15 numbers. The middle number is eighth. Since the eighth number is 4, the median is 4.

5. The arithmetic mean is:
$$\bar{x} = \frac{2+3+3+3+4+4+4+4+5+5+6+6+6+7+7}{15} = \frac{69}{15} = 4.6$$

9. The mode is the number that occurs most frequently, which is 4 since it occurs 4 times.

13. Arrange in numerical order; find the value of the 8th (middle) number.

18, 19, 19, 19, 20, 20, 21, 21, 21, 21, 22, 22, 22, 23, 25

The median value is 21.

17. $\overline{T} = (2(2.21) + 7(2.22) + 18(2.23) + 41(2.24) + 56(2.25) + 32(2.26) + 8(2.27) + 3(2.28) + 3(2.29)/170)$
$$= \frac{382.13}{170} = 2.248 \text{ s}$$

21. $\bar{d} = (3.83 + 3.90 + 3.96 + 4.09 + 4.15 + 4.18 + 4.21 + 4.23 + 4.25 + 4.26 + 4.27 + 4.29 + 4.33$
$$+ 4.34 + 2(4.36) + 4.37 + 4.41 + 4.44 + 4.51/20)$$
$$= 4.237 \text{ mR}$$

25. 25, 28, 28, 29, 29, 30, 30, 30, 30, 30, 31, 31, 31, 32, 32, 32, 33, 33, 34, 34, 34, 35, 36

The median value is the 12th number, which is 31 h.

29. Arrange the salaries in order: 375, 400, 425, 425, 450, 450, 475, 475, 500, 500, 500, 550, 550, 600
There are 14 numbers. The middle number is between the seventh and the eight, which are both 475. Therefore, the median is $475. The mode is $500, which occurs three times.

33. Arrange in order: 0.14, 0.15, 0.15, 0.17, 0.17, 0.18, 0.18, 0.18, 0.19, 0.20, 0.22, 0.22, 0.23, 0.23, 0.24, 0.26, 0.27, 0.32

The median is between the 9th and 10th.
Median = $(0.19 + 0.20) \div 2 = 0.195$
Mode = 0.18; $f = 3$

37. Add 100 to each number and arrange in order. 475, 500, 525, 525, 550, 550, 575, 575, 600, 600, 600, 650, 650, 700
The median is between the seventh and eighth number, both of which are $575. The mode is $600 since there are three of these. The mean is the sum divided by 14.
$$\bar{x} = \frac{8075}{14} = \$577$$

The mean, median, and mode are 100 more than the values in Exercise 29.

22.3 Standard Deviation

1.

x	$x - \overline{x}$	$(x - \overline{x})^2$
2	−260	6.76
3	−1.60	2.56
3	−1.60	2.56
3	−1.60	2.56
4	−0.60	0.36
4	−0.60	0.36
4	−0.60	0.36
4	−0.60	0.36
5	0.40	0.16
5	0.40	0.16
6	1.40	1.96
6	1.40	1.96
6	1.40	1.96
7	2.40	5.76
7	2.40	5.76
69		33.60

$\overline{x} = 69/15 = 4.60$

$33.60/14 = 2.4$

$s = \sqrt{2.4} = 1.55$

5.

x	x^2
2	4
3	9
3	9
3	9
4	16
4	16
4	16
4	16
5	25
5	25
6	36
6	36
6	36
7	49
7	49
69	351

$$s = \sqrt{\frac{15(351) - 69^2}{15(14)}}$$

$$s = 1.55$$

9.

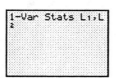

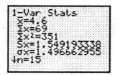

13.

x	x^2
18	324
19	361
19	361
19	361
20	400
20	400
21	441
21	441
21	441
21	441
22	484
22	484
22	484
23	529
25	625
313	6577

$$s = \sqrt{\frac{15(6577) - 313^2}{15(14)}}$$

$$s = 1.8$$

17.

x	f	fx	fx^2
2.21	2	4.42	9.7682
2.22	7	15.54	34.4988
2.23	18	40.14	89.5122
2.24	41	91.84	205.7216
2.25	56	126	283.5
2.26	32	72.32	163.4432
2.27	8	18.16	41.2232
2.28	3	6.84	15.5952
2.29	3	6.87	15.7323
	170	382.13	858.9947

$$\bar{x} = 382.13/170 = 2.2478$$

$$s = \sqrt{\frac{170(858.9947) - 382.13^2}{170(169)}}$$

$$s = 0.014 \text{ s}$$

22.4   Normal Distributions

1.

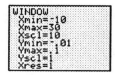 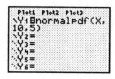 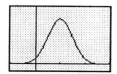

5. $200 \cdot (0.68) = 136$

9. As the calculator screens show, 68.3% or 342 of the batteries have voltages between 1.45 V and 1.55 V.

 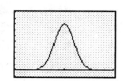 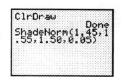 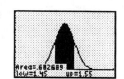

13. As the calculator screens show, 43.3% or 2165 of the tires will last between 85,000 miles and 100,000 km.

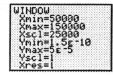 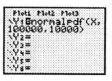 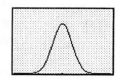 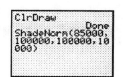 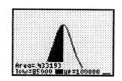

17. $\sigma_{\bar{x}} = \dfrac{\sigma}{\sqrt{n}} = \dfrac{10,000}{\sqrt{5000}} = 141$

21. $\bar{x} + s = 2.262$ and $\bar{x} - s = 2.234$. The readings within these bounds are 2.24, 2.25, and 2.26 with frequencies of $32 + 56 + 41 = 129$. Thus 129 of the 170 readings or 76% fall within these bounds as compared to a normal distribution with 68%.

22.5 Statistical Process Control

1.

Hour	Torques (in $N \cdot m$) of five engines					Mean x	Range R
1	366	352	354	360	362	358.8	14
2	370	374	362	366	356	365.6	18
3	358	357	365	372	361	362.6	15
4	360	368	367	359	363	363.4	9
5	352	356	354	348	350	352.0	8
6	366	361	372	370	363	366.4	11
7	365	366	361	370	362	364.8	9
8	354	363	360	361	364	360.4	10
9	361	358	356	364	364	360.6	8
10	368	366	368	358	360	364.0	10
11	355	360	359	362	353	357.8	9
12	365	364	357	367	370	364.6	13
13	360	364	372	358	365	363.8	14
14	348	360	352	360	354	354.8	12
15	358	364	362	372	361	363.4	14
16	360	361	371	366	346	360.8	25
17	354	359	358	366	366	360.6	12
18	362	366	367	361	357	362.6	10
19	363	373	364	360	358	363.6	15
20	372	362	360	365	367	365.2	12
					Sum	7235.7	248
					Mean	361.785	12.4

CL: $\bar{x} = 361.785$ N $\cdot$ m
$UCL(\bar{x}) = \bar{\bar{x}} + A_2 R = 361.785 + 0.577(12.4) = 368.9398$ N $\cdot$ m
$LCL(\bar{x}) = \bar{\bar{x}} - A_2 R = 361.785 + 0.577(12.4) = 354.6302$ N $\cdot$ m

5.

Subgroup	Output voltages of five adapters						
1	9.03	9.08	8.85	8.92	890	8.956	0.23
2	9.05	8.98	9.20	9.04	9.12	9.078	0.22
3	8.93	8.96	9.14	9.06	9.00	9.018	0.21
4	9.16	9.08	9.04	9.07	8.97	9.064	0.19
5	9.03	9.08	8.93	8.88	8.95	8.974	0.20
6	8.92	9.07	8.86	8.96	9.04	8.970	0.21
7	9.00	9.05	8.90	8.94	8.93	8.964	0.15
8	8.87	8.99	8.96	9.02	9.03	8.974	0.16
9	8.89	8.92	9.05	9.10	8.93	8.978	0.21
10	9.01	9.00	9.09	8.96	8.98	9.008	0.13
11	8.90	8.97	8.92	8.98	9.03	8.960	0.13
12	9.04	9.06	8.94	8.93	8.92	8.978	0.14
13	8.94	8.99	8.93	9.05	9.10	9.002	0.17
14	9.07	9.01	9.05	8.96	9.02	9.022	0.11
15	9.01	8.82	8.95	8.99	9.04	8.962	0.22
16	8.93	8.91	9.04	9.05	8.90	8.966	0.15
17	9.08	9.03	8.91	8.92	8.96	8.980	0.17
18	8.94	8.90	9.05	8.93	9.01	8.966	0.15
19	8.88	8.82	8.89	8.94	8.88	8.882	0.12
20	9.04	9.00	8.98	8.93	9.05	9.000	0.12
21	9.00	9.03	8.94	8.92	9.05	8.988	0.13
22	8.95	8.95	8.91	8.90	9.03	8.948	0.13
23	9.12	9.04	9.01	8.94	9.02	9.026	0.18
24	8.94	8.99	8.93	9.05	9.07	8.996	0.14
					Sum	215.66	3.97
					Mean	8.959	0.1654

CL: $\overline{\overline{x}} = 8.959$ V

$UCL(\overline{x}) = \overline{\overline{x}} + A_2 R = 8.959 + 0.577(0.1654) = 9.054$ V

$LCL(\overline{x}) = \overline{\overline{x}} - A_2 R = 8.959 - 0.577(0.1654) = 8.864$ V

9. CL: $\mu = \overline{\overline{x}} = 2.725$ in

$UCL(\overline{x}) = \mu + A\sigma = 2.725 + 1.342(0.0032) = 2.729$ in

$LCL(\overline{x}) = \mu - A\sigma = 2.725 + 1.342(0.0032) = 2.721$ in

13.

Week	Accounts with errors	Proportion with errors
1	52	0.052
2	36	0.036
3	27	0.027
4	58	0.058
5	44	0.044
6	21	0.021
7	48	0.048
8	63	0.063
9	32	0.032
10	38	0.038
11	27	0.027
12	43	0.043
13	22	0.022
14	35	0.035
15	41	0.041
16	20	0.020
17	28	0.028
18	37	0.037
19	24	0.024
20	42	0.042
Total	738	

$$CL: \quad \bar{p} = \frac{738}{1000(20)} = 0.0369$$

$$\sigma_p = \sqrt{\frac{\bar{p}(1 - \bar{p})}{n}}$$

$$= \sqrt{\frac{0.0369(1 - 0.0369)}{1000}}$$

$$= 0.00596$$

$$UCL(p) = 0.0369 + 3(0.00596) = 0.0548$$
$$LCL(p) = 0.0369 - 3(0.00596) = 0.0190$$

22.6   Linear Regression

1.

x	y	xy	x^2
4	1	4	16
6	4	24	36
8	5	40	64
10	8	80	100
12	9	108	144
40	27	256	360

$n = 5$

$$m = \frac{5(256) - (40)(27)}{5(360) - 40^2} = 1.0$$

$$b = \frac{360(27) - (256)(40)}{5(360) - 40^2} = -2.6$$

$$y = mx + b; y = 1.0x - 2.6$$

Plot points:

x	y
3	0.4
10	7.4

5.

i	V	iV	i^2
15.0	3.00	45.00	225.00
10.8	4.10	44.28	116.64
9.30	5.60	52.08	86.49
3.55	8.00	28.40	12.60
4.60	10.50	48.30	21.16
43.25	31.20	218.06	461.89

$n = 5$

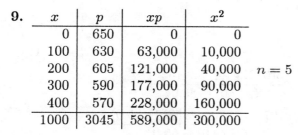

$$m = \frac{5(218.06) - (43.25)(31.20)}{5(461.89) - (43.25)^2} = -0.590$$

$$b = \frac{(461.89)(31.20) - (218.06)(43.25)}{5(461.89) - (43.25)^2} = 11.3$$

$V = mi + b; \; V = -0.590i + 11.3$

Plot points:

i	V
2	10.1
10	5.3

9.

x	p	xp	x^2
0	650	0	0
100	630	63,000	10,000
200	605	121,000	40,000
300	590	177,000	90,000
400	570	228,000	160,000
1000	3045	589,000	300,000

$n = 5$

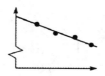

$$m = \frac{5(589,000) - (1000)(3045)}{5(300,000) - (1000)^2} = -0.200$$

$$b = \frac{(300,000)(3045) - (589,000)(1000)}{5(300,000) - (1000)^2} = 649$$

$p = mx + b; \; p = -0.200x + 649$

Plot points:

i	V
0	649
350	579

13.

x	y
4	1
6	4
8	5
10	8
12	9

x	x^2
4	16
6	36
8	64
10	100
12	144
40	360

y	y^2
1	1
4	16
5	25
8	64
9	81
27	187

$\overline{x} = \dfrac{40}{5} = 8; \; \overline{x}^2 = 64; \; \overline{x^2} = \dfrac{360}{5} = 72$

$s_x = \sqrt{72 - 64} = \sqrt{8}$

$\overline{y} = \dfrac{27}{5} = 5.4; \; \overline{y}^2 = 5.4^2 = 29.16;$

$\overline{y^2} = \dfrac{187}{5} = 37.4$

$s_y = \sqrt{37.4 - 29.16} = \sqrt{8.24}$

$r = m\left(\dfrac{s_x}{s_y}\right) = 1.0\left(\dfrac{\sqrt{8}}{\sqrt{8.24}}\right) = 0.985$

22.7   Nonlinear Regression

1.

x	y	x^2	yx^2	$(x^2)^2$
2	12	4	48	16
4	38	16	608	256
6	72	36	2 592	1 296
8	135	64	8 640	4 096
10	200	100	20 000	10 000
	457	220	31 888	15 664

$n = 5$

$$m = \frac{5(31\ 888) - (220)(457)}{5(15\ 664) - 220^2} = 1.97$$

$$b = \frac{(15\ 664)(457) - (31\ 888)(220)}{5(15\ 664) - 220^2}$$

$$= 4.8$$

Therefore, $y = 1.97x^2 + 4.8$

5. $y = mt^2 + b$

t	y	t^2	yt^2	$(t^2)^2$
1.0	6.0	1.0	6.0	1.0
2.0	23	4.0	92	16.0
3.0	55	9.0	495	81.0
4.0	98	16.0	1568	256
5.0	148	25.0	3700	625
	330	55.0	5861.0	979.0

$n = 5$

$$m = \frac{5(5861) - (55.0)(330)}{5(979.0) - 55.0^2} = 5.97$$

$$b = \frac{(979.0)(330) - (5861)(55.0)}{5(979.0) - 55.0^2}$$

$$= 0.38$$

$$y = 5.97t^2 + 0.38$$

9.

S	$\frac{1}{S}$	P	$\left(\frac{1}{S}\right)P$	$\left(\frac{1}{S}\right)2$
240	0.004 166 666 67	5.60	0.023 333 333 3	$1.736\ 111 \times 10^{-5}$
305	0.003 278 688 52	4.40	0.014 426 229 51	$1.074\ 980 \times 10^{-5}$
420	0.002 380 952 38	3.20	0.007 619 047 62	$5.668\ 935 \times 10^{-6}$
480	0.002 083 333 3	2.80	0.005 833 333 33	$4.340\ 279 \times 10^{-6}$
560	0.001 785 714 29	2.40	0.004 285 714 29	$3.188\ 776 \times 10^{-6}$
2005	0.013 695 36	18.40	0.055 497 66	$4.130\ 89 \times 10^{-5}$

$$m = \frac{5(0.55\ 497\ 66) - (0.013\ 695\ 36)(18.40)}{5(4.130\ 89 \times 10^{-5}) - (0.013\ 695\ 36)^2} = 1343$$

$$b = \frac{(4.130\ 89 \times 10^{-5})(18.40) - (0.055\ 497\ 66)(0.013\ 695\ 36)}{5(4.130\ 89 \times 10^{-5}) - (0.013\ 695\ 36)^2}$$

$$= 1.226\ 612 \times 10^{-3} \text{ or } 0$$

$$P = 1343 \left(\frac{1}{S}\right) + 0 = \frac{1343}{S}$$

Chapter 22 Review Exercises

1. Enter the 16 numbers as list L_1 in the calculator. Then $\boxed{\text{2nd}}$ $\boxed{\text{LIST}}$ $\boxed{\blacktriangleleft}$ to obtain

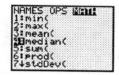   from which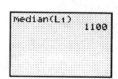

5.

Number	1093-1095	1096-1098	1099-1101	1102-1104	1105-1107
Frequency	3	4	3	4	2

9.

Number	1093-1095	1096-1098	1099-1101	1102-1104	1105-1107
CF	3	7	10	14	16

13. Enter the 12 numbers as list L_1 in the calculator. Then $\boxed{\text{STAT}}$ $\boxed{\blacktriangleright}$ to obtain

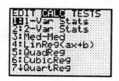   from which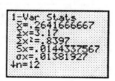

0.014 Pa · s is the standard deviation

17. Enter the 9 power numbers in the calculator as list L_1 and the corresponding frequencies as list L_2, then

     from which

700 W is the median

21. Using $L_1 \& L_2$ from problem 17,

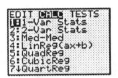   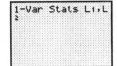   

17.3 W is the standard deviation

25. $\sum f = 200$; the median is the mean of the 100th and 101st entries.

Counts	0	1	2	3	4	5	6	7	8	9	10
Intervals	3	10	25	45	29	39	26	11	7	2	3

$\underbrace{\hspace{3cm}}_{83}$

$\underbrace{\hspace{4cm}}_{112}$ The 100th and 101st are both 4. Therefore, the median is 4.

29. Enter speeds in L_1 and no. cars in L_2, then

   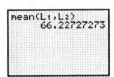

66.2 mi/h is the mean

33. CL; $\overline{p} = \dfrac{540}{500 \cdot 20} = 0.0540$

$\sigma_p = \sqrt{\dfrac{\overline{p}(1 - \overline{p})}{n}} = \sqrt{\dfrac{0.054(1 - 0.054)}{500}} = 0.01011$

$UCL(p) = 0.054 + 3(0.01011) = 0.0843$
$LCL(p) = 0.054 - 3(0.01011) = 0.0237$

37.

   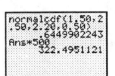

There are about 322 readings between 1.5 and 2.5 $\mu g/m^3$.

41.

T	R	TR	T^2
0.0	25.0	0	0
20.0	26.8	536	400
40.0	28.9	1156	1600
60.0	31.2	1872	3600
80.0	32.8	2624	6400
100	34.7	3470	10000
300	179.4	9658	22000

$$\overline{T} = \frac{300}{6} = 50.0 \qquad \left(\overline{T}\right)^2 = 2500$$

$$\overline{R} = \frac{179.4}{6} = 29.90 \qquad \overline{T}\,\overline{R} = 50.0(29.9) = 1495$$

$$\overline{TR} = \frac{9658}{6} = 1610 \qquad \overline{T^2} = \frac{22\,000}{6} = 3667$$

$$s_T^2 = \overline{T^2} - (\overline{T})^2 = 3667 - 2500 = 1167$$

$$m = \frac{\overline{TR} - \overline{T}\,\overline{R}}{s_T^2} = \frac{1610 - 1495}{1167} = 0.0985$$

$$b = \overline{R} - m\overline{T} = 29.90 - 0.0977(50.0) = 25.02$$

$$R = mT + b; \; R = 0.0985T + 25.0$$

(Answers may vary due to rounding.)

45.

	x	y	xy	x^2
t	t^2	s	$(t^2)(s)$	$(t^2)^2$
0.0	0.0	3000	0	0
3.0	9.0	2960	26640	81
6.0	36.0	2820	101520	1296
9.0	81.0	2600	210600	6561
12.0	144.0	2290	329760	20736
15.0	225.0	1900	427500	50625
18.0	324.0	1410	456840	104976
63.0	819.0	16980	1552860	184275

$$m = \frac{n\sum xy - (\sum x)(\sum y)}{n\sum x^2 - (\sum x)^2} = \frac{7(1552860) - (819)(16980)}{7(184275) - (819)^2} = -4.90$$

$$b = \frac{(\sum x^2)(\sum y) - (\sum xy)(\sum x)}{n\sum x^2 - (\sum x)^2} = \frac{(184275)(16980) - 1552860(819)}{7(184275) - (819)^2} = 3000$$

$$s = -4.90t^2 + 3000$$

49. There are 7 numbers. The geometric mean is

$$\sqrt[7]{(8.0)(8.2)(8.8)(9.5)(9.7)(10.0)(10.7)} = \sqrt[7]{5692009.7} = 5692009.7^{1/7} = 9.2 \text{ ppm}$$

53. Answers may vary.

THE DERIVATIVE

23.1 Limits

1. $f(x) = 3x - 2$ is continuous for all real x since it is defined for all x, and any small change in x will produce only a small change in $f(x)$.

5. $f(x) = \sqrt{\frac{x}{x-2}}$ is continuous for $x \leq 0$ and $x > 2$. The function is not defined for $0 < x \leq 2$. $0 < x < 2$ gives the square root of a negative and $x = 2$ gives division by zero.

9. The graph is not continuous at $x = 1$ since $f(0,9)$ is a value between -1 and -2, and $f(1.1)$ is a value greater than $+2$. This amount of change in y for a 0.2 change in x is not consistent with equivalent changes in x at other values in the function domain. A small change in x does not produce a small change in y at $x = 1$. The function is continuous for $x < 1$, and continuous for $x > 1$.

13. $f(x) = \begin{cases} x^2 & \text{for } x < 2 \\ 2 & \text{for } x \geq 2 \end{cases}$

Not continuous at $x = 2$. Small change in x around $x = 2$ produces a large change in $f(x)$.

17. $f(x) = 3x - 2$

x	2.900	2.990	2.999	3.001	3.010	3.100
$f(x)$	6.700	6.970	6.997	7.003	7.030	7.300

Therefore, $\lim_{x \to 3} (3x - 2) = 7$

21. $f(x) = \dfrac{2 - \sqrt{x+2}}{x-2}$

x	1.900	1.990	1.999
$f(x)$	-0.2516	-0.2502	$-0.250\,02$

x	2.001	2.010	2.100
$f(x)$	$-0.249\,98$	-0.2498	-0.2485

Therefore, $\lim_{x \to 2} f(x) = -0.25$

25. Since the function is continuous at $x = 3$, we may evaluate the limit by substitution. For $f(x) = 3x - 2$, $f(3) = 7$.

29. $\lim_{x \to 0} \dfrac{x^2 + x}{x} = \lim_{x \to 0} \dfrac{x(x+1)}{x}$
$$= \lim_{x \to 0} (x + 1)$$
$$= 0 + 1 = 1$$

33. $\lim_{x \to 1} \dfrac{x^3 - x}{x - 1} = \lim_{x \to 1} \dfrac{x(x^2 - 1)}{x - 1}$
$$= \lim_{x \to 1} \dfrac{x(x+1)(x-1)}{x - 1}$$
$$= \lim_{x \to 1} x(x + 1) = 1(2) = 2$$

37. $\lim\limits_{x\to -1}\sqrt{x}(x+1)$ does not exist since x cannot approach -1 without the function going through imaginary values. $f(0)$, which is real, is the only defined value for the function near $x = -1$.

41. $\lim\limits_{x\to\infty}\dfrac{3x^2+5}{x^2-2} = \lim\limits_{x\to\infty}\dfrac{(3x^2+5)\div x^2}{(x^2-2)\div x^2}$

$$= \lim\limits_{x\to\infty}\dfrac{3+\frac{5}{x^2}}{1-\frac{2}{x^2}}$$

$$= \dfrac{3+0}{1-0} = 3$$

45. $\lim\limits_{x\to 0}\dfrac{x^2-3x}{x}$

x	-0.1	-0.01	-0.001	0.001	0.01	0.1
$f(x)$	-3.10	-3.01	-3.001	-2.999	-2.99	-2.9

We see that $f(x) \to -3$ as $x \to 0$. We now find the limit by changing the algebraic form.

$$\lim\limits_{x\to 0}\dfrac{x^2-3x}{x} = \lim\limits_{x\to 0}\dfrac{x(x-3)}{x} = \lim\limits_{x\to 0} x-3 = 0-3 = -3$$

49. $\lim\limits_{t\to 0} v = 3 \text{ cm/s}$

d	0.480000	0.280000	0.029800	0.002998	0.00029998
t	0.200000	0.100000	0.010000	0.001000	0.00010000
$v = \frac{d}{t}$	2.40000	2.80000	2.98000	2.99800	2.99800

53. $\lim\limits_{x\to 0}(1+x)^{1/x} = 2.71827$ or e

x	0.1	0.01	0.001	0.0001	0.00001
$(1+x)^{1/x}$	2.6	2.70	2.716	2.718	2.71827

23.2 The Slope of a Tangent to a Curve

1. $y = x^2$; $P = (2,4)$

	Q_1	Q_2	Q_3	Q_4	P
x_2	1.5	1.9	1.99	1.99	2
y_2	2.25	3.61	3.9601	3.996001	4
$y_2 - 4$	-1.75	-0.39	-0.0399	-0.003999	
$x_2 - 2$	-0.5	-0.1	-0.01	-0.001	
$m = \frac{y_2-4}{x_2-2}$	3.5	3.9	3.99	3.999	
$m_{\tan} = 4$					

5. $y = x^2$; $P = (2,4)$
$4 + \Delta y = (2 + \Delta x)^2$; $4 + \Delta y = 4 + 4\Delta x + (\Delta x)^2$
$\Delta y = 4 - 4 + 4\Delta x + (\Delta x)^2$; $\Delta y = 4\Delta x + (\Delta x)^2$

$$m_{PQ} = \dfrac{\Delta y}{\Delta x} = \dfrac{4\Delta x + (\Delta x)^2}{\Delta x} = \dfrac{\Delta x(4 + \Delta x)}{\Delta x}$$

As Δx approaches zero, $m_{\tan} = 4 + 0 = 4$

9. $y = x^2$; $x = 2$, $x = -1$
$y_1 + \Delta y = (x_1 + \Delta x)^2$;
$y_1 + \Delta y = x_1^2 + 2x_1\Delta x + (\Delta x)^2$
$y_1 + \Delta y - y_1 = x_1^2 + 2x_1\Delta x + (\Delta x)^2 - x_1^2$
$\Delta y = 2x, \Delta x + (\Delta x)^2$

$$m_{\tan} = \frac{\Delta y}{\Delta x} = \frac{2x_1\Delta x + (\Delta x)^2}{\Delta x}$$
$$= \frac{(\Delta x)(2x_1 + \Delta x)}{\Delta x}$$
$$= 2x_1 + \Delta x$$
As $\Delta x \to 0$, $m_{\tan} = 2x_1$
$x_1 = 2$, $m_{\tan} = 4$
$x_1 = -1$, $m_{\tan} = -2$

13. $y = x^2 + 4x + 5$; $x = -3$, $x = 2$;
$y_1 = x_1^2 + 4x_1 + 5$
$y_1 + \Delta y = (x_1 + \Delta x)^2 + 4(x_1 + \Delta x) + 5$
$y_1 + \Delta y = x_1^2 + 2x_1\Delta x + (\Delta x)^2 + 4x_1 + 4\Delta x + 5$
$y_1 + \Delta y - y_1 = \Delta y = 2x_1\Delta x + (\Delta x)^2 + 4\Delta x$

$$\frac{\Delta y}{\Delta x} = \frac{\Delta x(2x_1 + \Delta x + 4)}{\Delta x} = 2x_1 + 4 + \Delta x$$
As $\Delta x \to 0$, $m_{\tan} = 2x_1 + 4$
$x_1 = -3$, $m_{\tan} = -2$
$x_1 = 2$, $m_{\tan} = 8$

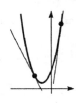

17. $y = x^4$; $x = 0$, $x = 0.5$, $x = 1$
$y_1 + \Delta y = (x_1 + \Delta x)^4$
$y_1 + \Delta y = x_1^4 + 4x_1^3\Delta x + 6x_1^2\Delta x^2 + 4x_1\Delta x^3 + x^4$
$\Delta y = 4x_1^3\Delta x + 6x_1^2\Delta x^2 + 4x_1\Delta x^3 + \Delta x^4$
$\Delta y = \Delta x(4x_1^3 + 6x_1^2\Delta x + 4x_1\Delta x^2 + \Delta x^3)$

$$m_{PQ} = \frac{\Delta y}{\Delta x} = 4x_1^3 + 6x_1^2\Delta x + 4x_1\Delta x^2 + \Delta x^3$$

$\Delta x \to 0$, $m_{PQ} \to m_{\tan} = 4x_1^3$
$x_1 = 0$, $m_{\tan} = 0$

$x_1 = 0.5$, $m_{\tan} = 0.5$
$x_1 = 1$, $m_{\tan} = 4$

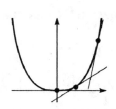

21. $y = 2x - 3x^2$; $x = 0$, $x = 0.5$
$y_1 + \Delta y = 2(x_1 + \Delta x) - 3(x_1 + \Delta x)^2$
$y_1 + \Delta y = 2x_1 + 2\Delta x - 3x_1^2 - 6x_1\Delta x - 3\Delta x^2$
$\Delta y = 2x_1 + 2\Delta x - 3x_1^2 - 6x_1\Delta x - 3\Delta x^2 - 2x_1 + 3x_1^2$
$\Delta y = 2\Delta x - 6x_1\Delta x - 3\Delta x^2 = \Delta x(2 - 6x_1 - 30\Delta x)$

$$m_{PQ} = \frac{\Delta y}{\Delta x} = 2 - 6x_1 - 3\Delta x$$

As $\Delta x \to 0$, $m_{PQ} \to m_{\tan} = 2 - 6x_1$
$x_1 = 0$, $m_{\tan} = 2$
$x_1 = 0.5$, $m_{\tan} = -1$

25. $y = x^2 + 2$; $P(2, 6), Q(2.1, 6.41)$

From P to Q, x changes by 0.1 units and Q by 0.41 units.

The average change in y for 1 unit change in x is

$\frac{0.41}{0.1} = 4.1$ units.

$y + \Delta y = (x + \Delta x)^2 + 2$
$y + \Delta y = x^2 + 2x\Delta x + \Delta x^2 + 2$
$y + \Delta y - y = (x^2 + 2x\Delta x + \Delta x^2 + 2) - (x^2 + 2)$
$\Delta y = 2x\Delta x + \Delta x^2$

$m_{PQ} = \dfrac{\Delta y}{\Delta x} = 2x + \Delta x$; $m_{\tan} = \lim\limits_{\Delta x \to 0} 2x + \Delta x = 2x$

$x = 2$, $m_{\tan} = 4$ (Instantaneous rate of change)

$\qquad m_{PQ} = 4.1$ (Average rate of change)

23.3   The Derivative

1. $y = 3x - 1$; $y + \Delta y = 3(x + \Delta x) - 1$
$\quad y + \Delta y = 3x + 3\Delta x - 1$
$\quad y + \Delta y - y = 3x + 3\Delta x - 1 - (3x - 1)$
$\qquad \Delta y = 3\Delta x$

$\qquad \dfrac{\Delta y}{\Delta x} = \dfrac{3\Delta x}{\Delta x}$;

$\qquad \lim\limits_{\Delta x \to 0} = 3$; $\dfrac{dy}{dx} = 3$

5. $y = x^2 - 1$; $y + \Delta y = (x + \Delta x)^2 - 1$
$\quad y + \Delta y = x^2 + 2x\Delta x + (\Delta x)^2 - 1$
$\quad y + \Delta y - y = x^2 + 2x\Delta x + (\Delta x)^2 - 1 - (x^2 - 1)$
$\qquad \Delta y = 2x\Delta x + (\Delta x)^2$;

$\qquad \dfrac{\Delta y}{\Delta x} = \dfrac{\Delta x(2x + \Delta x)}{\Delta x} = 2x + \Delta x$

$\qquad \lim\limits_{\Delta x \to 0} (2x + \Delta x) = 2x + 0 = 2x$; $\dfrac{dy}{dx} = 2x$

9. $y = x^2 - 7x$; $y + \Delta y = (x + \Delta x)^2 - 7(x + \Delta x)$
$\quad y + \Delta y = x^2 + 2x\Delta x + (\Delta x)^2 - 7x - 7\Delta x)$
$\quad y + \Delta y - y = x^2 + 2x\Delta x + (\Delta x)^2 - 7x - 7\Delta x - (x^2 - 7x)$
$\qquad \Delta y = 2x\Delta x + (\Delta x)^2 - 7\Delta x$

$\qquad \dfrac{\Delta y}{\Delta x} = \dfrac{\Delta x(2x + \Delta x - 7)}{\Delta x} = 2x + \Delta x - 7$

$\qquad \lim\limits_{\Delta x \to 0} (2x - \Delta x - 7) = 2x - 7$; $\dfrac{dy}{dx} = 2x - 7$

13. $y = x^3 + 4x - 6$; $y + \Delta y = (x + \Delta x)^3 + 4(x + \Delta x) - 6$
$\quad y + \Delta y = x^3 + 3x^2\Delta x + 3x(\Delta x)^2 + (\Delta x)^3 + 4x + 4\Delta x - 6$
$\quad y + \Delta y - y = x^3 + 3x^2\Delta x + 3x(\Delta x)^2 + (\Delta x)^3 + 4x + 4\Delta x$
$\quad -6 - (x^3 + 4x - 6)$
$\qquad \Delta y = 3x^2\Delta x + 3x(\Delta x)^2 + (\Delta x)^3 + 4\Delta x$

$\qquad \dfrac{\Delta y}{\Delta x} = \dfrac{\Delta x[3x^2 + 3x\Delta x + (\Delta x)^2 + 4]}{\Delta x}$

$\qquad\quad = 3x^2 + 3x\Delta x + (\Delta x)^2 + 4$

$\qquad \lim\limits_{\Delta x \to 0} [3x^2 + 3x\Delta x + (\Delta x)^2 + 4] = 3x^2 + 4$; $\dfrac{dy}{dx} = 3x^2 + 4$

17. $y = x + \dfrac{4}{3x};\ y + \Delta y = x + \Delta x + \dfrac{4}{3(x + \Delta x)}$

$\quad y + \Delta y - y = x + \Delta x + \dfrac{4}{3(x + \Delta x)} - x - \dfrac{4}{3x}$

$\quad \Delta y = \Delta x + \dfrac{4}{3(x + \Delta x)} - \dfrac{4}{3x}$

$\quad \Delta y = \dfrac{\Delta x(3x(x + \Delta x)) + 4x - 4(x + \Delta x)}{3x(x + \Delta x)}$

$\quad \dfrac{\Delta y}{\Delta x} = \dfrac{3x^2 + 3x\Delta x - 4}{3x(x + \Delta x)}$

$\quad \lim\limits_{\Delta x \to 0} \dfrac{3x^2 + 3x\Delta x - 4}{3x(x + \Delta x)} = \dfrac{3x^2 - 4}{3x^2} = 1 - \dfrac{4}{3x^2};\ \dfrac{dy}{dx} = 1 - \dfrac{4}{3x^2}$

21. $y = x^4 + x^3 + x^2 + x$

$\quad y + \Delta y = (x + \Delta x)^4 + (x + \Delta x)^3 + (x + \Delta x)^2 + (x + \Delta x)$

$\quad\quad = [x^4 + 4x^3\Delta x + 6x^2(\Delta x)^2 + 4x(\Delta x)^3 + (\Delta x)^4]$

$\quad\quad\quad + [x^3 + 3x^2\Delta x + 3x(\Delta x)^2 + (\Delta x)^3] + [x^2 + 2x\Delta x + (\Delta x)^2] + (x + \Delta x)$

$\quad y + \Delta y - y = 4x^3\Delta x + 6x^2(\Delta x)^2 + 4x(\Delta x)^3 + (\Delta x)^4 + 3x^2\Delta x + 3x(\Delta x)^2$

$\quad\quad\quad + (\Delta x)^3 + 2x\Delta x + (\Delta x)^2 + \Delta x$

$\quad \Delta y = (\Delta x)[4x^3 + 6x^2\Delta x + 4x(\Delta x)^2 + (\Delta x)^3 + 3x^2 + 3x(\Delta x) + (\Delta x)^2 + 2x + \Delta x + 1]$

$\quad \dfrac{\Delta y}{\Delta x} = (\Delta x)[4x^3 + 6x^2\Delta x + 4x(\Delta x)^2 + (\Delta x)^3 + 3x^2 + 3x(\Delta x) + (\Delta x)^2 + 2x + \Delta x + 1]/\Delta x$

$\quad \dfrac{\Delta y}{\Delta x} = 4x^3 + 6x^2\Delta x + 4x(\Delta x)^2 + (\Delta x)^3 + 3x^2 + 3x(\Delta x) + (\Delta x)^2 + 2x + \Delta x + 1$

$\quad \lim\limits_{\Delta x \to 0} \dfrac{\Delta y}{\Delta x} = 4x^3 + 3x^2 + 2x + 1;\ \dfrac{dy}{dx} = 4x^3 + 3x^2 + 2x + 1$

25. $y = 3x^2 - 2x;\ (-1, 5)$

$\quad y + \Delta y = 3(x + \Delta x)^2 - 2(x + \Delta x)$

$\quad y + \Delta y = 3x^2 + 6x\Delta x + 3\Delta x^2 - 2x - 2\Delta x$

$\quad y + \Delta y - y = 3x^2 + 6x\Delta x + 3\Delta x^2 - 2x - 2\Delta x - 3x^2 + 2x$

$\quad \Delta y = 6x\Delta x + 3\Delta x^2 - 2\Delta x$

$\quad \dfrac{\Delta y}{\Delta x} = 6x + 3\Delta x - 2;\ \dfrac{dy}{dx} = \lim\limits_{\Delta x \to 0} \dfrac{\Delta y}{\Delta x} = 6x - 2$

$\quad \dfrac{dy}{dx}\bigg|_{(-1.5)} = 6(-1) - 2 = -8$

29. $y = 1 + \dfrac{2}{x}$

$$y + \Delta y = 1 + \frac{2}{x + \Delta x} = \frac{x + \Delta x + 2}{x + \Delta x}$$

$$y + \Delta y - y = \frac{x + \Delta x + 2}{x + \Delta x} - 1 - \frac{2}{x} = \frac{x^2 + x\Delta x + 2x - x^2 - x\Delta x - 2x - 2\Delta x}{x(x + \Delta x)}$$

$$\Delta y = \frac{-2\Delta x}{x(x + \Delta x)}; \frac{\Delta y}{\Delta x} = \frac{-2}{x(x + \Delta x)}$$

$$\frac{dy}{dx} = \lim_{\Delta x \to 0} \frac{\Delta y}{\Delta x} = -\frac{2}{x^2}$$

Differentiable for all x, except $x = 0$.

33. $y = \sqrt{x + 1}$; $y^2 = x + 1$; $(y + \Delta y)^2 = x + \Delta x + 1$
$y^2 + 2y\Delta y + \Delta y^2 = x + \Delta x + 1$
$2y\Delta y + \Delta y^2 = x + \Delta x + 1 - x - 1 = \Delta x$
$\Delta y(2y + \Delta y) = \Delta x$

$$\Delta y = \frac{\Delta x}{2y + \Delta y}; \frac{\Delta y}{\Delta x} = \frac{1}{2y + \Delta y}; \left(\begin{array}{c} \Delta x \to 0 \\ \Delta y \to 0 \end{array} \right)$$

$$\frac{dy}{dx} = \lim_{\Delta x \to 0} \frac{\Delta y}{\Delta x} = \lim_{\Delta y \to 0} \frac{1}{2y + \Delta y} = \frac{1}{2y}$$

$$\frac{dy}{dx} = \frac{1}{2y} = \frac{1}{2\sqrt{x + 1}}$$

23.4   The Derivative as an Instantaneous Rate of Change

1. $y = x^2 - 1$; $(2, 3)$
$y + \Delta y = (x + \Delta x)^2 - 1$
$y + \Delta y = x^2 + 2x\Delta x + (\Delta x)^2 - 1$
$y + \Delta y - y = x^2 + 2x\Delta x + (\Delta x)^2 - 1 - (x^2 - 1)$
$\Delta y = 2x\Delta x + (\Delta x)^2$

$$\frac{\Delta y}{\Delta x} = 2x + \Delta x$$

$$\lim_{\Delta x \to 0} \frac{\Delta y}{\Delta x} = 2x;\ m_{\tan(2,3)} = 2(2) = 4$$

5. $s = 4t + 10$

	Q_1	Q_2	Q_3	Q_4	Q_5	P
t_2	2.0	2.5	2.9	2.99	2.999	3
s_2	18	20	21.6	21.96	21.996	22
t_{2-3}	−1	−0.5	−0.1	−0.01	−0.001	
$s_2 - 22$	−4	−2	−0.4	−0.04	−0.004	
$v = \frac{s_2-22}{t_2-3}$	4.00	4.00	4.00	4.00	4.00	

$$\lim_{t \to 3} \frac{\Delta s}{\Delta t} = 4 \text{ ft/s}$$

9. $s = 4t + 10; \ t = 3$

$s + \Delta s = 4(t + \Delta t) + 10 = 4t + 4\Delta t + 10$

$s + \Delta s - s = 4t + 4\Delta t + 10 - (4t + 10)$

$\Delta s = 4\Delta t; \ \dfrac{\Delta s}{\Delta t} = 4; \ \lim\limits_{\Delta t \to 0} \dfrac{\Delta s}{\Delta t} = \dfrac{ds}{dt} = 4$

$\dfrac{ds}{dt}\bigg|_{t=3} = 4 \text{ ft/s}$

13. $v = \lim\limits_{\Delta t \to 0} \dfrac{\Delta s}{\Delta t}; \ s + \Delta s = 3(t + \Delta t) - \dfrac{2}{5(t + \Delta t)}$

$s + \Delta s - s = 3t + 3\Delta t - \dfrac{2}{5(t + \Delta t)} - \left(3t - \dfrac{2}{5t}\right)$

$\Delta s = 3\Delta t - \dfrac{2}{5(t + \Delta t)} + \dfrac{2}{5t} = 3\Delta t + \dfrac{-2t + 2(t + \Delta t)}{5(t + \Delta t)(t)}$

$\Delta s = 3\Delta t + \dfrac{2\Delta t}{5(t + \Delta t)(t)}; \ \dfrac{\Delta s}{\Delta t} = \dfrac{3\Delta t}{\Delta t} + \dfrac{2\Delta t}{5(t + \Delta t)(t)} \times \dfrac{1}{\Delta t}$

$\dfrac{\Delta s}{\Delta t} = 3 + \dfrac{2}{5(t + \Delta t)(t)}$

$\lim\limits_{\Delta t \to 0} \dfrac{\Delta s}{\Delta t} = \dfrac{ds}{dt} = 3 + \dfrac{2}{5t^2} = v \text{ (instantaneous velocity)}$

17. $a = \lim\limits_{\Delta t \to 0} \dfrac{\Delta v}{\Delta t}; \ v + \Delta v = 6(t + \Delta t)^2 - 4(t + \Delta t) + 2$

$v + \Delta v = 6[t^2 + 2t\Delta t + (\Delta t)^2] - 4t - 4\Delta t + 2$

$v + \Delta v - v = 6t^2 + 12t\Delta t + 6(\Delta t)^2 - 4t - 4\Delta t + 2 - (6t^2 - 4t + 2)$

$\Delta v = 12t\Delta t + 6(\Delta t)^2 - 4\Delta t$

$\dfrac{\Delta v}{\Delta t} = \dfrac{\Delta t(12t + 6\Delta t - 4)}{\Delta t} = 12t + 6\Delta t - 4$

$\lim\limits_{\Delta t \to 0} \dfrac{\Delta v}{\Delta t} = \dfrac{dv}{dt} = 12t - 4 = a \text{ (instantaneous acceleration)}$

21. $q = 30 - 2t; \ q + \Delta q = 30 - 2(t + \Delta t)$

$q + \Delta q = 30 - 2t - 2\Delta t$

$q + \Delta q - q = (30 - 2t - 2\Delta t) - (30 - 2t) = 30 - 2t - 2\Delta t - 30 + 2t$

$\Delta q = -2\Delta t; \ \dfrac{\Delta q}{\Delta t} = -2$

Therefore, $i = \lim\limits_{\Delta t \to 0} \dfrac{\Delta q}{\Delta t} = -2$

25. $P = 500 + 250m^2$

$P + \Delta P = 500 + 250(m + \Delta m)^2$

$P + \Delta P = 500 + 250(m^2 + 2m\Delta m + (\Delta m)^2)$

$P + \Delta P = 500 + 250m^2 + 500m\Delta m + (\Delta m)^2$

$P + \Delta P - P = 500 + 250m^2 + 500m\Delta m + (\Delta m)^2 - (500 + 250m^2)$

$\Delta P = 500m\Delta m + (\Delta m)^2; \quad \dfrac{\Delta P}{\Delta m} = 500m + \Delta m$

$\lim\limits_{\Delta m \to 0} \dfrac{\Delta P}{\Delta m} = \dfrac{dP}{dm} = 500m$

$\dfrac{dP}{dm}\bigg|_{m=0.92} = 500(0.92) = 460 \text{ W}$

29. The volume of a cone $= \frac{1}{3}\pi r^2 h$. For this cone, the radius and height are equal to 4 cm. Due to the similarity of the figures, as the level of the oil decreases, the radius and height will still be equal; $r = h = d$. Therefore,

$V = \dfrac{1}{3}\pi d^2(d) = \dfrac{1}{3}\pi d^3$

$V + \Delta V = \dfrac{1}{3}\pi(d + \Delta d)^3 = \dfrac{1}{3}\pi(d^3 + 3d\Delta d^2 + 3d^2\Delta d + \Delta d^3)$

$V + \Delta V - V = \dfrac{1}{3}\pi d^3 + \pi d\Delta d^2 + \pi d^2\Delta d + \dfrac{1}{3}\pi\Delta d^3 - \dfrac{1}{3}\pi d^3$

$\Delta V = \pi d\Delta d^2 + \pi d^2\Delta d + \dfrac{1}{3}\pi\Delta d^3$

$\dfrac{\Delta V}{\Delta d} = \pi d\Delta d + \pi d^2 + \dfrac{1}{3}\pi\Delta d^2; \quad \lim\limits_{\Delta d \to 0} \dfrac{\Delta V}{\Delta d} = \dfrac{dV}{dd} = \pi d^2$

23.5   Derivatives of Polynomials

1. $y = x^5; \quad \dfrac{dy}{dx} = 5x^{5-1} = 5x^4$

5. $y = x^4 - 6; \quad \dfrac{dy}{dx} = 4x^{4-1} - 0 = 4x^3$

9. $p = 5r^3 - 2r + 1; \dfrac{dp}{dr} = 5(3r^2) - 2 + 0$

$= 15r^2 - 2$

13. $f(x) = -6x^7 + 5x^3 + \pi^2$

$\dfrac{f(x)}{dx} = -6(7x^6) + 5(3x^2) + 0 = -42x^6 + 15x^2$

17. $y = 6x^2 - 8x + 1; \dfrac{dy}{dx} = \dfrac{d(6x^2)}{dx} - \dfrac{d(8x)}{dx} + \dfrac{d(1)}{dx} = 12x - 8 + 0$

Since the derivative is a function of only x, we now evaluate it for $x = 2$.

$\dfrac{dy}{dx}\bigg|_{x=2} = 12(2) - 8 = 24 - 8 = 16$

21. $y = 2x^6 - 4x^2$; $m_{\tan} = \dfrac{dy}{dx} = 12x^5 - 8x$

$\dfrac{dy}{dx}\Big|_{x=-1} = m_{\tan} = 12(-1)^5 - 8(-1) = -12 + 8 = -4$

Move the trace to $x = -1$ and observe that the function is decreasing and that the slope is negative.

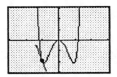

25. $s = 6t^5 - 5t + 2$; $v = \dfrac{ds}{dt} = 30t^4 - 5$

29. $s = 2t^3 - 4t^2$; $t = 4$

$v = \dfrac{ds}{dt} = 2(3t^2) - 4(2t) = 6t^2 - 8t$

$v\big|_{t=4} = 6(4^2) - 8(4) = 64$

33. $y = 3x^2 - 6x$; $m_{\tan} = \dfrac{dy}{dx} = 6x - 6$

Tangent is parallel where slope is zero.
Therefore $6x - 6 = 0$; $x = 1$

37. $y = 4x^2 + 3x$; $m_{\tan} = \dfrac{dy}{dx} = 8x + 3$;

$y = 5 - 2x^2$

$m_{\tan} = \dfrac{dy}{dx} = -4x$

The slopes are equal. Therefore, $8x + 3 = -4x$;
$12x = -3$; $x = -\frac{1}{4}$

41. $P = 16i^2 + 60i$;

$\dfrac{dP}{di} = 16(2i) + 60 = 32i + 60$

$\dfrac{dP}{di}\Big|_{i=0.75} = 32(0.75) + 60 = 84$ W/A

45. $h = 0.000104x^4 - 0.0417x^3 + 4.21x^2 - 8.33x$

$\dfrac{dh}{dx} = 0.000416x^3 - 0.1251x^2 + 8.42x - 8.33\big|_{x=120}$

$\dfrac{dh}{dx} = -80.5\ \dfrac{\text{m}}{\text{km}}$

23.6   Derivatives of Products and Quotients of Functions

1. $y = x^2(3x + 2)$; $u = x^2$; $\dfrac{du}{dx} = 2x$; $v = 3x + 2$; $\dfrac{dv}{dx} = 3$

$\dfrac{dy}{dx} = x^2(3) + (3x + 2)(2x) = 3x^2 + 6x^2 + 4x = 9x^2 + 4x$

5. $s = (3t + 2)(2t - 5)$
$u = 3t + 2, \ v = 2t - 5$

$$\frac{du}{dt} = 3 \qquad \frac{dv}{dt} = 2$$

$$\frac{ds}{dt} = (3t + 2)(2) + (2t - 5)(3)$$

$$= 6t + 4 + 6t - 15$$
$$= 12t - 11$$

9. $y = (2x - 7)(5 - 2x);$
$u = (2x - 7); \ v = (5 - 2x)$

$$\frac{dy}{dx} = (2x - 7)(-2) + (5 - 2x)(2)$$

$$= -4x + 14 + 10 - 4x = -8x + 24$$
$$y = (2x - 7)(5 - 2x)$$
$$= 10x - 4x^2 - 35 + 14x$$
$$= -4x^2 + 24x - 35$$

$$\frac{dy}{dx} = -8x + 24$$

13. $y = \dfrac{x}{2x + 3}; \ u = x; \ \dfrac{du}{dx} = 1;$

$v = 2x + 3; \ \dfrac{dv}{dx} = 2$

$$\frac{dy}{dx} = \frac{(2x + 3)(1) - x(2)}{(2x + 3)^2}$$

$$= \frac{2x + 3 - 2x}{(2x + 3)^2}$$

$$= \frac{3}{(2x + 3)^2}$$

17. $y = \dfrac{x^2}{3 - 2x}; \ u = x^2; \ \dfrac{du}{dx} = 2x;$

$v = 3 - 2x; \ \dfrac{dv}{dx} = -2$

$$\frac{dy}{dx} = \frac{(3 - 2x)(2x) - (x^2)(-2)}{(3 - 2x)^2}$$

$$= \frac{6x - 4x^2 + 2x^2}{(3 - 2x)^2}$$

$$= \frac{6 - 2x^2}{(3 - 2x)^2}$$

21. $f(x) = \dfrac{3x + 8}{x^2 + 4x + 2}$

$$\frac{df(x)}{dx} = \frac{(x^2 + 4x + 2)(3) - (3x + 8)(2x + 4)}{(x^2 + 4x + 2)^2}$$

$$= \frac{-3x^2 - 16x - 26}{(x^2 + 4x + 2)^2}$$

25. $y = (3x - 1)(4 - 7x)$

$$\frac{dy}{dx} = (3x - 1)(-7) + (4 - 7x)(3)$$

$$= -21x + 7 + 12 - 21x$$
$$= -42x + 19$$

$$\left. \frac{dy}{dx} \right|_{x=3} = -42(3) + 19 = -126 + 19 = -107$$

29. $y = \dfrac{3x - 5}{2x + 3}; \ y = 3x - 5; \ v = 2x + 3; \ du = 3dx; \ dv = 2dx$

$$\frac{dy}{dx} = \frac{(2x + 3)(3) - (3x - 5)(2)}{(2x + 3)^2}$$

$$= \frac{6x + 9 - 6x + 10}{(2x + 3)^2}$$

$$= \frac{19}{(2x + 3)^2}$$

$$\left. \frac{dy}{dx} \right|_{x=-2} = \frac{19}{(2(-2) + 3)^2} = \frac{19}{1} = 19$$

33. (1)
$$y = \frac{x^2(1 - 2x)}{3x - 7}$$

$$\frac{dy}{dx} = \frac{(3x - 7)[x^2(-2) + (1 - 2x)(2x)] - x^2(1 - 2x)(3)}{(3x - 7)^2}$$

$$= \frac{(3x - 7)(-6x^2 + 2x) - 3x^2 + 6x^3}{(3x - 7)^2} = \frac{-18x^3 + 6x^2 + 42x^2 - 14x - 3x^2 + 6x^3}{(3x - 7)^2}$$

$$= \frac{-12x^3 + 45x^2 - 14x}{(3x - 7)^2}$$

(2)
$$y = \frac{x^2 - 2x^3}{3x - 7}$$

$$\frac{dy}{dx} = \frac{(3x - 7)(2x - 6x^2) - (x^2 - 2x^3)(3)}{(3x - 7)^2} = \frac{-12x^3 + 45x^2 - 14x}{(3x - 7)^2}$$

37. $y = \dfrac{x}{x^2 + 1}; \ y = x; \ \dfrac{du}{dx} = 1; \ v = x^2 + 1; \ \dfrac{dv}{dx} = 2x$

$$\frac{dy}{dx} = \frac{(x^2 + 1)(1) - (x)(2x)}{(x^2 + 1)^2} = \frac{x^2 + 1 - 2x^2}{(x^2 + 1)^2} = \frac{-x^2 + 1}{(x^2 + 1)^2}$$

Therefore, $m_{\tan} = 0$ when $\dfrac{-x^2 + 1}{(x^2 + 1)^2} = 0$;

$-x^2 + 1 = 0; \ x^2 = 1; \ x = 1, -1$

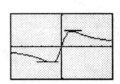

41.
$$V = \frac{6R + 25}{R + 3}$$

$$\frac{dV}{dR} = \frac{6(R + 3) - (6R + 25)}{(R + 3)^2} = \frac{6R + 18 - 6R - 25}{(R + 3)^2} = \frac{-7}{(R + 3)^2}$$

$$\left.\frac{dV}{dR}\right|_{R=7} = \frac{-7}{(7 + 3)^2} = \frac{-7}{100} = -0.07 \text{ V}/\Omega$$

45.
$$r_f = \frac{2(R^2 + Rr + r^2)}{3(R + r)}$$

$$\frac{dr_f}{dR} = \frac{2(2R + r)(3)(R + r) - 2(R^2 + Rr + r^2)(3)}{9(R + r)^2} \quad \frac{6(2R + r)(R + r) - 6(R^2 + Rr + r^2)}{9(R + r)^2}$$

$$= \frac{12R^2 + 18Rr + 6r^2 - 6R^2 - 6Rr - 6r^2}{9(R + r)^2} = \frac{6R^2 + 12Rr}{9(R + r)^2} = \frac{6R(R + 2)}{9(R + r)^2}$$

$$= \frac{2R(R + 2r)}{3(R + r)^2}$$

23.7   The Derivative of a Power of a Function

1. $y = \sqrt{x} = x^{1/2}$

$\dfrac{dy}{dx} = \dfrac{1}{2}x^{1/2-1} = \dfrac{1}{2}x^{-1/2}$

$\qquad = \dfrac{1}{2}\left(\dfrac{1}{x^{1/2}}\right) = \dfrac{1}{2x^{1/2}}$

$\qquad = \dfrac{1}{2x^{1/2}}$

5. $y = \dfrac{3}{\sqrt[3]{x}} = \dfrac{3}{x^{1/3}} = 3x^{-1/3};$

$\dfrac{dy}{dx} = 3\left(-\dfrac{1}{3}x^{-1/3-1}\right)$

$\qquad = -1x^{-4/3} = -1\left(\dfrac{1}{x^{4/3}}\right)$

$\qquad = -\dfrac{1}{x^{4/3}}$

9. $y = (x^2 + 1)^5;$

$\dfrac{dy}{dx} = 5(x^2 + 1)^4(2x)$

$\qquad = 10x(x^2 + 1)^4$

13. $y = (2x^3 - 3)^{1/3};$

$\dfrac{dy}{dx} = \dfrac{1}{3}(2x^3 - 3)^{-2/3}(6x^2)$

$\qquad = \dfrac{2x^2}{(2x^3 - 3)^{2/3}}$

17. $y = 4(2x^4 - 5)^{3/4}; \; u = 2x^4 - 5; \; \dfrac{du}{dx} = 8x^3$

$\dfrac{dy}{dx} = 4\left[\dfrac{3}{4}(2x^4 - 5)^{-1/4}(8x^3)\right] = \dfrac{24x^3}{(2x^4 - 5)^{1/4}}$

21. $y = x\sqrt{8x + 5} = x(8x + 5)^{1/2}$

$\dfrac{dy}{dx} = x\left(\dfrac{1}{2}\right)(8x + 5)^{-1/2}(8) + (8x + 5)^{1/2}(1) = 4x(8x + 5)^{-1/2} + (8x + 5)^{1/2}(1)$

$\qquad = \dfrac{4x}{(8x + 5)^{1/2}} + \dfrac{(8x + 5)}{(8x + 5)^{1/2}} = \dfrac{12x + 5}{(8x + 5)^{1/2}}$

25. $y = \sqrt{3x + 4}; \; x = 7$

$\qquad y = (3x + 4)^{1/2}; \; u = 3x + 4; \; n = \dfrac{1}{2}; \; \dfrac{du}{dx} = 3$

$\qquad \dfrac{dy}{dx} = \dfrac{1}{2}(3x + 4)^{-1/2}(3) = \dfrac{3}{2}(3x + 4)^{-1/2} = \dfrac{3}{2\sqrt{3x + 4}}$

$\left.\dfrac{dy}{dx}\right|_{x=7} = \dfrac{3}{2\sqrt{3(7) + 4}} = \dfrac{3}{2\sqrt{25}} = \dfrac{3}{2(5)} = \dfrac{3}{10}$

29. (a) $y = \dfrac{1}{x^3}; \; u = 1; \; \dfrac{du}{dx} = 0; \; v = x^3; \; \dfrac{dv}{dx} = 3x^2$

$\qquad \dfrac{dy}{dx} = \dfrac{x^3(0) - 1(3x^2)}{x^6} = \dfrac{-3x^2}{x^6} = \dfrac{-3}{x^4}$

(b) $y = x^{-3}; \; \dfrac{dy}{dx} = -3x^{-3-1} = -3x^{-4} = \dfrac{-3}{x^4}$

33. $y^2 = 4x$; $y = \sqrt{4x} = 2\sqrt{x} = 2x^{1/2}$

$$m_{\tan} = \frac{dy}{dx} = \frac{d(2x^{1/2})}{dx} = 2\left(\frac{1}{2}\right)x^{-1/2} = \frac{1}{\sqrt{x}}$$

$$m_{\tan}\big|_{x=1} = \frac{1}{\sqrt{1}} = \frac{1}{1} = 1$$

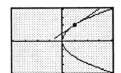

37. $P = \dfrac{k}{V^{3/2}}$; $P = 300$ kPa when $V = 100$ cm^3

$$300 = \frac{k}{100^{3/2}}$$

$$k = 300(100)^{3/2} = 300(10^3) = 300(1000) = 300\,000$$

$$P = \frac{300\,000}{V^{3/2}} = 300\,000V^{-3/2}; \quad \frac{dP}{dV} = 300\,000\left(-\frac{3}{2}\right)V^{-5/2}$$

$$\frac{dP}{dV} = -450\,000V^{-5/2} = \frac{-450\,000}{V^{5/2}}$$

$$\frac{dP}{dV}\bigg|_{V=100} = \frac{-450\,000}{100^{5/2}} = \frac{-450\,000}{100\,000} = -4.50 \text{ kPa/cm}^3$$

41. $\lambda_r = \dfrac{2a\lambda}{\sqrt{4a^2 - \lambda^2}} = \dfrac{2a\lambda}{(4a^2 - \lambda^2)^{1/2}}$

$$\frac{d\lambda_r}{d\lambda} = \frac{(4a^2 - \lambda^2)^{1/2}(2a) - 2a\lambda\left(\frac{1}{2}\right)(4a^2 - \lambda^2)^{-1/2}(-2\lambda)}{(4a^2 - \lambda^2)} = \frac{2a(4a^2 - \lambda^2)^{1/2} + 2a\lambda^2(4a^2 - \lambda^2)^{-1/2}}{(4a^2 - \lambda^2)}$$

$$= \frac{(4a^2 - \lambda^2)^{-1/2}[(2a)(4a^2 - \lambda^2) + 2a\lambda^2]}{(4a^2 - \lambda^2)} = \frac{8a^3}{(4a^2 - \lambda^2)^{3/2}}$$

23.8   Differentiation of Implicit Functions

1. $3x + 2y = 5$; $\dfrac{d(3x)}{dx} + \dfrac{d(2y)}{dx} = \dfrac{d(5)}{dx}$

$3 + \dfrac{2dy}{dx} = 0$; $\dfrac{2dy}{dx} = -3$; $\dfrac{dy}{dx} = -\dfrac{3}{2}$

5. $x^2 - 4y^2 - 9 = 0$; $\dfrac{d(x^2)}{dx} - \dfrac{d(4y^2)}{dx} - \dfrac{d(9)}{dx} = \dfrac{d(0)}{dx}$

$2x - \dfrac{8ydy}{dx} - 0 = 0$; $\dfrac{-8ydy}{dx} = -2x$; $\dfrac{dy}{dx} = \dfrac{x}{4y}$

9. $y^2 + y = x^2 - 4;$ $\dfrac{d(y^2)}{dx} + \dfrac{d(y)}{dx} = \dfrac{d(x^2)}{dx} - \dfrac{d(4)}{dx}$

$$\frac{2y\,dy}{dx} + \frac{dy}{dx} = 2x - 0$$

$$\frac{dy}{dx}(2y + 1) = 2x; \quad \frac{dy}{dx} = \frac{2x}{2y + 1}$$

13. $xy^3 + 3y + x^2 = 9;$

$$\frac{d(xy^3)}{dx} + \frac{d(3y)}{dx} + \frac{d(x^2)}{dx} = \frac{d(9)}{dx}$$

$$\frac{x\,dy^3}{dx} + \frac{y^3\,dx}{dx} + \frac{3\,dy}{dx} + 2x = 0$$

$$x(3y^2)\frac{dy}{dx} + y^3(1) + \frac{3\,dy}{dx} + 2x = 0$$

$$3xy^2\frac{dy}{dx} + y^3 + \frac{3\,dy}{dx} + 2x = 0$$

$$3xy^2\frac{dy}{dx} + \frac{3\,dy}{dx} = -y^3 - 2x$$

$$\frac{dy}{dx}(3xy^2 + 3) = -y^3 - 2x; \quad \frac{dy}{dx} = \frac{-2x - y^3}{3xy^2 + 3}$$

17. $(2y - x)^4 + x^2 = y + 3$

$$4(2y - x)^3\left(2\frac{dy}{dx} - 1\right) + 2x = \frac{dy}{dx} + 0$$

$$4(2y - x)^3\left(2\frac{dy}{dx}\right) - 4(2y - x)^3 + 2x = \frac{dy}{dx}$$

$$8\frac{dy}{dx}(2y - x)^3 - \frac{dy}{dx} = 4(2y - x)^3 - 2x$$

$$\frac{dy}{dx}[8(2y - x)^3 - 1] = 4(2y - x)^3 - 2x$$

$$\frac{dy}{dx} = \frac{4(2y - x)^3 - 2x}{8(2y - x)^3 - 1}$$

21. $3x^3y^2 - 2y^3 = -4;\ (1, 2)$

$$3x^3(2y)\frac{dy}{dx} + y^2(9x^2) - 6y^2\frac{dy}{dx} = 0$$

$$6x^3y\frac{dy}{dx} + 9x^2y^2 - 6y^2\frac{dy}{dx} = 0$$

$$6x^3y\frac{dy}{dx} - 6y^2\frac{dy}{dx} = -9x^2y^2$$

$$\frac{dy}{dx}(6x^3y - 6y^2) = -9x^2y^2$$

$$\frac{dy}{dx} = \frac{-9x^2y^2}{6x^3y - 6y^2} = \frac{-9x^2y^2}{3(2x^3y - 2y^2)}$$

$$\frac{dy}{dx} = \frac{-3x^2y^2}{2x^3y - 2y^2}$$

$$\frac{dy}{dx}\bigg|_{(1,2)} = \frac{-3(1^2)(2^2)}{2(1^3)(2) - 2(2^2)} = \frac{-12}{-4} = 3$$

25. $xy + y^2 + 2 = 0;$

$$\frac{d(xy)}{dx} + \frac{d(y^2)}{dx} + \frac{d(2)}{dx} = \frac{d(0)}{dx}$$

$$\frac{x\,dy}{dx} + \frac{y\,dx}{dx} + \frac{2y\,dy}{dx} + 0 = 0$$

$$\frac{x\,dy}{dx} + y(1) + \frac{2y\,dy}{dx} = 0; \quad \frac{x\,dy}{dx} + \frac{2y\,dy}{dx} = -y$$

$$\frac{dy}{dx}(x + 2y) = -y; \quad \frac{dy}{dx} = \frac{-y}{x + 2y}$$

$$\frac{dy}{dx}\bigg|_{(-3,1)} = \frac{-1}{-3 + 2(1)} = \frac{-1}{-1} = 1$$

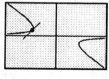

29. $r^2 = 2rR + 2R - 2r;$ $2r = 2R + \dfrac{dR}{dr}(2r) + 2\dfrac{dR}{dr} - 2$

$$2r - 2R + 2 = 2r\frac{dR}{dr} + 2\frac{dR}{dr} = \frac{dR}{dr}(2r + 2)$$

$$\frac{dR}{dr} = \frac{2(r - R + 1)}{2(r + 1)} = \frac{r - R + 1}{r + 1}$$

23.9   Higher Derivatives

1. $y = x^3 + x^2$; $y' = 3x^2 + 2x$; $y'' = 6x + 2$; $y''' = 6$; $y^{(4)} = 0$

5. $y = (1 - 2x)^4$;
$$y' = 4(1 - 2x)^3(-2) = -8(1 - 2x)^3$$
$$y'' = -24(1 - 2x)^2(-2) = 48(1 - 2x)^2;$$
$$y''' = 96(1 - 2x)(-2) = -192(1 - 2x)$$
$$y^{(4)} = -2(-192) = 384;\ y^{(5)} = 0$$

9. $y = 2x^7 - x^6 - 3x$;
$$y' = 14x^6 - 6x^5 - 3;$$
$$y'' = 84x^5 - 30x^4$$

13. $f(x) = \sqrt[4]{8x - 3} = (8x - 3)^{1/4}$
$$f'(x) = \frac{1}{4}(8x - 3)^{-3/4}(8)$$
$$= -12(8x - 3)^{-3/4}$$
$$f''(x) = -\frac{3}{2}(8x - 3)^{-7/4}(8)$$
$$= -12(8x - 3)^{-7/4}$$
$$= \frac{-12}{(8x - 3)^{7/4}}$$

17. $y = 2(2 - 5x)^4$;
$$y' = 8(2 - 5x)^3(-5)$$
$$= -40(2 - 5x)^3;$$
$$y'' = -120(2 - 5x)^2(-5)$$
$$= 600(2 - 5x)^2$$

21. $f(x) = \dfrac{2x}{1 - x}$; $f'(x) = \dfrac{(1 - x)(2) - (2x)(-1)}{(1 - x)^2} = \dfrac{2}{(1 - x)^2}$
$$f''(x) = \frac{(1 - x)^2(0) - 2(2)(1 - x)(-1)}{(1 - x)^4} = \frac{4(1 - x)}{(1 - x)^4} = \frac{4}{(1 - x)^3}$$

25. $x^2 - y^2 = 9$; $2x - 2yy' = 0$; $2x = 2yy'$; $\dfrac{2x}{2y} = \dfrac{x}{y} = y'$
$$y'' = \frac{y(1) - xy'}{y^2} = \frac{y - x\left(\frac{x}{y}\right)}{y^2} = \frac{\frac{y^2 - x^2}{y}}{y^2} = \frac{y^2 - x^2}{y^3} = \frac{-9}{y^3}$$

29. $f(x) = \sqrt{x^2 + 9} = (x^2 + 9)^{1/2}$
$$f'(x) = \frac{1}{2}(x^2 + 9)^{-1/2}(2x) = x(x^2 + 9)^{-1/2}$$
$$f''(x) = x\left[-\frac{1}{2}(x^2 + 9)^{-3/2}(2x)\right] + (x^2 + 9)^{-1/2} = -x^2(x^2 + 9)^{-3/2} + (x^2 + 9)^{-1/2}$$
$$= \frac{-x^2}{\sqrt{(x^2 + 9)^3}} + \frac{1}{\sqrt{x^2 + 9}}$$
$$f''(4) = \frac{-16}{(\sqrt{25})^3} + \frac{1}{\sqrt{25}} = \frac{-16}{5^3} + \frac{1}{5} = \frac{-16}{125} + \frac{25}{125} = \frac{9}{125}$$

33.
$$y = x(1-x)^5$$
$$y' = x[5(1-x)^4(-1)] + (1-x)^5(1) = -5x(1-x)^4 + (1-x)^5$$
$$y'' = (-5x)[4(1-x)^3(-1)] + (1-x)^4(-5) + 5(1-x)^4(-1)$$
$$= 20x(1-x)^3 - 10(1-x)^4$$
$$y''|_{x=2} = 20(2)(1-2)^3 - 10(1-2)^4 = -40 - 10 = -50$$

37. $s = 2250t - 16.1t^2;\ s' = 2250 - 32.2t$
$$s'' = a = -32.2 \text{ ft/s}^2$$

Chapter 23 Review Exercises

1. $\displaystyle\lim_{x \to 4} (8 - 3x) = 8 - 3(4) = -4$

5. $\displaystyle\lim_{x \to 2} \frac{4x - 8}{x^2 - 4} = \lim_{x \to 2} \frac{4(x-2)}{(x-2)(x+2)} = \lim_{x \to 2} \frac{4}{x+2} = \frac{4}{2+2} = 1$

9. $\displaystyle\lim_{x \to \infty} \frac{2 + \frac{1}{x+4}}{3 - \frac{1}{x^2}} = \frac{2+0}{3-0} = \frac{2}{3}$

13.
$$y = 7 + 5x$$
$$y + \Delta y = 7 + 5(x + \Delta x) = 7 + 5x + 5\Delta x$$
$$\Delta y = 7 + 5x + 5\Delta x - (7 + 5x) = 5\Delta x$$
$$\frac{\Delta y}{\Delta x} = \frac{5\Delta x}{\Delta x} = 5$$
$$\lim_{\Delta x \to 0} 5 = 5$$

17.
$$y = \frac{2}{x^2}$$
$$y + \Delta y = \frac{2}{(x + \Delta x)^2}$$
$$\Delta y = \frac{2}{(x + \Delta x)^2} - \frac{2}{x^2} = \frac{2x^2 - 2(x + \Delta x)^2}{x^2(x + \Delta x)^2} = \frac{2x^2 - 2x^2 - 4x\Delta x - 2(\Delta x)^2}{x^2(x + \Delta x)^2}$$
$$= \frac{\Delta x(-4x - 2\Delta x)}{x^2(x + \Delta x)^2}$$
$$\frac{\Delta y}{\Delta x} = \frac{-4x - 2\Delta x}{x^2(x + \Delta x)^2}$$
$$\lim_{\Delta x \to 0} \frac{-4x - 2\Delta x}{x^2(x + \Delta x)^2} = \frac{-4x - 0}{x^2(x + 0)^2} = \frac{-4x}{x^2(x^2)} = \frac{-4}{x^3}$$

21. $y = 2x^7 - 3x^2 + 5$
$$\frac{dy}{dx} = 2(7x^6) - 3(2x) + 0 = 14x^6 - 6x$$

25. $f(y) = \dfrac{3y}{1 - 5y}$

$\dfrac{df(y)}{dy} = \dfrac{(1 - 5y)(3) - 3y(-5)}{(1 - 5y)^2} = \dfrac{3 - 15y + 15y}{(1 - 5y)^2}$

$\dfrac{df(y)}{dy} = \dfrac{3}{(1 - 5y)^2}$

29. $y = \dfrac{3}{(5 - 2x^2)^{3/4}}; \ u = 3, \ \dfrac{du}{dx} = 0; \ v = (5 - 2x^2)^{3/4}$

$\dfrac{dy}{dx} = \dfrac{(5 - 2x^2)^{3/4}(0) - 3[-3x(5 - 2x^2)^{-1/4}]}{(5 - 2x^2)^{3/2}}$ $\dfrac{dv}{dx} = \dfrac{3}{4}(5 - 2x^2)^{-1/4}(-4x) = -3x(5 - 2x^2)^{-1/4}$

$\quad = \dfrac{9x(5 - 2x^2)^{-1/4}}{(5 - 2x^2)^{3/2}}$

$\quad = \dfrac{9x}{(5 - 2x^2)^{3/2}(5 - 2x^2)^{1/4}}$

$\quad = \dfrac{9x}{(5 - 2x^2)^{7/4}}$

33. $y = \dfrac{\sqrt{4x + 3}}{2x}; \ u = \sqrt{4x + 3} = (4x + 3)^{1/2}; \ \dfrac{du}{dx} = \dfrac{1}{2}(4x + 3)^{-1/2}(4)$

$\quad = 2(4x + 3)^{-1/2}$

$v = 2x, \ \dfrac{dv}{dx} = 2$

$\dfrac{dy}{dx} = \dfrac{2x(2)(4x + 3)^{-1/2} - (4x + 3)^{1/2}(2)}{(2x)^2}$

$\quad = \dfrac{(4x + 3)^{-1/2}[4x - (4x + 3)(2)]}{4x^2} = \dfrac{-4x - 6}{(4x + 3)^{1/2}(4x^2)} = \dfrac{2(-2x - 3)}{2(2x^2)(4x + 3)^{1/2}}$

$\quad = \dfrac{-2x - 3}{2x^2(4x + 3)^{1/2}}$

37. $y = \dfrac{4}{x} + 2\sqrt[3]{x}, \ x = 8$

$y = 4x^{-1} + 2x^{1/3}; \ \dfrac{dy}{dx} = 4(-1)x^{-2} + 2\left(\dfrac{1}{3}x^{-2/3}\right) = \dfrac{-4}{x^2} + \dfrac{2}{3x^{2/3}}$

$\dfrac{dy}{dx}\bigg|_{x=8} = \dfrac{-4}{8^2} + \dfrac{2}{3(8)^{2/3}} = \dfrac{-4}{64} + \dfrac{2}{3(4)} = \dfrac{-1}{16} + \dfrac{1}{6} = \dfrac{-3}{48} + \dfrac{8}{48} = \dfrac{5}{48}$

41. $y = 3x^4 - \dfrac{1}{x} = 3x^4 - x^{-1}$

$y' = 12x^3 + x^{-2}$

$y'' = 36x^2 - 2x^{-3}$

45. On a graphing calculator enter $y_1 = $ $\boxed{(}$ $\boxed{2}$ $\boxed{\times}$ $\boxed{(}$ $\boxed{x|t}$ $\boxed{x^2}$ $\boxed{-}$ $\boxed{4}$ $\boxed{)}$ $\boxed{)}$ $\boxed{\div}$ $\boxed{(}$ $\boxed{x|t}$ $\boxed{-}$ $\boxed{2}$ $\boxed{)}$

Using trace, when $x = 1.95$, $y = 7.9$; when $x = 2.05$, $y = 8.1$. Using zoom, when $x = 2.003$, $y = 8.006$. When $x = 1.997$, $y = 7.994$. The curve is discontinuous at $x = 2$; however, the accuracy of 3 significant digits may be achieved.

49. $y = 7x^4 - x^3$; $m_{\tan} = dy/dx = 28x^3 - 3x^2$; when $x = -1$, $m_{\tan} = 28(-1)^3 - 3(-1)^2 = -31$.

On a graphing calculator, enter $y_1 = 7$ $\boxed{x|t}$ $\boxed{\wedge}$ $\boxed{4}$ $\boxed{-}$ $\boxed{x|t}$ $\boxed{\wedge}$ $\boxed{3}$.
It is seen that the slope is decreasing rapidly when $x = -1$.

53. $R = 1 - kt + \dfrac{k^2 t^2}{2} - \dfrac{k^3 t^3}{6} = 1 - kt + \dfrac{1}{2}(k^2 t^2) - \dfrac{1}{6}(k^3 t^3)$

$R' = -k + k^2 t - \dfrac{1}{2}k^3 t^2$

57. $E = \dfrac{L\,dI}{dt}$; $I = t(0.01t + 1)^3$; $\dfrac{dI}{dt} = (0.01t + 1)^2(0.04t + 1)$; $L = 0.4H$

$E = 0.4\dfrac{dI}{dt}$; substituting the value for $\dfrac{dI}{dt}$, $E = 0.04(0.01t + 1)^2(0.04t + 1)$

61. $f = \dfrac{1}{2\pi\sqrt{C(L+2)}} = \dfrac{1}{2\pi}(CL + 2C)^{-1/2}$

$\dfrac{df}{dL} = -\dfrac{1}{2}\left(\dfrac{1}{2\pi}\right)(CL + 2C)^{-3/2}(C) = -\dfrac{C}{4\pi}(CL + 2C)^{-3/2}$

$= -\dfrac{C}{4\pi\sqrt{C^3(L+2)^3}} = -\dfrac{C}{4\pi C\sqrt{C(L+2)^3}} = -\dfrac{1}{4\pi\sqrt{C(L+2)^3}} = -\dfrac{1}{4\pi\sqrt{C}(L+2)^{3/2}}$

65. $y = \dfrac{W}{24EI}(6L^2 x^2 - 4Lx^3 + x^4)$

$y' = \dfrac{W}{24EI}(12L^2 x - 12Lx^2 + 4x^3) = \dfrac{W}{6EI}(3L^2 x - 3Lx^2 + x^3)$

$y'' = \dfrac{W}{6EI}(3L^2 - 6Lx + 3x^2) = \dfrac{W}{2EI}(L - x)^2$

$y''' = \dfrac{W}{2EI}(-2L + 2x) = \dfrac{W}{EI}(x - L)$

$y^{iv} = \dfrac{W}{EI}$

69. $A = xy = x(4 - x^2) = 4x - x^3$

$\dfrac{dA}{dx} = 4 - 3x^2$

73. The angle at which they cross is the angle between the tangent lines and can be found from the derivatives.

APPLICATIONS OF THE DERIVATIVE

24.1 Tangents and Normals

1. $y = x^2 + 2$ at $(2, 6)$

$\dfrac{dy}{dx} = 2x;\ \dfrac{dy}{dx}\bigg|_{(2,6)} = 4$

$m = 4;\ (x_1, y_1) = (2, 6)$
Eq. T.L.: $y - y_1 = m(x - x_1)$
$y - 6 = 4(x - 2) = 4x - 8$
$y = 4x - 8 + 6 = 4x - 2;$
$4x - y - 2 = 0$

5. $y = 6x - 2x^2$ at $(2, 4)$

$\dfrac{dy}{dx} = 6 - 4x;\ m_{\tan(2.4)} = -2$

Eq. of normal: $m_{\text{normal}} = \dfrac{1}{2}$

$y - y_1 = m(x - x_1)$

$y - 4 = \dfrac{1}{2}(x - 2);\ 2y - 8 = x - 2$

$x - 2y + 6 = 0$ or $y = \dfrac{1}{2}x + \dfrac{13}{2}$

9. $y = \dfrac{1}{\sqrt{x^2 + 1}}$ where $x = \sqrt{3},\ y = \dfrac{1}{2}$

$y = (x^2 + 1)^{-1/2};\ \dfrac{dy}{dx} = -\dfrac{1}{2}(x^2 + 1)^{-3/2}(2x) = -\dfrac{x}{(x^2 + 1)^{3/2}}$

$m_{\tan(\sqrt{3}, 1/2)} = \dfrac{-\sqrt{3}}{8};\ m_{\text{normal}} = \dfrac{8}{\sqrt{3}}$

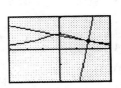

Eq. of T.L.:

$y - \dfrac{1}{2} = -\dfrac{\sqrt{3}}{8}(x - \sqrt{3});\ 8y - 4 = -\sqrt{3}x + 3$

$\sqrt{3}x + 8y - 7 = 0$

Eq. of N.L.:

$y - \dfrac{1}{2} = \dfrac{8}{\sqrt{3}}(x - \sqrt{3});\ 2\sqrt{3}y - \sqrt{3} = 16x - 16\sqrt{3}$

$16x - 2\sqrt{3}y - 15\sqrt{3} = 0$

13. $y = x^2 - 2x$; tangent line with slope of 2

$$m_{\tan} = \frac{dy}{dx} = 2x - 2; \; 2 = 2x - 2; \; x = 2; \; y = 2^2 - 2(2) = 0$$

Therefore, the point at which the slope is 2 is $(2, 0)$, Using the point slope formula for the equation of a line, $y - 0 = 2(x - 2)$; $y = 2x - 4$

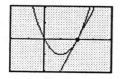

17. For the parabola $y^2 = 4x$, $2yy' = 4$, $y' = \left.\dfrac{2}{y}\right|_{(a,b)} = \dfrac{2}{b}$

$$m_{\text{parabola}}(a, b) = \frac{2}{b}$$

For the ellipse, $2x^2 + y^2 = 6$, $4x + 2yy' = 0$, $y' = \left.\dfrac{-2x}{y}\right|_{(a,b)} = \dfrac{-2a}{b}$

$$m_{ellipse}(a, b) = \frac{-2a}{b}$$

$$m_{\text{parabola}}(a, b) \cdot m_{ellipse}(a, b) = \frac{2}{b} \cdot \frac{-2a}{b} = \frac{-4a}{b^2} \text{ and since } b^2 = 4a$$

$$m_{\text{parabola}}(a, b) \cdot m_{ellipse}(a, b) = \frac{-4a}{-4a} = -1 \text{ which implies the TL's are perpendicular; they intersect at right angles.}$$

21. $y = \sqrt{2x^2 + 8} = (2x^2 + 8)^{1/2}$; $m = \tan 135° = -1$

$$\frac{dy}{dx} = \frac{1}{2}(2x^2 + 8)^{-1/2}(4x) = \frac{2x}{\sqrt{2x^2 + 8}} = 1$$

$2x = \sqrt{2x^2 + 8}$; $4x^2 = 2x^2 + 8$; $2x^2 = 8$; $x^2 = 4$

$x = \pm 2$; $y = \sqrt{2(4) + 8} = \sqrt{16} = 4$

$y - y_1 = m(x - x_1)$

$y - 4 = -1(x - 2)$

$y + x - 6 = 0$

24.2   Newtons Method for Solving Equations

1. $x^2 - 2x - 5 = 0$ (between 3 and 4)
   $f(x) = x^2 - 2x - 5;\ f'(x) = 2x - 2$
   $f(3) = 3^2 - 2(3) - 5 = -2;\ f(4) = 4^2 - 2(4) - 5 = 3$

   The root is possibly closer to 3 than 4. Thus, let $x_1 = 3.3$

   $f(x_1) = 3.3^2 - 2(3.3) - 5 = -0.71;\ f'(x_1) = 2(3.3) - 2 = 4.6$

   $x_2 = x_1 - \dfrac{f(x_1)}{f'(x_1)} = 3.3 - \dfrac{(-0.71)}{4.6} = 3.454\ 347\ 8$

   $f(x_2) = 3.454\ 347\ 8^2 - 2(3.454\ 347\ 8) - 5 = 0.023\ 762\ 7$

   $f'(x_2) = 2(3.454\ 347\ 8 - \dfrac{0.023\ 762\ 7}{4.908\ 695\ 6} = 3.449\ 506\ 9$

   Using the quadratic formula,

   $x = \dfrac{-(-2) \pm \sqrt{(-2)^2 - 4(1)(-5)}}{2} = \dfrac{2 \pm \sqrt{24}}{2}$

   The positive root is the one between 3 and 4

   $x = \dfrac{2 + \sqrt{24}}{2} = 3.449\ 489\ 7$

   The results agree to four (rounded off) decimal places.

5. $x^3 - 6x^2 + 10x - 4 = 0$ (between 0 and 1)
   $f(x) = x^3 - 6x^2 + 10x - 4;\ f'(x) = 3x^2 - 12x + 10;\ f(0) = -4;\ f(1) = 1$

   The root is probably closer to 1. Let $x_1 = 0.7$

n	x_n	$f(x_n)$	$f'(x_n)$	$x_n - \frac{f(x_n)}{f'(x_n)}$
1	0.7	0.403	3.07	0.568 729 6
2	0.568 729 6	−0.069 466 6	4.145 604 9	0.585 486 3
3	0.585 486 3	−0.001 200 9	4.002 547	0.585 786 3
4	0.585 786 3	−0.000 000 5	4.000 001 2	0.585 786 4

   $x_4 = x_3 = 0.585\ 786\ 4$ to seven decimal places

9. $x^4 - x^3 - 3x^2 - x - 4 = 0$; (between 2 and 3)
   $f(x) = x^4 - x^3 - 3x^2 - x - 4;\ f'(x) = 4x^3 - 3x^2 - 6x - 1$
   $f(2) = -10;\ f(3) = 20$; the root is possibly closer to 2.
   Let $x_1 = 2.3$

n	x_n	$f(x_n)$	$f'(x_n)$	$x_n - \frac{f(x_n)}{f'(x_n)}$
1	2.3	−6.352 9	17.998	2.652 978 1
2	2.652 978 1	3.097 272 5	36.657 001	2.568 484 8
3	2.568 484 8	0.217 500 7	31.576 097	2.561 596 7
4	2.561 596 7	0.001 368 3	−31.179 599	2.561 552 8

   $x_3 = x_4 = 2.561\ 552\ 8$ to seven decimal places.

13. $2x^2 = \sqrt{2x+1}$ or $2x^2 - \sqrt{2x+1} = 0$
$4x^4 - 2x - 1 = 0$ (Square both sides.)
$f(x) = 4x^4 - 2x - 1$

From sketch, the intersections lie between 0 and 1, and between 0 and -1. Approximate the positive root at 0.8.

n	x_n	$f(x_n)$	$f'(x_n)$	$x_n - \frac{f(x_n)}{f'(x_n)}$
1	0.8	$-0.961\ 6$	6.192	0.955 297 2
2	0.955 297 2	0.420 707 1	11.948 757	0.920 087 9
3	0.920 087 9	0.026 491 4	10.462 579	0.917 555 9
4	0.917 555 9	0.000 130 1	10.360 1	0.917 543 3

The positive root is approximately 0.917 543 3.

17. $f(x) = x^3 - 2x^2 - 5x + 4$. From sketch, one root lies between -1 and -2, and the other between 0 and 1.

$f'(x) = 3x^2 - 4x - 5$
$f(-1) = 6$ and $f(-2) = -2$

One root is possibly closer to -2.
Let $x_1 = -1.7$
$f(0) = 4$; $f(1) = -2$. The second root is possibly closer to 1. Let $x_1 = 0.7$
$f(3) = -2$; $f(4) = 16$. The third root is closer to 3. Let $x_1 = 3.1$

n	x_n	$f(x_n)$	$f'(x_n)$	$x_n - \frac{f(x_n)}{f'(x_n)}$
1	-1.7	1.807	10.47	$-1.872\ 588\ 3$
2	$-1.872\ 588\ 3$	$-0.216\ 626\ 1$	13.010 114	$-1.855\ 937\ 7$
3	$-1.855\ 937\ 7$	$-0.002\ 107\ 2$	12.757 265	$-1.855\ 772\ 5$
4	$-1.855\ 772\ 5$	$-0.000\ 000\ 057$	11.562 575	$-1.855\ 772\ 5$

1	0.7	-0.137	-6.33	0.678 357
2	0.678 357	0.000 036 9	$-6.332\ 923$	0.678 362 8

1	3.1	-0.929	11.43	3.181 277 3
2	3.1181 277 3	0.048 761 7	12.636 467	3.177 418 5
3	3.177 418 5	0.000 111	12.578 291	3.177 409 7

The roots are $-1.855\ 772\ 5$, $0.678\ 362\ 8$, and $3.177\ 409\ 7$.

21. $V = \frac{1}{6}\pi h(h^2 + 3r^2) = \frac{1}{6}\pi h^3 + \frac{1}{2}\pi r^2 h$

$180\ 000 = \frac{1}{6}\pi h[h^2 + 3(60)^2] = \frac{1}{6}\pi h^3 + 1800\pi h$

$f(h) = \frac{1}{6}\pi h^3 + 1800\pi h - 180\ 000$

$f'(h) = \frac{1}{2}\pi h^2 + \frac{1}{2}\pi r^2 = \frac{1}{2}\pi h^2 + 1800\pi$

n	h_n	$f(h_n)$	$f'(h_n)$	$h_n - \dfrac{f(h_n)}{f'(h_n)}$
1	29	-3238.813	6975.906	29.464 286
2	29.464 286	9.874 670	7018.544	29.462 879
3	29.462 879			

$h_3 = 29.462\ 879;\ h = 29.5$ m

24.3   Curvilinear Motion

1. $x = 3t;\ y = 1 - t;\ t = 4$

$$v_x = \frac{dx}{dt} = 3;\ v_y = \frac{dy}{dt} = -1$$

$$v_x|_{t=4} = 3;\ v_y|_{t=4} = -1$$

$$v|_{t=4} = \sqrt{3^2 + (-1)^2} = \sqrt{9+1} = \sqrt{10} = 3.16$$

$$\tan\theta = \frac{-1}{3} = -0.3333;\ \theta = -18.4° = 341.6°$$

5. $x = 3t;\ y = 1 - t;$

$$t = 4;\ v_x = \frac{dx}{dt} = 3$$

$$a_x = \frac{d^2x}{dt^2} = \frac{d(3)}{dt} = 0;$$

$$a_x|_{t=4} = 0$$

$$v_y = \frac{dy}{dt} = -1;\ a_x = \frac{d^2y}{dt^2}$$

$$= \frac{d}{dt}(-1) = 0$$

$$a_y|_{t=4} = 0$$

The particle is not accelerating since
$a = \sqrt{0^2 + 0^2} = 0$.

9. $y = 4.0 - 0.20x^2;$

$$v_x = \frac{dx}{dt} = 5.0 \text{ m/s}$$

$$v_y = \frac{dy}{dt} = -0.20\left(2x\frac{dx}{dt}\right)$$

$$= -0.40x\frac{dx}{dt}$$

$$v_y|_{x=4.0} = -0.40(4.0)(5.0) = -8.0 \text{ m/s}$$

$$v = \sqrt{(5.0)^2 + (-8.0)^2}$$
$$= \sqrt{25 + 64} = \sqrt{89} = 9.4 \text{ m/s}$$

$$\tan\theta = \frac{v_y}{v_x} = \frac{-8.0}{5.0} = -1.60;$$

$$\theta = -58° = 302°$$

13. $x = 96t;\ \dfrac{dx}{dt} = 96 \text{ ft/s} = v_x;\ y = 120t - 16t^2$

$$\frac{dy}{dt} = 120 - 32t;\ \left.\frac{dy}{dt}\right|_{t=6.0} = -72 \text{ ft/s} = v_y$$

$$v = \sqrt{96^2 + (-72)^2} = 120 \text{ ft/s}$$

$$\tan\theta = \frac{-72}{96};\ \theta = 323°$$

$$\frac{d^2x}{dt^2} = 0 = a_x;\ \frac{d^2y}{dt^2} = -32 = a_y;\ a = \sqrt{0^2 + (-32)^2} = 32 \text{ ft/s}^2$$

$\tan\theta$ is undefined; $\theta = 270°$

17. $x = 10(\sqrt{1 + t^4} - 1) = 10(1 + t^4)^{1/2} - 10; \; y = 40t^{3/2}$

$\dfrac{dx}{dt} = 5(1 + t^4)^{-1/2}(4t^3) = 20t^3(1 + t^4)^{-1/2} = \dfrac{20t^3}{\sqrt{1 + t^4}}$

$\dfrac{dy}{dt} = 60t^{1/2} = 60\sqrt{t}$

$ax = \dfrac{(1 + t^4)^{1/2}(60t^2) - (20t^3)\left(\frac{1}{2}\right)(1 + t^4)^{-1/2}(4t^3)}{1 + t^4} = \dfrac{(1 + t^4)^{-1/2}[60t^2(1 + t^4) - 40t^6]}{(1 + t^4)}$

$\qquad = \dfrac{60t^2(1 + t^4) - 40t^6}{(1 + t^4)^{3/2}}$

$ay = 30t^{-1/2}$

$a_x\big|_{t=10.0} = \dfrac{6000(10,000) - 40,000,000}{1,000,000} = 20.0$

$a_y\big|_{t=10.0} = 30(10)^{-1/2} = \dfrac{30}{\sqrt{10}} = 9.5$

$a = \sqrt{(20)^2 + (9.5)^2} = 22.1 \text{ m/s}^2$

$\tan\theta = \dfrac{9.5}{20.0} = 0.475; \; \theta = 25.4°$

$a_x\big|_{t=100} = \dfrac{60(10^4)(10^8) - 40(10^{12})}{(10^8)^{3/2}} = \dfrac{6 \times 10^{13} - 4 \times 10^{13}}{10^{12}} = \dfrac{2 \times 10^{13}}{10^{12}} = 20.0$

$a_y\big|_{t=100} = 30(10^2)^{-1/2} = 30(10^{-1}) = 3.0$

$a = \sqrt{(20.0)^2 + (3.0)^2} = 20.2 \text{ m/s}^2$

$\tan\theta = \dfrac{3.0}{20.0} = 0.150; \; \theta = 8.5°$

21. $d = 3.50$ in; $r = 1.75$ in; $x^2 + y^2 = 1.75^2$; $\dfrac{dy}{dx} = -\dfrac{x}{y}$

3600 r/min $= 7200\pi$ rad/min $= \omega$

$v = \omega r = 7200\pi(1.75) = 12,600\pi$ in/min

$x^2 + y^2 = 1.75^2$

$\quad y^2 = 1.75^2 - x^2 = 1.75^2 - 1.20^2 = 3.062 - 1.44 = 1.622$

$\quad y = 1.274$ in

$\dfrac{dy}{dx} = -\dfrac{x}{y} = -\dfrac{1.20}{1.274} = -0.942 = \dfrac{v_y}{v_x}; \; v_y = -0.942v_x$

$v = 12,600\pi = \sqrt{v_x^2 + v_y^2} = \sqrt{(-0.942v_x)^2 + v_x^2}$

$\quad = \sqrt{1.893v_x^2} = 1.376v_x$

$v_x = 9158\pi = 28,800$ in/min

$v_y = -0.942v_x = 8654\pi = -27,100$ in/min

24.4   Related Rates

1. $R = 4.000 + 0.003T^2$; $\dfrac{dT}{dt} = 0.100°C/s$

$$\frac{dR}{dt} = 0 + 0.006T\frac{dT}{dt}$$

$$\left.\frac{dR}{dt}\right|_{T=150°C} = 0 + 0.006(150)(0.1) = 0.0900\ \Omega/s$$

5. $r = \sqrt{0.4\lambda}$; $\dfrac{d\lambda}{dt} = 0.10 \times 10^{-7}$; $r = (0.4\lambda)^{1/2}$

$$\frac{dr}{dt} = \frac{1}{2}(0.4\lambda)^{-1/2}(0.4)\frac{d\lambda}{dt} = 0.2(0.4\lambda)^{-1/2}\frac{d\lambda}{dt}$$

$$\left.\frac{dr}{dt}\right|_{\lambda=6.0\times10^{-7}} = 0.2[0.4(6.0 \times 10^{-7})]^{-1/2}(0.10 \times 10^{-7}) = \frac{2 \times 10^{-9}}{\sqrt{24} \times 10^{-4}} = 4.1 \times 10^{-6}\ \text{m/s}$$

9. $A = \pi r^2$; $\dfrac{dr}{dt} = 0.020$ mm/mo; $\dfrac{dA}{dt} = 2\pi r\dfrac{dr}{dt}$; $\left.\dfrac{dA}{dt}\right|_{r=1.2} = 2\pi(1.2)(0.020) = 0.15\ \text{mm}^2/\text{mo}$

13. $p = \dfrac{k}{v}$; $\dfrac{dv}{dt} = 20$ cm^3/min; $v = 810$ cm^3

$$230 = \frac{k}{650}; \quad k = 1.495 \times 10^5\ \text{kPa} \times \text{cm}^3$$

$$p = \frac{149\,500}{v} = 149\,500 v^{-1}; \quad \frac{dp}{dt} = -149\,500 v^{-2}\frac{dv}{dt}$$

$$\left.\frac{dp}{dt}\right|_{v=810} = -149\,500(810)^{-2}(20) = -4.56\ \text{kPa/min}$$

17.
$$\frac{r}{h} = \frac{1.15}{3.6}$$

$$V = \frac{1}{3}\pi r^2 h = \frac{1}{3}\pi\left(\frac{1.15}{3.6}\right)^2 h^3$$

$$\frac{dV}{dt} = \frac{dV}{dh} \cdot \frac{dh}{dt}$$

$$0.50 = \pi \cdot \left(\frac{1.15}{3.6}\right)^2 \cdot h^2 \cdot \frac{dh}{dt}$$

$$0.50 = \pi\left(\frac{1.15}{3.6}\right)^2 (1.8)^2 \cdot \frac{dh}{dt}$$

$$\frac{dh}{dt} = 0.48\ \text{m/min}$$

21. Let x be the distance traveled by the jet going due east, and y be the distance traveled by the jet going north of east.

Since the second jet remains due north of the first jet, we have a right triangle and can use the Pythagorean theorem. $x^2 + z^2 = y^2$

Taking the derivative of this expression,

$$2x\frac{dx}{dt} + 2z\frac{dz}{dt} = 2y\frac{dy}{dt}$$

$$x\big|_{t=(1/2)} = 1600\left(\frac{1}{2}\right) = 800 \text{ mi}; \; y\big|_{t=(1/2)} = 1800\left(\frac{1}{2}\right) = 900 \text{ mi}$$

$$z = \sqrt{y^2 - x^2} = \sqrt{900^2 - 800^2} = 412.3 \text{ mi}$$

$$\frac{dx}{dt} = 1600; \; \frac{dy}{dt} = 1800 \text{ mi/h}$$

Substituting, $2(800)(1600) + 2(412.3)\dfrac{dz}{dt} = 2(900)(1800)$

$$\frac{dz}{dt} = 820 \text{ mi/h}$$

24.5   Using Derivatives in Curve Sketching

1. $y = x^2 + 2x$; $y' = 2x + 2$; $2x + 2 > 0$
$2x > -2$; $x > -1$; $f(x)$ increases.
$2x + 2 < 0$; $2x < -2$; $x < -1$; $f(x)$ decreases.

5. $y = x^2 + 2x$; $y' = 2x + 2$; $y' = 0$ at $x = 1$
$y'' = 2 > 0$ at $x = -1$ and $(-1, -1)$
is a relative minimum.

9. $y = x^2 + 2x$; $y' = 2x + 2$; $y'' = 2$
Thus, $y'' > 0$ for all x. The graph is concave up for all x and has no points of inflection.

13. $y = x^2 + 2x$

17. $y = 12x - 2x^2$; $y' = 12 - 4x$
$y' = 0$ at $x = 3$; for $x = 3$, $y = 12(3) - 2(3)^2 = 18$
and $(3, 18)$ is a critical point. $12 - 4x < 0$ for $x > 3$ and the function decreases; $12 - 4x > 0$ for $x < 3$ and the function increases; $y'' = -4$; thus $y'' < 0$ for all x. There are no inflections; the graph is concave down for all x, and $(3, 18)$ is a maximum point.

21. $y = x^3 + 3x^2 + 3x + 2$

$y' = 3x^2 + 6x = 3(x^2 + 2x + 1) = 3(x + 1)(x + 1)$

$3(x + 1)(x + 1) = 0$ for $x = -1$

$(-1, 1)$ is a critical point.

$3(x + 1)(x + 1) > 0$ for $x < -1$ and the slope is positive.

$3(x + 1)(x + 1) > 0$ for $x > -1$ and the slope is positive.

$y'' = 6x + 6$; $6x + 6 = 0$ for $x = 1$, and $(-1, 1)$ is an inflection point.

$6x + 6 < 0$ for $x < -1$ and the graph is concave down.

$6x + 6 > 0$ for $x > -1$ and the graph is concave up.

Since there is no change in slope from positive to negative or vice versa, there are no maximum or minimum points.

25. $y = 4x^3 - 3x^4$; $y' = 12x^2 - 12x^3 = 12x^2(1 - x) = 0$

$12x^2(1 - x) = 0$ for $x = 0$ and $x = 1$

$(0, 0)$ and $(1, 1)$ are critical points.

$12x^2 - 12x^3 > 0$ for $x < 0$ and the slope is positive.

$12x - 12x^3 > 0$ for $0 < x < 1$ and the slope is positive.

$12x - 12x^3 < 0$ for $x > 1$ and the slope is negative.

$y'' = 24x - 36x^2$; $24x - 36x^2 = 12x(2 - 3x) = 0$ for $x = 0$, $x = \frac{2}{3}$

$(0, 0)$ and $\left(\frac{2}{3}, \frac{16}{27}\right)$ are possible inflection points.

$24x - 36x^2 < 0$ for $x < 0$ and the graph is concave down.

$24x - 36x^2 > 0$ for $0 < x < \frac{2}{3}$ and the graph is concave up.

$24x - 36x^2 < 0$ for $x > \frac{2}{3}$ and the graph is concave down.

$(1,1)$ is a relative maximum point since $y' = 0$ at $(1, 1)$ and the slope is positive for $x < 1$ and negative for $x > 1$. $(0, 0)$ and $\left(\frac{2}{3}, \frac{16}{27}\right)$ are inflection points since there is a concavity change.

29. $y = x^3 - 12x$; $y' = 3x^2 - 12$; $y'' = 6x$. On graphing calculator with $x_{\min} = -5$, $x_{\max} = 5$, $y_{\min} = -20$, $y_{\max} = 20$, enter $y_1 = x^3 - 12x$; $y_2 = 3x^2 - 12$; $y_3 = 6x$. From the graph is observed that the maximum and minimum values of y occur when y' is zero.

A maximum value for y occurs when $x = -2$, and a minimum value occurs when $x = 2$. An inflection point (change in curvature) occurs when y'' is zero. x is also zero at this point.

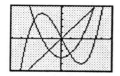

33. $R = 75 - 18i^2 + 8i^3 - i^4$
$R' = -36i + 24i^2 - 4i^3 = -4i(i^2 - 6i + 9) = -4i(i-3)^2$
$R' = 0$ for $i = 0$ and $i = 3$
$(0, 75)$ and $(3, 48)$ are critical points.
$R' > 0$ for $i < 0$, $R' < 0$ for $0 < i < 3$
$R' < 0$ for $i > 3$
Max. at $(0, 75)$, no max. or min. at $(3, 48)$
$R'' = -36 + 48i - 12i^2 = -12(i-1)(i-3)$
$(1, 64)$ and $(3, 48)$ are possible inflection points.
$R'' < 0$ for $i < 1$, concave down
$R'' > 0$ for $1 < i < 3$, concave up
$R'' < 0$ for $i > 3$, concave down
$(1, 64)$ and $(3, 48)$ are inflection points.
(From calculator graph, $R = 0$ for $i = -1.5$ and $i = 5.0$)

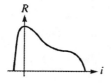

37. $f(1) = 0$; therefore $(1, 0)$ is an x-intercept
$f'(x) > 0$ for all x; therefore curve rises left to right
$f''(x) < 0$ for all x; therefore concave down

24.6   More on Curve Sketching

1. $y = \dfrac{4}{x^2}$

Intercepts:

(1) y is defined for $x = 0$, so the graph is not continuous at $x = 0$ and cannot cross the y-axis.

(2) Since 4 is positive and x^2 is positive for all non-zero x, $\frac{4}{x^2}$ is always positive, and the graph does not cross into quadrants III or IV.

(3) Since $\frac{4}{x^2} > 0$ for any x, $y \neq 0$ and the graph does not intersect the x-axis at any point.

Symmetry:

(4) Symmetrical about y-axis since replacing x with $-x$ produces no change. Not symmetrical about x-axis.

Behavior as x becomes large:

(5) As $x \to -\infty$, $\frac{4}{x^2}$ approaches 0, and the negative x-axis is an asymptote.

(6) As $x \to +\infty$, $\frac{4}{x^2}$ approaches 0, and the positive x-axis is an asymptote.

Derivatives:

(7) $y' = -8x^{-3} = \frac{-8}{x^3}$. For negative x, $\frac{-8}{x^3}$ is positive and the graph rises. For positive x, $\frac{-8}{x^3}$ is negative and the graph falls.

(8) $y'' = 24x^{-4} = \frac{24}{x^4}$. y'' is positive for all x. There are no inflection points.
Inc. $x < 0$, dec. $x > 0$
Concave up $x < 0$, $x > 0$
Asym. $x = 0$, $y = 0$

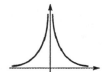

5. $y = x^2 + \dfrac{2}{x} = \dfrac{x^3 + 2}{x}$

(1) $\frac{2}{x}$ is undefined for $x = 0$, so the graph is not continuous at the y-axis; i.e., no y-intercept exists.

(2) $\frac{x^3+2}{x} = 0$ at $x = \sqrt[3]{-2} = -\sqrt[3]{2}$. There is an x-intercept at $(-\sqrt[3]{2}, 0)$.

(3) As $x \to \infty$, $x^2 \to \infty$ and $\frac{2}{x} \to 0$, so $x^2 + \frac{2}{x} \to \infty$.

(4) As $x \to 0$ through positive x, $x^2 \to 0$ and $\frac{2}{x} \to \infty$, so $x^2 + \frac{2}{x} \to \infty$.

(5) As $x \to -\infty$, $x^2 \to \infty$ and $\frac{2}{x} \to 0$ so $x^2 + \frac{2}{x} \to \infty$.

(6) As $x \to 0$ through negative numbers, $x^2 \to 0$ and $\frac{2}{x} \to -\infty$, so $x^2 + \frac{2}{x} \to -\infty$.

(7) $y' = 2x - 2x^{-2} = 0$ at $x = 1$ and the slope is zero at $(1, 3)$.

(8) $y'' = 2 + 4x^{-3} = 0$ at $x = -\sqrt[3]{2}$ and $(-\sqrt[3]{2}, 0)$ is an inflection point.

(9) $y'' > 0$ at $x = 1$, so the graph is concave up and $(1, 3)$ is a relative minimum.

(10) Since $(-\sqrt[3]{2}, 0)$ is an inflection, $f''(-1) < 0$ and the graph is concave down. $f''(-2) > 0$ and the graph is concave up.

(11) Not symmetrical about the x- or y-axis.

9. $y = \dfrac{x^2}{x+1}$

Intercepts:

(1) Function undefined at $x = -1$; not continuous at $x = -1$.

(2) At $x = 0$, $y = 0$. The origin is the only intercept.

Behavior as x becomes large:

(3) As $x \to \infty$, $y \to x$, so $y = x$ is an asymptote. As $x \to \infty$, $y = -\infty$.

Vertical asymptotes:

(4) As $x \to -1$ from the left, $x + 1 \to 0$ through negative values and $\frac{x^2}{x+1} \to -\infty$ since $x^2 > 0$ for all x. As $x \to -1$ from the right, $x + 1 \to 0$ through positive values and $\frac{x^2}{x+1} \to +\infty$. $x = -1$ is an asymptote.

Symmetry:

(5) The graph is not symmetrical about the y-axis or the x-axis.

Derivatives:

(6) $y' = \frac{x^2+2x}{(x+1)^2}$; $y' = 0$ at $x = -2$, $x = 0$. $(-2, -4)$ and $(0, 0)$ are critical points. Checking the derivative at $x = -3$, the slope is positive, and at $x = -1.5$ the slope is negative. $(-2, -4)$ is a relative maximum point. Checking the derivative at $x = -0.5$, the slope is negative, and at $x = 1$ the slope is positive, so $(0, 0)$ is a relative minimum point.
Int. $(0, 0)$, max. $(-2, -4)$, min $(0, 0)$, asym. $x = -1$

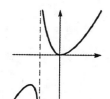

13. $y = \dfrac{4}{x} - \dfrac{4}{x^2}$

Intercepts:

(1) There are no y intercepts since $x = 0$ is undefined.

(2) $y = 0$ when $x = 1$ so $(1, 0)$ is an x-intercept.

Asymptotes:

(3) $x = 0$ is an asymptote; the denominator is 0.

Symmetry:

(4) Not symmetrical about the y-axis since $\frac{4}{x} - \frac{4}{x^2}$ is different from $\frac{4}{(-x)} - \frac{4}{(-x)^2}$

(5) Not symmetrical about the x-axis since $y = \frac{4}{x} - \frac{4}{x^2}$ is different from $-y = \frac{4}{x} - \frac{4}{x^2}$

(6) Not symmetrical about the origin since $y = \frac{4}{x} - \frac{4}{x^2}$ is different from $-y = \frac{4}{-x} - \frac{4}{(-x)^2}$

Derivatives:

(7) $y' = -4x^{-2} + 8x^{-3} = 0$ at $x = 2$; $(2, 1)$ is a relative maximum.

(8) $y'' = 8x^{-3} - 24x^{-4} = 0$ at $x = 3$ so $\left(3, \frac{8}{9}\right)$ is a possible inflection.
$y'' < 0$ (concave down) for $x < 3$ and 0 (concave up) for $x > 3$ so $\left(3, \frac{8}{9}\right)$ is an inflection.

Behavior as x becomes large:

(9) As $x \to \infty$ or $-\infty$, $\frac{4}{x}$ and $-\frac{4}{x^2}$ each approach 0.

As $x \to 0$, $\frac{4}{x} - \frac{4}{x^2} = \frac{4x-4}{x^2}$ approaches $-\infty$, through positive or negative values of x.

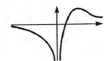

17. $y = \dfrac{9x}{9 - x^2}$

Intercept:

(1) Intercept at $x = 0, y = 0$ only

(2) Asymptotes at $x = -3, x = 3$

Derivatives:

(3) $y' = \dfrac{(9 - x^2)(9) - (9x)(-2x)}{(9 - x^2)^2} = \dfrac{81 + 9x^2}{(9 - x^2)^2}$

$81 + 9x^2 = 0$; $9x^2 = -81$; $x^2 = \sqrt{-9}$ (imaginary)
No real value max. or min., ± 3 are critical values.

(4) $y'' = \dfrac{(9 - x^2)^2(18x) - (81 + 9x^2)(2)(9 - x^2)(-2x)}{(9 - x^2)^2}$

$\qquad = \dfrac{-18x^5 - 324x^3 + 4374x}{(9 - x^2)^4}$

$-18x^5 - 324x^3 + 4374x = 0$; $-18x(x^4 + 18x^2 - 243) = 0$
$-18x = 0$; $x = 0$
$x^4 + 18x^2 - 243 = 0$; $(x^2 + 27)(x^2 - 9) = 0$
(imaginary) $x^2 = 9$; $x = \pm 3$ (these are asymptotes)

$y|_{x=0} = 0$; a possible inflection is $(0, 0)$

at $\left(-1, -\frac{9}{8}\right)$, $y'' = -4032$; concave down at $\left(1, \frac{9}{8}\right)$, $y'' = 4032$; concave up, and $(0, 0)$ is an inflection point

Symmetry:

(5) There is symmetry to the origin.

(6) As $x \to +\infty$ and as $x \to -\infty$, $y \to 0$. Therefore, $y = 0$ is an asymptote.

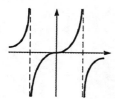

21. $R = \dfrac{200}{\sqrt{t^2 + 40,000}}$

Intercepts:

(1) Not continuous at $t = 0$ (not defined at $t < 0$).

(2) No t-intercept; R-intercept at $(0, 1)$

Symmetry:

(3) No symmetry about either axis (R is undefined for $t < 0$).

Derivatives:

(4) $R = 200(t^2 + 40,000)^{-1/2}$
$R' = -100(t^2 + 40,000)^{-3/2}(2t) = -200t(t^2 + 40,000)^{-3/2}$;

$\dfrac{-200t}{(t^2 + 40,000)^{3/2}} = 0$; $-200t = 0$; $t = 0$ is a max. since $R|_{t=0} = 1$ and $R|_{t=1} < 1$ (R is undefined for $t < 0$)

(5) $R'' = \dfrac{(t^2 + 40,000)^{3/2}(-200) - (-200t)\left(\frac{3}{2}\right)(t^2 + 40,000)^{1/2}(2t)}{[(t^2 + 40,000)^{3/2}]^2}$

$= -200(t^2 + 40,000)^{3/2} + 600t^2(t^2 + 40,000)^{1/2} = 0$

$(t^2 + 40,000)^{1/2}[-200(t^2 + 40,000) + 600t^2] = 0$
$(t^2 + 40,000)^{1/2} = 0$; $t^2 + 40,000 = 0$
$t^2 = -40,000$ (imaginary)
$-200(t^2 + 40,000) + 600t^2 = 0$
$-200[(t^2 + 40,000) - 3t^2] = 0$; $t^2 + 40,000 - 3t^2 = 0$
$-2t^2 = -40,000$; $t^2 = 20,000$
$t = 141$ possible inflection

$R''|_{t=140} \le 0$; $R''|_{t=142} > 0$; $R|_{t=141} = 0.82$ is an inflection

As x becomes large:

(6) As $x \to \infty$, $\sqrt{t^2 + 40,000}$ becomes infinitely large and $\dfrac{200}{\sqrt{t^2 + 40,000}}$ is a positive value that becomes infinitely small but never zero.

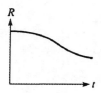

24.7   Applied Maximum and Minimum Problems

1. $s = 112t - 16.0t^2$; find maximum s.
$s' = 112 - 32.0t = 0$; $-32.0t = -112$; $t = 3.50\ s$
$s'' = -32.0 < 0$ for all t, so the graph is concave down and
$t = 3.50$ is a maximum.
$s = 112(3.50) - 16.0(3.50)^2 = 196$ ft

5. $S = 360A - 0.1A^2$, find maximum S.
$S' = 360 - 0.3A^2$; $A^2 = 1200$; $A = 35\ \mathrm{m^2}$
$S'' = -0.6A < 0$ for all valid (positive) A so the graph is concave down
and $A = 34.6\ \mathrm{m^2}$ is a max. Maximum savings are $S = 360(35) - 0.1(35)^2 = \$8300$.

9.

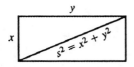

$P = 2x + 2y = 48$
$x + y = 24$

Diagonal will be a minimum if $l = s^2$ is a minimum.

$l = x^2 + y^2$
$l = x^2 + (24 - x)^2$
$l = x^2 + 24x^2 - 48x + x^2$
$l = 2x^2 - 48x + 24^2$

$\dfrac{dl}{dx} = 4x - 48 = 0$
$x = 12$
from which $y = 12$

Dimensions are 12 in by 12 in, a square will minimize the diagonal.

13. Distance traveled from B is 16.0t; $40.0 - 16.0t$ is side of a right triangle. Distance traveled from A is 18.0t.

$d = \sqrt{(40.0 - 16.0t)^2 + (18.0t)^2} = \sqrt{1600 - 1280t + 580t^2}$
$y = d^2 = 1600 - 1280t + 580t^2$

$\dfrac{dy}{dt} = -1280 + 106t$

Minimum will occur when $-1280 + 1060t = 0$; $t = 1.1$ h

17. $A = \dfrac{1}{2}(a+b)h; \; a = 6.00 \text{ ft}$

$A = \dfrac{1}{2}(6+2x)(\sqrt{9-x^2}) = (3+x)(9-x^2)^{1/2}$

$A' = (3+x)\left(\dfrac{1}{2}\right)(9-x^2)^{-1/2}(-2x) + (1)(9-x^2)^{1/2} = (-3x-x^2)(9-x^2)^{-1/2} + (9-x^2)^{1/2}$

$\quad = \dfrac{-3x-x^2}{(9-x^2)^{1/2}} + \dfrac{9-x^2}{(9-x^2)^{1/2}} = \dfrac{9-x^2-3x-x^2}{(9-x^2)^{1/2}} = \dfrac{-2x^2-3x+9}{(9-x^2)^{1/2}}$

$-2x^2 - 3x + 9 = 0; \; 2x^2 + 3x - 9 = 0; \; (2x-3)(x+3) = 0$

$x + 3 = 0; \; x = -3 \text{ (not valid)}; \; 2x - 3 = 0; \; x = 1.50$

$b = 2x = 3.00 \text{ ft}$

21. $y = k(2x^4 - 5Lx^3 + 3L^2x^2) = 2kx^4 - 5kLx^3 + 3kL^2x^2$

$y' = 8kx^3 - 15kLx^2 + 6kL^2x = 0$

$kx(8x^2 - 15Lx + 6L^2) = 0$

$kx = 0; \; x = 0$

$8x^2 - 15Lx + 6L^2 = 0$

$x = \dfrac{-(-15) \pm \sqrt{(-15)^2 - 4(8)(6)}}{2(8)} = \dfrac{15 \pm \sqrt{33}}{16} = \dfrac{15 \pm 5.75}{16}$

$x = 0.58L, \; 1.30L \text{ (not valid—this distance is greater than } L, \text{ the length of the beam)}$

25. $2x + \pi d = 400; \; \pi d = 400 - 2x; \; d = \dfrac{400 - 2x}{\pi}$

$A = x(d) = x\left(\dfrac{400 - 2x}{\pi}\right) = \dfrac{400x - 2x^2}{\pi}$

$A' = \dfrac{400 - 4x}{\pi} = 0; \; 400 - 4x = 0; \; x = 100 \text{ m}$

29. Let C = total cost

$C = 50\,000(10 - x) + 80\,000(\sqrt{x^2 + 2.5^2}) = 500\,000 - 50\,000x + 80\,000(x^2 + 6.25)^{1/2}$

$C' = -50\,000 + 40\,000(x^2 + 6.25)^{-1/2}(2x) = -50\,000 + 80\,000x(x^2 + 6.25)^{-1/2}$

$\quad = -50\,000 + \dfrac{80\,000x}{\sqrt{x^2 + 6.25}} = \dfrac{-50\,000\sqrt{x^2 + 6.25} + 80\,000x}{\sqrt{x^2 + 6.25}} = 0$

$-50\,000\sqrt{x^2 + 6.25} + 80\,000x = 0$

$\sqrt{x^2 + 6.25} = \dfrac{-80\,000x}{-50\,000} = \dfrac{8x}{5}; \; x^2 + 6.25 = \dfrac{64}{25}x^2$

$6.25 = \dfrac{64}{25}x^2 = \dfrac{39}{25}x^2; \; x^2 = 6.25\left(\dfrac{25}{39}\right) = 4.00$

$x = 2.00 \text{ mi}; \; 10 - x = 8.00 \text{ mi}$

24.8   Differentials and Linear Approximations

1. $y = f(x) = x^5 + x$
$dy = f'(x)dx = (5x^4 + 1)dx$

5. $s = f(t) = 2(3t^2 - 5)^4$
$ds = f'(t)dt = 8(3t^2 - 5)^3(6t)dt$
$ds = 48t(3t^2 - 5)^3 dt$

9. $y = f(x) = x^2(1-x)^3$
$dy = f'(x)dx = [x^2 \cdot 3(1-x)^2(-1) + (1-x)^3 \cdot 2x]dx$
$dy = (-3x^2(1-x)^2 + 2x(1-x)^3)dx$
$dy = (1-x)^2(-3x^2 + 2x(1-x))dx$
$dy = (1-x)^2(-5x^2 + 2x)dx$
$dy = x(1-x)^2(-5x + 2)dx$

13. $y = f(x) = 7x^2 + 4x, dy = f'(x)dx = (14x + 4)dx$
$\Delta y = f(x + \Delta x) - f(x) = 7(4.2)^2 + 4(4.2) - (7 \cdot 4^2 + 4 \cdot 4) = 12.28$
$dy = (14 \cdot 4 + 4)(0.2) = 12$

17. $y = f(x) = (1-3x)^5$
$dy = f'(x)dx = 5(1-3x)^4(-3)dx = -15(1-3x)^4 dx$
$dy = -15(1-3(1))^4(0.01) = -2.4$
$f(x + \Delta x) - f(x) = f(1.01) - f(1) = (1-3(1.01))^5 - (1-3(1))^5 = -2.47$

21. $f(x) = x^2 + 2x; f'(x) = 2x + 2$
$L(x) = f(a) + f'(a)(x-a) = f(0) + f'(0)(x-0)$
$L(x) = 0^2 + 2 \cdot 0 + (2 \cdot 0 + 2)(x-0)$
$L(x) = 2x$

25. $A = f(x) = x^2$
$dA = f'(x)dx = 2xdx$
$dA = 2 \cdot (0.950)(.002) = 0.0038 \text{ cm}^2$

29. $r = \sqrt{\lambda}$
$\dfrac{dr}{d\lambda} = \dfrac{1}{2\sqrt{\lambda}}; \dfrac{dr}{r} = \dfrac{1}{r} \cdot \dfrac{d\lambda}{2\sqrt{\lambda}} = \dfrac{1}{2} \cdot \dfrac{1}{\sqrt{\lambda}} \cdot \dfrac{d\lambda}{\sqrt{\lambda}}$
$\dfrac{dr}{r} = \dfrac{1}{2} \cdot \dfrac{d\lambda}{\lambda}$

33. $f(x) = \sqrt{2-x}; f'(x) = \dfrac{-1}{2\sqrt{2-x}}$

$L(x) = f(a) + f'(a)(x-a)$

$L(x) = f(1) + f'(1)(x-1) = \sqrt{2-1} + \dfrac{-1}{2\sqrt{2-1}}(x-1)$

$L(x) = 1 - \dfrac{1}{2}(x-1)$

$\sqrt{1.9} = f(0.1) \approx L(0.1) = 1 - \dfrac{1}{2}(0.1-1) = 1.45$

Chapter 24 Review Exercises

1. $y = 3x - x^2$ at $(-1, -4)$; $y' = 3 - 2x$

$y'|_{x=-1} = 3 - 2(-1) = 5$

$m = 5$ for tangent line

$y - y_1 = 5(x - x_1)$

$y - (-4) = 5[x - (-1)]$; $y + 4 = 5x + 5$

$5x - y + 1 = 0$

5. $y = \sqrt{x^2 + 3}$; $m = \dfrac{1}{2}$

$y = (x^2 + 3)^{1/2}$; $\dfrac{dy}{dx} = \dfrac{1}{2}(x^2 + 3)^{-1/2}(2x) = \dfrac{x}{\sqrt{x^2 + 3}}$

$m_{\text{tan}} = \dfrac{dy}{dx} = \dfrac{x}{\sqrt{x^2 + 3}} = \dfrac{1}{2}$

$2x = \sqrt{x^2 + 3}$

Squaring both sides, $4x^2 = x^2 + 3$; $3x^2 = 3$; $x^2 = 1$; $x = 1$ and $x = -1$. Therefore, the abscissa of the point at which $m = \frac{1}{2}$ is 1 or -1. If $x = 1, y = \sqrt{1^2 + 3} = \sqrt{4} = 2$ or -2. If $x = -1, y = \sqrt{(-1)^2 + 3} = 2$ or -2. The possible points where the slope of the tangent line is $\frac{1}{2}$ are $(1, 2)$, $(1, -2)$, $(-1, 2)$, $(-1, -2)$. A sketch of the curve shows that the only relative maximum or minimum point is at m(0,1.7). Therefore the point is $(1, 2)$.

$y - 2 = \dfrac{1}{2}(x - 1)$; $y = \dfrac{1}{2}x + \dfrac{3}{2}$ is the equation of the tangent line.

9. $y = 0.5x^2 + x$; $V_y = \dfrac{dy}{dt} = \dfrac{x\,dx}{dt} + \dfrac{dx}{dt}$; $V_x = 0.5\sqrt{x}$

Substituting, $V_y = x(0.5\sqrt{x}) + 0.5\sqrt{x}$

Find V_y at $(2, 4)$:

$V_y|_{x=2} = 2(0.5\sqrt{2}) + 0.5\sqrt{2} = \sqrt{2} + 0.5\sqrt{2} = 1.5\sqrt{2} = 2.12$

13. $x^3 - 3x^2 - x + 2 = 0$ (between 0 and 1)

$f(x) = x^3 - 3x^2 - x + 2$; $f'(x) = 3x^2 - 6x - 1$

$f(0) = 0^3 - 3(0^2) - 0 + 2 = 2$; $f(1) = 1^3 - 3(1^2) - 1 + 2 = -1$

The root is possibly closer to 1 than 0. Let $x_1 = 0.6$:

n	x_n	$f(x_n)$	$f'(x_n)$	$x_n - \dfrac{f(x_n)}{f'(x_n)}$	
1	0.6	0.536	−3.52	0.7522727	
2	0.7522727	−0.0242935	−3.8158936	0.7459063	$x_4 = x_3 = 0.7458983$
3	0.7459063	−0.0000304	−3.8063092	0.7458983	

17. $y = 4x^2 + 16x$

    (1) The graph is continuous for all x.

    (2) The intercepts are $(0,0)$ and $(-4,0)$.

    (3) As $x \to +\infty$ and $-\infty, y \to +\infty$.

    (4) The graph is not symmetrical about either axis or the origin.

    (5) $y' = 8x + 16; y' = 0$ at $x = -2$. $(-2, -16)$ is a critical point.

    (6) $y'' = 8 > 0$ for all x; the graph is concave up and $(-2, -16)$ is a minimum.

21. $y = x^4 - 32x$

    (1) The graph is continuous for all x.

    (2) The intercepts are $(0,0)$ and $(2\sqrt[3]{4}, 0)$.

    (3) As $x \to -\infty, y \to +\infty$; as $x \to +\infty, y \to +\infty$.

    (4) The graph is not symmetrical about either axis or the origin.

    (5) $y' = 4x^3 - 32 = 0$ for $x = 2$

    (6) $y'' = 12x^2; y'' = 0$ at $x = 0$; $(0,0)$ is a possible point of inflection. Since $f''(x) > 0$; the graph is concave up everywhere and $(0,0)$ is not an inflection point. $(2, -48)$ is a minimum.

25. $y = f(x) = 4x^3 + \dfrac{1}{x}$

    $dy = f'(x)dx$

    $= \left(12x^2 - \dfrac{1}{x^2}\right)dx$

29. $y = f(x) = x^3, x = 2, \Delta x = 0.1$
$$\begin{aligned}\Delta y - dy &= f(x + \Delta x) - f(x) - f'(x)dx \\ &= (x + \Delta x)^3 - x^3 - 3x^2 dx \\ &= 2.1^3 - 2^3 - 3 \cdot 2^2(0.1) \\ &= 0.061\end{aligned}$$

33. $V = f(r) = \dfrac{4}{3}\pi r^3, r = 3.500, \Delta r = 0.012$

    $dV = f'(r)dr = 4\pi r^2 \cdot dr = 4\pi(3.500)^2(0.012)$

    $dV = 1.847 \text{ m}^3$

37. $y = x^2 + 2$ and $y = 4x - x^2$
$y' = 2x; y' = 4 - 2x$
$2x = 4 - 2x; 4x = 4; x = 1$

The point $(1, 3)$ belongs to both graphs; the slope of the tangent line is 2.

$y - y_1 = 2(x - x_1); y - 3 = 2(x - 1); y - 3 = 2x - 2$
$2x - y + 1 = 0$ is the equation of the tangent line.

41.

$$x = 8t \qquad y = -0.15t^2 \qquad v = \sqrt{8^2 + (-3.6)^2}$$

$$\frac{dx}{dt} = 8 \qquad \frac{dy}{dt} = -0.30t \qquad v = \sqrt{64 + 12.96}$$

$$v_x|_{t=12} = 8 \qquad v_y|_{t=12} = -3.6 \qquad v|_{t=12} = \sqrt{76.96} = 8.8 \text{ m/s}$$

$$\tan\theta = \frac{-3.6}{8} = -0.45;\ \theta = 336°$$

45.

$$P = 0.030r^3 - 2.6r^2 + 71r - 200,\ 6 \le r \le 30 \text{ m}^3/\text{s}$$

$$\frac{dP}{dr} = 0.09r^2 - 5.2r + 71 = 0$$

$$r = 22.1,\ r = 35.6\ (\text{reject},\ 6 \le r \le 30)$$

$$\frac{d^2P}{dr^2} = 0.18r - 5.2|_{22.1} = -1.222 < 0,\ \text{maximum}$$

P is a maximum when rate is $22.1 \text{ m}^3/\text{s}$

49.

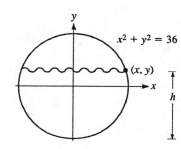

$$w = 2x,\ h = 6 + y$$

$$w = 2\sqrt{36 - y^2} = 2 \cdot \sqrt{36 - (h-6)^2}$$

$$\frac{dw}{dt} = \frac{dw}{dh} \cdot \frac{dh}{dt} = \frac{-2(h-6)}{\sqrt{36 - (h-6)^2}} \cdot \frac{dh}{dt}$$

$$\frac{dw}{dt} = \frac{-2(1.5 - 6)}{\sqrt{36 - (1.5 - 6)^2}} \cdot (-0.250)$$

$$= -0.567 \text{ ft/s}$$

53. $y + 2x = 200;\ y = -2x + 200$

$$A = xy = x(-2x + 200) = -2x^2 + 200x$$

$$A' = -4x + 200 = 0$$

$$x = \frac{-200}{-4} = 50;\ y = 200 - 2(50) = 100$$

$$A = 50(100) = 5000 \text{ cm}^2$$

57. 8000 ft = 1.52 mi

$$\frac{dz}{dt} = 680;\ \frac{dy}{dt} = 0;\ \text{find } \frac{dx}{dt}$$

$$z^2 = x^2 + y^2;\ z = \sqrt{5.00^2 + 1.52^2} = \sqrt{27.31} = 5.23 \text{ mi}$$

$$2z\frac{dz}{dt} = 2x\frac{dx}{dt} + 2y\frac{dy}{dt}$$

$$2(5.23)(680) = 2(5.00)\frac{dx}{dt} + 2(1.52)(0)$$

$$7107 = 10.00\frac{dx}{dt}$$

$$\frac{dx}{dt} = 711 \text{ mi/h}$$

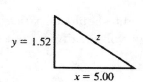

61. $V = x^2 y;\ x^2 + 4xy = 27,\ y = \dfrac{27 - x^2}{4x}$

$\quad\quad V = x^2 \cdot \dfrac{27 - x^2}{4x} = \dfrac{27x - x^3}{4}$

$\quad\quad V' = \dfrac{27}{4} - \dfrac{3x^2}{4} = 0 \text{ for } x = 3$

$\quad\quad V'' = \dfrac{-64}{4} < 0 \text{ for } x > 0,\ V \text{ is concave down}$

$\quad\quad V = \dfrac{27 \cdot 3 - 3^2}{4} = 13.5 \text{ ft}^3 \text{ is the maximum volume.}$

65. $V = \dfrac{1}{3}\pi r^2 h = \dfrac{1}{3}\pi r^2 \cdot r = \dfrac{1}{3}\pi r^3$

$\quad\quad \dfrac{dV}{dt} = \pi r^2 \dfrac{dr}{dt}$

$\quad\quad 100 = \pi(10.0)^2 \dfrac{dr}{dt}$

$\quad\quad \dfrac{dr}{dt} = 0.318 \text{ ft/min}$

69.

The amount of plastic used is determined by the surface area $S = 2 \cdot \pi r^2 + 2\pi r \cdot h$. $V = \pi r^2 h = \text{constant}$ from which $h = \dfrac{\text{constant}}{\pi r^2}$.

$\quad\quad S = 2\pi r^2 + 2\pi r \cdot \dfrac{\text{constant}}{\pi r^2}$

$\quad\quad S(r) = 2\pi r^2 + \dfrac{2 \cdot \text{constant}}{r}$

$\quad\quad \dfrac{dS}{dr} = 4\pi r - \dfrac{2 \cdot \text{constant}}{r^2} = 0 \text{ for minimum}$

$\quad\quad 4\pi r^3 = 2 \cdot \text{constant}$

$\quad\quad r^3 = \dfrac{\text{constant}}{2\pi}$

$\quad\quad \dfrac{h}{r} = \dfrac{\text{constant}}{\pi \cdot r^3} = \dfrac{\text{constant}}{\pi \frac{\text{constant}}{2\pi}} = 2$

The height should be twice the radius to minimize the surface area. In finding the $\frac{h}{r}$ ratio, the constant volume divides out so it is not necessary to specify the volume.

INTEGRATION

25.1 Antiderivatives

1. $3x^2$; the power of x required in the antiderivative is 3. Therefore, we must multiply by $\frac{1}{3}$. The antiderivative of $3x^2$ is $\frac{1}{3}(3x^3) = x^3$. $a = 1$.

5. The power of x required in the antiderivative of $f(x) = 9\sqrt{x}$ is $\frac{3}{2}$. Multiply by $\frac{2}{3}$. The antiderivative of $9\sqrt{x}$ is $\frac{2}{3} \cdot 9x^{3/2}$. $a = 6$.

9. The power of x required in the antiderivative of $\frac{5}{2}x^{3/2}$ is $\frac{5}{2}$. Multiply by $\frac{2}{5}$. The antiderivative of $\frac{5}{2}x^{3/2}$ is $\frac{2}{5}\left(\frac{5}{2}\right)x^{5/2} = x^{5/2}$.

13. $f(x) = 2x^2 - x$; $2x^2 \rightarrow kx^3$; $\dfrac{d}{dx}(kx^3) = k3x^2$

    $3k = 2$; $k = \frac{2}{3}$, therefore $\frac{2}{3}x^3$; $x \rightarrow kx^2$; $\dfrac{d}{dx}(kx^2) = k2x$

    $2k = 1$; $k = \frac{1}{2}$, therefore $\frac{1}{2}x^2$

    Antiderivative of $2x^2 - x$ is $\frac{2}{3}x^3 - \frac{1}{2}x^2$.

17. $f(x) = \dfrac{-7}{x^6}$; $f(x) = -7x^{-6}$; power required is -5, therefore x^{-5};

    $\dfrac{d}{dx}kx^{-5} = -5kx^{-6}$; $-5k = -7$; $k = \frac{7}{5}$, therefore $\frac{7}{5}x^{-5}$

    Therefore, antiderivative is $\dfrac{7}{5x^5}$.

21. The power of x required for the antiderivative of x^2 is 3, so it will be multiplied by $\frac{1}{3}$, the power of x required for $2 = 2x^0$ is 1, and it will be multiplied by 1. The power of x required for x^{-2} is -1, and it will be multiplied by -1. The antiderivative of $x^2 + 2 + x^{-2}$ is $\frac{1}{3}x^3 + 2x - \frac{1}{x}$.

25. The antiderivative requires $(p^2 - 1)^4$. We multiply by $\frac{1}{4}$. Thus, we have $\frac{1}{4}[4(p^2 - 1)^4]$. The derivative of $(p^2 - 1)^4$ is $4(p^2 - 1)^3(2p)$. The antiderivative of $4(p^2 - 1)^3(2p)$ is $(p^2 - 1)^4$.

29. The antiderivative requires $(6x + 1)^{3/2}$. We multiply by $\frac{2}{3}$. Thus we have $\frac{2}{3}\left(\frac{3}{2}\right)(6x + 1)^{3/2}$. The derivative of $(6x + 1)^{3/2}$ is $\frac{3}{2}(6x + 1)^{1/2}(6)$. The antiderivative of $\frac{3}{2}(6x + 1)^{1/2}(6)$ is $(6x + 1)^{3/2}$.

25.2 The Indefinite Integral

1. $\displaystyle\int 2x\,dx = 2\int x\,dx$; $u = x$; $du = dx$; $n = 1$

    $\displaystyle 2\int x\,dx = 2\left(\frac{x^{1+1}}{1+1}\right) + C = x^2 + C$

5. $\int 2x^{3/2}dx$; $u = x$; $du = dx$; $n = \frac{3}{2}$

$$\int 2x^{3/2}dx = \frac{2x^{(3/2)+1}}{\frac{3}{2}+1} + C$$

$$= \frac{2x^{5/2}}{\frac{5}{2}} + C$$

$$= \frac{4}{5}x^{5/2} + C$$

13. $\int \left(\frac{t^2}{2} - \frac{2}{t^2}\right)dt = \frac{t^{2+1}}{2(2+1)} - \frac{2t^{-2+1}}{-2+1} + C$

$$= \frac{t^3}{6} + \frac{2}{t} + C$$

17. $\int (2x^{-2/3} + 3^{-2})dx = \int 2x^{-2/3}dx + \int 3^{-2}x^0 dx$

$$= 2\int x^{-2/3}dx + 3^{-2}\int x^0 dx$$

$$= 2 + \frac{1}{3}(3x^{1/3}) + 3^{-2}(x^1)$$

$$= 6x^{1/3} + \frac{1}{9}x + C$$

21. $\int (x^2 - 1)^5(2xdx)$; $u = x^2 - 1$; $du = 2xdx$; $n = 5$

$$\int (x^2 - 1)^5(2xdx) = \frac{(x^2 - 1)^6}{6} + C$$

$$= \frac{1}{6}(x^2 - 1)^6 + C$$

25. $\int (2\theta^5 + 5)^7\theta^4 d\theta = \frac{1}{10}\int (2\theta^5 + 5)^7 \cdot (10\theta^4)d\theta$

$$= \frac{1}{10} \cdot \frac{(2\theta^5 + 5)^8}{8} + C$$

$$= \frac{(2\theta^5 + 5)^8}{80} + C$$

29. $\int \frac{xdx}{\sqrt{6x^2 + 1}} = \int (6x^2 + 1)^{-1/2}xdx$

$u = 6x^2 + 1$; $du = 12x$; $n = -\frac{1}{2}$

$$\int (6x^2 + 1)^{-1/2}xdx = \frac{1}{12}\int (6x^2 + 1)^{-1/2}(12xdx)$$

$$= \frac{1}{12}\frac{(6x^2 + 1)^{1/2}}{\frac{1}{2}} + C$$

$$= \frac{1}{6}\sqrt{6x^2 + 1} + C$$

9. $\int (x^2 - x^5)dx = \int x^2 d - x\int x^5 dx$

$$= \frac{x^3}{3} - \frac{x^6}{6} + C$$

$$= \frac{1}{3}x^3 - \frac{1}{6}x^6 + C$$

33. $\dfrac{dy}{dx} = 6x^2$; $dy = 6x^2 dx$

$$y = \int 6x^2 dx = 6\int x^2 dx = \frac{6x^3}{3} + C = 2x^3 + C$$

The curve passes through $(0, 2)$. $2 = 2(0^3) + C$, $C = 2$; $y = 2x^3 + 2$

37. Slope: $\dfrac{dy}{dx} = -x\sqrt{1 - 4x^2}$

$dy = -x\sqrt{1 - 4x^2}\, dx$

$$y = \int -x\sqrt{1 - 4x^2}\, dx = \int (1 - 4x^2)^{1/2}(-x\, dx);$$

$u = 1 - 4x^2$; $du = -8x$; $n = \dfrac{1}{2}$

$$y = \frac{1}{8}\int (1 - 4x^2)^{1/2}(-8x\, dx)$$

$$= \frac{1}{8}\frac{(1 - 4x^2)^{3/2}}{\frac{3}{2}} + C$$

$$= \frac{1}{12}(1 - 4x^2)^{3/2} + C$$

The curve passes through $(0, 7)$.

$$7 = \frac{1}{12}[1 - 4(0^2)]^{3/2} + C$$

$$7 = \frac{1}{12}(1)^{3/2} + C; \ 7 = \frac{1}{12} + C; \ C = \frac{83}{12}$$

$$y = \frac{1}{12}(1 - 4x^2)^{3/2} + \frac{83}{12}; \ 12y = 83 + (1 - 4x^2)^{3/2}$$

41. $\dfrac{df}{dA} = \dfrac{0.005}{\sqrt{0.01A + 1}} = 0.005(0.01A + 1)^{-1/2}$

$f(A) = \int 0.005(0.01A + 1)^{-1/2} dA + C$

$$= \frac{1}{2}\int (0.01A + 1)^{-1/2}(0.01dA)$$

$$= (0.01A + 1)^{1/2} + C$$

$f = 0$ for $A = 0 \text{ m}^2$

$f(0) = 0 = (0.01(0) + 1)^{1/2} + C; \ C = -1$

$f(A) = (0.01A + 1)^{1/2} - 1 = \sqrt{0.01A + 1} - 1$

25.3 The Area Under a Curve

1. $y = 3x$, between $x = 0$ and $x = 3$

(a)

x	y
1	3
2	6
3	9

$n = 3$
$\Delta x = 1$

$A = 1(0 + 3 + 6) = 9$; (first rectangle has 0 height)

(b)

x	y
0	0
0.3	0.9
0.6	1.8
0.9	2.7
1.2	3.6
1.5	4.5
1.8	5.4
2.1	6.3
2.4	7.2
2.7	8.1
3.0	9.0

$n = 10$
$\Delta x = 0.3$

$A = 0.3(0 + 0.9 + 1.8 + 2.7 + 3.6 + 4.5 + 5.4 + 6.3 + 7.2 + 8.1)$
$= 0.3(40.5) = 12.15$

5. $y = 4x - x^2$, between $x = 1$ and $x = 4$

(a) $n = 6, \Delta x = 0.5$
$A = 0.5(3.00 + 3.75 + 3.75 + 3.00 + 1.75 + 0.00)$
$A = 7.625$

x	y
1.0	3.00
1.5	3.75
2.0	4.00
2.5	3.75
3.0	3.00
3.5	1.75
4.0	0.00

($y = 4.00$ is not the height of any inscribed rectangle)

(b) $n = 10, \Delta x = 0.3$
$A = 0.3(3.00 + 3.51 + 3.84 + 3.96 + 3.75 + 3.36 + 2.79 + 2.04 + 1.11)$
$A = 8.208$

x	y
1.0	3.00
1.3	3.51
1.6	3.84
1.9	3.99
2.2	3.96
2.5	3.75
2.8	3.36
3.1	2.79
3.4	2.04
3.7	1.11
4.0	0.00

($y = 3.99$ is not the height of any inscribed rectangle)

9. $y = \dfrac{1}{\sqrt{x+1}}$, between $x = 3$ and $x = 8$

(a) $n = 5, \Delta x = \dfrac{8-3}{5} = 1$

x	y
3	0.5
4	0.447
5	0.408
6	0.378
7	0.355
8	0.333

$A = \sum\limits_{i=1}^{5} A_i = \sum\limits_{i=1}^{5} y_i \Delta x$

$y_1 = f(4)$
$A = (0.447 + 0.408 + \cdots + 0.354 + 0.333)(1)$
$A = 1.92$

(b) $n = 10, \Delta x = \dfrac{8-3}{10} = 0.5$

x	y
3	0.5
3.5	0.471
4	0.447
4.5	0.426
5	0.408
5.5	0.392
6	0.378
6.5	0.365
7	0.354
7.5	0.343
8	0.333

$A = \sum\limits_{i=1}^{10} A_i = \sum\limits_{i=1}^{10} y_i \Delta x$

$y_1 = f(3.5)$
$A = (0.471 + 0.447 + \cdots + 0.343 + 0.333)(0.5)$
$A = 1.96$

13. $y = x^2$, between $x = 0$ and $x = 2$

$$A_{0.2} = \left[\int x^2\,dx\right]_0^2 = \left.\frac{x^3}{3}\right|_0^2 = \frac{8}{3} - 0 = \frac{8}{3}$$

17. $y = \dfrac{1}{x^2} = x^{-2}$, between $x = 1$ and $x = 5$

$$A_{1.5} = \left[\int x^{-2}dx\right]_1^5 = \frac{x^{-1}}{-1}\Big|_1^5 = \frac{-1}{x}\Big|_1^5 = -\frac{1}{5} - (-1) = \frac{4}{5}$$

25.4   The Definite Integral

1. $\displaystyle\int_0^1 2x\,dx = \frac{2x^2}{2}\Big|_0^1 = x^2\Big|_0^1 = 1^2 - 0^2 = 1$

5. $\displaystyle\int_3^6\left(\frac{1}{\sqrt{x}}+2\right)dx = \int_3^6\left(\frac{1}{\sqrt{x}}\right)dx + \int_3^6 2\,dx$

$$= \int_3^6 x^{-1/2}dx + \int_3^6 2x^0 dx$$
$$= 2x^{1/2}\Big|_3^6 + 2x\Big|_3^6$$
$$= [2(6)^{1/2} - 2(3)^{1/2}] + [2(6) - 2(3)]$$
$$= 6 + 2\sqrt{6} - 2\sqrt{3}$$

9. $\displaystyle\int_{-2}^2 (T-2)(T+2)dT = \int_{-2}^2 (T^2 - 4)dT = \frac{1}{3}T^3 - 4T\Big|_{-2}^2$

$$= \frac{1}{3}(2)^3 - 4(2) - \left(\frac{1}{3}(-2)^3 - 4(-2)\right)$$
$$= -\frac{32}{3}$$

13. $\displaystyle\int_0^4 (1-\sqrt{x})^2 dx = \int_0^4 (1 - 2\sqrt{x} + x)dx$

$$= \int_0^4 1\,dx - \int_0^4 2x^{1/2}dx + \int_0^4 x\,dx$$
$$= x - \frac{2x^{3/2}}{\frac{3}{2}} + \frac{x^2}{2}\Big|_0^4$$
$$= x - \frac{4}{3}x^{3/2} + \frac{x^2}{2}\Big|_0^4$$
$$= \left[4 - \frac{4}{3}(4^{3/2}) + \frac{4^2}{2}\right] - 0$$
$$= 4 - \frac{4}{3}(8) + 8 - 0$$
$$= \frac{12}{3} - \frac{32}{3} + \frac{24}{3} = \frac{4}{3}$$

17. $\displaystyle\int_0^4 \frac{x\,dx}{\sqrt{x^2+9}} = \int_0^4 (x^2+9)^{-1/2}x\,dx$

$$= \frac{1}{2}\int_0^4 (x^2+9)^{-1/2}2x\,dx$$
$$= \frac{1}{2} \times \frac{(x^2+9)^{1/2}}{\frac{1}{2}}\Big|_0^4$$
$$= (x^2+9)^{1/2}\Big|_0^4$$
$$= (4^2+9)^{1/2} - (0^2+9)^{1/2}$$
$$= 25^{1/2} - 9^{1/2} = 5 - 3 = 2$$

21. If $u = (2x^2 + 1)$, $n = 3$, and $du = 4x\,dx$, then

$$\int_1^3 \frac{2x\,dx}{(2x^2 + 1)^3} = \frac{1}{2}\int_1^3 \frac{4x\,dx}{(2x^2 + 1)^3}$$

$$= \frac{1}{2}\left[-\frac{1}{2}(2x^2 + 1)^{-2}\right]\Big|_1^3$$

$$= \frac{1}{2}\left[-\frac{1}{2}(2(3)^2 + 1)^{-2}\right] - \frac{1}{2}\left[-\frac{1}{2}(2(1)^2 + 1)^{-2}\right]$$

$$= -\frac{1}{4}\left(\frac{1}{19^2}\right) + \frac{1}{4}\left(\frac{1}{3^2}\right) = -\frac{1}{1444} + \frac{1}{36} = \frac{88}{3249}$$

$$= 0.0271$$

25. $\int_0^2 2x(9 - 2x^2)^2\,dx$; $u = (9 - 2x^2)$; $du = -4x$; $n = 2$

$$-\frac{1}{2}\int_0^2 (9 - 2x^2)^2(-2)(2x\,dx) = -\frac{1}{2}\int_0^2 (9 - 2x^2)^2(-4x\,dx)$$

$$= -\frac{1}{2}\frac{(9 - 2x^2)^3}{3}\Big|_0^2 = -\frac{1}{6}(9 - 2x^2)^3\Big|_0^2$$

$$= -\frac{1}{6}[9 - 2(2)^2]^3 - \left(\frac{1}{6}\right)[9 - 2(0)^2]^3$$

$$= -\frac{1}{6}(1)^3 - \left(-\frac{1}{6}\right)(9)^3 = -\frac{1}{6}(1) - \left(-\frac{1}{6}\right)(729)$$

$$= -\frac{1}{6} + \frac{729}{6} = \frac{364}{3}$$

29. $\int_{-1}^2 \frac{8x - 2}{(2x^2 - x + 1)^3}\,dx = \int_{-1}^2 (8x - 2)(2x^2 - x + 1)^{-1}\,dx$

$$= 2\int_{-1}^2 (4x - 1)(2x^2 - x + 1)^{-3}\,dx = \frac{2(2x^2 - x + 1)^{-2}}{-2}\Big|_{-1}^2$$

$$= -(2x^2 - x + 1)^{-2}\big|_{-1}^2 = -\frac{1}{7^2} - \left(-\frac{1}{4^2}\right) = 0.0421$$

33. $W = \int_0^{80} (1000 - 5x)\,dx = \left(1000x - \frac{5}{2}x^2\right)\Big|_0^{80}$

$$= 1000(80) - \frac{5}{2}(80)^2 - [0 - 0] = 80\,000 - 16\,000$$

$$= 64,000 \text{ ft} \cdot \text{lb}$$

25.5   Numerical Integration: the Trapezoidal Rule

1. $\int_0^2 2x^2\,dx$; $n = 4$; $\Delta x = \dfrac{2-0}{4} = \dfrac{1}{2}$; $\dfrac{\Delta x}{2} = \dfrac{1}{4}$

n	x_n	y_n
0	0	0
1	$\frac{1}{2}$	$\frac{1}{2}$
2	1	2
3	$\frac{3}{2}$	$\frac{9}{2}$
4	2	8

$$A_T = \frac{1}{4}\left[0 + 2\left(\frac{1}{2}\right) + 2(2) + 2\left(\frac{9}{2}\right) + 8\right] = \frac{11}{2} = 5.50$$

$$\int_0^2 2x^2\,dx = \left.\frac{2x^3}{3}\right|_0^2 = \frac{16}{3} - 0 = \frac{16}{3} = 5.33$$

5. $\int_2^3 \dfrac{1}{2x}\,dx$; $n = 2$; $\Delta x = \dfrac{3-2}{2} = \dfrac{1}{2}$; $\dfrac{\Delta x}{2} = \dfrac{1}{4}$

n	x_n	y_n
0	2	$\frac{1}{4}$
1	$\frac{5}{2}$	$\frac{1}{5}$
2	3	$\frac{1}{6}$

$$A_T = \frac{1}{4}\left[\frac{1}{4} + 2\left(\frac{1}{5}\right) + \frac{1}{6}\right] = \frac{1}{4}\left(\frac{49}{60}\right) = \frac{49}{240} = 0.2042$$

9. $\int_1^5 \dfrac{1}{x^2 + x}\,dx$; $n = 10$; $\Delta x = \dfrac{5-1}{10} = 0.4$; $\dfrac{\Delta x}{2} = 0.2$

n	x_n	y_n
0	1	0.5000
1	1.4	0.2976
2	1.8	0.1984
3	2.2	0.1420
4	2.6	0.1068
5	3	0.0833
6	3.4	0.0668
7	3.8	0.0548
8	4.2	0.0458
9	4.6	0.0388
10	5	0.3333

$$A = 0.2[0.5000 + 2(0.2976) + 2(0.1984) + 2(0.1420) + 2(0.1068) + 2(0.0833)$$
$$+2(0.0668) + 2(0.0548) + 2(0.0458) + 2(0.0388) + 0.0333]$$
$$= 0.5205$$

13. $\int_2^{14} y\,dx$; $\Delta x = 2$; $\dfrac{\Delta x}{2} = 1$

x	y
2	0.67
4	2.34
6	4.56
8	3.67
10	3.56
12	4.78
14	6.87

$A = 1[0.67 + 2(2.34) + 2(4.56) + 2(3.67) + 2(3.56) + 2(4.78) + 6.87]$
$\quad = 45.36$

25.6   Simpson's Rule

1. $\int_0^2 (1 + x^3)\,dx$; $n = 2$; $\Delta x = 1$; $\dfrac{\Delta x}{3} = \dfrac{1}{3}$

$A_s = \dfrac{1}{3}[1 + 4(2) + 9]$

$A_s = 6$

$A = \int_0^2 (1 + x^3)\,dx$

$\quad = x + \dfrac{1}{4}x^4 \Big|_0^2$

$\quad = 2 + \dfrac{1}{4}(16) = 6$

n	x_n	y_n
0	0	1
1	1	2
2	2	9

5. $\int_2^3 \dfrac{1}{2x}\,dx$; $n = 2$; $\Delta x = \dfrac{1}{2}$; $\dfrac{\Delta x}{3} = \dfrac{1}{6}$

$A_s = \dfrac{1}{6}\left[\dfrac{1}{4} + 4\left(\dfrac{1}{5}\right) + \dfrac{1}{6}\right]$

$\quad = \dfrac{1}{6}\left(\dfrac{73}{60}\right) = \dfrac{73}{360} = 0.2028$

n	x_n	y_n
0	2	$\dfrac{1}{4}$
1	$\dfrac{5}{2}$	$\dfrac{1}{5}$
2	3	$\dfrac{1}{6}$

9. $\int_1^5 \dfrac{dx}{x^2 + x}$; $n = 10$; $\Delta x = 0.4$; $\dfrac{\Delta x}{3} = \dfrac{0.4}{3}$

$A_s = \dfrac{0.4}{3}[0.5000 + 4(0.2976) + 2(0.1984) + 4(0.1420) + 2(0.1068) + 4(0.0833) + 2(0.0668)$

$\qquad + 4(0.0548) + 2(0.0458) + 4(0.0388) + 0.0333]$

$\quad = \dfrac{0.4}{3}(3.8349) = 0.5114$

13. $\Delta x = 2; \dfrac{\Delta x}{3} = \dfrac{2}{3}$

x	y
2	0.67
4	2.34
6	4.56
8	3.67
10	3.56
12	4.78
14	6.87

$$\int_2^{14} y\, dx = \frac{2}{3}[0.67 + 4(2.34) + 2(4.56) + 4(3.67) + 2(3.56) + 4(4.78) + 6.87]$$

$$= 44.63$$

Chapter 25 Review Exercises

1. $\displaystyle\int (4x^3 - x)\,dx = \int 4x^3\,dx - \int x\,dx = \frac{4x^4}{4} - \frac{x^2}{2} + C$

$$= x^4 - \frac{1}{2}x^2 + C$$

5. $\displaystyle\int_1^4 \left(\frac{\sqrt{x}}{2} + \frac{2}{\sqrt{x}}\right)dx = \frac{1}{2}\int_1^4 x^{1/2}\,dx + 2\int_1^4 x^{-1/2}\,dx$

$$= \frac{1}{2}\frac{x^{3/2}}{\frac{3}{2}} + \frac{2x^{1/2}}{\frac{1}{2}}\bigg|_1^4 = \frac{1}{3}x^{3/2} + 4x^{1/2}\bigg|_1^4$$

$$= \left[\frac{1}{3}(4)^{3/2} + 4(4)^{1/2}\right] - \left[\frac{1}{3}(1)^{3/2} + 4(1)^{1/2}\right] = \frac{19}{3}$$

9. $\displaystyle\int \left(3 + \frac{2}{x^3}\right)dx = \int 3\,dx + \int \frac{2}{x^3}\,dx$

$$= \int 3\,dx + \int 2x^{-3}\,dx$$

$$= (3x) + \left(-\frac{2}{2}x^{-2}\right)$$

$$= 3x - x^{-2} = 3x - \frac{1}{x^2} + C$$

13. $\displaystyle\int \frac{dn}{(2-5n)^3} = \int (2-5n)^{-3}\,dn$

$$-\frac{1}{5}\int (2-5n)^{-3}(-5\,dn) = -\frac{1}{5} \times \frac{(2-5n)^{-2}}{-2} + C$$

$$= \frac{1}{10} \times \frac{1}{(2-5n)^2} + C$$

$$= \frac{1}{10(2-5n)^2} + C$$

17. $\displaystyle\int_0^2 \frac{3x\,dx}{\sqrt[3]{1+2x^2}} = \int_0^2 (1+2x^2)^{-1/3}(3x)\,dx$

$$u = 1 + 2x^2;\ du = 4x\,dx;\ n = -\frac{1}{3}$$

$$\frac{3}{4}\int_0^2 (1+2x^2)^{-1/3}(4x)\,dx = \frac{3}{4} \times \frac{(1+2x^2)^{2/3}}{\frac{2}{3}}\Bigg|_0^2$$

$$= \frac{9}{8}(1+2x^2)^{2/3}\Bigg|_0^2$$

$$= \frac{9}{8}[1+2(2^2)]^{2/3} - \frac{9}{8}[1+2(0)^2]^{2/3}$$

$$= \frac{9}{8}(9)^{2/3} - \frac{9}{8}(1)^{2/3}$$

$$= \frac{9}{8}(\sqrt[3]{81} - 1) = \frac{9}{8}(3\sqrt[3]{3} - 1)$$

21. $\displaystyle\int \frac{(2-3x^2)\,dx}{(2x-x^3)^2} = \int (2x-x^3)^{-2}(2-3x^2)\,dx;$

$$u = 2x - x^3;\ du = 2 - 3x^2;\ n = -2$$

$$\int (2x-x^3)^{-2}(2-3x^2)\,dx = \frac{(2x-x^3)^{-1}}{-1} + C$$

$$= -\frac{1}{(2x-x^3)} + C$$

25. $\displaystyle\frac{dy}{dx} = 3 - x^2$

$$y = 3x - \frac{x^3}{3} + C$$

$$3 = 3(-1) - \frac{(-1)^3}{3} + C$$

$$C = \frac{17}{3}$$

$$y = 3x - \frac{x^3}{3} + \frac{17}{3}$$

29. $A = \int_1^3 (6x - 1)\,dx$

$A = 3x^2 - x\big|_1^3$

$A = 3(3^2) - 3 - [3 \cdot 1^2 - 1]$

$A = 22$

33. $\Delta x = \dfrac{3-1}{4} = \dfrac{1}{2}; \; x_0 = 1, x_1 = 1.5, x_2 = 2, x_3 = 2.5, x_4 = 3.0$

$$\int_1^3 \frac{dx}{2x-1} \approx \frac{\Delta x}{3}[y_0 + 4y_1 + 2y_2 + 4y_3 + y_4]$$

$$\approx \frac{\frac{1}{2}}{3}\left[\frac{1}{2.1-1} + \frac{4}{2(1.5)-1} + \frac{2}{2(2)-1} + \frac{4}{2(2.5)-1} + \frac{1}{2(3)-1}\right]$$

$$\approx \frac{73}{90} = 0.811$$

37. $y = x\sqrt[3]{2x^2+1}, a = 1, b = 4, n = 3.$

$\Delta x = \dfrac{b-a}{n} = \dfrac{4-1}{3} = 1$

$x_0 = 1, y_0 = 1\sqrt[3]{2\cdot1^2+1} = \sqrt[3]{3}$

$x_1 = 2, y_1 = 2\sqrt[3]{2\cdot2^2+1} = 2\sqrt[3]{9}$

$x_3 = 3, y_2 = 3\sqrt[3]{2\cdot3^2+1} = 3\sqrt[3]{19}$

$x_3 = 4, y_3 = 4\sqrt[3]{2\cdot4^2+1} = 4\sqrt[3]{33}$

$$\int_1^4 x\sqrt[3]{2x^2+1}\,dx \approx \frac{\Delta x}{2}[y_0 + 2y_1 + 2y_2 + y_3] \approx \frac{1}{2}[\sqrt[3]{3} + 4\sqrt[3]{9} + 6\sqrt[3]{19} + 4\sqrt[3]{33}] \approx 19.30156604$$

41. $A = \dfrac{1}{2}\left[y(0) + y\left(\dfrac{1}{2}\right) + y(1) + y\left(\dfrac{3}{2}\right) + y\left(\dfrac{5}{2}\right) + y(3) + y\left(\dfrac{7}{2}\right) + y(4)\right]$

$A = 24.68 \text{ m}^2$

45. $\dfrac{dy}{dx} = k(2L^3 - 12Lx + 2x^4)$

$dy = k(2L^3x^0 - 12Lx + 2x^4)\,dx$

$y = \int k(2L^3x^0 - 12Lx + 2x^4)\,dx$

$= k\int (2L^3x^0 - 12Lx + 2x^4)\,dx$

$= k\left(2L^3x^0 - \dfrac{12Lx}{2} + \dfrac{2x^5}{5}\right) + C$

$y = 0$ for $x = 0$; $0 = k(0 - 0 + 0) + C$

$C = 0; \; y = k\left(2L^3x - 6Lx^2 + \dfrac{2}{5}x^5\right)$

49. The area of a quarter circle is $\frac{1}{4} \cdot \pi r^2$, thus $\pi = \frac{4 \cdot A}{r^2}$. The area of a quarter circle of radius one may be found from $\int_0^1 \sqrt{1-x^2}\, dx$. This may be evaluated using rectangles. See problem 41.

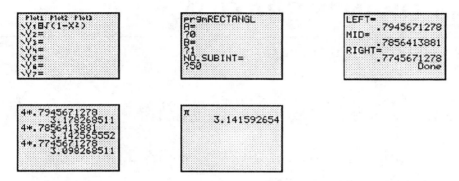

Multiplying each of these values by four gives an approximation for π. Increasing the number of subintervals will increase the accuracy of the approximation.

APPLICATIONS OF INTEGRATION

26.1 Applications of the Indefinite Integral

1. $v = \int a\,dt = \int -32\,dt = -32t + C$; $v = 0, t = 0$; $0 = -32(0) + C$; $C = 0$
$v = -32t$; find v for $t = 2.5$ s; $v = -32(2.5) = -80$ ft/s; 80 ft/s downward

5. $a = 90(1 - 4t)^{1/2}$; $v = \int 90(1 - 4t)^{1/2}dt = 90 \int (1 - 4t)^{1/2}dt = -\dfrac{1}{4}(90) \int (1 - 4t)^{1/2}(-4\,dt)$

$= \dfrac{-45}{2}\dfrac{(1 - 4t)^{3/2}}{\frac{3}{2}} + C = -15(1 - 4t)^{3/2} + C$; $v = 0$ for $t = 0$

$0 = -15(1 - 0)^{3/2} + C$; $0 = -15 + C$; $C = 15$
$v = 15(1 - 4t)^{3/2} + 15$; for $t = 0.25$ s; $v = -15(1 - 1)^{3/2} + 15$; $v = 15$ ft/s

9. $v = \int -32\,dt = -32t + C$; $v = v_0$; $t = 0, C = v_0$; $v = -32t + v_0$;

$s = \int (-32t + v_0)\,dt = -16t^2 + v_0 t + C_1$

$s = 0, t = 0$; $C_1 = 0$; $s = -16t^2 + v_0)$; $s = 90$ ft when $v = 0$

$90 = -16t^2 + v_0 t$; $0 = -32t + v_0$; $t = \dfrac{v_0}{32}$; $90 = -16\left(\dfrac{v_0}{32}\right)^2 + v_0\left(\dfrac{v_0}{32}\right)$; $\dfrac{v_0^2}{64} = 90$

$v_0 = \sqrt{64(90)} = 76$ ft/s

13. $q = \int i\,dt = \int 0.230 \times 10^{-6}dt = 0.230 \times 10^{-6}t + C$; $q = 0, t = 0$; $C = 0$; $q = 0.230 \times 10^{-6}t$

Find q for $t = 1.50 \times 10^{-3}$s: $q = 0.230 \times 10^{-6}(1.50 \times 10^{-3}) = 0.345 \times 10^{-9} = 0.345$ nC

17. $V_c = \dfrac{1}{C} \displaystyle\int i\,dt = \dfrac{1}{2.5 \times 10^{-6}} \int 0.025\,dt = \dfrac{1}{2.5 \times 10^{-6}}(0.025)t = 1.0 \times 10^4 t + C$

$v_c = 0, t = 0$; $C = 0$; $v_c = 1.0 \times 10^4 t$

Find v_c for $t = 0.012$ s: $v_c = 1.0 \times 10^4(0.012) = 120$ V

21. $\omega = \dfrac{d\theta}{dt} = 16t + 0.5t^2$; $d\theta = (16t + 0.50t^2)\,dt$

$\theta = 8t^2 + \dfrac{0.50t^3}{3} + C$; $\theta = 0, t = 0$; $C = 0$; $\theta = 8t^2 + \dfrac{0.50t^3}{3}$; find θ for $t = 10.0$ s

$\theta = 8(10.0)^2 + \dfrac{0.50}{3}(10.0)^3 = 970$ rad

25. $\dfrac{dV}{dx} = \dfrac{-k}{x^2}$; $V = -\displaystyle\int \dfrac{k}{x^2}\,dx = kx^{-1} + C = \dfrac{k}{x} + C$

$\displaystyle\lim_{V \to 0} V = \lim_{x \to \infty} \dfrac{k}{x} + C$; $0 = 0 + C$; $C = 0$; therefore, $V|_{x=x_1} = \dfrac{k}{x_1}$

26.2   Areas by Integration

1. $y = 4x; y = 0, x = 1$

Using vertical elements,

$$A = \int_0^1 y\, dx = \int_0^1 4x\, dx = 2x^2\big|_0^1 = 2(1)^2 - 2(0) = 2$$

5. $y = 6 - 4x; x = 0, y = 0, y = 3$

$$y - 6 = -4x, x = -\frac{1}{4}y + \frac{3}{2}$$

$$A = \int_0^3 x\, dy = \int_0^3 \left(-\frac{1}{4}y + \frac{3}{2}\right) dy = -\frac{1}{8}y^2 + \frac{3}{2}y\Big|_0^3 = -\frac{1}{8}(3)^2 + \frac{3}{2}(3) + \frac{1}{8}(0)^2 - \frac{3}{2}(0)$$

$$= -\frac{9}{8} + \frac{9}{2} = -\frac{9}{8} + \frac{36}{8} = \frac{27}{8}$$

9. $y = x^{-2}; y = 0, x = 2, x = 3$

$$A = \int_2^3 x^{-2} dy = -x^{-1}\big|_2^3 = -\frac{1}{x}\Big|_2^3 = -\frac{1}{3} - \left(-\frac{1}{2}\right) = \frac{1}{6}$$

13. $y = \dfrac{2}{\sqrt{x}}; x = 0, y = 1, y = 4$

$$\sqrt{x} = \frac{2}{y}; x = \frac{4}{y^2}$$

$$A = \int_1^4 x\, dy;\ A = 4\int_1^4 y^{-2} dy = -4y^{-1}\Big|_1^4 = -\frac{4}{y}\Big|_1^4 = -\frac{4}{4} - \left(-\frac{4}{1}\right) = -1 + 4 = 3$$

17. $y_2 = x^2; y_1 = 2 - x, x = 0\ (x \geq 0)$
$dA = l\, dx; l = y_1 - y_2 = 2 - x - x^2$
$y_1 = y_2; 2 - x = x^2; x^2 + x - 2 = 0; (x + 2)(x - 1) = 0; x = 1$

$$A = \int_0^1 (2 - x - x^2)\, dx = \left(2x - \frac{1}{2}x^2 - \frac{1}{3}x^3\right)\Big|_0^1; A = 2 - \frac{1}{2} - \frac{1}{3} - 0 = \frac{7}{6}$$

21. $y_1 = x^2 + 5x, y_2 = 3 - x^2$
$dA = l\, dx; l = y_2 - y_1$
$l = 3 - x^2 - (x^2 + 5x) = 3 - x^2 - x^2 - 5x = 3 - 5x - 2x^2$
$y_1 = y_2$
$x^2 + 5x = 3 - x^2; 2x^2 + 5x - 3 = 0$

$$(2x - 1)(x + 3) = 0; x = \frac{1}{2}, x = -3$$

$$A = \int_{-3}^{1/2} (3 - 5x - 2x^2)\, dx = \left(3x - \frac{5}{2}x^2 - \frac{2}{3}x^3\right)\Big|_{-3}^{1/2}$$

$$A = \left(\frac{3}{2} - \frac{5}{2} \cdot \frac{1}{4} - \frac{2}{3} \cdot \frac{1}{8}\right) - \left(-9 - \frac{45}{2} + 18\right) = \frac{343}{24}$$

25. $y = 8x; x = 0, y = 4$

(a) Using horizontal elements,

$dA = x\,dy, y = 8x, x = \dfrac{1}{8}y$

$A = \displaystyle\int_0^4 x\,dy = \dfrac{1}{8}\int_0^4 y\,dy = \dfrac{1}{8}\dfrac{y^2}{2}\Big|_0^4 = \dfrac{1}{16}y^2\Big|_0^4 = 1 - 0 = 1$

(b) Using vertical elements,

$A = \displaystyle\int_0^{1/2} (4 - 8x)\,dx = 4x - 4x^2\big|_0^{1/2} = 1$

29. $dw = p\,dt; p = 12t - 4t^2$

$w = \displaystyle\int_0^3 (12t - 4t^2)\,dt = 6t^2 - \dfrac{4}{3}t^3\Big|_0^3 = 6(3)^2 - \dfrac{4}{3}(3)^3 - 0 = 54 - 36 = 18.0 \text{ J}$

33. $y = x^3 - 2x^2 - x + 2$ and $y = x^2 - 1$

Find the points of intersection: $x^3 - 2x^2 - x + 2 = x^2 - 1$
$x^3 - 3x^2 - x + 3 = 0; (x^2 - 1)(x - 3) = 0$
$x^2 = 1; x = 3; x = \pm 1$

$A_1 = \displaystyle\int_{-1}^1 [(x^3 - 2x^2 - x + 2) - (x^2 - 1)]\,dx = \int_{-1}^1 (x^3 - 3x^2 - x + 3)\,dx$

$= \dfrac{1}{4}x^4 - x^3 - \dfrac{1}{2}x^2 + 3x\Big|_{-1}^1 = \left(\dfrac{1}{4} - 1 - \dfrac{1}{2} + 3\right) - \left(\dfrac{1}{4} + 1 - \dfrac{1}{2} - 3\right)$

$= 4 \text{ cm}^2$

26.3 Volumes by Integration

1. $y = 2 - x, x = 0, y = 0$

Disk: $dV = \pi y^2\,dx$

$V = \pi \displaystyle\int_0^2 y^2\,dx = \pi \int_0^2 (2 - x)^2\,dx = \pi\left[-\dfrac{1}{3}(2 - 2)^3 + \dfrac{1}{3}(2 - 0)^3\right] = \dfrac{8}{3}\pi$

5. $y = x, y = 0, x = 2$

Disk: $dV = \pi y^2\,dx$

$V = \pi \displaystyle\int_0^2 x^2\,dx = \dfrac{\pi}{3}x^3\Big|_0^2 = \dfrac{8\pi}{3}$

9. $y = x^3, y = 8, x = 0$
$y = x^3$; therefore $x = \sqrt[3]{y}$
Shell: $dV = 2\pi yx\,dy$

$V = 2\pi \displaystyle\int_0^8 y\sqrt[3]{y}\,dy = 2\pi \int_0^8 y^{4/3}\,dy = 2\pi\dfrac{3}{7}y^{7/3} = \dfrac{6\pi}{7}y^{7/3}\Big|_0^8 = \dfrac{768\pi}{7}$

13. $x = 4y - y^2 - 3, x = 0$
Shell: $dV = 2\pi yx\, dy$

$$V = 2\pi \int_1^3 y(4y - y^2 - 3)\, dy = 2\pi \int_1^3 (4y^2 - y^3 - 3y)\, dy = 2\pi \left(\frac{4}{3}y^3 - \frac{1}{4}y^4 - \frac{3}{2}y^2 \right)\bigg|_1^3$$

$$= 2\pi \left[36 - \frac{81}{4} - \frac{27}{2} - \left(\frac{4}{3} - \frac{1}{4} - \frac{3}{2} \right) \right] = \frac{16\pi}{3}$$

17. $y = 2\sqrt{x}, x = 0, y = 2$

Disk: $dV = \pi x^2 dy$; $x = \frac{1}{4}y^2$

$$V = \pi \int_0^2 \left(\frac{1}{4}y^2 \right) dy = \pi \int_0^2 \frac{1}{16}y^4 dy = \pi \left(\frac{1}{80}y^5 \right)\bigg|_0^2 = \pi \left(\frac{32}{80} - 0 \right) = \frac{2}{5}\pi$$

21. $x = 6y - y^2, x = 0$
Disk: $dV = \pi x^2 dy$

$$V = \pi \int_0^6 (36y^2 - 12y^3 + y^4)\, dy = \pi \left(12y^3 - 3y^4 + \frac{y^5}{5} \right)\bigg|_0^6 = \pi \left(2592 - 3888 + \frac{7776}{5} \right) = \frac{1296\pi}{5}$$

25. $y = 2x - x^2$, $y = 0$, rotated around
$x = 2$, using shells $r = 2 - x$
$h = y, t = dx$
$dV = 2\pi(2 - x)y\, dx$

$$V = 2\pi \int_0^2 (2 - x)(2 - x^2)\, dx = 2\pi \int_0^2 (4x - 4x^2 + x^3)\, dx$$

$$= 2\pi \left[2x^2 - \frac{4}{3}x^3 + \frac{1}{4}x^4 \right]\bigg|_0^2 = 2\pi \left[8 - \frac{32}{3} + 4 \right] = \frac{8}{3}\pi$$

29. $y = x^4 + 1.5$; $x = 0, x = 1.1$
Shell: $dV = 2\pi xy\, dx$

$$V = 2\pi \int_0^{1.1} xy\, dx = 2\pi \int_0^{1.1} x(x^4 + 1.5)\, dx = 2\pi \int_0^{1.1} (x^5 + 1.5x)\, dx$$

$$= 2\pi \left(\frac{x^6}{6} + \frac{3x^2}{4} \right)\bigg|_0^{1.1} = 2\pi \left(\frac{1.1^6}{6} + \frac{3(1.1)^2}{4} \right) = 7.56 \text{ mm}^3$$

26.4 Centroids

1. $m_1 = 5.0, d_1 = 1.0, m_2 = 8.5, d_2 = 4.2$
$m_3 = 3.6, d_3 = 2.5$
$m_1 d_1 + m_2 d_2 + m_3 d_3 = (m_1 + m_2 + m_3)\overline{d}$
$$5.0(1.0) + (8.5)(4.2) + (3.6)(2.5) = 17\overline{d}$$
$$\overline{d} = 2.9 \text{ cm}$$

5. The area of the left rectangle is $4(2) = 8$.
The center is $(-2, 0)$. The area of the right rectangle is $2(4) = 8$. The center is $(1, 1)$;
$$8(-2) + 8(1) = (8 + 8)\overline{x}$$

$$-8 = 16\overline{x};\ \overline{x} = -\frac{1}{2}$$

$$8(0) + 8(1) = (8 + 8)\overline{y};\ 8 = 16\overline{y} = \frac{1}{2}$$

The centroid is $(\overline{x}, \overline{y}) = (-0.5 \text{ in},\ 0.5 \text{ in})$.

9. $y = x^2,\ y = 2$

The curve is symmetrical to the y-axis.
Therefore, $\overline{x} = 0$

$$\overline{y} = \frac{\int_0^2 y(2x)\,dy}{\int_0^2 2x\,dy} = \frac{\int_0^2 y(2\sqrt{y})\,dy}{\int_0^2 2\sqrt{y}\,dy} = \frac{\int_0^2 2y^{3/2}\,dy}{\int_0^2 2y^{1/2}\,dy} = \frac{\frac{4}{5}y^{5/2}\Big|_0^2}{\frac{4}{3}y^{3/2}\Big|_0^2} = \frac{\frac{4}{5}\sqrt{32}}{\frac{4}{3}\sqrt{8}} = \frac{6}{5}$$

The centroid is $\left(0, \frac{6}{5}\right)$.

13. $y = x^2,\ y = x^3$

$$\overline{x} = \frac{\int_0^1 x(x^2 - x^3)\,dx}{\int_0^1 (x^2 - x^3)\,dx} = \frac{\int_0^1 (x^3 - x^4)\,dx}{\int_0^1 (x^2 - x^3)\,dx} = \frac{\frac{1}{4}x^4 - \frac{1}{5}x^5\Big|_0^1}{\frac{1}{3}x^3 - \frac{1}{4}x^4\Big|_0^1} = \frac{\frac{1}{4} - \frac{1}{5}}{\frac{1}{3} - \frac{1}{4}} = \frac{\frac{1}{20}}{\frac{1}{12}} = \frac{3}{5}$$

$$\overline{y} = \frac{\int_0^1 y(y^{1/3} - y^{1/2})\,dy}{\int_0^1 (y^{1/3} - y^{1/2})\,dy} = \frac{\int_0^1 (y^{4/3} - y^{3/2})\,dy}{\int_0^1 (y^{1/3} - y^{1/2})\,dy} = \frac{\frac{3}{7}y^{7/3} - \frac{2}{5}y^{5/2}\Big|_0^1}{\frac{3}{4}y^{4/3} - \frac{2}{3}y^{3/2}\Big|_0^1} = \frac{\frac{3}{7} - \frac{2}{5}}{\frac{3}{4} - \frac{2}{3}} = \frac{\frac{1}{35}}{\frac{1}{12}} = \frac{12}{35}$$

The centroid is $\left(\frac{3}{5}, \frac{12}{35}\right)$.

17. $y = x^3,\ y = 0,\ x = 1$
Rotated about x-axis, $\overline{y} = 0$

$$\overline{x} = \frac{\int_a^b xy^2\,dx}{\int_a^b y^2\,dx} = \frac{\int_0^1 x(x^3)^2\,dx}{\int_0^1 (x^3)^2\,dx} = \frac{\int_0^1 x^7\,dx}{\int_0^1 x^6\,dx} = \frac{\frac{x^8}{8}\Big|_0^1}{\frac{x^7}{7}\Big|_0^1} = \frac{\frac{1}{8}}{\frac{1}{7}} = \frac{7}{8}$$

Centroid is $\left(\frac{7}{8}, 0\right)$.

21. $y^2 = 4x;\ x = 1$
Rotated about x-axis, $\overline{y} = 0$

$$\overline{x} = \frac{\int_a^b xy^2\,dx}{\int_a^b y^2\,dx} = \frac{\int_0^1 x(4x)\,dx}{\int_0^1 4x\,dx} = \frac{\frac{4x^3}{3}\Big|_0^1}{2x^2\Big|_0^1} = \frac{\frac{4}{3}}{2} = \frac{2}{3}$$

Centroid is $\left(\frac{2}{3}, 0\right)$.

25. $\dfrac{x^2}{5.00^2} + \dfrac{y^2}{1.00^2} = 1;\ y^2 = 1 - \dfrac{x^2}{25.0};\ x^2 = 25.0 - 25.0y^2$

$$\int_0^{1.00}(yx^2)\,dy = \int_0^{1.00} y(25.0 - 25.0y^2)\,dy = \int_0^{1.00}(25.0y - 25.0y^3)\,dy = \left(\dfrac{25.0}{2}y^2 - \dfrac{25.0}{4}y^4\right)\Big|_0^{1.00}$$

$$= \dfrac{25.0}{2} - \dfrac{25.0}{4} = \dfrac{25.0}{4}$$

$$\int_0^{1.00} x^2\,dy = \int_0^{1.00}(25.0 - 25.0y^2)\,dy = \left(25.0y - \dfrac{25.0}{3}y^3\right) dy\Big|_0^{1.00} = \left(25.0 - \dfrac{25.0}{3}\right) = \dfrac{50.0}{3}$$

$$\bar{y} = \dfrac{25.0}{4} \div \dfrac{50.0}{3} = \dfrac{25.0}{4} \cdot \dfrac{3}{50.0} = 0.375 \text{ cm above center of base}$$

26.5   Moments of Inertia

1. $5.0(2.4)^2 + 3.2(3.5)^2 = 68 \text{ g} \cdot \text{cm}^2$
$(5.0 + 3.2)R^2 = 68;\ R = 2.9 \text{ cm}$

5. $y^2 = x, x = 4, x$-axis, with respect to the x-axis

$$I_x = k\int_0^2 y^2(4 - y^2)\,dy = k\int_0^2 (4y^2 - y^4)\,dy = k\left(\dfrac{4}{3}y^3 - \dfrac{1}{5}y^5\right)\Big|_0^2 = k\left(\dfrac{32}{3} - \dfrac{32}{5}\right) = \dfrac{64}{15}k$$

9. $y = \dfrac{b}{a}x;\ x = a, y = 0$

$$I_x = k\int_0^b y^2(a - x)\,dy = k\int_0^b y^2\left(a - \dfrac{ay}{b}\right)\,dy = ka\int_0^b \left(y^2 - \dfrac{y^3}{b}\right)\,dy = ka\left(\dfrac{1}{3}y^3 - \dfrac{1}{4b}y^4\right)\Big|_0^b$$

$$= \dfrac{kab^3}{12}$$

For $k = 1;\ m = \dfrac{1}{2}ab;\ I_x = ab\left(\dfrac{b^2}{12}\right) = 2m\left(\dfrac{b^2}{12}\right) = \dfrac{1}{6}mb^2$

13. $y^2 = x^3, y = 8, y$-axis, with respect to the x-axis

$$I_x = k\int_0^8 y^2 x\,dy = k\int_0^8 y^2(y^{2/3})\,dy = k\int_0^8 y^{8/3}\,dy = k\dfrac{3}{11}y^{11/3}\Big|_0^8 = \dfrac{3}{11}(8)^{11/3}k = \dfrac{3}{11}(2)^{11}k$$

$$= \dfrac{6144}{11}k$$

$$m = k\int_0^8 x\,dy = k\int_0^8 y^{2/3}\,dy = k\left(\dfrac{3}{5}y^{5/3}\right)\Big|_0^8 = \dfrac{3}{5}(8)^{5/3}k = \dfrac{3}{5}(2)^5k = \dfrac{96k}{5}$$

$$R^2 = \dfrac{I_x}{m} = \dfrac{6144}{11} \div \dfrac{96k}{5} = \dfrac{64(5)}{11}$$

$$R = \sqrt{\dfrac{64(5)}{11}} = \dfrac{8}{11}\sqrt{55}$$

17.　$y = 2x - x^2, y = 0$, rotated about y-axis

$$I_y = 2\pi k \int_0^2 (2x - x^2)(x^3)\,dx = 2\pi k \int_0^2 (2x^4 - x^5)\,dx = 2\pi k \left[\frac{2}{5}x^5 - \frac{1}{6}x^6 \right]\Big|_0^2 = 2\pi k \left[\frac{2}{5}(2)^5 - \frac{1}{6}(2)^6 \right]$$

$$= 2\pi k \left[\frac{64}{5} - \frac{64}{6} \right] = \frac{64\pi k}{15}$$

$$m = 2\pi k \int_0^2 (2x - x^2)(x)\,dx = 2\pi k \int_0^2 (2x^2 - x^3)\,dx = 2\pi k \left[\frac{2}{3}x^3 - \frac{1}{4}x^4 \right]\Big|_0^2$$

$$= 2\pi k \left[\frac{2}{3}(2)^3 - \frac{1}{4}(2)^4 \right] = 2\pi k \left[\frac{16}{3} - \frac{16}{4} \right] = \frac{8\pi k}{3}$$

$$R_y^2 = \frac{I_y}{m} = \frac{65\pi k}{15} \div \frac{8\pi k}{3} = \frac{8}{5}; \; R_y = \sqrt{\frac{8}{5}\left(\frac{5}{5}\right)} = \frac{2}{5}\sqrt{10}$$

21.　$r = 0.600$ cm, $h = 0.800$ cm, $m = 3.00$ g

$$y = \frac{0.600}{0.800}x = 0.750x; \; x = 1.333y$$

$$I_x = 2\pi k \int_0^{0.600} (0.800 - 1.333y)y^3\,dy = 2\pi k (0.200y^4 - 0.2667y^5)\big|_0^{0.600} = 2\pi k (0.005\,181)$$

$$m = \frac{k}{3}\pi r^2 h, 2\pi k = \frac{6m}{r^2 h} = \frac{6(3.00)}{(0.600^2)(0.800)} = 62.5 \text{ g/cm}^3$$

$$I_x = (62.5)(0.005\,181) = 0.324 \text{ g} \cdot \text{cm}^2$$

26.6　Other Applications

1. $f(x) = kx; 6.0 = k(1.5); k = 4.0$ lb/in

$$w = \int_0^{2.0} 4.0x\,dx = 2.0x^2\big|_0^{2.0} = 2.0(2.0^2) - 0 = 8.0 \text{ lb} \cdot \text{in}$$

5. $f(x) = \dfrac{9.0 \times 10^9 q_1 q_2}{x^2} = \dfrac{9.0 \times 10^9 \times 1.6 \times 10^{-19} \times 1.3 \times 10^{-18}}{x^2} = \dfrac{1.87 \times 10^{-27}}{x^2}$

$$W = \int_{2.0 \times 10^{-6}}^{1.0} (1.87 \times 10^{-27} x^{-2})\,dx = -1.9 \times 10^{-27} x^{-1}\big|_{2.0 \times 10^{-6}}^{1.0} = -1.9 \times 10^{-27} - \frac{-1.9 \times 10^{-27}}{2.0 \times 10^{-6}}$$

$$= -1.9 \times 10^{-27} + 0.94 \times 10^{-21} = -1.9 \times 10^{-27} + 9.4 \times 10^{-22} = 9.4 \times 10^{-22} \text{ N} \cdot \text{m}$$

9. $f(x) = 32.5 - (1.25 \times 10^{-3})x$

$$W = \int_0^{12000} [32.5 - (1.25 \times 10^{-3})x]\,dx = 32.5x - 0.63 \times 10^{-3}x^2\big|_0^{12000} = 300,000 - 0$$

$$= 3.00 \times 10^5 \text{ ft} \cdot \text{ton}$$

13. $F = 62.4 \int_0^{2.50} (12.0x)\,dx = 62.4 \left(6.00x^2\right)\big|_0^{2.50} = 62.4(37.5 - 0) = 2340 \text{ lb}$

17. $F = 9800 \int_0^{1.0} (xy)(dy); \quad y = \frac{1.0}{2.0}x; \quad x = 2.0y$

$$F = 9800 \int_0^{1.0} (2.0y)(y)\, dy = 9800 \left[\frac{2.0}{3}y^3\right]\Big|_0^{1.0} = 9800\left(\frac{2.0}{3}\right) = 6500 \text{ N}$$

21. $i = 4t - t^2$. Find i_{av} with respect to time for $t = 0$ to $t = 4$.

$$i_{av} = \frac{\int_0^{4.0} i\, dt}{4.0 - 0} = \frac{\int_0^{4.0} (4t - t^2)\, dt}{4.0} = \frac{2t^2 - \frac{1}{3}t^3 \Big|_0^{4.0}}{4.0} = \frac{2(16) - \frac{1}{3}(64)}{4.0} = \frac{10.7}{4.0} = 2.7 \text{ A}$$

25. $s = \int_a^b \sqrt{1 + \left(\frac{dy}{dx}\right)^2}\, dx; \quad y = 0.4x^{3/2}; \quad \frac{dy}{dx} = 0.06x^{1/2}$

$$s = \int_0^{100} \sqrt{1 + (0.06x^{1/2})^2}\, dx = \int_0^{100} \sqrt{(1 + 0.0036x)}\, dx = \int_0^{100} (1 + 0.0036x)^{1/2}\, dx$$

$$= \frac{1}{0.0036} \int_0^{100} (1 + 0.0036x)^{1/2}(0.0036)(dx) = \frac{1}{0.0036}\left[\frac{2}{3}(1 + 0.0036x)^{3/2}\right]\Big|_0^{100}$$

$$= \frac{1}{0.0036}\left[\frac{2}{3}(1.586) - \frac{2}{3}(1)^{3/2}\right] = \frac{1}{0.0036}\left[\frac{2}{3}(0.586)\right] = \frac{1}{0.0036}[0.391]$$

$$= 109 \text{ ft}$$

Chapter 26 Review Exercises

1. $v = \int a\, dt; \quad a = 32 \text{ ft/s}^2; \quad v = at + c, \text{ at } t = 0, v = 0 \Rightarrow c = 0$

$v = at = 32t$

$140 = 32t$

$t = 4.4s$

5. $i = 0.25(2\sqrt{t} - t); \quad q = \int i\, dt$

$$q = \int_0^2 0.25(2\sqrt{t} - t)\, dt = \int_0^2 0.25(2t^{1/2} - t)\, dt = \int_0^2 (0.50t^{1/2} - 0.25t)\, dt$$

$$= \left(\frac{1}{3}t^{3/2} - \frac{1}{8}t^2\right)\Big|_0^2 = \frac{\sqrt{8}}{3} - \frac{1}{2} = 0.44 \text{ C}$$

9. $dy = \left(20 + \frac{1}{40}x^2\right) dx; \quad y = \int \left(20 + \frac{1}{40}x^2\right) dx = 20x + \frac{1}{120}x^3 + C$

$y = 0$ when $x = 0$

$20(0) + \frac{1}{120}(0)^3 + C = 0; \quad C = 0$

$y = 20x + \frac{1}{120}x^3$

13. $y^2 = 2x, y = x - 4$

$x_1 = \dfrac{1}{2}y^2; \; x_2 = y + 4;$

$A = \displaystyle\int_{-2}^{4} (x_2 - x_1)\, dy = \int_{-2}^{4} \left(y + 4 - \frac{1}{2}y^2 \right) dy = -\frac{1}{6}y^3 + \frac{1}{2}y^2 + 4y \bigg|_{-2}^{4}$

$\qquad = -\dfrac{1}{6}(64) + \dfrac{1}{2}(16) + 4(4) + \dfrac{1}{6}(-8) - \dfrac{1}{2}(4) - 4(-2) = 18$

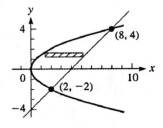

17.

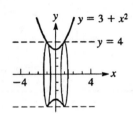

$V = \displaystyle\pi \int_{-1}^{1} 4^2\, dx - \pi \int_{-1}^{1} y^2\, dx = \pi \int_{-1}^{1} (4^2 - y^2)\, dx = \pi \int_{-1}^{1} [16 - (3 + x^2)^2]\, dx$

$\qquad = \pi \displaystyle\int_{-1}^{1} (16 - 9 - 6x^2 - x^4)\, dx = \pi \int_{-1}^{1} (7 - 6x^2 - x^4)\, dx = \pi \left(7x - 2x^3 - \frac{1}{5}x^5 \right) \bigg|_{-1}^{1}$

$\qquad = \pi \left(7 - 2 - \dfrac{1}{5} \right) - \pi \left(-7 + 2 + \dfrac{1}{5} \right) = \dfrac{24\pi}{5} + \dfrac{24\pi}{5} = \dfrac{48\pi}{5}$

21.

$\dfrac{x^2}{a^2} - \dfrac{y^2}{b^2} = 1$

$b^2 x^2 + a^2 y^2 = a^2 b^2$

$y = \sqrt{\dfrac{a^2 b^2 - b^2 x^2}{a^2}}$

$V = \displaystyle\pi \int_{-a}^{a} y^2\, dx = \pi \int_{-a}^{a} \left(\frac{a^2 b^2 - b^2 x^2}{a^2} \right) dx = \pi \int_{-a}^{a} \left(b^2 - \frac{b^2}{a^2}x^2 \right) dx = \pi \left(b^2 x - \frac{b^2}{3a^2}x^3 \right) \bigg|_{-a}^{a}$

$\qquad = \pi \left[\left(ab^2 - \dfrac{ab^2}{3} \right) - \left(-ab^2 + \dfrac{ab^2}{3} \right) \right] = \pi \left(2ab^2 - \dfrac{2ab^2}{3} \right) = \pi \left(\dfrac{4ab^2}{3} \right) = \dfrac{4\pi ab^2}{3} = \dfrac{4}{3}\pi ab^2$

25. $y = \sqrt{x}, x = 1, x = 4, y = 0, x$-axis; $y = 0$

$$\bar{x} = \frac{\int_1^4 xy^2\,dx}{\int_1^4 y^2\,dx} = \frac{\int_1^4 x(x)\,dx}{\int_1^4 x\,dx} = \frac{\int_1^4 x^2\,dx}{\int_1^4 x\,dx} = \frac{\frac{1}{3}x^3\Big|_1^4}{\frac{1}{2}x^2\Big|_1^4} = \frac{\frac{63}{3}}{\frac{15}{2}} = \frac{14}{5}$$

Thus, $(\bar{x}, \bar{y})$ is $\left(\frac{14}{5}, 0\right)$.

29. $I_x = 2\pi k \int_c^d (x_2 - x_1)y^3\,dy$

$$I_x = 2\pi(0.0114) \int_0^{3.00(20.0)^{0.10}} \left(20.0 - \left(\frac{y}{3.00}\right)^{10}\right) y^3\,dy$$

$$I_x = 68.7 \text{ g} \cdot \text{mm}^2$$

33. $y_2 = 3x^2 - x^3;\ y_1 = 0;$

$$\bar{x} = \frac{\int_0^3 x(y_2^2 - y_1)\,dx}{\int_0^3 x(y_2^2 - y_1)\,dx} = \frac{\int_0^3 x(3x^2 - x^3)\,dx}{\int_0^3 (3x^2 - x^3)\,dx} = \frac{\int_0^3 (3x^3 - x^4)\,dx}{\int_0^3 (3x^2 - x^3)\,dx} = \frac{\left(\frac{3}{4}x^4 - \frac{1}{5}x^5\right)\Big|_0^3}{\left(x^3 - \frac{1}{4}x^4\right)\Big|_0^3}$$

$$= \frac{\frac{243}{4} - \frac{243}{5}}{27 - \frac{81}{4}} = \frac{12.15}{6.75} = 1.8 \text{ m}$$

37. The circumference of the bottom, $c = 2\pi r = 9\pi$, equates to l, the length of the vertical surface area.

$$F = 68.0 \int_0^{3.25} (9\pi h)\,dh = 68.0\left[4.50\pi h^2\right]\Big|_0^{3.25}$$

$$= 68.0[4.50\pi(3.25)^2 - 0] = 10,200 \text{ lb}$$

41. The formula is $V = 2\int_0^a (2 \cdot \sqrt{a^2 - x^2})^2\,dx$ where x is the distance from the center of the circular top to the cross section.

DIFFERENTIATION OF TRANSCENDENTAL FUNCTIONS

27.1 Derivatives of the Sine and Cosine Functions

1. $y = \sin(x + 2);$

$$\frac{dy}{dx} = \cos(x + 2)\frac{d(x + 2)}{dx}$$

$$= \cos(x + 2)$$

5. $y = 6\cos\left(\frac{1}{2}x\right);$

$$\frac{dy}{dx} = -6\left[\sin\left(\frac{1}{2}x\right)\right]\left[\frac{1}{2}\right]$$

$$= -3\sin\left(\frac{1}{2}x\right)$$

9. $r = \sin^2(3\pi\theta)$

$$\frac{dr}{d\theta} = 2\sin(3\pi\theta) \cdot \cos(3\pi\theta) \cdot 3\pi$$

$$\frac{dr}{d\theta} = 6\pi\sin(3\pi\theta) \cdot \cos(3\pi\theta)$$

$$= 3\pi\sin 6\pi\theta$$

13. $y = x\sin 3x;$

$$\frac{dy}{dx} = x(3\cos 3x) + (\sin 3x)(1)$$

$$\frac{dy}{dx} = \sin 3x + 3x\cos 3x$$

17. $y = \sin x^2\cos 2x;$

$$\frac{dy}{dx} = \sin x^2(-2\sin 2x) + \cos 2x(2x\cos x^2)$$

$$\frac{dy}{dx} = 2x\cos x^2\cos 2x - 2\sin x^2\sin 2x$$

21. $r = \dfrac{\sin\left(3t - \frac{\pi}{3}\right)}{2t}$

$$\frac{dr}{dt} = \frac{2t \cdot \cos\left(3t - \frac{\pi}{3}\right) \cdot 3 - \sin\left(3t - \frac{\pi}{3}\right) \cdot 2}{(2t)^2}$$

$$\frac{dr}{dt} = \frac{3t\cos\left(3t - \frac{\pi}{3}\right) - \sin\left(3t - \frac{\pi}{3}\right)}{2t^2}$$

25. $y = 2\sin^2 3x\cos 2x$

$$\frac{dy}{dx} = (2\sin^2 3x)(-\sin 2x)(2) + (\cos 2x)(2)(2\sin 3x)(\cos 3x)(3)$$

$$\frac{dy}{dx} = -4\sin^2 3x\sin 2x + 12\cos 2x\sin 3x\cos 3x$$

$$\frac{dy}{dx} = 4\sin 3x(3\cos 3x\cos 2x - \sin 3x\sin 2x)$$

29. $y = \sin^3 x - \cos 2x;$

$$\frac{dy}{dx} = 3\sin^2 x(\cos x) - (-\sin 2x)(2); \quad \frac{dy}{dx} = 3\sin^2 x\cos x + 2\sin 2x$$

33. (a) Set mode to radian. Set range to $x_{\min} = -4$, $x_{\max} = 4$, $y_{\min} = -2$, $y_{\max} = 2$, and enter $y_1 = \sin x \div x$.

Construct zoom box and use trace. $x = 0.001\ 994\ 46$, $y = 0.999\ 999\ 34$.

(b) Checking the values on *p.* 775, using a zoom box, when $x = 0.503\ 047\ 09$, $y = 0.954\ 900\ 48$; when $x = 0.103\ 268\ 7$, $y = 0.998\ 223\ 54$; when $x = 0.050\ 969\ 53$, $y = 0.999\ 567\ 07$; when $x = 0.01108033$, $y = 0.999\ 979\ 54$; when $0.002\ 216\ 07$, $y = 0.999\ 999\ 18$.

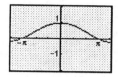

37. $y = \sin x$

$m_{\tan}\big|_{x=0} = 1.0$	$m_{\tan}\big|_{x=\pi/4} = 0.7$	$m_{\tan}\big|_{x=\pi/2} = 0.0$
$m_{\tan}\big|_{x=3\pi/4} = -0.7$	$m_{\tan}\big|_{x=\pi} = -1.0$	$m_{\tan}\big|_{x=5\pi/4} = -0.7$
$m_{\tan}\big|_{x=3\pi/2} = 0.0$	$m_{\tan}\big|_{x=7\pi/4} = 0.7$	$m_{\tan}\big|_{x=2\pi} = 1.0$

Plot points: $(0, 1.0)$, $\left(\frac{\pi}{4}, 0.7\right)$, $\left(\frac{\pi}{2}, 0.0\right)$, $\left(\frac{3\pi}{4}, -0.7\right)$, $(\pi, -1.0)$, $\left(\frac{5\pi}{4}, -0.7\right)$, $\left(\frac{3\pi}{2}, 0.0\right)$, $\left(\frac{7\pi}{4}, 0.7\right)$, $(2\pi, 1.0)$.

Resulting curve is $y = \cos x$.

41. $\dfrac{d}{dx}\sin x = \cos x;$

$\dfrac{d^2}{dx^2}\sin x = -\sin x$

$\dfrac{d^3}{dx^3}\sin x = -\cos x;$

$\dfrac{d^4}{dx^4}\sin x = \sin x$

45. $y = 3\sin 2x;$

$\dfrac{dy}{dx} = 3(\cos 2x)(2x) = 6\cos 2x;$

$dy = (6\cos 2x)dx = \left(6\cos\dfrac{\pi}{4}\right)(0.02)$

$= \dfrac{6\sqrt{2}}{2}(0.02) = 0.085$

49. $y = 1.85\sin 36\pi t;$ $v = \dfrac{dy}{dx} = (1.85\cos 36\pi t)(36\pi)$

$v\big|_{t=0.0250} = [1.85\cos(36\pi \cdot 0.025)][36\pi] = -199 \text{ cm/s}$

27.2 Derivatives of the Other Trigonometric Functions

1. $y = \tan 5x;$

$\dfrac{dy}{dx} = \sec^2 5x\,\dfrac{d\,5x}{dx} = 5\sec^2 5x$

5. $y = 3\sec 2x;$

$\dfrac{dy}{dx} = 3\sec 2x\tan 2x\,\dfrac{d\,2x}{dx} = 6\sec 2x\tan 2x$

9. $y = 5\tan^2 3x$

$$\frac{dy}{dx} = 5(2\tan 3x)(\sec^2 3x)\frac{d3x}{dx}$$

$$= 30\tan 3x \sec^2 3x$$

13. $y = \sqrt{\sec 4x};$

$$\frac{dy}{dx} = \frac{1}{2}(\sec 4x^{-1/2})\sec 4x \tan 4x \frac{d4x}{dx}$$

$$\frac{dy}{dx} = \frac{2\sec 4x \tan 4x}{\sqrt{\sec 4x}} = 2\tan 4x\sqrt{\sec 4x}$$

17. $r = t^2 \tan(0.5t)$

$$\frac{dr}{dt} = t^2 \cdot \sec^2(0.5t) \cdot 0.5 + \tan(0.5t) \cdot 2t$$

$$\frac{dr}{dt} = \frac{t^2}{2}\sec^2(0.5t) + 2t\tan(0.5t)$$

21. $y = \dfrac{\csc}{x};$

$$\frac{dy}{dx} = \frac{x\left(\frac{d\,\csc x}{dx}\right) - \csc x\left(\frac{dx}{dx}\right)}{x^2}$$

$$\frac{dy}{dx} = \frac{x(-\csc x \cot x) - \csc x}{x^2}$$

$$= -\frac{\csc x(x\cot x + 1)}{x^2}$$

25. $y = \dfrac{1}{3}\tan 3x - \tan x;$

$$\frac{dy}{dx} = 3\left(\frac{1}{3}\right)\tan^2 x\frac{d(\tan x)}{dx} - \sec^2 x$$

$$\frac{dy}{dx} = \tan^2 x \sec^2 x - \sec^2 x$$

$$= \sec^2 x(\tan^2 x - 1)$$

29. $y = \sqrt{2x + \tan 4x} = (2x + \tan 4x)^{1/2}$

$$\frac{dy}{dx} = \frac{1}{2}(2x + \tan 4x)^{-1/2}(2 + 4\sec^2 4x)$$

$$\frac{dy}{dx} = \frac{1 + 2\sec^2 4x}{\sqrt{2x + \tan 4x}}$$

33. $y = 4\tan^2 3x$

$$\frac{dy}{dx} = 4(2\tan 3x)(\sec^2 3x)(3) = 24\tan 3x \sec^2 3x; \; dy = 24\;\tan 3x \sec^2 3x\,dx$$

37. (a) $\sec^2 1.000 = \frac{1}{\cos^2 1.0000} = 3.425\,518\,8$. This is the slope of a tangent line to the curve $f(x) = \tan x$
at $x = 1.0000$. It is the value of $f'(x) = \sec^2 x$ at $x = 1.0000$ since $\frac{d(\tan x)}{dx} = \sec^2 x$.

(b) $\frac{\tan 1.0001 - \tan 1.0000}{0.0001} = 3.426\,052\,4$. This is the slope of a secant line through the curve $f(x) = \tan x$
at $x = 1.000$, where $\Delta x = 0.0001$.

$$\lim_{\Delta x \to 0}\frac{\tan(x + \Delta x) - \tan x}{\Delta x} = \frac{d\tan x}{dx} = \sec^2 x$$

For $\Delta x = 0.0001$, the slope of the tangent line is approximately equal to the slope of the secant line.
(Final digits may vary.)

41. $y = 2\cot 3x; \; x = \dfrac{\pi}{12}; \; \dfrac{dy}{dx} = 2(-\csc^2 3x)(3) = -6\csc^2 3x; \; \left.\dfrac{dy}{dx}\right|_{x=\pi/12} = -6\csc^2\dfrac{\pi}{4} = -6(\sqrt{2})^2 = -12$

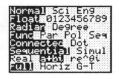
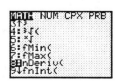
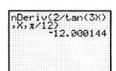

45. $y = 2t^{1.5} - \tan 0.1t$; $v = \dfrac{dy}{dt} = 3t^{0.5} - 0.1\sec^2 0.1t$; $v|_{t=15} = 3(15)^{0.5} - 0.1\sec^2[0.1(15)] = -8.4$ cm/s

27.3 Derivatives of the Inverse Trigonometric Functions

1. $y = \sin^{-1}(x^2)$; $\dfrac{dy}{dx} = \dfrac{1}{\sqrt{1-(x^2)^2}}\dfrac{dx^2}{dx} = \dfrac{2x}{\sqrt{1-x^4}}$

5. $y = 3.6\cos^{-1} 0.5s$

$\dfrac{dy}{ds} = 3.6\dfrac{-1}{\sqrt{1-(-.5s)^2}}\cdot 0.5$

$\dfrac{dy}{ds} = \dfrac{-1.8}{\sqrt{1-\frac{s^2}{4}}}$

$\dfrac{dy}{ds} = \dfrac{-3.6}{\sqrt{4-s^2}}$

9. $y = \tan^{-1}\sqrt{x} = \tan^{-1} x^{1/2}$;

$\dfrac{dy}{dx} = \dfrac{1}{1+(\sqrt{x})^2}\dfrac{dx^{1/2}}{dx}$

$\dfrac{dy}{dx} = \dfrac{1}{1+x}\left(\dfrac{1}{2}x^{-1/2}\right)$

$= \dfrac{1}{2\sqrt{x}(1+x)}$

13. $y = 5x\sin^{-1} x$;

$\dfrac{dy}{dx} = 5x\left(\dfrac{1}{\sqrt{1-x^2}}\right)(1) + (\sin^{-1})(5)$

$\dfrac{dy}{dx} = \dfrac{5x}{\sqrt{1-x^2}} + 5\sin^{-1} x$

17. $y = \dfrac{3x-1}{\sin^{-1} 2x}$

$\dfrac{dy}{dx} = \dfrac{(\sin^{-1} 2x)(3) - (3x-1)\frac{1}{\sqrt{1-4x^2}}(2)}{(\sin^{-1} 2x)^2}$

$\dfrac{dy}{dx} = \left(3\sin^{-1} 2x - \dfrac{6x-2}{\sqrt{1-4x^2}}\right) \times \dfrac{1}{(\sin^{-1} 2x)^2}$

$\dfrac{dy}{dx} = \dfrac{3\sqrt{1-4x^2}\sin^{-1} 2x - 6x + 2}{\sqrt{1-4x^2}(\sin^{-1} 2x)^2}$

21. $y = 2(\cos^{-1} 4x)^3$

$\dfrac{dy}{dx} = 6(\cos^{-1} 4x)^2\dfrac{d(\cos^{-1} 4x)}{dx}$

$\dfrac{dy}{dx} = 6(\cos^{-1} 4x)^2\left(-\dfrac{1}{\sqrt{1-(4x)^2}}\right)(4)$

$\dfrac{dy}{dx} = -\dfrac{24(\cos^{-1} 4x)^2}{\sqrt{1-16x^2}}$

25. $y = \tan^{-1}\left(\dfrac{1-t}{1+t}\right)$

$\dfrac{dy}{dt} = \dfrac{1}{1+\left(\frac{1-t}{1+t}\right)^2}\cdot\dfrac{(1+t)(-1)-(1-t)(1)}{(1+t)^2}$

$\dfrac{dy}{dt} = \dfrac{-1-t-1+t}{(1+t)^2+(1-t)^2}$

$\dfrac{dy}{dt} = \dfrac{-2}{1+2t+t^2+1-2t+t^2}$

$\dfrac{dy}{dt} = \dfrac{-2}{2+2t^2}; \dfrac{dy}{dt} = \dfrac{-1}{1+t^2}$

29. $y = 3(4-\cos^{-1} 2x)^3$

$\dfrac{dy}{dx} = 9(4-\cos^{-1} 2x)^2\left(\dfrac{1}{\sqrt{1-(2x)^2}}\right)(2) = \dfrac{18(4-\cos^{-1} 2x)^2}{\sqrt{1-4x^2}}$

33. (a) $\dfrac{1}{\sqrt{1 - 0.5^2}} = 1.154\ 700\ 5;$ **(b)** $\dfrac{\sin^{-1} 0.5001 - \sin^{-1} 0.5000}{0.0001} = 1.154\ 739\ 0;$

value of derivative   slope of secant line

37. $y = \dfrac{x}{\tan^{-1} x}; m_{\text{tan}} = \dfrac{dy}{dx} = \dfrac{\tan^{-1} x(1) - x\left[\frac{1}{1+x^2}\right]}{(\tan^{-1} x)^2} = \dfrac{\tan^{-1} x - \frac{x}{1+x^2}}{(\tan^{-1} x)^2}$

When $x = 0.80, m_{\text{tan}} = \dfrac{\tan^{-1}(0.8) - \frac{0.8}{1+0.8^2}}{(\tan^{-1} 0.8)^2} = 0.41$

 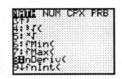 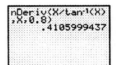

41. $y = \tan^{-1} 2x; \dfrac{dy}{dx} = \dfrac{1}{1 + (2x)^2}(2) = \dfrac{2}{1 + 4x^2}; \dfrac{d^2y}{dx^2} = \dfrac{(1 + 4x^2)(0) - 2(8x)}{(1 + 4x^2)^2} = \dfrac{-16x}{(1 + 4x^2)^2}$

45. $t = \dfrac{1}{\omega} \sin^{-1} \dfrac{A - E}{mE} = \dfrac{1}{\omega} \sin^{-1} \left(\dfrac{A - E}{E}\right)\left(\dfrac{1}{m}\right) = \dfrac{1}{\omega} \sin^{-1} \left(\dfrac{A - E}{E}\right) m^{-1}$

$u = \left(\dfrac{A - E}{E}\right) m^{-1}; \dfrac{du}{dm} = -\left(\dfrac{A - E}{E}\right) m^{-2}$

$\dfrac{dt}{dm} = \dfrac{1}{\omega\sqrt{1 - \left(\frac{A-E}{E}\right)^2 m^{-2}}}\left(\dfrac{-A + E}{Em^2}\right) = \dfrac{E - A}{\omega Em^2\sqrt{1 - \left(\frac{A-E}{E^2m^2}\right)^2}} = \dfrac{E - A}{\omega Em^2\sqrt{\frac{E^2m^2 - (A-E)^2}{E^2m^2}}}$

$\dfrac{dt}{dm} = \dfrac{E - A}{\omega m\sqrt{E^2m^2 - (A - E)^2}}$

27.4   Applications

1. Points of intersection occur when $\sin x = \cos x$.

$y_1 = \sin x; y_2 = \cos x; \dfrac{dy_1}{dx} = \cos x; \dfrac{dy_2}{dx} = -\sin x$

At points of intersection, $\dfrac{dy_1}{dx} = -\dfrac{dy_2}{dx}.$

5. $y = x - \tan x$; $\left(-\dfrac{\pi}{2} < x < \dfrac{\pi}{2}\right)$; $\dfrac{dy}{dx} = 1 - \sec^2 x$; $\dfrac{d^2y}{dx^2} = 0 - 2\sec x(\sec x \tan x) = -2\sec^2 x \tan x$.

(1) $\dfrac{dy}{dx} < 0$ for $\dfrac{-\pi}{2} < x < 0$; $\dfrac{dy}{dx} = 0$ for $x = 0$; $\dfrac{dy}{dx} < 0$ for $0 < x < \dfrac{\pi}{2}$. The function decreases from $-\dfrac{\pi}{2}$ to $\dfrac{\pi}{2}$ $(x \neq 0)$ and has one critical point at $x = 0$.

(2) Since $0 - \tan = 0$, there is an intercept at $(0, 0)$.

(3) $\dfrac{d^2y}{dx^2} = 0$ when $\sec^2 x = 0$ or $\tan x = 0$. Since $|\sec x| \geq 1$ for all x, $\sec^2 x \geq 1$ for all x, and $\sec^2 x = 0$ has no solution. Between $-\dfrac{\pi}{2}$ and $\dfrac{\pi}{2}$, $\tan x = 0$ when $x = 0$.

(4) $-2\sec^2 x \tan x > 0$ when $\tan x < 0$. $\tan x < 0$ for $-\dfrac{\pi}{2} < x < 0$. The graph is concave up in this region. $-2\sec^2 x \tan x < 0$ when $\tan x > 0$ or x is between 0 and $\dfrac{\pi}{2}$. The graph is concave down in this region. Since there is a change of sign in the second derivative, the only critical point is an inflection point, and there are no maximum or minimum points.

(5) As $x \to -\dfrac{\pi}{2}$ from the right, $x - \tan x \to -\dfrac{\pi}{2} - (-\infty)$ or $+\infty$; $x = -\dfrac{\pi}{2}$ is an asymptote.

Summarizing, the function decreases, intersects the x-axis at 0, is concave up for $-\dfrac{\pi}{2} < x < 0$, concave down for $0 < x < \dfrac{\pi}{2}$, has point of inflection at $x = 0$, and asymptotes at $x = -\dfrac{\pi}{2}$ and $\dfrac{\pi}{2}$.
Dec., $x > 0, x < 0$
Infl. $(0, 0)$

Asym., $x = \dfrac{\pi}{2}, x = -\dfrac{\pi}{2}$

9. $x^2 - 4\sin x = 0$

By a rough sketch of the graph using $f(0) = 0$, $f(1) = -2.37$, and $f(2) = 0.36$, the root is between 1 and 2 and is possibly closer to 2. Let $x_1 = 1.7$:

n	x_n	$f(x_n)$	$f'(x_n)$	$x_n - \dfrac{f(x_n)}{f'(x_n)}$
1	1.7	−1.076 659 2	3.915 378	1.974 982 2
2	1.974 982 2	0.222 863 2	5.523 045 9	1.934 630 7
3	1.934 630 7	0.004 639 1	5.292 702 3	1.933 754 2
4	1.933 754 2	0.000 002 3	5.287 672 2	1.933 753 8

$x_4 = 1.933\ 753\ 8$

13. $y = 0.50\sin 2t + 0.30\cos t$; $v = \dfrac{dy}{dt} = 1.00\cos 2t - 0.30\sin t$

$v|_{t=0.40s} = 1.00\cos 0.80 - 0.30\sin 0.40 = 0.58$ ft/s

$a = \dfrac{d^2y}{dt^2} = -2.00\sin 2t - 0.30\cos t$

$a|_{t=0.40s} = -200\sin 0.80 - 0.30\cos 0.40 = -1.7$ ft/s²

17. $x = 2.625 \cos 12\pi t$; $\dfrac{dx}{dt} = -2.625(12\pi) \sin 12\pi t = -31.50\pi \sin 12\pi t$;

$y = 2.625 \sin 12\pi t$; $\dfrac{dy}{dt} = 2.625(12\pi) \cos 12\pi t = 31.50\pi \cos 12\pi t$

$\dfrac{dx}{dy}\bigg|_{t=1.250} = -31.50\pi \sin 47.12 = 0$

$\dfrac{dy}{dt}\bigg|_{t=1.250} = -31.50\pi \cos 47.12 = 31.50\pi(-1.000) = -98.96$ in/s

$v = \sqrt{(0)^2 + (-98.96)^2} = 98.96$ in/s; $\theta = 270° (v_x = 0, v_y < 0)$

21. $s = 16t^2$

$\tan\theta = \dfrac{200 - s}{100} = \dfrac{200 - 16t^2}{100} = 2 - 0.16t^2$

$\theta = \tan^{-1}(2 - 0.16t^2)$

$\dfrac{d\theta}{dt} = \dfrac{1}{1 + (2 - 0.16t^2)^2}(-0.32t) = \dfrac{-0.32t}{5 - 0.64(t)^2 + 0.0256(t)^4}$

$\dfrac{d\theta}{dt}\bigg|_{t=1.0} = \dfrac{-0.32(1.0)}{5 - 0.64(1.0)^2 + 0.0256(1.0)^4} = -0.073$ rad/s

25. $\mu = \tan\theta$; $\theta = 20° = \dfrac{\pi}{9} = 0.349$ rad

$d\theta = 1° = \dfrac{\pi}{180} = 0.0175$ rad; $\dfrac{d\mu}{d\theta} = \sec^2\theta$

$d\mu = \sec^2\theta \, d\theta$

$d\mu|_{\theta=20°} = 0.349 = (\sec^2 0.349)(0.0175) = (1.064)^2(0.0175) = 0.020$

29. $s = kwd^2$; $d^2 = 256 - w^2$; $\cos\theta = \dfrac{w}{16.0}$; $w = 16.0\cos\theta$; $S = k(16.0\cos\theta)[256 - (16.0\cos\theta)^2]$;

$S = 4100k[\cos\theta - \cos^3\theta]$; $\dfrac{dS}{d\theta} = 4100k[-\sin\theta - 3\cos^2\theta(-\sin\theta)]$; $\dfrac{dS}{d\theta} = 4100k(-\sin\theta + 3\sin\theta\cos^2\theta)$

Maximum occurs when $\dfrac{dS}{d\theta} = 0$.

$4100k(-\sin\theta + 3\sin\theta\cos^2\theta) = 0$; $-\sin\theta(1 - 3\cos^2\theta) = 0$

$-\sin\theta = 0$; $\sin\theta = 0$; $\theta = 0, \theta = \pi$; $1 - 3\cos^2\theta = 0$; $\cos^2\theta = \dfrac{1}{3}$; $\cos\theta = \sqrt{\dfrac{1}{3}}$

$w = 16.0\sqrt{\dfrac{1}{3}} = 9.24$ in

$l = \sqrt{16.0^2 - 16.0^2\left(\dfrac{1}{3}\right)} = 13.1$ in

27.5   Derivative of the Logarithmic Function

1. $y = \log x^2;\ u = x^2;\ \dfrac{du}{dx} = 2x$

$\dfrac{dy}{dx} = \dfrac{1}{x^2}(\log e)(2x) = \dfrac{2\log e}{x}$

5. $y = 0.2\ln(1 - 3x)$

$\dfrac{dy}{dx} = \dfrac{0.2}{1 - 3x}(-3) = \dfrac{-0.6}{1 - 3x}$

9. $y = \ln\sqrt{x} = \ln(x^{1/2}) = \dfrac{1}{2}\ln x;\ \dfrac{dy}{dx} = \dfrac{1}{2}\left(\dfrac{1}{x}\right) = \dfrac{1}{2x}$

13. $y = 3[t + \ln t^2]^2$

$\dfrac{dy}{dt} = 6[t + \ln t^2]\cdot\left[1 + \dfrac{1}{t^2}\cdot 2t\right]$

$\dfrac{dy}{dt} = 6[t + \ln t^2]\cdot\left[1 + \dfrac{2}{t}\right]$

17. $y = \ln(\ln x);$ let $u = \ln x$

$\dfrac{dy}{dx} = \dfrac{1}{\ln x}\left(\dfrac{d(\ln x)}{dx}\right)$

$= \dfrac{1}{\ln x}\left(\dfrac{1}{x}\right) = \dfrac{1}{x\ln x}$

21. $y = \sin(\ln x)$

$\dfrac{dy}{dx} = \cos(\ln x)\dfrac{d(\ln x)}{dx} = \cos(\ln x)\left(\dfrac{1}{x}\right)$

$= \dfrac{\cos(\ln x)}{x}$

25. $y = \ln(x\tan x)$

$\dfrac{dy}{dx} = \dfrac{1}{x\tan x}[x(\sec^2 x) + \tan x]$

$= \dfrac{x\sec^2 x + \tan x}{x\tan x}$

29. $y = \sqrt{x^2 + 1} - \ln\dfrac{1 + \sqrt{x^2 + 1}}{x} = (x^2 + 1)^{1/2} - \ln(1 + \sqrt{x^2 + 1}) + \ln x$

$\dfrac{dy}{dx} = \dfrac{1}{2}(x^2 + 1)^{-1/2}(2x) - \dfrac{\frac{1}{2}(x^2 + 1)^{-1/2}(2x)}{1 + \sqrt{x^2 + 1}} + \dfrac{1}{x} = \dfrac{x}{(x^2 + 1)^{1/2}} - \dfrac{x}{(x^2 + 1)^{1/2}(1 + \sqrt{x^2 + 1})} + \dfrac{1}{x}$

$= \dfrac{x^2(1 + \sqrt{x^2 + 1}) - x^2 + (x^2 + 1)^{1/2}(1 + \sqrt{x^2 + 1})}{x(x^2 + 1)^{1/2}(1 + \sqrt{x^2 + 1})} = \dfrac{x^2\sqrt{x^2 + 1} + \sqrt{x^2 + 1} + (x^2 + 1)}{x(x^2 + 1)^{1/2}(1 + \sqrt{x^2 + 1})}$

$= \dfrac{(x^2 + 1)\sqrt{x^2 + 1} + (x^2 + 1)}{x(x^2 + 1)^{1/2}(1 + \sqrt{x^2 + 1})} = \dfrac{(x^2 + 1)(\sqrt{x^2 + 1} + 1)}{x(x^2 + 1)^{1/2}(1 + \sqrt{x^2 + 1})} = \dfrac{\sqrt{x^2 + 1}}{x}$

33. $\dfrac{\ln 2.0001 - \ln 2.0000}{0.0001} = 0.499\,987\,5$

Slope of secant line through $(2.0000, \ln 2.0000)$ and $(2.0001, \ln 2.0001)$

$0.5 = \dfrac{d\ln x}{dx}$ for $x = 2$ and is slope of tangent line through $(2.0000, \ln 2.0000)$

37. $y = \sin^{-1} 2x + \sqrt{1 - 4x^2}$; $x = 0.250$; $\dfrac{dy}{dx} = \dfrac{2}{\sqrt{1 - (2x)^2}} + \dfrac{1}{2}(1 - 4x^2)^{-1/2}(-8x) = \dfrac{2 - 4x}{\sqrt{1 - 4x^2}}$

$\dfrac{dy}{dx}\bigg|_{x=0.250} = \dfrac{2 - 4(0.250)}{\sqrt{1 - 4(0.250)^2}} = 1.15$

41. $y = \tan^{-1} 2x + \ln(4x^2 + 1)$; $m_{\text{tan}} = \dfrac{dy}{dx} = \dfrac{1}{4x^2 + 1}(2) + \dfrac{1}{4x^2 + 1}(8x) = \dfrac{2}{1 + 4x^2} + \dfrac{8x}{4x^2 + 1} = \dfrac{2 + 8x}{1 + 4x^2}$

When $x = 0.625$, $m_{\text{tan}} = \dfrac{2 + 8(0.625)}{1 + 4(0.625)^2} = \dfrac{7}{2.5625} = 2.73$

45. $b = 10\log\left(\dfrac{I}{I_0}\right) = 10\log(I_0^{-1}I)$; $u = (I_0^{-1}I)$; $\dfrac{db}{dt} = 10\left(\dfrac{I_0}{I}\right)(\log e)\left(\dfrac{I}{I_0}\right)\dfrac{dI}{dt} = \dfrac{10}{I}\log e \dfrac{dI}{dt}$

27.6   Derivative of the Exponential Function

1. $y = 3^{2x}$;

$\dfrac{dy}{dx} = 3^{2x}\ln 3 \dfrac{d\, 2x}{dx} = (2\ln 3)3^{2x}$

5. $y = e^{\sqrt{x}}$;

$\dfrac{dy}{dx} = e^{\sqrt{x}}\dfrac{1}{2}x^{-1/2} = \dfrac{e^{\sqrt{x}}}{2\sqrt{x}}$

9. $y = xe^{-x}$

$\dfrac{dy}{dx} = x(e^{-x})(-1) + (1)(e^{-x})$

$\quad = e^{-x} - xe^{-x} = e^{-x}(1 - x)$

13. $r = \dfrac{2(e^{2s} - e^{-2s})}{e^{2s}}$

$r = 2(1 - e^{-4s})$

$\dfrac{dr}{ds} = 2(4e^{-4s})$

$\dfrac{dr}{ds} = 8e^{-4s}$

17. $y = \dfrac{2e^{3x}}{4x + 3}$;

$\dfrac{dy}{dx} = \dfrac{(4x + 3)(2e^{3x})(3) - (2e^{3x})(4)}{(4x + 3)^2}$

$\dfrac{dy}{dx} = \dfrac{(12x + 9)(2e^{3x}) - 8e^{3x}}{(4x + 3)^2}$

$\quad = \dfrac{2e^{3x}(12x + 5)}{(4x + 3)^2}$

21. $y = (2e^{2x})^3 \sin x^2$
$\quad = 8e^{6x}\sin x^2$

$\dfrac{dy}{dx} = 8e^{6x}(\cos x^2)(2x) + \sin x^2(8e^{6x})(6)$

$\quad = 16e^{6x}(x\cos x^2 + 3\sin x^2)$

25. $y = xe^{xy} + \sin y$

$$\frac{dy}{dx} = x(e^{xy})\left(x\frac{dy}{dx} + y\right) + (1)e^{xy} + \cos y\frac{dy}{dx} = x(e^{xy})\left(x\frac{dy}{dx}\right) + x(e^{xy})(y) + e^{xy} + \cos y\frac{dy}{dx}$$

$$\frac{dy}{dx} - x(e^{xy})\left(x\frac{dy}{dx}\right) - \cos y\frac{dy}{dx} = x(e^{xy})y + e^{xy}$$

$$\frac{dy}{dx}(1 - x(e^{xy})(x) - \cos y) = x(e^{xy})y + e^{xy}$$

$$\frac{dy}{dx} = \frac{xy(e^{xy}) + e^{xy}}{1 - x^2e^{xy} - \cos y} = \frac{e^{xy}(xy+1)}{1 - x^2e^{xy} - \cos y}$$

29. $y = \ln \sin 2e^{6x}$; $\dfrac{dy}{dx} = \dfrac{1}{\sin 2e^{6x}}(\cos 2e^{6x})(2e^{6x})(6)$; $\dfrac{dy}{dx} = \dfrac{12e^{6x}\cos 2e^{6x}}{\sin 2e^{6x}} = 12e^{6x}\cot 2e^{6x}$

33. (a) $e = e^x = 2.718\,281\,8$ when $x = 1.0000$. This is the slope of a tangent line to the curve $f(x) = e^x$ when $x = 1.0000$. It is the value of $f'(x) = e^x$, since $\frac{d\,e^x}{dx} = e^x$.

(b) $\dfrac{e^{1.0001} - e^{1.0000}}{0.0001} = 2.718\,417\,8$ This is the slope of a secant line through the curve $f(x) = e^x$ at

$x = 1.0000$, where $\Delta x = 0.0001$.

$$\lim_{\Delta x \to 0} \frac{e^{(x+\Delta x)} - e^x}{\Delta x} = \frac{d\,e^x}{dx} = e^x$$

For $\Delta x = 0.0001$, the slope of the tangent line is approximately equal to the slope of the secant line.

37. $y = \dfrac{2e^{4x}}{(x+2)}$;

$$\frac{dy}{dx} = \frac{(x+2)8e^{4x} - 2e^{4x}}{(x+2)^2}; \frac{dy}{dx} = 2e^{4x}\frac{[4(x+2) - 1]}{(x+2)^2} = \frac{2e^{4x}(4x+7)}{(x+2)^2}; dy = \frac{2e^{4x}(4x+7)}{(x+2)^2}dx$$

41. $y = xe^{-x}$; $\dfrac{dy}{dx} = x(e^{-x})(-1) + (e^{-x})(1) = -xe^{-x} + e^{-x}$

Substituting, $\dfrac{dy}{dx} + y = (-xe^{-x} + e^{-x}) + (xe^{-x}) = e^{-x}$

45. $R = e^{-0.002t}$; $t = 100$ h: $R(0 \le R \le 1)$;

$$\frac{dR}{dt} = e^{-0.002t}(-0.002); \frac{dR}{dt} = -0.002e^{-0.002t}; \left.\frac{dR}{dt}\right|_{t=100} = -0.001\,64/\text{h}$$

49. $\cosh^2 u - \sinh^2 u = \dfrac{1}{4}(e^u + e^{-u})^2 - \dfrac{1}{4}(e^u - e^{-u})^2 = \dfrac{1}{4}(e^{2u} + 2e^0 + e^{-2u}) - \dfrac{1}{4}(e^{2u} - 2e^0 + e^{-2u})$

$$= \frac{1}{4}[0 + 2e^0 + 2e^0 + 0] = \frac{1}{4}(4e^0) = \frac{1}{4}(4) = 1$$

27.7 Applications

1. $y = \ln \cos x$; $\dfrac{dy}{dx} = \dfrac{1}{\cos x}(-\sin x) = \dfrac{-\sin x}{\cos x} = -\tan x$; $\dfrac{d^2 y}{dy^2} = -\sec^2 x$

(1) Since $\ln(1) = 0$ and $\cos 0 = 1$, there is an intercept at $(0,0)$.

(2) Since ln functions are not defined for negatives, y is undefined for $\cos x < 0$. The function is defined for x between $-\frac{\pi}{2}$ and $\frac{\pi}{2}$, etc.

(3) As $\cos x$ approaches 0, $\ln \cos x$ approaches negative infinity.
$\cos x = 0$ when $x = -\frac{\pi}{2}, \frac{\pi}{2}$, and their odd multiples, so $x = -\frac{\pi}{2}$, $x = \frac{\pi}{2}$, $x = \frac{3\pi}{2}$, etc., are asymptotes.

(4) Critical points exist where $-\tan x = 0$; i.e., where $x = 0, 2\pi, 4\pi$, etc.

(5) $-\sec^2 x$ is negative at all critical points so the graph is concave down, and all critical points are maximum points.

(6) Maximum points are $(0,0), (2\pi, 0), (4\pi, 0)$, etc.

Summary:
Int. $(0,0)$, max. $(0,0)$, not defined for $\cos x < 0$, asym. $x = -\frac{1}{2}\pi, \frac{1}{2}\pi, \ldots$

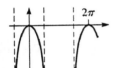

5. $y = \ln \dfrac{1}{x^2 + 1} = -\ln(x^2 + 1)$; $\dfrac{dy}{dx} = -\dfrac{1}{x^2 + 1}(2x) = \dfrac{-2x}{x^2 + 1}$;

$\dfrac{d^2 y}{dx^2} = \dfrac{(x^2 + 1)(-2) - (-2x)(2x)}{(x^2 + 1)^2} = \dfrac{2x^2 - 2}{(x^2 + 1)^2}$

(1) Since $\frac{1}{x^2+1} > 0$ for all numbers, $\ln \frac{1}{x^2+1}$ is defined for all numbers; there are no asymptotes.

(2) When $x = 0$, $y = \ln \frac{1}{0+1} = \ln 1 = 0$; $(0,0)$ is an intercept.

(3) Critical points; $\frac{dy}{dx} = \frac{-2x}{x^2+1} = 0$ when $-2x = 0$; $(0,0)$ is a critical point.

(4) $\frac{d^2 y}{dx^2} = \frac{2x^2-2}{(x^2+1)^2} = 0$ when $2x^2 - 2 = 0$; $x^2 = 1, x = -1, x = 1$.

Inflections occur at $x = -1$.

$y = \ln \left(\dfrac{1}{(-1)^2 + 1} \right) = \ln \dfrac{1}{2} = \ln(2^{-1}) = -\ln 2$ and $x = 1$, $y = \ln \left(\dfrac{1}{1^2 + 1} \right) = -\ln 2$

(5) Since $\frac{d^2 y}{dx^2}$ is negative at the critical point $(0,0)$, it is a maximum. The graph is concave down between the points $(-1, -\ln 2)$ and $(1, -\ln 2)$. It is concave up for $x < -1$ and $x > 1$. Since the second derivative goes through a change of sign at $x = -1$ and again at $x = 1$, $(-1, -\ln 2)$ and $(1, -\ln 2)$ are inflection points.

Summary:
Int. $(0,0)$, max. $(0,0)$, infl. $(-1, -\ln 2), (1, -\ln 2)$

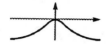

9. $y = \ln x - x;\ \dfrac{dy}{dx} = \dfrac{1}{x} - 1;\ \dfrac{d^2y}{dx^2} = -x^{-2} = -\dfrac{1}{x^2}$

(1) $x \not< 0$ since ln is undefined for those values. $y \ne 0$ since $\ln x \ne x$ for any number.

(2) There is an asymptote at $x = 0$.

(3) Critical points occur at $\frac{1}{x} - 1 = 0$ or $x = 1, y = -1$.

(4) Since $\frac{1}{x^2} \ne 0$ for any x, there is no point of inflection.

(5) Since $-\frac{1}{x^2} < 0$, for all x, the graph is concave down, and the critical point $(1, -1)$ is a maximum.

Max. $(1, -1)$, asymptote $x = 0$

13. $y = x^2 \ln x;\ \dfrac{dy}{dx} = x^2 \left(\dfrac{1}{x}\right) + (\ln 2)(2x) = x + 2x \ln x;\ \dfrac{dy}{dx}\bigg|_{x=1} = 1 + 2 \ln 1 = 1 + 2(0) = 1$

Slope is 1, $x = 1, y = 0$; using slope intercept form of the equation and substituting gives $0 = 1(1) + b$ or $b = 1$. The equation is $y = (1)x - 1$ or $y = x - 1$.

17. $f(x) = x^2 - 2 + \ln x;\ f'(x) = 2x + \dfrac{1}{x};\ f(1) = 1^2 - 2 + \ln 1 = -1;\ f(2) = 2^2 - 4 + \ln 2 = 0.69$

Therefore we choose $x_1 = 1.5$

n	x_n	$f(x_n)$	$f'(x_n)$	$x_n - \frac{f(x_n)}{f'(x_n)}$
1	1.5	0.655 465 1	3.666 666 7	1.321 236 8
2	1.321 236 8	0.024 234 9	3.399 340 2	1.314 107 5
3	1.314 107 5	0.000 036 2	3.389 187 8	1.314 096 8

Therefore, the root is 1.314 096 8, which is correct to the number of decimal places shown.

21. $\ln p = \dfrac{a}{T} + b \ln T + c;\ p = e^{(a/T + b \ln T + c)}$

$$\dfrac{dp}{dT} = e^{(a/T + b\ln T + c)}\left(-aT^{-2} + \dfrac{b}{T}\right) = e^{(a/T + b\ln T + c)}\left(\dfrac{-a}{T^2} + \dfrac{b}{T}\right) = p\left(\dfrac{-a}{T^2} + \dfrac{bT}{T^2}\right) = p\left(\dfrac{-a + bT}{T^2}\right)$$

25. $y = \ln \sec x$; $-1.5 \le x \le 1.5$; $u = \sec x$; $\frac{dy}{dx} = \frac{1}{\sec x} \cdot \sec x \tan x = \tan x = 0$ at $x = 0$; $x = 0$ is a critical value; also, multiples of 2π. $\frac{d^2y}{dx^2} = \sec^2 x$; $\sec^2(0) = 1$ so the curve is concave up and there is a minimum point at $x = 0$, $y = \ln \sec 0 = \ln 1 = 0$ recurring at multiples of $x = 2\pi$; $(2\pi, 0), (4\pi, 0), \ldots (0, 0)$ is an intercept.

Asymptotes occur where $\frac{dy}{dx} = \tan x$ is undefined.

These values are odd multiples of $\frac{\pi}{2}$; $-\frac{\pi}{2}, \frac{\pi}{2}, \frac{3\pi}{2} \ldots$

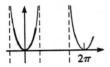

29.
$$y = e^{-0.5t}(0.4 \cos 6t - 0.2 \sin 6t);$$

$$v = \frac{dy}{dt} = e^{-0.5t}(-2.4 \sin 6t - 1.2 \cos 6t) + (0.4 \cos 6t - 0.2 \sin 6t) \times (e^{-0.5t})(-0.5)$$

$$\frac{dy}{dt} = -2.4e^{-0.5t} \sin 6t - 1.2e^{-0.5t} \cos 6t + 0.1e^{-0.5t} \sin 6t - 0.2e^{-0.5t} \cos 6t$$

$$\frac{dy}{dt} = -2.3e^{-0.5t} \sin 6t - 1.4e^{-0.5t} \cos 6t; \quad \frac{dy}{dt} = -e^{-0.5t}(1.4 \cos 6t + 2.3 \sin 6t)$$

$$\left. \frac{dy}{dt} \right|_{t=-026} = -e^{-0.5(0.26)}[1.4 \cos 6(0.26) + 2.3 \sin 6(0.26)] = -e^{-0.13}[0.011 + 2.3(1.00)] = -2.03 \text{ cm/s}$$

Chapter 27 Review Exercises

1. $y = 3 \cos(4x - 1)$; $\frac{dy}{dx} = [-3 \sin(4x - 1)][4] = -12 \sin(4x - 1)$

5. $y = \csc^2(3x + 2)$; $\frac{dy}{dx} = 2 \csc(3x + 2)[- \csc(3x + 2) \cot(3x + 2)](3)$
$$= -6 \csc^2(3x + 2) \cot(3x + 2)$$

9. $y = (e^{x-3})^2$; $\frac{dy}{dx} = 2(e^{x-3})(e^{x-3})(1) = 2e^{2(x-3)}$

13. $y = 3 \tan^{-1}\left(\frac{x}{3}\right)$; $\frac{dy}{dx} = 3\left[\frac{1}{1 + \left(\frac{x}{3}\right)^2}\right]\frac{1}{3} = \frac{1}{1 + \left(\frac{x}{3}\right)^2} = \frac{1}{1 + \frac{x^2}{9}} = \frac{9}{9 + x^2}$

17. $y = \sqrt{\csc 4x + \cot 4x} = (\csc 4x + \cot 4x)^{1/2}$

$$\frac{dy}{dx} = \frac{1}{2}(\csc 4x + \cot 4x)^{-1/2}(-4 \csc 4x \cot 4x - 4 \csc^2 4x)$$

$$= \frac{1}{2}(\csc 4x + \cot 4x)^{-1/2}(-4 \csc 4x)(\csc 4x + \cot 4x)$$

$$= -2 \csc 4x(\csc 4x + \cot 4x)^{1/2} = (-2 \csc 4x)\sqrt{\csc 4x + \cot 4x}$$

21. $y = \dfrac{\cos^2 x}{e^{3x} + 1}$

$$\frac{dy}{dx} = \frac{(e^{3x} + 1)[2\cos x(-\sin x)] - (\cos^2 x)(e^{3x})(3)}{(e^{3x} + 1)^2} = \frac{(e^{3x} + 1)[-2\sin x \cos x] - 3e^{3x}\cos^2 x}{(e^{3x} + 1)^2}$$

$$= \frac{-\cos x[(e^{3x} + 1)(2\sin x) + 3e^{3x}\cos x]}{(e^{3x} + 1)^2} = \frac{-\cos x[2e^{3x}\sin x + 2\sin x + 3e^{3x}\cos x]}{(e^{3x} + 1)^2}$$

$$= \frac{-\cos x(2e^{3x}\sin x + 3e^{3x}\cos x + 2\sin x)}{(e^{3x} + 1)^2}$$

25. $y = \ln(\csc x^2); \quad \dfrac{dy}{dx} = \dfrac{1}{\csc x^2}(-\csc x^2 \cot x^2)(2x) = -2x\cot x^2$

29. $L = 0.1e^{-2t}\sec(\pi t)$

$$\frac{dL}{dt} = 0.1e^{-2t}\sec(\pi t)\tan(\pi t) \cdot \pi + \sec(\pi t) \cdot 0.1(-2)e^{-2t}$$

$$\frac{dL}{dt} = 0.1\pi e^{-2t}\sec(\pi t)\tan(\pi t) - 0.2e^{-2t}\sec(\pi t)$$

$$\frac{dL}{dt} = e^{-2t}\sec(\pi t) \cdot [0.1\pi\tan(\pi t) - 0.2]$$

33. $\tan^{-1}\dfrac{y}{x} = x^2 e^y; \quad u = \dfrac{y}{x} = yx^{-1}; \quad \dfrac{du}{dx} = -yx^{-2} + x^{-1}\dfrac{dy}{dx}$

$$\frac{1}{1 + (yx^{-1})^2}\left(-yx^{-2} + x^{-1}\right)\frac{dy}{dx} = x^2 e^y \frac{dy}{dx} + 2xe^y$$

$$\frac{\frac{-y}{x^2} + \frac{1}{x}\frac{dy}{dx}}{1 + y^2 x^{-2}} = x^2 e^y \frac{dy}{dx} + 2xe^y$$

$$\frac{-y}{x^2} + \frac{1}{x}\frac{dy}{dx} = \left(x^2 e^y \frac{dy}{dx} + 2xe^y\right)(1 + y^2 x^{-2})$$

$$\frac{-y}{x^2} + \frac{1}{x}\frac{dy}{dx} = x^2 e^y \frac{dy}{dx} + 2xe^y + y^2 e^y \frac{dy}{dx} + 2x^{-1}y^2 e^y$$

$$\frac{1}{x}\frac{dy}{dx} - x^2 e^y \frac{dy}{dx} - y^2 e^y \frac{dy}{dx} = 2xe^y + 2x^{-1}y^2 e^y + \frac{y}{x^2}$$

$$\frac{dy}{dx}\left(\frac{1}{x} - x^2 e^y - y^2 e^y\right) = 2xe^y + \frac{2y^2 ey}{x} + \frac{y}{x^2}$$

$$\frac{dy}{dx}\left(\frac{1 - x^3 e^y - xy^2 e^y}{x}\right) = \frac{2xe^y + 2xy^2 e^y + y}{x^2}$$

$$\frac{dy}{dx} = \frac{2x^3 e^y + 2xy^2 e^y + y}{x^2} \cdot \frac{x}{1 - x^3 e^y - xy^2 e^y}$$

$$\frac{dy}{dx} = \frac{2x^3 e^y + 2xy^2 e^y + y}{x - x^4 e^y - x^2 y^2 e^y}$$

37. $\ln xy + ye^{-x} = 1$

Using implicit differentiation, $\dfrac{d\ln xy}{dx} + \dfrac{dye^{-x}}{dx} = \dfrac{d(1)}{dx}$

$$\frac{1}{xy}\left(x\frac{dy}{dx} + y\frac{dx}{dy}\right) + y\frac{de^{-x}}{dx} + e^{-x}\frac{dy}{dx} = 0$$

$$\frac{1}{xy}\left(x\frac{dy}{dx} + y\right) + ye^{-x}(-1) + e^{-x}\frac{dy}{dx} = 0$$

$$\frac{1}{y}\frac{dy}{dx} + \frac{1}{x} - ye^{-x} + e^{-x}\frac{dy}{dx} = 0$$

$$\frac{1}{y}\frac{dy}{dx} + e^{-x}\frac{dy}{dx} = ye^{-x} - \frac{1}{x}$$

$$\frac{dy}{dx}\left(\frac{1}{y} + e^{-x}\right) = ye^{-x} - \frac{1}{x}$$

$$\frac{dy}{dx}\left(\frac{1 + ye^{-x}}{y}\right) = \frac{xye^{-x} - 1}{x}$$

$$\frac{dy}{dx} = \left(\frac{xye^{-x} - 1}{x}\right)\left(\frac{y}{1 + ye^{-x}}\right) = \frac{y(xye^{-x} - 1)}{x(1 + ye^{-x})}$$

41. $y = x - \cos x$; $\dfrac{dy}{dx} = 1 + \sin x$; $\dfrac{d^2y}{dx^2} = \cos x$ Infl.: $\left(\dfrac{1}{2}\pi, \dfrac{1}{2}\pi\right), \left(\dfrac{3}{2}\pi, \dfrac{3}{2}\pi\right)$

(a) $x = 0, y = 0 - \cos 0 = 0 - 1 = -1$; $(0, -1)$ is an intercept. $x - \cos x = 0$ when $x = \cos x$; $x = 0.74$ (see Table 3); $(0.74, 0)$ is an intercept.

(b) y is defined for all x; no asymptotes.

(c) Critical points occur at $1 + \sin x = 0$, $\sin x = -1$, $x = -\dfrac{\pi}{2}, \dfrac{3\pi}{2}, \dfrac{7\pi}{2}$, etc.

(d) Inflections occur at $\cos x = 0$, $x = -\dfrac{\pi}{2}, \dfrac{\pi}{2}, \dfrac{3\pi}{2}, \dfrac{5\pi}{2}$, etc., since the second derivative undergoes a change of sign at each of these points.

(e) All critical points are inflections; no maximum or minimum points.

(f) Checking concavity at $x = 0, -\cos 0 = -1$; the graph is concave up at $(0, -1)$ and on each side of this point up to the inflection points at $\left(-\dfrac{\pi}{2}, -\dfrac{\pi}{2}\right)$ and $\left(\dfrac{\pi}{2}, \dfrac{\pi}{2}\right)$. It will switch concavity again at each subsequent inflection point.

x	$-\dfrac{3\pi}{2}$	$-\pi$	$-\dfrac{\pi}{2}$	0	0.7	$\dfrac{\pi}{2}$	π	$\dfrac{3\pi}{2}$
y	-4.7	-4.1	-1.6	-1	0	1.6	4.1	4.7

 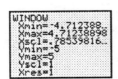 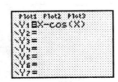 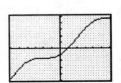

45. $y = 4\cos^2(x^2)$; slope $= \dfrac{dy}{dx} = 2[4\cos(x^2)][-\sin(x^2)](2x) = -16x\cos x^2 \sin x^2$

$\left.\dfrac{dy}{dx}\right|_{x=1} = -16\cos(1^2)\sin(1^2) = -16(0.5403)(0.8415) = -7.27$

$f(1) = 4\cos^2(1^2) = 4(0.5403)^2 = 1.168$

By Eq. (21-9), $y = -7.27x + b$; $1.168 = -7.27(1) + b$, $b = 8.44$; $y = -7.27x + 8.44$; $7.27x + y - 8.44 = 0$

49.

$$\sin^2 x + \cos^2 x = 1$$

$$\dfrac{d(\sin^2 x + \cos^2 x)}{dx} = \dfrac{d(1)}{dx}$$

$$\dfrac{d\sin^2 x}{dx} + \dfrac{d\cos^2 x}{dx} = 0$$

$$2\sin x \cos x + 2\cos x(-\sin x) = 0$$

$$2\sin x \cos x - 2\cos x \sin x = 0; \quad 0 = 0$$

53. $T = 17.2 + 5.2\cos\left[\dfrac{\pi}{6}(x - 0.50)\right]$

$\dfrac{dT}{dx} = -5.2\sin\left[\dfrac{\pi}{6}(x - 0.50)\right] \cdot \left.\dfrac{\pi}{6}\right|_{x=2}$

$\quad = -1.93\dfrac{^\circ C}{mo}$

$\dfrac{dT}{dx} = \dfrac{dT}{dx} \cdot \dfrac{dx}{dt} = -1.93\dfrac{^\circ C}{mo} \cdot \dfrac{12\ mo}{365\ day}$

$\dfrac{dT}{dx} = -0.064\dfrac{^\circ C}{day}$

57. For $N = 8$, $n = xN\log_x N$ becomes $n = 8x\log_x 8$ with $1 < x < 10$. For graph, use $y = 8x\dfrac{\ln 8}{\ln x}$.

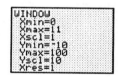  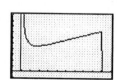

The VL's may be eliminated by using the Dot mode rather than Connected.

61. $\theta = \sin^{-1}\dfrac{Ff}{R}$; $u = \dfrac{Ff}{R} = \dfrac{f}{R}F$

$\dfrac{d\theta}{dF} = \dfrac{1}{\sqrt{1 - u^2}}\dfrac{du}{dx}; \dfrac{du}{dF} = \dfrac{f}{R}$

$\quad = \dfrac{1}{\sqrt{1 - \left(\dfrac{Ff}{R}\right)^2}} \cdot \dfrac{f}{R} = \dfrac{1 \cdot f}{\sqrt{1 - \left(\dfrac{Ff}{R}\right)^2}\sqrt{R^2}} = \dfrac{f}{\sqrt{R^2 - F^2 f^2}}$

65.
$$T = 80 + 120(0.5)^{0.2t}; \quad u = 0.2t, \quad \frac{du}{dt} = 0.2$$

$$T = 80 + 120(0.5)^u$$

$$\frac{dT}{dt} = 120(0.5)^{0.2t}(\ln(0.5))(0.2)$$

$$\left.\frac{dT}{dt}\right|_{t=5.00} = 120(0.5)^{0.2(5.00)}(-0.693)(0.2) = 60(-0.693)(0.2) = -8.32°\text{F/min}$$

$$L(t) = -8.32(t - 5.00) + 140$$

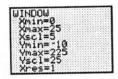  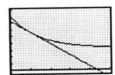

69. $A = xy = 3xe^{-0.5x^2}$ (Working with $\frac{1}{2}$ the actual area)

$$\frac{dA}{dx} = 3x\,d(e^{-0.5x^2}) + e^{-0.5x^2}\left(\frac{d3x}{dx}\right) = 3x(e^{-0.5x^2})(-1.0x) + e^{-0.5x^2}(3) = -3x^2e^{-0.5x^2} + 3e^{-0.5x^2}$$

The maximum value will occur when $\frac{dA}{dx} = 0$

$$-3x^2e^{-0.5x^2} + 3e^{-0.5x^2} = 0$$
$$(e^{-0.5x^2})(-3x^2 + 3) = 0$$
$$-3x^2 + 3 = 0; \quad x^2 = 1, x = 1.00$$
$$e^{-0.5x^2} = 0 \text{ has no real solution}$$
$$y = 3e^{-0.5(1)^2} = 3e^{-0.5} = 1.82$$
$$W = 2x = 2.00; \quad H = 1.82$$

73. $x = r(\theta - \sin\theta); \; y = r(1 - \cos\theta); \; r = 5.00 \text{ cm}; \; \dfrac{d\theta}{dt} = 0.12 \text{ rad/s}; \; \theta = 35°$

horizontal component of velocity,

$$v_x = \frac{dx}{dt} = \frac{d[5.5(\theta - \sin\theta)]}{dt} = 5.5\left(\frac{d\theta}{dt} - \cos\theta\frac{d\theta}{dt}\right) = 5.5(0.12 - 0.12\cos 35°) = 0.119 \text{ cm/s}$$

vertical component of velocity,

$$v_y = \frac{dy}{dt} = \frac{d[5.5(1 - \cos\theta)]}{dt} = 5.5\sin\theta\frac{d\theta}{dt} = (5.5)(0.12)\sin 35° = 0.379$$

$$v = \sqrt{0.119^2 + 0.379^2} = 0.4 \text{ cm/s}. \quad \theta = \tan^{-1}\frac{0.379}{0.119} = 72.5°$$

77.

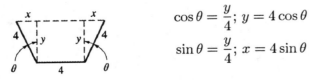

$$\cos\theta = \frac{y}{4}; \; y = 4\cos\theta$$

$$\sin\theta = \frac{y}{4}; \; x = 4\sin\theta$$

$$A = (4 + x)y = 4y + xy = 16\cos\theta + 16\sin\theta\cos\theta = 16\cos\theta(1 + \sin\theta)$$

$$\frac{dA}{d\theta} = -16\sin\theta + 16\sin\theta(-\sin\theta) + 16\cos\theta(\cos\theta) = -16\sin\theta - 16\sin^2\theta + 16\cos^2\theta$$
$$= 16(\cos^2\theta - \sin^2\theta - \sin\theta) = 16(1 - 2\sin^2\theta - \sin\theta)$$

(1) Not valid for negative θ or A. Domain and range are positive real numbers.

(2) A-intercept at $\theta = 0$, $A = 16$; To find θ-intercept, $A = 0$

$$16\cos\theta(1 + \sin\theta) = 0$$
$$16\cos\theta = 0 \qquad 1 + \sin\theta = 0$$
$$\theta = \frac{\pi}{2} \qquad \sin\theta = -1, \theta = -\frac{\pi}{2} \quad \text{(not in domain)}$$

(3) Critical value is

$$16(1 - 2\sin^2\theta - \sin\theta) = 0$$
$$1 - 2\sin^2\theta - \sin\theta = 0$$
$$-1 + 2\sin^2\theta + \sin\theta = 0$$
$$(2\sin\theta - 1)(\sin\theta + 1) = 0$$
$$2\sin\theta = 1 \qquad \sin\theta = -1$$

$$\sin\theta = \frac{1}{2} \qquad \theta = -\frac{\pi}{2} \quad \text{(not in domain)}$$

$$\theta = \frac{\pi}{6}$$

(4) $\dfrac{d^2A}{d\theta^2} = 16(-4\sin\theta\cos\theta - \cos\theta) = -16\cos\theta(4\sin\theta + 1)$
$$= -16(4\sin\theta + 1)|_{\theta=\pi/6} = -16[4(0.5) + 1] = -16(3) = -48$$

Curve is concave down at $\frac{\pi}{6}$ and $\theta = \frac{\pi}{6}$ is a maximum.

81.

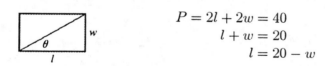

$$P = 2l + 2w = 40$$
$$l + w = 20$$
$$l = 20 - w$$

(a) Using algebra, $\quad A = l \cdot w = (20 - w) \cdot w = 20w - w^2$

$$\frac{dA}{dw} = 20 - 2w = 0 \Rightarrow w = 10 \text{ and } l = 10$$

The area is maximum when the rectangle is a square.

(b) Using trigonometry $\tan\theta = \dfrac{w}{l} = \dfrac{w}{20 - w}$ from which $w = \dfrac{20\tan\theta}{1 + \tan\theta}$, then

$$A = \frac{20^2\tan\theta}{1 + \tan\theta} - \frac{20^2\tan^2\theta}{(1 + \tan\theta)^2} \quad \text{and taking} \quad \frac{dA}{d\theta} = 0$$

gives a maximum for $\theta = 45°$; again, a square.

METHODS OF INTEGRATION

28.1   The General Power Formula

1. $u = \sin x; \; du = \cos x \, dx$

$$\int \sin^4 x \cos x \, dx = \frac{1}{5} \sin^5 x + C$$

5. $\int 4 \tan^2 x \sec^2 x \, dx = 4 \int \tan^2 x \sec^2 x \, dx$

$u = \tan x; \; du = \sec^2 x \, dx$

$$4 \int \tan^2 x \sec^2 x \, dx = 4 \left(\frac{1}{3} \tan^3 x + c \right) = \frac{4}{3} \tan^3 x + c$$

9. $u = \sin^{-1} x; \; du = \dfrac{1}{\sqrt{1 - x^2}} \, dx = \dfrac{dx}{\sqrt{1 - x^2}}$

$$\int (\sin^{-1} x)^3 \left(\frac{dx}{\sqrt{1 - x^2}} \right) = \frac{1}{4} (\sin^{-1} x)^4 + C$$

13. $u = \ln(x + 1); \; du = \dfrac{1}{x + 1}(1) dx = \dfrac{dx}{x + 1}$

$$\int [\ln(x + 1)^2 \frac{dx}{x + 1} = \frac{1}{3} [\ln(x + 1)]^3 + C$$

17. $u = 4 + e^x; \; du = e^x dx; \; \int (4 + e^x)^3 e^x \, dx = \dfrac{1}{4}(4 + e^x)^4 + C$

21. $u = 1 + \sec^2 x; \; \dfrac{du}{dx} = 2 \sec x \dfrac{d(\sec x)}{dx} = 2 \sec x \sec x \tan x; \; du = 2 \sec^2 x \tan x \, dx$

$$\int (1 + \sec^2 x)^4 (\sec^2 x \tan x \, dx) = \frac{1}{2} \int (1 + \sec^2 x)^4 2 \sec^2 x \tan x \, dx = \frac{1}{2} \times \frac{1}{5}(1 + \sec^2 x)^5 + C$$

$$= \frac{1}{10}(1 + \sec^2 x)^5 + C$$

25. $A = \displaystyle\int_0^2 \dfrac{1 + \tan^{-1} 2x}{1 + 4x^2} \, dx; \; u = \tan^{-1} 2x; \; du = \dfrac{1}{1 + (2x)^2} \times 2$

$$A = \frac{1}{2} \int_0^2 1 + u \, du = \frac{1}{2} \left(u + \frac{1}{2}u^2 \right) \Big|_0^2 = \frac{1}{2} \left[\tan^{-1} 2x + \frac{1}{2}(\tan^{-1} 2x)^2 \right] \Big|_0^2$$

$$= \frac{1}{2} \left[\tan^{-1} 4 + \frac{1}{2}(\tan^{-1} 2x)^2 - \tan^{-1} 0 - \frac{1}{2}(\tan^{-1} 0)^2 \right] = \frac{1}{2}[1.326 + 0.879 - 0 - 0] = 1.102$$

29. $P = mnv^2 \displaystyle\int_0^{\pi/2} \sin \theta \cos^2 \theta \, d\theta; \; n = 2; \; \mu = \cos \theta; \; du = -\sin \theta \, d\theta$

$$P = mnv^2 \int_0^{\pi/2} \cos^2 \theta (-\sin \theta \, d\theta) = -mnv^2 \left[\frac{\cos^3 \theta}{3} \right] \Big|_0^{\pi/2} = -mnv^2 \left[\frac{1}{3} \left(\cos^3 \frac{\pi}{2} - \cos^3 0 \right) \right]$$

$$= -mnv^2 \left[\frac{1}{3}(0 - 1) \right] = -mnv^2 \left(-\frac{1}{3} \right) = \frac{1}{3}mnv^2$$

28.2 The Basic Logarithmic Form

1. $u = 1 + 4x$; $du = 4\,dx$. Introduce a factor of 4.

$$\int \frac{dx}{1 + 4x} = \frac{1}{4} \int \frac{4\,dx}{1 + 4x} = \frac{1}{4} \ln |1 + 4x| + C$$

5. $u = 8 - 3x$; $du = -3\,dx$

$$\int_0^2 \frac{dx}{8 - 3x} = -\frac{1}{3} \int_0^2 \frac{-3\,dx}{8 - 3x} = -\frac{1}{3} \ln |8 - 3x| \Big|_0^2 = -\frac{1}{3} \ln 2 + \frac{1}{3} \ln 8 = 0.462$$

9. $u = 1 + \sin x$; $du = \cos x\,dx$

$$\int_0^{\pi/2} \frac{\cos x\,dx}{1 + \sin x} = \ln |\sin x| \big|_0^{\pi/2} = \ln \left|1 + \sin \frac{\pi}{2}\right| - \ln |1 + \sin 0| = \ln |2| - \ln |1| = 0.693 - 0 = 0.693$$

13. $u = x + e^x$; $du = (1 + e^x)\,dx$; $\displaystyle\int \frac{1 + e^x}{x + e^x}\,dx = \ln |x + e^x| + C$

17. $u = 4x + 2x^2$; $du = (4 + 4x)\,dx$

$$\int_1^3 \frac{1 + x}{4x + 2x^2}\,dx = \frac{1}{4} \int_1^3 \frac{4 + 4x}{4x + 2x^2}\,dx = \frac{1}{4} \ln |4x + 2x^2| \Big|_1^3 = \frac{1}{4} \ln 30 - \frac{1}{4} \ln 6 = 0.402$$

21. $u = 2x + \tan x$; $du = (2 + \sec^2 x)\,dx$; $\displaystyle\int \frac{2 + \sec^2 x}{2x + \tan x}\,dx = \ln |2x + \tan x| + C$

25. $\displaystyle\int \frac{x + 2}{x^2}\,dx = \int \frac{1}{x}\,dx + \int \frac{2}{x^2}\,dx = \int \frac{1}{x}\,dx + \int (2x^{-2})\,dx = \ln |x| - 2x^{-1} + C = \ln |x| - \frac{2}{x} + C$

29. $y = \dfrac{1}{x + 1}$

$$A = \int_0^2 \frac{1}{x + 1}\,dx = \ln(x + 1) \Big|_0^2$$

$$A = \ln 3 - \ln 1 = 1.10 - 0 = 1.10$$

33. $m = \dfrac{dy}{dx} = \dfrac{\sin x}{3 + \cos x}$; $y = \displaystyle\int \frac{1}{3 + \cos x} \times \sin x\,dx$

$$y = -\int \frac{1}{3 + \cos x}(-\sin x)\,dx; \text{ let } u = 3 + \cos x;\ du = -\sin x\,dx$$

$$y = -\int \frac{1}{u}\,du = -\ln |u| + C = -\ln(3 + \cos x) + C$$

$$2 = -\ln\left(3 + \cos \frac{\pi}{3}\right) + C; \text{ substitute values of } x \text{ and } y$$

$$2 = -\ln(3 + 0.5) + C;\ C = 2 + \ln 3.5$$

$$y = -\ln(3 + \cos x) + \ln 3.5 + 2; \text{ substituting for } C$$

$$y = \ln \frac{3.5}{3 + \cos x} + 2$$

37. $t = L \int \dfrac{di}{E - iR}; \; u = E - iR; \; du = -R\,di$

$t = -\dfrac{L}{R} \int \dfrac{-R\,di}{E - iR} = \dfrac{-L}{R} \ln|E - iR| + C; \; t = 0 \text{ for } i = 0$

$0 = -\dfrac{L}{R} \ln E + C; \; C = \dfrac{L}{R} \ln|E|$

$t = \dfrac{L}{R}(-\ln|E - iR| + \ln|E|)$

$t = \dfrac{L}{R} \ln \dfrac{E}{E - iR}; \; \dfrac{R}{L}t = \ln \dfrac{E}{E - iR}; \; e^{Rt/L} = \dfrac{E}{E - iR}$

$i = \dfrac{E}{R} - \dfrac{E}{R}e^{-Rt/L}; \; i = \dfrac{E}{R}(1 - e^{-Rt/L})$

28.3 The Exponential Form

1. $u = 7x; \; du = 7\,dx; \; \int e^{7x}(7\,dx) = e^{7x} + C$

5. $\displaystyle\int_{-2}^{2} 6e^{s/2}ds = \int_{-2}^{2} 12 \cdot e^{s/2} \cdot \dfrac{1}{2}\,ds = 12 \cdot e^{s/2}\Big|_{-2}^{2} = 12(e^{2/2} - e^{-2/2}) = 12\left(e - \dfrac{1}{e}\right) = 28.2$

9. $u = \sqrt{x} = x^{1/2}; \; du = \dfrac{1}{2}x^{-1/2}dx = \dfrac{dx}{2\sqrt{x}}; \; \displaystyle\int_{1}^{4} \dfrac{e^{\sqrt{x}}}{\sqrt{x}}\,dx = 2\int_{1}^{4} e^{\sqrt{x}}\dfrac{dx}{2\sqrt{x}} = 2e^{\sqrt{x}}\Big|_{1}^{4} = 2e^2 - 2e = 9.34$

13. $\displaystyle\int \sqrt{e^{2y} + e^{3y}}\,dy = \int \sqrt{e^{2y}(1 + e^y)}\,dy = \int \sqrt{1 + e^y} \cdot e^y\,dy = \dfrac{2}{3}(1 + e^y)^{3/2} + C$

17. $\displaystyle\int \dfrac{2\,dx}{\sqrt{x}e^{\sqrt{x}}} = 2\int \dfrac{dx}{x^{1/2}e^{x^{1/2}}} = 2\int e^{-x^{1/2}}x^{-1/2}dx$

$u = -x^{1/2}; \; du = -\dfrac{1}{2}x^{-1/2}dx$

$2\displaystyle\int e^{-x^{1/2}}x^{-1/2}dx = 2(-2)\int e^{-x^{1/2}}\left(-\dfrac{1}{2}x^{-1/2}\right) = -4e^{-x^{1/2}} + C = -\dfrac{4}{e^{x^{1/2}}} + C = -\dfrac{4}{e^{\sqrt{x}}} + C$

21. $u = \cos 3x\,dx; \; du = -\sin 3x\,(3\,dx) = \dfrac{-3}{\csc 3x}\,dx$

$\displaystyle\int \dfrac{e^{\cos 3x}\,dx}{\csc 3x} = -\dfrac{1}{3}\int e^{\cos 3x}[-3\sin 3x(3\,dx)] = -\dfrac{1}{3}e^{\cos 3x} + C$

25. $A = \displaystyle\int_{0}^{2} 3e^x\,dx = 3e^x\Big|_{0}^{2} = 3e^2 - 3e^0 = 19.2$

29. $y_{av} = \dfrac{\displaystyle\int_{0}^{4} e^{2x}\,dx}{4 - 0} = \dfrac{\dfrac{1}{2}\displaystyle\int e^{2x}(2\,dx)}{4} = \dfrac{\dfrac{1}{2}e^{2x}\Big|_{0}^{4}}{4} = \dfrac{1}{8}e^{2x}\Big|_{0}^{4} = \dfrac{1}{8}(e^8 - 1) = 372$

33. $qe^{t/RC} = \dfrac{E}{R} \displaystyle\int e^{t/RC} dt; \; u = \dfrac{t}{RC}; \; du = \dfrac{1}{RC} dt$

$qe^{t/RC} = RC \cdot \dfrac{E}{R} \displaystyle\int e^{t/RC} \left(\dfrac{1}{RC}\right) dt$

$qe^{t/RC} = EC(e^{t/RC}) + C_1$, where C_1 is the constant of integration.

$q = 0$ for $t = 0$; $0 = EC + C_1$; $C_1 = -EC$;

$qe^{t/RC} = EC(r^{t/RC}) - EC$

$q = EC - \dfrac{EC}{e^{t/RC}}; \; q = EC(1 - e^{-t/RC})$

28.4   Basic Trigonometric Forms

1. $u = 2x; \; du = 2\,dx$

$\displaystyle\int \cos 2x\, dx = \dfrac{1}{2}\int \cos 2x\,(dx)$

$= \dfrac{1}{2}\sin 2x + C$

5. $u = \dfrac{1}{2}x; \; du = \dfrac{1}{2}\,dx$

$\displaystyle\int \sec \dfrac{1}{2}x \tan \dfrac{1}{2}x\, dx = 2\int \sec \dfrac{1}{2}x \tan \dfrac{1}{2}x \left(\dfrac{1}{2}dx\right)$

$= 2\sec \dfrac{1}{2}x + C$

9. $\displaystyle\int 3\phi \sec^2 \phi^2 \cos \phi^2 d\phi = \dfrac{3}{2}\int \sec \phi^2 \cdot 2\phi\,d\phi = \dfrac{3}{2}\cdot \ln|\sec \phi^2 + \tan \phi^2| + C$

13. $u = 2x; \; du = 2\,dx$

$\displaystyle\int_0^{\pi/6} \dfrac{dx}{\cos^2 2x} = \dfrac{1}{2}\int_0^{\pi/6} \sec^2 2x (2\,dx) = \dfrac{1}{2}\tan 2x \Big|_0^{\pi/6} = \dfrac{1}{2}\left(\tan \dfrac{\pi}{3} - \tan 0\right) = \dfrac{1}{2}(\sqrt{3} - 0) = \dfrac{1}{2}\sqrt{3}$

17. $\displaystyle\int \sqrt{\tan^2 2x + 1}\,dx = \int \sqrt{\sec^2 2x}\,dx = \int \sec 2x\, dx$

$u = 2x; \; du = 2\,dx$

$\displaystyle\int \sec 2x\, dx = \dfrac{1}{2}\int \sec 2x (2\,dx) = \dfrac{1}{2}\ln|\sec 2x + \tan 2x| + C$

21. $\displaystyle\int \dfrac{1 - \sin x}{1 + \cos x}\, dx = \int \dfrac{1 - \sin x}{1 + \cos x} \times \dfrac{1 - \cos x}{1 - \cos x}\, dx = \int \dfrac{1 - \sin x - \cos x + \sin x \cos x}{1 - \cos^2 x}\, dx$

$= \displaystyle\int \dfrac{1 - \sin x - \cos x + \sin x \cos x}{\sin^2 x}\, dx = \int \left(\dfrac{1}{\sin^2 x} - \dfrac{\sin x}{\sin^2 x} - \dfrac{\cos x}{\sin^2 x} + \dfrac{\sin x \cos x}{\sin^2 x}\right) dx$

$= \displaystyle\int (\csc^2 x - \csc x - \cot x \csc x + \cot x)\, dx$

$= -\cot x - \ln|\csc x - \cot x| - (-\csc x) + \ln|\sin x| + C$

$= \csc x - \cot x - \ln|\csc x - \cot x| + \ln|\sin x| + C$

25. $A = \displaystyle\int_0^{\pi/4} y\, dx = \int_0^{\pi/4} 2\tan x\, dx = 2(-\ln)\,|\cos x|\big|_0^{\pi/4} = (-2)\left[\ln\left|\cos \dfrac{\pi}{4}\right| - 2\ln|\cos 0|\right]$

$= -2(\ln 0.7071 - \ln 1) = 0.693$

29. $\omega = -0.25\sin 2.5t$

$$\theta = \int -0.25\sin 2.5t\,dt; \; u = 2.5t; \; du = 2.5\,dt$$

$$\theta = -0.10\int \sin 2.5t(2.5\,dt) = -0.10(-\cos 2.5t) + C$$

$\theta = 0.10\cos 2.5t + C$
$0.10 = 0.10\cos 0 + C; \; C = 0; \; \theta = 0.10\cos 2.5t$

28.5   Other Trigonometric Forms

1. $u = \sin x; \; du = \cos x\,dx; \; \displaystyle\int \sin^2 x\cos x\,dx = \frac{1}{3}\sin^3 x + C$

5. $\displaystyle\int 4(\cos^4\theta - \sin^4\theta)\,d\theta = \int 4(\cos^2\theta + \sin^2\theta)(\cos^2\theta - \sin^{-2}\theta)\,d\theta = \int 4\cdot 1\cdot(1-\sin^2\theta)\,d\theta$

$$= \int 4\cdot\left(1 - 2\frac{1-\cos 2\theta}{2}\right)d\theta = \int 4\cdot(1 - 1 + \cos 2\theta)\,d\theta = \int \frac{4}{2}\cdot\cos 2\theta\cdot 2\,d\theta$$

$$= 2\sin(2\theta) + C$$

9. $\displaystyle\int \sin^2 x\,dx = \int\left[\frac{1}{2}(1-\cos 2x)\right]dx = \frac{1}{2}\int dx - \frac{1}{2}\int\cos 2x\,dx = \frac{1}{2}\int dx - \frac{1}{4}\int\cos 2x(2\,dx)$

$$= \frac{1}{2}x - \frac{1}{4}\sin 2x + C$$

13. $\displaystyle\int \tan x^3\,dx = \int \tan^2 x\tan x\,dx = \int(\sec^2 x - 1)\tan x\,dx = \int \tan x\,sex^2 x\,dx - \int \tan x\,dx$

$$= \frac{1}{2}\tan^2 x - (-\ln|\cos x|) + C = \frac{1}{2}\tan^2 x + \ln|\cos x| + C$$

17. $\displaystyle\int \tan^4 2x\,dx = \int(\tan^2 2x)(\tan^2 2x)\,dx = \int(\tan^2 2x)(\sec^2 2x - 1)\,dx = \int(\tan^2 2x\sec^2 2x - \tan^2 2x)\,dx$

$$= \int \tan^2 2x\sec^2 2x\,dx - \int \tan^2 2x\,dx = \int \tan^2 2x\sec^2 2x\,dx - \int(\sec^2 2x - 1)dx$$

$$= \frac{1}{2}\int(\tan^2 2x\sec^2 2x)(2\,dx) - \int \sec^2 2x\,dx + \int 1\,dx$$

$$= \frac{1}{2}\int(\tan^2 2x\sec^2 2x)(2\,dx) - \frac{1}{2}\int \sec^2 2x(2\,dx) + \int 1\,dx$$

$$= \frac{1}{2}\times\frac{1}{3}\tan^3 2x - \frac{1}{2}\tan 2x + x + C = \frac{1}{6}\tan^3 2x - \frac{1}{2}\tan 2x + x + C$$

21. $\displaystyle\int(\sin x + \cos x)^2\,dx = = \int(\sin^2 x + \cos^2 x)\,dx + \int 2\sin x\cos x\,dx = \int 1\,dx + \int \sin 2x\,dx$

$$= \int 1\,dx + \frac{1}{2}\int \sin 2x\,(2\,dx) = x - \frac{1}{2}\cos 2x + C$$

25. $\displaystyle\int_{\pi/6}^{\pi/4} \cot^5 x \, dx = \int_{\pi/6}^{\pi/4} \cot^3 x(\csc^2 x - 1)\, dx = \int_{\pi/6}^{\pi/4} \cot^3 x \csc^2 x \, dx - \int_{\pi/6}^{\pi/4} \cot^3 x \, dx$

$\displaystyle\qquad = -\int_{\pi/6}^{\pi/4} \cot^3 x(-\csc^2 x \, dx) - \int_{\pi/6}^{\pi/4} \cot x(\csc^2 x - 1)\, dx$

$\displaystyle\qquad = -\frac{1}{4}\cot^4 x \Big|_{\pi/6}^{\pi/4} - \int_{\pi/6}^{\pi/4} \cot x \csc^2 x \, dx + \int_{\pi/6}^{\pi/4} \cot x \, dx$

$\displaystyle\qquad = -\frac{1}{4}\cot^4 x + \frac{1}{2}\cot^2 x + \ln|\sin x| \Big|_{\pi/6}^{\pi/4}$

$\displaystyle\qquad = -\frac{1}{4}\cot^4 \frac{\pi}{4} + \frac{1}{2}\cot^2 \frac{\pi}{4} + \ln\left|\sin\frac{\pi}{4}\right| - \left(-\frac{1}{4}\cot^4 \frac{\pi}{6} + \frac{1}{2}\cot^2 \frac{\pi}{6} + \ln\left|\sin\frac{\pi}{6}\right|\right)$

$\displaystyle\qquad = -\frac{1}{4} + \frac{1}{2} + \ln\frac{1}{2}\sqrt{2} + \frac{1}{4}(\sqrt{3})^4 - \frac{1}{2}(\sqrt{3})^2 + \ln\frac{1}{2} = 1.347$

29. Rotate about x-axis, disks

$\displaystyle V = \pi\int_0^\pi y^2 \, dx = \pi\int_0^\pi \sin^2 x \, dx = \pi\int_0^\pi \frac{1}{2}(1 - \cos 2x)\, dx = \frac{\pi}{2}\int_0^\pi dx - \frac{\pi}{2}\int_0^\pi \cos 2x \, dx$

$\displaystyle\qquad = \frac{\pi}{2}x - \frac{\pi}{2}\times\frac{1}{2}\sin 2x \Big|_0^\pi = \frac{\pi^2}{2} - \frac{\pi}{4}\sin 2\pi - 0 + \frac{\pi}{4}\sin 0 = \frac{1}{2}\pi^2 = 4.935$

33. $\displaystyle\int \sin x \cos x \, dx; \; u = \sin x; \; du = \cos x \, dx$

$\displaystyle\int u\, du = \frac{1}{2}u^2 + C = \frac{1}{2}\sin^2 x + C_1$

Let $u = \cos x; \; du = -\sin x \, dx$

$\displaystyle -\int \cos x(-\sin x)\, dx = -\frac{1}{2}\cos^2 x + C_2$

$\displaystyle \frac{1}{2}\sin^2 x + C_1 = \frac{1}{2}(1 - \cos^2 x + C_1) = \frac{1}{2} - \frac{1}{2}\cos^2 x + C_1$

$\displaystyle -\frac{1}{2}\cos^2 x + C_2 = \frac{1}{2} - \frac{1}{2}\cos^2 x + C_1$

$\displaystyle C_2 = C_1 + \frac{1}{2}$

37. $\displaystyle V_{rms} = \sqrt{\frac{1}{1/60.0}\int_0^{1/60.0} (340\sin 120\pi t)^2 dt} = 240 \text{ V}$

28.6   Inverse Trigonometric Forms

1. $a = 2; \; u = x; \; \displaystyle\int \frac{dx}{\sqrt{4 - x^2}} = \int \frac{dx}{\sqrt{2^2 - x^2}} = \sin^{-1}\frac{x}{2} + C$

5. $a = 1; \; u = 4x; \; du = 4\, dx$

$\displaystyle\int \frac{dx}{\sqrt{1 - (4x)^2}} = \frac{1}{4}\int \frac{4\, dx}{\sqrt{1 - (4x)^2}} = \frac{1}{4}\sin^{-1}\frac{4x}{1} + C = \frac{1}{4}\sin^{-1} 4x + C$

9. $\displaystyle \frac{2}{\sqrt{5}}\int_0^{0.4} \frac{\sqrt{5}\, dx}{\sqrt{4 - 5x^2}} = \frac{2}{\sqrt{5}}\left(\sin^{-1}\frac{\sqrt{5}x}{2}\right)\Big|_0^{0.4} = \frac{2\sqrt{5}}{5}[\sin^{-1}(0.2\sqrt{5}) - \sin^{-1} 0] = 0.415$

13. $\int_1^e \dfrac{3\,du}{u[1+(\ln u)^2]} = 3\int_1^e \dfrac{1}{1+(\ln u)^2}\cdot\dfrac{du}{u} = 3\cdot\tan^{-1}(\ln u)\Big|_1^e = 3[\tan^{-1}(\ln e)-\tan^{-1}(\ln 1)]$

$$= 3\cdot[\tan^{-1}1-\tan^{-1}0] = 3\cdot\left[\dfrac{\pi}{4}-0\right] = \dfrac{3\pi}{4} = 2.356$$

17. $a=1$; $u=x+1$; $du=dx$

$$\int\dfrac{dx}{x^2+2x+2} = \int\dfrac{dx}{(x^2+2x+1)+1} = \int\dfrac{dx}{(x+1)^2+1^2} = \dfrac{1}{1}\tan^{-1}\dfrac{(x+1)}{1}+C = \tan^{-1}(x+1)+C$$

21. $a=1$; $u=\sin 2\theta$; $du=\cos 2\theta(2\,d\theta)$

$$\int_{\pi/6}^{\pi/2}\dfrac{2\cos 2\theta}{1+\sin^2 2\theta}\,d\theta = \int_{\pi/6}^{\pi/2}\dfrac{\cos 2\theta\,(2d\theta)}{1+\sin^2 2\theta} = \tan^{-1}\dfrac{\sin 2\theta}{1}\Big|_{\pi/6}^{\pi/2} = \left(\tan^{-1}\sin\pi-\tan^{-1}\sin\dfrac{\pi}{3}\right)$$

$$= -0.714$$

25. (a) Inverse tangent, $\int\dfrac{du}{a^2+u^2}$ where $u=3x$, $du=3\,dx$, $a=2$; numerator cannot fit du of denominator. Positive $9x^2$ leads to inverse tangent form.

(b) $\int\dfrac{2\,dx}{4+9x} = 2\int\dfrac{dx}{4+9x}$; $u=4+9x$; $du=9\,dx$

Therefore, the integral is logarithmic.

(c) $\int\dfrac{2x\,dx}{\sqrt{4+9x^2}} = 2\int(4+9x^2)^{-1/2}(x\,dx)$; $u=4+9x^2$; $du=18x\,dx$

Therefore, the form of the integral is general power.

29. $y=\dfrac{1}{1+x^2}$; $A=\displaystyle\int_0^2\dfrac{1}{1+x^2}\,dx$; $a=1$; $u=x$; $du=dx$

$$A = \dfrac{1}{1}\tan^{-1}\dfrac{x}{1}\Big|_0^2 = \tan^{-1}2-\tan^{-1}0 = 1.11$$

33. $\int\dfrac{dx}{\sqrt{A^2-x^2}} = \int\sqrt{\dfrac{k}{m}}\,dt$; $\sin^{-1}\dfrac{x}{A} = \sqrt{\dfrac{k}{m}}\,t+C$

Solve for C by letting $x=x_0$ and $t=0$.

$\sin^{-1}\dfrac{x_0}{A} = \sqrt{\dfrac{k}{m}}(0)+C$; $C=\sin^{-1}\dfrac{x_0}{A}$; therefore, $\sin^{-1}\dfrac{x}{A} = \sqrt{\dfrac{k}{m}}\,t+\sin^{-1}\dfrac{x_0}{A}$

28.7   Integrations by Parts

1. $u=\theta$; $du=d\theta$; $dv=\cos\theta\,d\theta$; $v=\displaystyle\int\cos\theta\,d\theta = \sin\theta$

$$\int(\theta)(\cos\theta\,d\theta) = \theta(\sin\theta)-\int\sin\theta\,d\theta = \theta(\sin\theta)-(-\cos\theta)+C = \cos\theta+\theta\sin\theta+C$$

5. $u=x$; $du=dx$; $dv=\sec^2 x\,dx$; $v=\displaystyle\int\sec^2 x\,dx = \tan x$

$$\int(x)(\sec^2 x\,dx) = x\tan x-\int\tan x\,dx = x\tan x-(-\ln|\cos x|+C) = x\tan x+\ln|\cos x|+C$$

9. $\displaystyle\int_{-3}^{0} \frac{4t\,dt}{\sqrt{1-t}} = 4\int \frac{t\,dt}{\sqrt{1-t}}; \; u = t; \; du = dt$

$\displaystyle dv = \frac{1}{\sqrt{1-t}}\,dt; \; v = \int (1-t)^{-1/2}dt$

$\displaystyle v = -\int (1-x)^{-1/2}(-dt) = -(1-t)^{1/2}(2) = -2(1-t)^{1/2}$

$\displaystyle 4\int (t)(1-t)^{-1/2}dt = 4t[-2(1-t)^{1/2}] - 4\int -2(1-t)^{1/2}dt = -8t(1-t)^{1/2} + 8\int (1-x)^{1/2}dt$

$\displaystyle = -8t(1-t)^{1/2} - 8\int (1-t)^{1/2}(-dt) = -8t(1-t)^{1/2} - 8(1-t)^{3/2}\left(\frac{2}{3}\right)\Big|_{-3}^{0} = -10\frac{2}{3}$

13. $\displaystyle\int 2\phi^2 \sin\phi\cos\phi\,d\phi = \int \phi^2 \sin(2\phi)\,d\phi$

let $\quad u = \phi^2 \qquad dv = \sin(2\phi)\,d\phi$

$\qquad du = 2\phi\,d\phi$

$\displaystyle \qquad\qquad v = \int \sin(2\phi)\,d\phi = \frac{1}{2}\sin(2\phi)\,2\,d\phi; \; v = -\frac{1}{2}\cos(2\phi)$

$\displaystyle \int 2\phi^2 \sin\phi\cos\phi\,d\phi = \int \phi^2 \sin(2\phi)\,d\phi = -\frac{1}{2}\cdot\phi^2\cos(2\phi) + \int \phi\cos(2\phi)\,d\phi$

let $\quad u = \phi \quad dv = \cos(2\phi)\,d\phi$

$\qquad du = d\phi$

$\displaystyle \qquad\qquad v = \int \cos(2\phi)\,d\phi = \frac{1}{2}\sin(2\phi)$

$\displaystyle \int 2\phi^2 \sin\phi\cos\phi\,d\phi = -\frac{1}{2}\phi^2\cos(2\phi) + \frac{\phi}{2}\sin(2\phi) - \frac{1}{2}\int \sin(2\phi)\,d\phi$

$\displaystyle \qquad\qquad = -\frac{\phi^2}{2}\cos(2\phi) + \frac{\phi}{2}\sin(2\phi) + \frac{1}{4}\cos(2\phi) + C$

$\displaystyle \qquad\qquad = \frac{\phi}{2}\sin(2\phi) - \frac{1}{4}(2\phi^2 - 1)\cos(2\phi) + C$

17. $\displaystyle A = \int_0^2 xe^{-x}dx; \; u = x; \; du = dx; \; dv = e^{-x}dx; \; v = \int e^{-x}dx = -e^{-x}$

$\displaystyle A = -xe^{-x}\Big|_0^2 = \int_0^2 -e^{-x}dx = -xe^{-x} - e^{-x}\Big|_0^2 = -2e^{-2} - e^{-2} - (0-1) = 1 - \frac{3}{e^2} = 0.594$

21. $\displaystyle \overline{x} = \frac{\displaystyle\int_0^{\pi/2} x(\cos x)\,dx}{\displaystyle\int_0^{\pi/2} \cos x\,dx}$

Let $u = x; \; du = dx; \; dv = \cos x; \; v = \sin x$

$\displaystyle \overline{x} = \frac{x\sin x\big|_0^{\pi/2} - \displaystyle\int_0^{\pi/2} \sin x\,dx}{\sin x\big|_0^{\pi/2}} = \frac{x\sin x\big|_0^{\pi/2} - (-\cos x)\big|_0^{\pi/2}}{1} = x\sin x + \cos x\big|_0^{\pi/2} = \frac{\pi}{2} - 1 = 0.571$

25. $v = \dfrac{ds}{dt} = \dfrac{t^3}{\sqrt{t^2+1}};\ s = \displaystyle\int \dfrac{t^3\,dt}{\sqrt{t^2+1}}$

Let $u = t^2;\ du = 2t\,dt;\ dv = \dfrac{t\,dt}{(t^2+1)^{1/2}}$

$v = \dfrac{1}{2}\displaystyle\int \dfrac{2t\,dt}{(t^2+)^{1/2}} = \dfrac{1}{2}(2)(t^2+1)^{1/2} = (t^2+1)^{1/2}$

$s = t^2(t^2+1)^{1/2} - \displaystyle\int (t^2+1)^{1/2}(2t\,dt) = t^2(t^2+1)^{1/2} - \dfrac{2}{3}(t^2+1)^{3/2} + C$

$s = 0$ for $t = 0;\ 0 = -\dfrac{2}{3} + C;\ C = \dfrac{2}{3}$

$s = \dfrac{1}{3}[3t^2(t^2+1)^{1/2} - 2(t^2+1)^{3/2} + 2] = \dfrac{1}{3}[(t^2-2)(t^2+1)^{1/2} + 2]$

28.8   Integration by Trigonometric Substitution

1. Let $x = \sin\theta;\ dx = \cos\theta\,d\theta$

$\displaystyle\int \dfrac{\sqrt{1-x^2}}{x^2}\,dx = \int \dfrac{\sqrt{1-\sin^2\theta}}{\sin^2\theta}\cos\theta\,d\theta = \int \dfrac{\cos^2\theta}{\sin^2\theta}\,d\theta = \int \cot^2\theta\,d\theta = \int (\csc^2\theta - 1)\,d\theta$

$\qquad = \displaystyle\int \csc^2\theta\,d\theta - \int d\theta = -\cot\theta - \theta + C = \dfrac{-\sqrt{1-x^2}}{x} - \sin^{-1}x + C$

5. Let $z = 3\tan\theta;\ dz = 3\sec^2\theta\,d\theta$

$\displaystyle\int \dfrac{6\,dz}{z^2\sqrt{z^2+9}} = 6\int \dfrac{3\sec^2\theta\,d\theta}{9\tan^2\theta\sqrt{9\tan^2\theta+9}} = 6\int \dfrac{3\sec^2\theta\,d\theta}{27\tan^2\theta\sqrt{\tan^2\theta+1}} = \dfrac{6}{9}\int \dfrac{\sec\theta\,d\theta}{\tan^2\theta}$

$\qquad = \dfrac{6}{9}\displaystyle\int \dfrac{\cos\theta\,d\theta}{\sin^2\theta} = \dfrac{6}{9}\int \csc\theta\cot\theta\,d\theta = -\dfrac{6}{9}\csc\theta + C = \dfrac{6}{9\sin\theta} + C$

$\tan\theta = \dfrac{z}{3};\ \sin\theta = \dfrac{z}{\sqrt{9+z^2}}$

$\dfrac{-6}{9\sin\theta} + C = \dfrac{-6}{\dfrac{9z}{\sqrt{9+z^2}}} + C = -\dfrac{2\sqrt{z^2+9}}{3z} + C$

9. $\displaystyle\int_0^{0.5} \dfrac{x^3\,dx}{\sqrt{1-x^2}},\ x = \sin\theta;\ dx = \cos\theta\,d\theta$

$\displaystyle\int \dfrac{\sin^3\theta\cos\theta\,d\theta}{\sqrt{1-\sin^2\theta}} = \int \sin^3\theta\,d\theta = \int \sin\theta\sin^2\theta\,d\theta = \int \sin\theta(1-\cos^2\theta)\,d\theta = \int \sin\theta\,d\theta - \int \cos^2\theta\sin\theta\,d\theta$

$\qquad = -\cos\theta + \dfrac{\cos^3\theta}{3}$

$\cos\theta = \sqrt{1-x^2};\ -\sqrt{1-x^2} + \dfrac{1}{3}(\sqrt{1-x^2})\Big|_0^{0.5} = -\sqrt{1-0.5^2} + \dfrac{1}{3}(\sqrt{1-0.5^2})^3 + \sqrt{1} - \dfrac{1}{3}\sqrt{1} = 0.017$

13. $\int \dfrac{dy}{y\sqrt{4y^2-9}}$; $2y=3\sec\theta$; $y=\dfrac{3}{2}\sec\theta$; $dy=\dfrac{3}{2}\sec\theta\tan\theta\,d\theta$

$$\int \dfrac{\frac{3}{2}\sec\theta\tan\theta\,d\theta}{\frac{3}{2}\sec\theta\sqrt{4\left(\frac{3}{2}\sec\theta\right)^2-9}}=\int\dfrac{\tan\theta\,d\theta}{\sqrt{9\sec^2\theta-9}}=\int\dfrac{\tan\theta\,d\theta}{3\sqrt{\sec^2\theta-1}}=\int\dfrac{\tan\theta\,d\theta}{3\tan\theta}=\dfrac{1}{3}\int d\theta$$

$$=\dfrac{1}{3}\theta+C=\dfrac{1}{3}\sec^{-1}\dfrac{2}{3}y+C$$

$$\int_{2.5}^{3}\dfrac{dy}{y\sqrt{4y^2-9}}=\dfrac{1}{3}\sec^{-1}\left(\dfrac{2y}{3}\right)\Big|_{2.5}^{3}=\dfrac{1}{3}\cos^{-1}\left(\dfrac{3}{2y}\right)\Big|_{2.5}^{3}=0.039967$$

17. $A=4\displaystyle\int_0^1 y\,dx=4\int_0^1\sqrt{1-x^2}\,dx$

Let $x=\sin\theta$, $dx=\cos\theta\,d\theta$

$$\int\sqrt{1-x^2}\,dx=\int\sqrt{1-\sin^2\theta}\cos\theta\,d\theta=\int\cos^2\theta\,d\theta=\dfrac{1}{2}\int(1+\cos2\theta)\,d\theta$$

$$=\dfrac{1}{2}\theta+\dfrac{1}{4}\sin2\theta+C=\dfrac{1}{2}\theta+\dfrac{1}{2}\sin\theta\cos\theta+C$$

$$=\dfrac{1}{2}\sin^{-1}x+\dfrac{1}{2}x\sqrt{1-x^2}+C$$

$$A=4\int_0^1\sqrt{1-x^2}\,dx=4\left(\dfrac{1}{2}\sin^{-1}x+\dfrac{1}{2}x\sqrt{1-x^2}\right)\Big|_0^1=2\sin^{-1}1+2(1)\sqrt{0}-[2\sin^{-1}0+2(0)]$$

$$=2\sin^{-1}1=2\left(\dfrac{\pi}{2}\right)=\pi$$

21. $V=2\pi\displaystyle\int_4^5\dfrac{x(\sqrt{x^2-16})}{x^2}\,dx$, disks

(The limits of integration are $x=4$ since $x=4$ when $y=0$; and $x=5$.)
Let $x=4\sec\theta$; $dx=4\sec\theta\tan\theta\,d\theta$

$$V=2\pi\int_4^5\dfrac{\sqrt{x^2-16}}{x}\,dx$$

$$2\pi\int\dfrac{\sqrt{x^2-16}}{x}\,dx=2\pi\int\dfrac{\sqrt{16\sec^2\theta-16}}{4\sec\theta}(4\sec\theta\tan\theta\,d\theta)=2\pi\int\sqrt{16\sec^2\theta-16}\tan\theta\,d\theta$$

$$=2\pi\int4\sqrt{\sec^2\theta-1}\tan\theta\,d\theta=8\pi\int(\tan\theta)\tan\theta\,d\theta=8\pi\int\tan^2\theta\,d\theta$$

$$=8\pi\int(\sec^2\theta-1)\,d\theta=8\pi(\tan\theta-\theta)$$

Since $x=4\sec\theta$; $\sec=\dfrac{x}{4}$; and $\tan\theta=\dfrac{\sqrt{x^2-16}}{4}$; and

$$2\pi\int\dfrac{\sqrt{x^2-16}}{x}\,dx=8\pi(\tan\theta-\theta)=8\pi\left(\dfrac{\sqrt{x^2-16}}{4}-\sec^{-1}\dfrac{x}{4}\right)$$

$$V=2\pi\int_4^5\dfrac{x\sqrt{x^2-16}}{x^2}\,dx=8\pi\left(\dfrac{\sqrt{x^2-16}}{4}-\sec^{-1}\dfrac{x}{4}\right)\Big|_4^5=8\pi\left(\dfrac{3}{4}-\sec^{-1}\dfrac{5}{4}-0\right)=2.68$$

28.9   Integration by Partial Fractions: Nonrepeated Linear Factors

1. $\dfrac{x+3}{(x+1)(x+2)} = \dfrac{A}{x+1} + \dfrac{B}{x+2} = \dfrac{A(x+2)+B(x+1)}{(x+1)(x+2)}$

$$x+3 = A(x+2) + B(x+1)$$
$$x = -2, \quad 1 = -B, \quad B = -1$$
$$x = -1, \quad 2 = A$$

$$\int \frac{x+3}{(x+1)(x+2)} = \int \frac{2}{x+1}\,dx + \int \frac{-1}{x+2}\,dx = 2\ln|x+1| - \ln|x+2| + C$$

$$= \ln(x+1)^2 - \ln|x+2| = \ln\frac{(x+1)^2}{|x+2|} + C$$

5.

$$
\begin{array}{r}
1 \\[2pt]
x^2 + 3x\overline{)x^2 + 0x + 3} \\
\underline{x^2 + 3x} \\
-3x + 3
\end{array}
$$

$$\frac{x^2+3}{x^2+3x} = 1 + \frac{-3x+3}{x(x+3)}$$

$$\frac{-3x+3}{x(x+3)} = \frac{A}{x} + \frac{B}{x+3} = \frac{A(x+3)+Bx}{x(x+3)}$$

$$-3x+3 = A(x+3) + Bx$$
$$x = 0, 3 = 3A, = A = 1$$
$$x = -3, 12 = -3B = B = -4$$

$$\int \frac{x^2+3}{x^2+3x} = \int dx + \int \frac{dx}{x} + \int \frac{-4}{x+3}\,dx = x + \ln|x| - 4\ln|x+3| + C = x + \ln\frac{|x|}{(x+3)^4} + C$$

9. $\dfrac{4x^2-10}{x(x+1)(x-5)} = \dfrac{A}{x} + \dfrac{B}{x+1} + \dfrac{C}{x-5} = \dfrac{A(x+1)(x-5) + B\cdot x(x-5) + Cx(x+1)}{x(x+1)(x-5)}$

$$4x^2 - 10 = A(x+1)(x-5) + Bx(x-5) + Cx(x+1)$$
$$x = 0, -10 = -5A, A = 2$$
$$x = 5, \quad 90 = 30C, C = 3$$
$$4x^2 - 10 = 2(x+1)(x-5) + Bx(x-5) + 3x(x+1)$$
$$x = 1, -6 = -16 - 4B + 6, B = -1$$

$$\int \frac{4x^2-10}{x(x+1)(x-5)}\,dx = \int \frac{2}{x}\,dx - \int \frac{dx}{x+1} + \int \frac{3}{x-5}\,dx = 2\ln|x| - \ln|x+1| + 3\ln|x-5| + C$$

$$= \ln\frac{x^2 \cdot |x-5|^3}{|x+1|} + C$$

13. $\dfrac{x^3 + 7x^2 + 9x + 2}{x(x^2 + 3x + 2)} = \dfrac{x^3 + 7x^2 + 9x + 2}{x^3 + 3x^2 + 2x}$

$$
\begin{array}{r}
1 \\
x^3 + 3x^2 + 2x\overline{)x^3 + 7x^2 + 9x + 2} \\
\underline{x^3 + 3x^2 + 2x} \\
4x^2 + 7x + 2
\end{array}
$$

$\dfrac{x^3 + 7x^2 + 9x + 2}{x(x^2 + 3x + 2)} = 1 + \dfrac{4x^2 + 7x + 2}{x(x^2 + 3x + 2)} = 1 + \dfrac{4x^2 + 7x + 2}{x(x + 1)(x + 2)}$

$\dfrac{4x^2 + 7x + 2}{x(x + 1)(x + 2)} = \dfrac{A}{x} + \dfrac{B}{x + 1} + \dfrac{C}{x + 2} = \dfrac{A(x + 1)(x + 2) + Bx(x + 2) + Cx(x + 1)}{x(x + 1)(x + 2)}$

$4x^2 + 7x + 2 = A(x + 1)(x + 2) + Bx(x + 2) + Cx(x + 1)$

$x = 0, \qquad 2 = 2A, \ A = 1$

$x = -1, \quad -1 = -B, \ B = 1$

$x = -2, \qquad 4 = 2C, \ C = 2$

$\displaystyle\int_1^2 \dfrac{x^3 + 7x^2 + 9x + 2}{x(x^2 + 3x + 2)}\,dx = \int_1^2 dx + \int_1^2 \dfrac{dx}{x} + \int_1^2 \dfrac{dx}{x + 1} + \int_1^2 \dfrac{2}{x + 2}\,dx$

$$= x + \ln|x| + \ln|x + 1| + 2\ln|x + 2| \ \Big|_1^2$$

$$= 2.674$$

17. $A = \displaystyle\int_2^4 \dfrac{x - 16}{x^2 - 5x - 14}\,dx = 1.322$

21. $\qquad \dfrac{3x + 5}{x^2 + 5x} = \dfrac{3x + 5}{x(x + 5)} = \dfrac{A}{x} + \dfrac{B}{x + 5} = \dfrac{A(x + 5) + Bx}{x(x + 5)}$

$\qquad\qquad 3x + 5 = A(x + 5) + Bx$

$x = 0, \qquad 5 = 5A, \quad A = 1$

$x = -5, \ -10 = -5B, \ B = 2$

$$y = \int \dfrac{3x + 5}{x^2 + 5x}\,dx = \int \dfrac{dx}{x} + \int \dfrac{2}{x + 5}\,dx$$

$$y = \ln|x| + 2\ln|x + 5| + C$$

$$0 = \ln|1| + 2\ln|6| + C, \ C = -2\ln 6$$

$$y = \ln|x| + 2\ln|x + 5| - 2\ln 6$$

$$y = \ln|x| + \ln|x + 5|^2 - \ln 36$$

$$y = \ln \dfrac{|x| \cdot (x + 5)^2}{36}$$

28.10   Integration by Partial Fractions: Other Cases

1. $\dfrac{x-8}{x^3-4x^2+4x} = \dfrac{x-8}{x(x^2-4x+4)} = \dfrac{x-8}{x(x-2)(x-2)} = \dfrac{A}{x} + \dfrac{B}{x-2} + \dfrac{C}{(x-2)^2}$

$$x-8 = A(x-2)^2 + Bx(x-2) + Cx$$
$$x=0, \ -8 = 4A, \ A = -2$$
$$x=2, \ -6 = 2C, \ C = -3$$
$$x-8 = -2(x-2)^2 + Bx(x-2) - 3x$$
$$x=1, \ -7 = -2 - B - 3, \ B = 2$$

$$\int \frac{x-8}{x^3-4x^2+4x} = \int \frac{-2}{x}\,dx + \int \frac{2}{x-2}\,dx - \int \frac{3}{(x-2)^2}\,dx = -2\ln|x| + 2\ln|x-2| + \frac{3}{x-2} + C$$

$$= \ln\left(\frac{x-2}{x}\right)^2 + \frac{3}{x-2} + C$$

5. $\dfrac{2s}{(s-3)^3} = \dfrac{A}{s-3} + \dfrac{B}{(s-3)^2} + \dfrac{C}{(s-3)^3} = \dfrac{A(s-3)^2 + B(s-3) + C}{(s-3)^3}$

$$2s = A(s-3)^2 + B(s-3) + C$$
$$s=3, \quad 6 = C$$
$$2s = A(s-3)^2 + B(s-3) + 6$$

$$\left.\begin{array}{ll} s=1, & 2 = 4A - 2B + 6 \\ s=2, & 4 = A - B + 6 \end{array}\right\} A = 0, \ B = 2$$

$$\int_1^2 \frac{2s}{(s-3)^3}\,ds = \int_1^2 \frac{2}{(s-3)^3}\,ds + \int_1^2 \frac{6}{(s-3)^3}\,ds = 2\cdot\frac{(s-3)^{-2+1}}{-2+1} + 6\cdot\frac{(s-3)^{-3+1}}{-3+1}\bigg|_1^2$$

$$= \frac{-2}{(s-3)} - \frac{3}{(s-3)^2}\bigg|_1^2 = -\frac{5}{4}$$

9. $\dfrac{x^2+x+5}{(x+1)(x^2+4)} = \dfrac{A}{x+1} + \dfrac{Bx+C}{x^2+4} = \dfrac{A(x^2+4) + (Bx+C)(x+1)}{(x+1)(x^2+4)}$

$$x^2 + x + 5 = A(x^2+4) + (Bx+C)(x+1)$$

$$\left.\begin{array}{lll} x=0, & 5 = 4A + & C \\ x=1, & 7 = 5A + 2B + 2C \\ x=2, & 11 = 8A + 6B + 3C \end{array}\right\} A = 1, \ B = 0, \ C = 1$$

$$\int_0^2 \frac{x^2+x+5}{(x+1)(x^2+4)}\,dx = \int_0^2 \frac{dx}{x+1} + \int_1^2 \frac{1}{x^2+4}\,dx \ \ln|x+1| + \frac{\tan^{-1}\frac{x}{2}}{2}\bigg|_0^2 = 1.49$$

13. $\dfrac{10x^3 + 40x^2 + 22x + 7}{(4x^2+1)(x^2+6x+10)} = \dfrac{Ax+B}{4x^2+1} + \dfrac{Cx+D}{x^2+6x+10}$

$$10x^3 + 40x^2 + 22x + 7 = (Ax+B)(x^2+6x+10) + (Cx+D)(4x^2+1)$$
$$10x^3 + 40x^2 + 22x + 7 = Ax^3 + 6Ax^2 + 10Ax + Bx^2 + 6Bx + 10B + 4Cx^3 + Cx + 4Dx^2 + D$$
$$10x^3 + 40x^2 + 22x + 7 = (A+4C)x^3 + (6A+B+4D)x^2 + (10A+6B+C)x + 10B+D$$

$$\left.\begin{array}{l}\text{(1)} \quad A+4C \qquad\qquad = 10 \\ \text{(2)} \quad 6A+B+4D \quad = 40 \\ \text{(3)} \quad 10A+6B+C = 22 \\ \text{(4)} \quad 10B+D \qquad\quad = 7\end{array}\right\} A=2, B=0, C=2, D=7$$

$$\int \frac{10x^3+40x^2+22x+7}{(4x^2+1)(x^2+6x+10)}\,dx = \int \frac{2x}{4x^2+1}\,dx + \int \frac{2x+7}{x^2+6x+10}\,dx$$

$$\int \frac{2x}{4x^2+1}\,dx = \frac{1}{4}\int \frac{8x}{4x^2+1}\,dx = \frac{\ln(4x^2+1)}{4}$$

$$\int \frac{2x+7}{x^2+6x+10}\,dx = \int \frac{2x+7}{x^2+6x+9+1}\,dx = \int \frac{2x+7}{(x+3)^2+1}\,dx \quad \text{let} \quad u=x+3,\ x=u-3$$
$$du=dx$$

$$\int \frac{2x+7}{x^2+6x+10}\,dx = \int \frac{2(u-3)+7}{u^2+1}\,du$$

$$\int \frac{2x+7}{x^2+6x+10}\,dx = \int \frac{2u-6+7}{u^2+1}\,du$$

$$\int \frac{2x+7}{x^2+6x+10}\,dx = \int \frac{2u}{u^2+1}\,du + \int \frac{1}{u^2+1}\,du$$

$$\int \frac{2x+7}{x^2+6x+10}\,dx = \ln(u^2+1) + \tan^{-1}u + C$$

$$\int \frac{2x+7}{x^2+6x+10}\,dx = \ln(x^2+6x+10) + \tan^{-1}(x+3) + C$$

$$\int \frac{10x^3+40x^2+22x+7}{(4x^2+1)(x^2+6x+10)}\,dx = \frac{\ln(4x^2+1)}{4} + \ln(x^2+6x+10) + \tan^{-1}(x+3) + C$$

$$\int \frac{10x^3+40x^2+22x+7}{(4x^2+1)(x^2+6x+10)}\,dx = \ln(\sqrt[4]{4x^2+1}\cdot(x^2+6x+10)) + \tan^{-1}(x+3) + C$$

17.
$$A = -\int_1^3 \frac{x-3}{x^3+x^2}\,dx = -\int_1^3 \frac{x-3}{x^2(x+1)}\,dx$$

$$\frac{x-3}{x^2(x+1)} = \frac{Ax+B}{x^2} + \frac{C}{x+1} = \frac{(Ax+B)(x+1)+Cx^2}{x^2(x+1)} = \frac{Ax^2+Ax+Bx+B+Cx^2}{x^2(x+1)}$$
$$x-3 = (A+C)x^2 + (A+B)x + B$$

(1) $A+C=0,\ C=-4$
(2) $A+B=1,\ A=4$
(3) $\qquad B=-3$

$$\frac{x-3}{x^2(x+1)} = \frac{4x-3}{x^2} - \frac{4}{x+1} = \frac{4}{x} - \frac{3}{x^2} - \frac{4}{x+1}$$

$$\int_1^3 \frac{x-3}{x^3+x^2}\,dx = \int_1^3 \frac{4}{x}\,dx - 3\int_1^3 \frac{dx}{x^2} - 4\int_1^3 \frac{dx}{x+1} = 4\ln|x|\Big|_1^3 + \frac{3}{x}\Big|_1^3 - 4\ln(x+1)\Big|_1^3$$

$$= 4(\ln 3 - \ln 1) + \frac{3}{3} - \frac{3}{1} - 4\ln(4) + 4\ln 2 = \ln 3^4 + 1 - 3 - \ln 4^4 + \ln 2^4$$

$$= -2 + \ln 81 - \ln 256 + \ln 16 = -2 + \ln \frac{81\cdot 16}{256} = -2 + \ln \frac{81}{16} = -0.378$$

$$A = 2 - \ln \frac{81}{16} = 0.378$$

21. $\dfrac{t^2 + 14t + 27}{(2t+1)(t+5)^2} = \dfrac{A}{2t+1} + \dfrac{B}{t+5} + \dfrac{C}{(t+5)^2}$

$\dfrac{t^2 + 14t + 27}{(2t+1)(t+5)^2} = \dfrac{A(t+5)^2 + B(t+5)(2t+1) + C(2t+1)}{(2t+1)(t+5)^2}$

$t^2 + 14t + 27 = At^2 + 10At + 25A + 2Bt^2 + 11Bt + 5B + 2Ct + C$

$t^2 + 14t + 27 = (A+2B)t^2 + (10A + 11B + 2C)t + 25A + 5B + C$

$\left.\begin{array}{ll}(1) & A + 2B = 1 \\ (2) & 10A + 11B + 2C = 14 \\ (3) & 25A + 5B + C = 27\end{array}\right\}\ A = 1,\ B = 0,\ C = 2$

$\dfrac{ds}{dt} = \dfrac{t^2 + 14t + 27}{(2t+1)(t+5)^2} = \dfrac{1}{2t+1} + \dfrac{2}{(t+5)^2}$

$s = \dfrac{1}{2}\displaystyle\int_0^{2.00} \dfrac{2}{2t+1}\,dt + \int_0^{2.00} \dfrac{2}{(t+5)^2}\,dt;\ s = \dfrac{1}{2}\ln|2t+1|\,\Big|_0^{2.00} - 2\cdot \dfrac{1}{(t+5)}\,\Big|_0^{2.00}$

$s = \dfrac{1}{2}\ln 5.00 - \dfrac{1}{2}\ln 1 - \dfrac{2}{7.00} + \dfrac{2}{5.00} = 0.919\ \text{m}$

28.11   Integration by Use of Tables

1. Formula #1; $u = x$; $a = 2$; $b = 5$; $du = dx$

$\displaystyle\int \dfrac{3x\,dx}{2+5x} = 3\int \dfrac{3x\,dx}{2+5x} = 3\left\{\dfrac{1}{25}[(2+5x) - 2\ln|2+5x|]\right\} + C = \dfrac{3}{25}[2 + 5x - 2\ln|2+5x|] + C$

5. Formula #24; $u = y$, $a = 2$

$\displaystyle\int \dfrac{dy}{(y^2+4)^{3/2}} = \dfrac{y}{4\sqrt{y^2+4}} + C$

9. Formula #17; $u = 2x$; $du = 2\,dx$; $a = 3$

$\displaystyle\int \dfrac{\sqrt{4x^2-9}}{x}\,dx = \int \dfrac{\sqrt{(2x)^2 - 3^2}}{2x}\,dx$

$\qquad = \sqrt{4x^2-9} - 3\sec^{-1}\left(\dfrac{2x}{3}\right) + C$

13. Formula #52; $u = r^2$; $du = 2r\,dr$

$6\displaystyle\int \tan^{-1}r^2(r\,dr) = 3\int \tan^{-1}r^2(2r\,dr) = 3\left[r^2\tan^{-1}r^2 - \dfrac{1}{2}\ln(1+r^4)\right] + C$

$\qquad = 3r^2\tan^{-1}r^2 - \dfrac{3}{2}\ln(1+r^4) + C$

17. Formula #11; $u = 2x$; $du = 2\,dx$; $a = 1$

$\displaystyle\int \dfrac{dx}{x\sqrt{4x^2+1}} = \int \dfrac{2\,dx}{2x\sqrt{(2x)^2 + 1^2}} = -\ln\left(\dfrac{1+\sqrt{4x^2+1}}{2x}\right) + C$

21. Formula #40; $a = 1$; $u = x$; $du = dx$; $b = 5$

$$\int_0^{\pi/12} \sin\theta \cos 5\theta\, d\theta = -\frac{\cos(-4\theta)}{2(-4)} - \frac{\cos 6\theta}{12} = \frac{1}{8}\cos 4\theta - \frac{1}{12}\cos 6\theta \Big|_0^{\pi/12} = 0.0208$$

25. let $u = x^2$, $du = 2x\, dx$
$u^2 = x^4$

$$\int \frac{2x\, dx}{(1 - x^4)^{3/2}} = \int \frac{du}{(1 - u^2)^{3/2}}$$

Formula 25: $a = 1$

$$\int \frac{2x\, dx}{(1 - x^4)^{3/2}} = \frac{u}{\sqrt{1 - u^2}} + C;$$
$$\int \frac{2x\, dx}{(1 - x^4)^{3/2}} = \frac{x^2}{\sqrt{1 - x^4}} + C$$

29. Formula #46; $u = x^2$; $du = 2x\, dx$; $n = 1$

$$\int x^3 \ln x^2\, dx = \frac{1}{2}\int x^2 \ln x^2 (2x\, dx) = \frac{1}{2}\left[(x^2)^2\left(\frac{\ln x^2}{2} - \frac{1}{4}\right)\right]$$

$$= \frac{1}{2}\left[\frac{x^4}{2}\left(\ln x^2 - \frac{1}{2}\right)\right] = \frac{1}{4}x^4\left(\ln x^2 - \frac{1}{2}\right) + C$$

33. From Exercise 17 of Section 26-6,

$$s = \int_a^b \sqrt{1 + \left(\frac{dy}{dx}\right)^2}\, dx; \quad y = x^2; \quad \frac{dy}{dx} = 2x; \quad s = \int_0^1 \sqrt{1 + (2x)^2}\, dx = \frac{1}{2}\int_0^1 \sqrt{(2x)^2 + 1}(2dx)$$

Formula #14; $u = 2x$; $du = 2\, dx$

$$s = \frac{1}{2}\left[\frac{2x}{2}\sqrt{4x^2 + 1} + \frac{1}{2}\ln(2x + \sqrt{4x^2 + 1})\right]\Big|_0^1 = \frac{1}{2}\left[\left(1\sqrt{5} + \frac{1}{2}\ln(2 + \sqrt{5})\right) - \frac{1}{2}\ln 1\right]$$

$$= \frac{1}{4}[2\sqrt{5} + \ln(2 + \sqrt{5})] = 1.479$$

37. $F = w \int_0^3 lh\, dh = w \int_0^3 x(3 - y)\, dy = w \int_0^3 \frac{3 - y}{\sqrt{1 + y}}\, dy$

(Formula #6)

$$\int \frac{3 - y}{\sqrt{1 + y}}\, dy = 3\int \frac{dy}{\sqrt{1 + y}} - \int \frac{y\, dy}{\sqrt{1 + y}} = 3\frac{(1 + y)^{1/2}}{\frac{1}{2}} - \left[\frac{-2(2 - y)\sqrt{1 + y}}{3(1)^2}\right] + C$$

$$F = w \int_0^3 \frac{3 - y}{\sqrt{1 + y}}\, dy = w\left[6(1 + y)^{1/2} + \frac{2}{3}(2 - y)(1 + y)^{1/2}\right]\Big|_0^3$$

$$= w\left[6(2) + \frac{2}{3}(-1)(2) - 6(1) - \frac{2}{3}(2)(1)\right]$$

$$F = w\left(12 - \frac{4}{3} - 6 - \frac{4}{3}\right) = \frac{10w}{3} = \frac{10(62.4)}{3} = 208 \text{ lb}$$

Chapter 28 Review Exercises

1. $u = -2x$, $du = -2\,dx$

$$\int e^{-2x}\,dx = -\frac{1}{2}\int e^{-2x}(-2\,dx) = -\frac{1}{2}e^{-2x} + C$$

5. $\displaystyle\int_0^{\pi/2} \frac{4\cos\theta\,d\theta}{1+\sin\theta} = 4\int_0^{\pi/2} \frac{\cos\theta\,d\theta}{1+\sin\theta};\ u = 1+\sin\theta,\ du = \cos\theta = 4\ln(1+\sin\theta)\Big|_0^{\pi/2} = 2.77$

9. $\displaystyle\int_0^{\pi/2}\cos^3 2x\,dx = \int_0^{\pi/2}\cos^2 2x\cos 2x\,dx = \int_0^{\pi/2}(1-\sin^2 2x)\cos 2x\,dx$

$$= \int_0^{\pi/2}\cos 2x\,dx - \int_0^{\pi/2}\sin^2 2x\cos 2x\,dx$$

$$= \frac{1}{2}\int_0^{\pi/2}\cos 2x(2\,dx) - \frac{1}{2}\int_0^{\pi/2}\sin^2 2x\cos 2x(2\,dx)$$

$$= \frac{1}{2}\left[\sin 2x - \frac{1}{3}\sin^3 2x\right]\Big|_0^{\pi/2}$$

$$= \frac{1}{2}\left[\left(\sin\pi - \frac{1}{3}\sin^3\pi\right) = \left(\sin 0 - \frac{1}{3}\sin^3 0\right)\right] = \frac{1}{2}(0) = 0$$

13. $\displaystyle\int (\sin t + \cos t)^2 \cdot \sin t\,dt = \int (\sin^2 t + 2\sin t\cos t + \cos^2 t)\cdot\sin t\,dt$

$$= \int (1 + 2\sin t\cos t)\cdot\sin t\,dt$$

$$= \int (\sin t + 2\sin^2 t\cos t)dt$$

$$= \int \sin t\,dt + 2\int\sin^2 t(\cos t\,dt)$$

$$= -\cos t + \frac{2\sin^3 t}{3} + C$$

17. $\displaystyle\int\sec^4 3x\,dx = \int\sec^2 3x\sec^2 3x\,dx = \int(1+\tan^2 3x)\sec^2 3x\,dx$

$$= \frac{1}{3}\int\sec^2 3x(3\,dx) + \frac{1}{3}\int\tan^2 3x\sec^2 3x(3\,dx)$$

$$= \frac{1}{3}\tan 3x + \frac{1}{3}\frac{\tan^3 3x}{3} + C = \frac{1}{9}\tan^3 3x + \frac{1}{3}\tan 3x + C$$

21. $\displaystyle\int\frac{3x\,dx}{4+x^4} = 3\int\frac{x\,dx}{4+x^4} = 3\int\frac{1}{2^2+(x^2)^2}x\,dx = \frac{3}{2}\int\frac{1}{2^2+(x^2)^2}2x\,dx$

$$= \frac{3}{2}\left(\frac{1}{2}\tan^{-1}\frac{x^2}{2} + C_1\right) = \frac{3}{4}\tan^{-1}\frac{x^2}{2} + C \text{ where } C = \frac{3}{2}C_1.$$

25. $u = e^{2x}, du = e^{2x}(2\,dx)$

$$\int \frac{e^{2x}\,dx}{\sqrt{e^{2x}+1}} = \frac{1}{2}\int (e^{2x}+1)^{-1/2}e^{2x}(2\,dx) = \frac{1}{2}(e^{2x}+1)^{1/2}(2) + C = \sqrt{e^{2x}+1} + C$$

29. $\displaystyle\int_0^{\pi/6} 3\sin^2 3\theta\,d\theta = \int_0^{\pi/6} 3\cdot\frac{(1-\cos 6\theta)}{2}\,d\theta = \int_0^{\pi/6}\frac{3}{2}\,d\theta - \frac{1}{4}\int_0^{\pi/6}\cos 6\theta(6\,d\theta)$

$$= \frac{3}{2}\theta\Big|_0^{\pi/6} - \frac{1}{4}\sin 6\theta\Big|_0^{\pi/6} = \frac{3}{2}\left[\frac{\pi}{6}-0\right] - \frac{1}{4}[\sin\pi - \sin 0] = \frac{\pi}{4}$$

33. $\dfrac{3u^2 - 6u - 2}{u^2(3u+1)} = \dfrac{Au+B}{u^2} + \dfrac{C}{3u+1} = \dfrac{(Au+B)(3u+1) + Cu^2}{u^2(3u+1)}$

$3u^2 - 6u - 2 = 3Au^2 + Au + 3Bu + B + Cu^2$

$3u^2 - 6u - 2 = (3A+C)u^2 + (A+3B)u + B$

   (1) $3A + C = 3$, $3(0) + C = 3$, $C = 3$

   (2) $A + 3B = -6$; $A + 3(-2) = -6$, $A = 0$

   (3) $B = -2$

$$\int \frac{3u^2 - 6u - 2}{u^2(3u+1)}\,du = \int \frac{-2}{u^2}\,du + \int \frac{3}{3u+1}\,du = \frac{2}{u} + \ln|3u+1| + C$$

37. $\displaystyle\int_1^e 3\cos(\ln x)\cdot\frac{dx}{x} = 3\sin(\ln x)\Big|_1^e = 3\sin(\ln e) - 3\sin(\ln 1) = 3\sin(1) - 3\sin(0)$

$$= 3\sin 1 - 3\cdot 0 = 3\sin 1 \approx 2.52$$

41. Use the general power formula. Let $u = e^x + 1$, $du = e^x\,dx$, $n = 2$.

$$\int e^x(e^x+1)^2\,dx = \int (e^x+1)^2 e^x\,dx = \frac{(e^x+1)^3}{3} + C_1 = \frac{e^{3x} + 3e^{2x} + 3e^x + 1}{3} + C_1$$

The second method used is to multiply the factors before integrating:

$$\int e^x(e^x+1)^2\,dx = \int e^x(e^{2x} + 2e^x + 1)\,dx = \int (e^{3x} + 2e^{2x} + e^x)\,dx = \int e^{3x}\,dx + \int 2e^{2x}\,dx + \int e^x\,dx$$

$$= \frac{1}{3}\int e^{3x}(3\,dx) + 2\left(\frac{1}{2}\right)\int e^{2x}(2\,dx) + \int e^x\,dx$$

$$= \frac{1}{3}e^{3x} + e^{2x} + e^x + C_2;\ C_2 = C_1 + \frac{1}{3}$$

45. $\displaystyle\int \sec^4 x\,dx = \frac{\sec^2\tan x}{3} + \frac{2}{3}\int\sec^2 x\,dx$ Formula 37 in table of integrals.

$$y = \frac{\sec^2 x\tan x}{3} + \frac{2}{3}\tan x + C = \frac{1}{3}(1 + \tan^2 x)(\tan x) + \frac{2}{3}\tan x + C$$

$$= \frac{1}{3}\tan x + \frac{1}{3}\tan^3 x + \frac{2}{3}\tan x + C = \frac{1}{3}\tan^3 x + \frac{2}{3}\tan x + C$$

$$0 = \frac{1}{3}\tan^3(0) + \frac{2}{3}\tan 0 + C$$

$$0 = 0 + 0 + C;\ C = 0;\ y = \frac{1}{3}\tan^3 x + \tan x$$

49. $x^2 + y^2 = 5^2$; $y = \sqrt{25 - x^2}$; $A = 2\int_3^5 \sqrt{25 - x^2}\,dx$

$$A = 2\left[\frac{x}{2}\sqrt{25 - x^2} + \frac{25}{2}\sin^{-1}\frac{x}{5}\right]\Big|_3^5 \quad \text{Formula 15 in table of integrals.}$$

$$A = 2\left[\frac{5}{2}\sqrt{0} + \frac{25}{2}\sin^{-1}1\right] - 2\left[\frac{3}{2}\sqrt{16} + \frac{25}{2}\sin^{-1}\frac{3}{5}\right]$$

$$= 2\left[\frac{25}{2}\left(\frac{\pi}{2}\right)\right] - 2\left[6 + \frac{25}{2}(0.6435)\right]$$

$$= 2[19.63] - 2[14.04] = 11.18$$

53. $y = e^x\sin x$; $x = 0$; $x = \pi$

$$V = \pi\int \sin^2 x\, e^{2x}\,dx = \frac{\pi}{2}\int 2\sin^2 x\, e^{2x}\,dx = \frac{\pi}{2}\int(1 - \cos 2x)e^{2x}\,dx$$

$$= \frac{\pi}{2}\int(e^{2x} - e^{2x}\cos 2x)dx = \frac{\pi}{2}\left[\int e^{2x}\,dx - \int e^{2x}\cos 2x\,dx\right]$$

$$= \frac{\pi}{2}\left[\frac{1}{2}\int e^{2x}\,dx - \int e^{2x}\cos 2x\,dx\right] \quad \text{Formula 50 in table of integrals.}$$

$$= \frac{\pi}{2}\left[\frac{1}{2}e^{2x} - \frac{e^{2x}(2\cos 2x + 2\sin 2x)}{2^2 + 2^2}\right]\Big|_0^\pi = \frac{\pi}{2}\left[\frac{1}{2}e^{2x} - e^{2x}\frac{(\cos 2x + \sin 2x)}{4}\right]\Big|_0^\pi$$

$$= \frac{\pi}{2}\left[\frac{2e^{2x} - e^{2x}(\cos 2x + \sin 2x)}{4}\right]\Big|_0^\pi = \frac{\pi}{8}\left[2e^{2x} - e^{2x}(\cos 2x + \sin 2x)\right]\Big|_0^\pi$$

$$= \frac{\pi}{8}\left[2e^{2\pi} - e^{2\pi}(\cos 2\pi + \sin 2\pi) - 2e^0 + e^0(\cos 0 + \sin 0)\right]\Big|_0^\pi$$

$$= \frac{\pi}{8}\left[2e^{2\pi} - e^{2\pi}(1) - 2 + 1\right] = \frac{\pi}{8}[e^{2\pi} - 1] = 209.9$$

57. $\Delta s = \int(C_v/T)dT = \int\frac{a + bT + cT^2}{T}\,dT = \int\frac{a}{T}\,dT + \int b\,dT + \int cT\,dT$

$$= a\ln T + bT + \frac{1}{2}cT^2 + C$$

61.
$$y_{rms} = \sqrt{\frac{1}{T}\int i^2 dt} = \sqrt{\frac{1}{T}\int(2\sin t)^2 dt}$$

$$\int_0^{2\pi}(2\sin t)^2 dt = \int_0^{2\pi} 4\sin^2 t\,dt = 4\int_0^{2\pi}\sin^2 t\,dt$$

$$= 4\left(\frac{t}{2} - \frac{1}{2}\sin t\cos t\right)\Big|_0^{2\pi} \quad \text{Formula 29 in table of integrals.}$$

$$= 2t - 2\sin t\cos t\Big|_0^{2\pi} = 4\pi; \quad T = 2\pi$$

$$y_{rms} = \sqrt{\frac{1}{2\pi}(4\pi)} = \sqrt{2}$$

65. $V = \pi \int_{2.00}^{4.00} y^2 \, dx = \pi \int_{2.00}^{4.00} e^{-0.2x} \, dx$; and $u = -0.2x \; du = -0.2 \, dx$

$$= -\frac{\pi}{0.2} \int_{2.00}^{4.00} e^{-0.2x}(-0.2) \, dx = \frac{-\pi}{0.2} e^{-0.2x} \Big|_{2.00}^{4.00}$$

$$= -\frac{\pi}{2}[e^{-0.8} - e^{-0.4}] = 3.47 \text{ cm}^3$$

69. (a) $\int_0^{2.8} \frac{4}{1 + e^x} \, dx = \int_0^{2.8} \frac{-4e^{-x}}{e^{-x} + 1} \cdot (-e^{-x} \, dx) = -4 \ln(e^{-x} + 1) \Big|_0^{2.8} = 2.536$

  (b) let $\quad u = e^x$

  $\qquad du = e^x \, dx$

  $\qquad du = u \, dx$

  $\qquad \dfrac{1}{u} du = dx$

  $\displaystyle \int \frac{4}{1 + e^x} \, dx = \int \frac{4}{1 + u} \cdot \frac{du}{u} = \int \frac{4}{u(u + 1)} \, du$

  $\displaystyle \int \frac{4}{1 + e^x} \, dx = \int \left(\frac{4}{u} - \frac{4}{u + 1} \right) du$

  $\displaystyle \int \frac{4}{1 + e^x} \, dx = 4 \ln |u| - 4 \ln |u + 1|$

  $\displaystyle \int \frac{4}{1 + e^x} \, dx = 4 \ln \left(\frac{u}{u + 1} \right) = 4 \cdot \ln \left(\frac{e^x}{e^x + 1} \right)$

  $\displaystyle \int \frac{4}{1 + e^x} \, dx = 4 \ln \left(\frac{e^x}{e^x + 1} \right) \Big|_0^{2.8} = 2.536$

In (a) the integrand was multiplied by $\dfrac{e^{-x}}{e^{-x}}$.

In (b) a u-substitution led to an integral which was evaluated using partial fractions.

EXPANSION OF FUNCTIONS IN SERIES

29.1   Infinite Series

1. $a_n = n^2;\ n = 1, 2, 3 \ldots a_1 = (1)^1 = 1;\ a_2 = (2)^2 = 4;\ a_3 = (3)^2 = 9;\ a_4 = (4)^2 = 16$

5. $a_n = \left(-\dfrac{2}{5}\right)^n;\ n = 0, 1, 2, 3, \ldots$

(a) $a_1 = \left(-\dfrac{2}{5}\right)^1 = -\dfrac{2}{5}$ $\qquad\qquad a_3 = \left(-\dfrac{2}{5}\right)^3 = -\dfrac{8}{125}$

$\quad\ a_2 = \left(-\dfrac{2}{5}\right)^2 = \dfrac{4}{25}$ $\qquad\qquad a_4 = \left(-\dfrac{2}{5}\right)^4 = \dfrac{16}{625}$

(b) $S = -\dfrac{2}{5} + \dfrac{4}{25} - \dfrac{8}{125} + \dfrac{16}{625}$

9. $\dfrac{1}{2} + \dfrac{1}{3} + \dfrac{1}{4} + \dfrac{1}{5} + \cdots$

$n = 1,\ a_1 = \dfrac{1}{2} = \dfrac{1}{1+1};\ n = 2,\ a_2 = \dfrac{1}{3} = \dfrac{1}{2+1}$

$n^{\text{th}} \text{ term} = \dfrac{1}{n+1}$

13. $1 + \dfrac{1}{8} + \dfrac{1}{27} + \dfrac{1}{64} + \dfrac{1}{15} + \cdots$

$S_1 = 1;\ S_2 = 1 + \dfrac{1}{8} = 1.125;\ S_3 = 1 + \dfrac{1}{8} + \dfrac{1}{27} = 1.162\ 037\ 0;\ S_4 = 1 + \dfrac{1}{8} + \dfrac{1}{27} + \dfrac{1}{64} = 1.77\ 662\ 0;$

$S_5 = 1 + \dfrac{1}{8} + \dfrac{1}{27} + \dfrac{1}{64} + \dfrac{1}{125} = 1.185\ 662\ 0$

Values appear to approach 1.2. Convergent to approx. 1.2.

17. $\displaystyle\sum_{n=0}^{\infty} \sqrt{n} = \sqrt{0} + \sqrt{1} + \sqrt{2} + \sqrt{3} + \cdots$

$S_1 = 0$
$S_2 = 1$
$S_3 = 2.414213562$
$S_4 = 4.14626437$
$S_5 = 6.14626437$

Series appears divergent

21. $1 + 2 + 4 + \cdots + 2^n + \cdots;\ n = 0, 1, 2, 3, \cdots$

$S_0 = 1$
$S_1 = 3$
$S_2 = 7$
$S_3 = 15$
$\vdots$
$S_n = 2^{n+1} - 1$

$\displaystyle\lim_{n \to \infty} S_n = \lim_{n \to \infty} (2^{n+1} - 1) = \infty,\ \text{divergent}$

25. $10+9+8.1+7.29+6.561+\cdots+10(0.9)^n+\cdots$; $n=0,1,2,3,\ldots$ $S_n = 100-100(0.9)^{n+1}$; $n=0,1,2,3,\ldots$

$\displaystyle\lim_{n\to\infty} S_n = \lim_{n\to\infty} 100 - 100(0.9)^{n+1} = 100$, convergent

29. Successive square roots of 2 are approximately: 1.414 214, 1.189 207, 1.090 508, 1.044 274, 1.021 897, 1.010 889, 1.005 430, 1.002 711, 1.001 355, 1.000 677, 1.000 339, 1.000 169, 1.000 085, 1.000 042, 1.000 021, 1.000 011, 1.000 005, 1.000 003, 1.000 001, 1.000 000

(a) It appears that the value of $\displaystyle\lim_{n\to\infty} 2^{(1/2)n}$ is 1.

(b) The infinite series will diverge since each successive term will increase the sum by approximately 1.

33. $a_1 = 105, r = 1.05$; $S = \dfrac{105(1-1.05^n)}{1-1.05} = 2100(1.05^n - 1) \to \infty$, diverges

On a graphing calculator, let $y_1 = 2100(1.05^x - 1)$, $x_{\min} = 0$, $x_{\max} = 10$, $y_{\min} = 0$, $y_{\max} = 1500$.

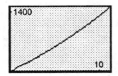

29.2   Maclaurin's Series

1. $f(x) = e^x$ $\qquad$ $f(0) = e^0 = 1$
 $f'(x) = e^x$ $\qquad$ $f'(0) = e^0 = 1$
 $f''(x) = e^x$ $\qquad$ $f''(0) = e^0 = 1$

$$f(x) = 1 + (1)x + \frac{(1)x^2}{2!} + \cdots$$
$$= 1 + x + \frac{1}{2}x^2 + \cdots$$

5. $f(x) = (1+x)^{1/2}$ $\qquad$ $f(0) = 1$

$f'(x) = \dfrac{1}{2}(1+x)^{-1/2}$ $\qquad$ $f'(0) = \dfrac{1}{2}$

$f''(x) = -\dfrac{1}{4}(1+x)^{-3/2}$ $\qquad$ $f''(0) = -\dfrac{1}{4}$

$$f(x) = 1 + \frac{1}{2}x - \frac{1}{4}\frac{x^2}{2!} + \cdots = 1 + \frac{1}{2}x - \frac{1}{8}x^2 + \cdots$$

9. $f(x) = \cos 4\pi x$ $\qquad$ $f(0) = 1$
 $f'(x) = -4\pi \sin 4\pi x$ $\qquad$ $f'(0) = 0$
 $f''(x) = -16\pi^2 \cos 4\pi x$ $\qquad$ $f''(0) = -16\pi^2$
 $f'''(x) = 64\pi^3 \sin 4\pi x$ $\qquad$ $f'''(0) = 0$
 $f^{iv}(x) = 256\pi^4 \cos 4\pi x$ $\qquad$ $f^{iv}(0) = 256\pi^4$

$$f(x) = 1 + 0 \cdot x - \frac{16\pi^2}{2!} \cdot x^2 + 0 \cdot \frac{x^3}{3!} + \frac{256\pi^4}{4!}x^4 + \cdots = f(x) = 1 - 8\pi^2 x^2 + \frac{32\pi^4}{3} \cdot x^4 - \cdots$$

13. $f(x) = \ln(1 - 2x)$ $f(0) = \ln 1 = 0$

$$f'(x) = \frac{1}{1 - 2x}(-2) = \frac{-2}{1 - 2x}$$ $f'(0) = -2$

$$f''(x) = \frac{0 - (-2)(-2)}{(1 - 2x)^2} = \frac{-4}{(1 - 2x)^2}$$ $f''(0) = -4$

$$f'''(x) = \frac{0 - (-4)2(1 - 2x)(-2)}{(1 - 2x)^4} = \frac{-16(1 - 2x)}{(1 - 2x)^4}$$ $f'''(0) = -16$

$$\ln(1 - 2x) = 0 + (-2)x + (-4)\frac{x^2}{2!} + (-16)\frac{x^3}{3!} + \cdots = -2x - 2x^2 - \frac{8}{3}x^3 - \cdots$$

17. $f(x) = \tan^{-1} x$ $f(0) = 0$

$$f'(x) = \frac{1}{1 + x^2} = (1 + x^2)^{-1}$$ $f'(0) = 1$

$$f''(x) = -(1 + x^2)^{-2}2x = -2x(1 + x^2)^{-2}$$ $f''(0) = 0$
$$f'''(x) = -2x[-2(1 + x^2)^{-3}(2x)] + (1 + x^2)^{-2}(-2)$$ $f'''(0) = -2$

$$f(x) = 0 + 1x + \frac{0x^2}{2!} - \frac{2x^3}{3!} + \cdots = x - \frac{1}{3}x^3 + \cdots$$

21. $f(x) = \ln \cos x$ $f(0) = \ln 1 = 0$

$$f'(x) = -\frac{1}{\cos x}\sin x = -\tan x$$ $f'(0) = 0$

$$f''(x) = -\sec^2 x$$ $f''(0) = -1$
$$f'''(x) = -2\sec x \sec x \tan x = -2\sec^2 x \tan x$$ $f'''(0) = 0$
$$f^{iv}(x) = -2\sec^2 x \sec^2 x - 2\tan x(2\sec x \sec x \tan x)$$ $f^{iv}(0) = -2 - 0 = -2$

$$f(x) = 0 + 0x - \frac{1x^2}{2!} + \frac{0x^3}{3!} - \frac{2x^4}{4!} + \cdots = -\frac{1}{2}x^2 - \frac{1}{12}x^4 - \cdots$$

25. (a) It is not possible to find a Maclaurin's expansion for $f(x) = \csc x$ since the formula is not defined when $x = 0$.

(b) $f(x) = \ln x$ is not defined when $x = 0$.

29. The Maclaurin's expansion of $f(x) = e^{3x}$ is $f(x) = 1 + 3x + \frac{9}{2}x^2 + \frac{9}{2}x^3 \cdots$. The linearization is $L(x) = 1 + 3x$, the first two terms of the expansion.

29.3 Certain Operations with Series

1. $f(x) = e^{3x}; \quad e^x = 1 + x + \dfrac{x^2}{2!} + \dfrac{x^3}{3!} + \cdots$

$$f(x) = e^{3x} = 1 + 3x + \dfrac{(3x)^2}{2!} + \dfrac{(3x)^3}{3!} + \cdots = 1 + 3x + \dfrac{9}{2}x^2 + \dfrac{9}{2}x^3 + \cdots$$

5. $f(x) = 1 - \dfrac{x^2}{2!} + \dfrac{x^4}{4!} - \dfrac{x^6}{6!} + \cdots$

$$\cos 4x = f(4x) = 1 - \dfrac{(4x)^2}{2!} + \dfrac{(4x)^4}{4!} - \dfrac{4x)^6}{6!} + \cdots = 1 - 8x^2 + \dfrac{32}{3}x^4 - \dfrac{256}{45}x^6 + \cdots$$

$x \cos 4x = x f(4x)$

$$= x\left(1 - 8x^2 + \dfrac{32}{3}x^4 - \dfrac{256}{45}x^6 + \cdots\right) = x - 8x^3 + \dfrac{32}{3}x^5 - \dfrac{256}{45}x^7 + \cdots$$

9. $\displaystyle\int_0^1 \sin x^2 \, dx = \int_0^1 \left(x^2 - \dfrac{(x^2)^3}{3!} + \dfrac{(x^2)^5}{5!}\right) dx = -\int_0^1 \left(x^2 - \dfrac{x^6}{6} + \dfrac{x^{10}}{120}\right) dx = \left(\dfrac{1}{3}x^3 - \dfrac{1}{42}x^7 + \dfrac{1}{1320}x^{11}\right)\Big|_0^1$

$$= \dfrac{1}{3} - \dfrac{1}{42} + \dfrac{1}{1320} = 0.3103$$

13. $f(x) = \dfrac{2}{1 - x^2} = \dfrac{1}{1 + x} + \dfrac{1}{1 - x}$

$f(x) = 1 - x + x^2 - x^3 + \cdots + 1 + x + x^2 + x^3 + \cdots$

$f(x) = 1 - x + x^2 - x^3 + x^4 - x^5 + x^6 - x^7 + x^8 + \cdots + 1 + x + x^2 + x^3 + x^4 + x^5 + x^6 + x^7 + x^8 + \cdots$

$f(x) = 2(1 + x^2 + x^4 + x^6 + \cdots)$

17. $\dfrac{d(\sin x)}{dx} = \dfrac{d}{dx}\left(x - \dfrac{x^3}{3!} + \dfrac{x^5}{5!} + \cdots\right) = 1 - \dfrac{3x^2}{3!} + \dfrac{5x^4}{4!} + \cdots = 1 - \dfrac{x^2}{2!} + \dfrac{x^4}{4!} + \cdots = \cos x$

21. $\displaystyle\int_0^1 e^x \, dx = e^x \Big|_0^1 = e - e^0 = e - 1 = 2.718 - 1 = 1.718$

$$\begin{aligned}
f(x) &= e^x & f(0) &= e^0 = 1 \\
f'(x) &= e^x & f'(0) &= 1 \\
f''(x) &= e^x & f''(0) &= 1 \\
f'''(x) &= e^x & f'''(0) &= 1
\end{aligned}$$

$$e^x = 1 + x + \dfrac{x^2}{2!} + \dfrac{x^3}{3!} + \cdots$$

$$\int_0^1 \left(1 + x + \dfrac{x^2}{2} + \dfrac{x^3}{6}\right) dx = x + \dfrac{x^2}{2} + \dfrac{x^3}{6} + \dfrac{x^4}{24}\Big|_0^1 = 1 + \dfrac{1}{2} + \dfrac{1}{6} + \dfrac{1}{24} = 1.708\ 333\ 3$$

25.
$$\int_0^x \cos t^2\, dt = \int_0^{0.2} \left[1 - \frac{(t^2)^2}{2!}\right] dt$$
$$= \int_0^{0.2} \left(1 - \frac{t^4}{2}\right) dt$$

$$\left(t - \frac{t^5}{10}\right)\Bigg|_0^{0.2} = 0.2 - 0.000\,032$$
$$= 1.999\,968$$

29. $y_1 = e^x,\ y_2 = 1,\ y_3 = 1 + x,\ y_4 = 1 + x + \frac{1}{2}x^2;$
$x_{\min} = -5,\ x_{\max} = 5,\ y_{\min} = -1,\ y_{\max} = 3$

29.4   Computations by Use of Series Expansions

1. $e^x = 1 + x + \frac{x^2}{2!} + \cdots$

$e^{0.2} = 1 + 0.2 + \frac{(0.2)^2}{2!} = 1.22$

(1.221 402 8 calculator)

5. $e^x = 1 + x + \frac{x}{2!} + \frac{x}{3!} + \frac{x}{4!} + \frac{x}{5!} + \frac{x}{6!} + \cdots$

$e^1 = 1 + 1 + \frac{1}{2} + \frac{1}{6} + \frac{1}{24} + \frac{1}{120} + \frac{1}{720} + \cdots$

$= 2.718\,055\,6;\ (2.718\,281\,8\ \text{calculator})$

9. $\ln(1+x) = x - \frac{x^2}{2} + \frac{x^3}{3} - \frac{x^4}{4} + \cdots$

$\ln(1 + 0.4) = 0.4 - \frac{(0.4)^2}{2} + \frac{(0.4)^3}{3} - \frac{(0.4)^4}{4} + \cdots$
$$= 0.334\,933\,3;\ (0.336\,472\,2\ \text{calculator})$$

13. $\ln 0.9861$

$\ln(1+x) = x - \frac{x^2}{2} + \frac{x^3}{3} - \frac{x^4}{4} + \cdots$

Let $x = -0.0139$

$\ln[1 + (-0.0139)] = -0.0139 - \frac{(-0.0139)^2}{2} + \frac{(-0.0139)^3}{3} + \cdots$
$$= -0.0139 - 0.000\,096\,6 - 0.000\,000\,9 + \cdots$$
$$= -0.013\,997\,5;\ (-0.013\,997\,5\ \text{calculator})$$

17. $\sqrt{1.1076} = 1.1076^{1/2} = (1 + 0.1076)^{1/2}$

$(1+x)^n = 1 + nx + \frac{n(n-1)x^2}{2!} + \cdots$

$x = 0.1076$ and $n = \frac{1}{2}$

$\sqrt{1.1076} = 1 + \frac{1}{2}(0.1076) + \frac{\frac{1}{2}\left(-\frac{1}{2}\right)(0.1076)^2}{2} + \cdots$
$$= 1 + 0.053\,800\,0 - 0.001\,447\,2 + \cdots$$
$$= 1.052\,352\,8$$

21. From Exercise 3, $\sin(0.1) = 0.1 + \frac{0.1^3}{6} = 0.100\ 166\ 7$

The maximum possible error is the value of the first term omitted, $\dfrac{x^5}{5!} = \left|\dfrac{0.1^5}{120}\right| = 8.3 \times 10^{-8}$

25. $\tan^{-1}\dfrac{1}{2} = \dfrac{1}{2} - \dfrac{1}{3}\left(\dfrac{1}{2}\right)^3 + \dfrac{1}{5}\left(\dfrac{1}{2}\right)^5 = 0.4646$

$\tan^{-1}\dfrac{1}{3} = \dfrac{1}{3} - \dfrac{1}{3}\left(\dfrac{1}{3}\right)^3 + \dfrac{1}{5}\left(\dfrac{1}{3}\right)^5 = 0.3218$

$\pi = 4(0.4646 + 0.3218) = 3.146$

29. $f(t) = \dfrac{E}{R}(1 - e^{-Rt/L}); \quad e^x = 1 + x + \dfrac{x^2}{2} + \cdots$

$e^{-Rt/L} = 1 - \dfrac{Rt}{L} + \dfrac{R^2 t^2}{2L^2} + \cdots$

$i = \dfrac{E}{R}\left[1 - \left(1 - \dfrac{Rt}{L} + \dfrac{R^2 t^2}{2L^2}\right)\right] = \dfrac{E}{L}\left(t - \dfrac{Rt^2}{2L}\right)$

The approximation will be valid for small values of t.

29.5   Taylor Series

1. $e^x = e\left[1 + (x-1) + \dfrac{(x-1)^2}{2} + \dfrac{(x-1)^3}{6} + \cdots\right]$

Let $x = 1.2$; $e^{1.2} = 2.7183\left[1 + (1.2-1) + \dfrac{(1.2-1)^2}{2} + \dfrac{(1.2-1)^3}{6} + \cdots\right] = 2.7183(1.2227) = 3.32$

5. Let $a = 30° = \dfrac{\pi}{6}$; from Example 4,

$\sin x = \dfrac{1}{2} + \dfrac{\sqrt{3}}{2}\left(x - \dfrac{\pi}{6}\right) - \dfrac{1}{4}\left(x - \dfrac{\pi}{6}\right)^2 - \cdots$

$31° = \dfrac{\pi}{6} + \dfrac{\pi}{180}$

$\sin 31° = \dfrac{1}{2} + \dfrac{\sqrt{3}}{2}\left(\dfrac{\pi}{6} + \dfrac{\pi}{180} - \dfrac{\pi}{6}\right) - \dfrac{1}{4}\left(\dfrac{\pi}{6} + \dfrac{\pi}{180} - \dfrac{\pi}{6}\right)^2 = \dfrac{1}{2} + \dfrac{\sqrt{3}}{2}\left(\dfrac{\pi}{180}\right) - \dfrac{1}{4}\left(\dfrac{\pi}{180}\right)^2 = 0.5150$

9. $f(x) = e^{-x}$ $\qquad\qquad\qquad f(2) = e^{-2}$
$f'(x) = -e^{-x}$ $\qquad\qquad\quad f'(2) = -e^{-2}$
$f''(x) = e^{-x}$ $\qquad\qquad\quad f''(2) = e^{-2}$

$f(x) = e^{-2} - e^{-2}(x-2) + \dfrac{e^{-2}(x-2)^2}{2!} + \cdots = e^{-2}\left[1 - (x-2) + \dfrac{(x-2)^2}{2!} - \cdots\right]$

13. $f(x) = x^{1/3}$ $f(8) = 8^{1/3} = 2$

$f'(x) = \dfrac{1}{3}x^{-2/3}$ $f'(8) = \dfrac{1}{3(8)^{2/3}} = \dfrac{1}{12}$

$f''(x) = -\dfrac{2}{9}x^{-5/3}$ $f''(8) = \dfrac{-2}{9(8^{5/3})} = -\dfrac{1}{144}$

$f(x) = 2 + \dfrac{1}{12}(x - 8) - \dfrac{1}{288}(x - 8)^2 + \cdots$

17. $e^x = e^3\left[1 + (x - 3) + \dfrac{(x-3)^2}{2}\right]$

$e^\pi = e^3\left[1 + (\pi - 3) + \dfrac{(\pi-3)^2}{2}\right]$

$e^\pi = 23.13084363$ as compared with $e^\pi = 23.14069263$ on a calculator.

21. $\sqrt[3]{8.3};\ f(x) = x^{1/3}$

$f'(x) = \dfrac{1}{3}x^{-2/3}$ $f(8) = 2$

$f''(x) = -\dfrac{2}{9}x^{-5/3}$ $f'(8) = \dfrac{1}{3\sqrt[3]{8^2}} = \dfrac{1}{12} = 0.083\,333$

$f''(8) = -\dfrac{2}{9\sqrt[3]{8^5}} = -\dfrac{2}{9(32)} = -0.006\,944\,4$

$a = 8$

$\sqrt[3]{8.3} = 2 + 0.083333(8.3 - 8) - 0.0069444(8.3 - 8)^2 = 2.02437$

25. $f(x) = c_0 + c_1(x - a) + c_2(x - a)^2 + c_3(x - a)^3 + c_4(x - a)^4 + c_5(x - a)^5 + \cdots + c_n(x - a)^n$
$f'(x) = c_1 + 2c_2(x - a) + 3c_3(x - a)^2 + 4c_4(x - a)^3 + 5c_5(x - a)^4 + \cdots + nc_n(x - a)^{n-1}$
$f''(x) = 2c_2 + 2 \times 3c_3(x - a) + 3 \times 4c_4(x - a)^2 + 4 \times 5c_5(x - a)^3 + \cdots + (n - 1)nc_n(x - a)^{n-2}$
$f'''(x) = 2 \times 3c_3 + 2 \times 3 \times 4c_4(x - a) + 3 \times 4 \times 5c_5(x - a)^2 + \cdots + (n - 2)(n - 1)nc_n(x - a)^{n-3}$
$f^{iv}(x) = 2 \times 3 \times 4c_4 + 2 \times 3 \times 4 \times 5c_5(x - a) + \cdots + (n - 3)(n - 2)(n - 1)nc_n(x - a)^{n-4}$

Let $x = a$; $f(a) = c_0$; $f'(a) = c_1$; $f''(a) = 2c_2$; $c_2 = \dfrac{f''(a)}{2!}$

$f'''(a) = 2 \times 3c_3$; $c_3 = \dfrac{f'''(a)}{3!}$

$f^{iv}(a) = 2 \times 3 \times 4c_4$; $c_4 = \dfrac{f^{iv}(a)}{4!}$

$f(x) = f(a) + f'(a)(x - a) + \dfrac{f''(a)(x - a)^2}{2!} + \dfrac{f'''(a)(x - a)^3}{3!} + \dfrac{f^{iv}(a)(x - a)^4}{4!} + \cdots$

29. $f(x) = \sin x$; $x = 0$ to $x = 2$

    (a) $y_1 = \sin x$

    (b) $y_2 = \dfrac{\sqrt{3}}{2} + \dfrac{1}{2}\left(x - \dfrac{\pi}{3}\right)$

        $x_{\min} = 0$, $x_{\max} = 2$, $y_{\min} = 0$, $y_{\max} = 1.2$

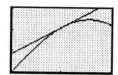

The series gives a good approximation near $x = \frac{\pi}{3}$ and deteriorates as x moves away from $x = \frac{\pi}{3}$.

29.6 Introduction to Fourier Series

1. $a_0 = \dfrac{1}{2\pi}\displaystyle\int_{-\pi}^{0} 1\,dx + \dfrac{1}{2\pi}\displaystyle\int_{0}^{\pi} 0\,dx = \dfrac{1}{2\pi}x\bigg|_{-\pi}^{0} = 0 - \dfrac{1}{2\pi}(-\pi) = \dfrac{1}{2}$

    $a_1 = \dfrac{1}{\pi}\displaystyle\int_{-\pi}^{0} (1)\cos x\,dx + \dfrac{1}{\pi}\displaystyle\int_{0}^{\pi} 0\,dx = \dfrac{1}{\pi}\sin x\bigg|_{-\pi}^{0} = 0$

    $a_n = 0$ for all values of n since $\sin n\pi = 0$

    $b_1 = \dfrac{1}{\pi}\displaystyle\int_{-\pi}^{0} 1\sin x\,dx + \displaystyle\int_{0}^{\pi} 0\,dx = \dfrac{1}{\pi}(-\cos x)\bigg|_{-\pi}^{0} = \dfrac{1}{\pi}(-1-1) = -\dfrac{2}{\pi}$

    $b_2 = \dfrac{1}{\pi}\displaystyle\int_{-\pi}^{0} 1\sin 2x\,dx + \displaystyle\int_{0}^{\pi} 0\,dx = \dfrac{1}{2\pi}\displaystyle\int_{-\pi}^{0} \sin 2x(2\,dx) = \dfrac{1}{2\pi}\displaystyle\int_{-\pi}^{0} \sin 2x(2\,dx) = \dfrac{1}{2\pi}(-\cos 2x)\bigg|_{-\pi}^{0}$

        $= \dfrac{1}{2\pi}(-1+1) = 0$

    $b_3 = \dfrac{1}{\pi}\displaystyle\int_{-\pi}^{0} 1\sin 3x\,dx + \displaystyle\int_{0}^{\pi} 0\,dx = \dfrac{1}{3\pi}\displaystyle\int_{-\pi}^{0} \sin 3x(3\,dx) = \dfrac{1}{3\pi}(-\cos 3x)\bigg|_{-\pi}^{0} = \dfrac{1}{3\pi}(-1-1) = \dfrac{-2}{3\pi}$

$f(x) = \dfrac{1}{2} - \dfrac{2}{\pi}\sin x - \dfrac{2}{3\pi}\sin 3x - \cdots$

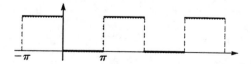

5. $a_0 = \dfrac{1}{2\pi} \displaystyle\int_{-\pi}^{0} 0 \, dx + \dfrac{1}{2\pi} \int_{0}^{\pi} x \, dx = \dfrac{1}{2\pi} \left(\dfrac{x^2}{2} \right)\Big|_{0}^{\pi} = \dfrac{\pi}{4};$ Eq. (29-25)

$a_1 = \dfrac{1}{\pi} \displaystyle\int_{-\pi}^{0} 0 \cos x \, dx + \dfrac{1}{\pi} \int_{0}^{\pi} x \cos x \, dx = \dfrac{1}{\pi}(\cos x + x \sin x)\Big|_{0}^{\pi} = \dfrac{1}{\pi}(-1 - 1) = -\dfrac{2}{\pi}$

$a_2 = \dfrac{1}{\pi} \displaystyle\int_{-\pi}^{0} 0 \, dx + \dfrac{1}{\pi} \int_{0}^{\pi} x \cos 2x \, dx = \dfrac{1}{4\pi} \int_{0}^{\pi} 2x \cos 2x (2 \, dx) = \dfrac{1}{4\pi}(\cos 2x + 2x \sin 2x)\Big|_{0}^{\pi}$

$\quad = \dfrac{1}{4\pi}(\cos 2\pi - \cos 0) = 0$

$a_3 = \dfrac{1}{\pi} \displaystyle\int_{-\pi}^{0} 0 \, dx + \dfrac{1}{\pi} \int_{0}^{\pi} x \cos 3x \, dx = \dfrac{1}{9\pi} \int_{0}^{\pi} 3x \cos 3x (3 \, dx) = \dfrac{1}{9\pi}(\cos 3x + 3x \sin 3x)\Big|_{0}^{\pi}$

$\quad = \dfrac{1}{9\pi}[(\cos 3\pi + 3\pi \sin 3\pi) - (\cos 0 + 0)] = \dfrac{1}{9\pi}(-1 - 1) = -\dfrac{2}{9\pi}$

$b_1 = \dfrac{1}{\pi} \displaystyle\int_{-\pi}^{0} 0 \, dx + \dfrac{1}{\pi} \int_{0}^{\pi} x \sin x \, dx = \dfrac{1}{\pi}(\sin x - x \cos x)\Big|_{0}^{\pi} = \dfrac{1}{\pi}[(\sin \pi - \pi \cos \pi) - (\sin 0 - 0)]$

$\quad = \dfrac{1}{\pi}(-\pi)(-1) = 1$

$b_2 = \dfrac{1}{\pi} \displaystyle\int_{-\pi}^{0} 0 \, dx + \dfrac{1}{\pi} \int_{0}^{\pi} x \sin 2x \, dx = \dfrac{1}{4\pi} \int_{0}^{\pi} 2x \sin 2x (2 \, dx) = \dfrac{1}{4\pi}(\sin 2x - 2x \cos 2x)\Big|_{0}^{\pi}$

$\quad = -\dfrac{1}{4\pi}[(\sin 2\pi - 2\pi \cos 2\pi) - (\sin 0 - 0)] = \dfrac{1}{4\pi}(0 - 2\pi - 0) = -\dfrac{1}{2}$

$f(x) = \dfrac{\pi}{4} - \dfrac{2}{\pi} \cos x + 0 \cos 2x - \dfrac{2}{9\pi} \cos 3x + \cdots + \sin x - \dfrac{1}{2} \sin 2x + \cdots$

$\quad = \dfrac{\pi}{4} - \dfrac{2}{\pi} \left(\cos x + \dfrac{1}{9} \cos 3x + \cdots \right) + \left(\sin x - \dfrac{1}{2} \sin 2x + \cdots \right)$

9. $a_0 = \dfrac{1}{2\pi} \displaystyle\int_{-\pi}^{0} -x \, dx + \dfrac{1}{2\pi} \int_{0}^{\pi} x \, dx = -\dfrac{1}{2\pi} \left(\dfrac{x^2}{2} \right)\Big|_{0}^{\pi} + \dfrac{1}{2\pi} \left(\dfrac{x^2}{2} \right)\Big|_{0}^{\pi} = 0 - \left(-\dfrac{\pi}{4} \right) + \dfrac{\pi^2}{4\pi} - 0$

$\quad = \dfrac{\pi}{4} + \dfrac{\pi}{4} = \dfrac{\pi}{2}$

$a_1 = \dfrac{1}{\pi} \displaystyle\int_{-\pi}^{0} -x \cos x \, dx + \dfrac{1}{\pi} \int_{0}^{\pi} x \cos x \, dx = -\dfrac{1}{\pi}(\cos x + x \sin x)\Big|_{-\pi}^{0} + \dfrac{1}{\pi}(\cos x + x \sin x)\Big|_{0}^{\pi}$

$\quad = -\dfrac{1}{\pi}[\cos 0 + 0 \sin 0 - \cos(-\pi) + \pi \sin(-\pi)] + \dfrac{1}{\pi}(\cos \pi + \pi \sin \pi - \cos 0 - 0 \sin 0) = -\dfrac{4}{\pi}$

$a_2 = \dfrac{1}{\pi} \displaystyle\int_{-\pi}^{0} -x \cos 2x \, dx + \dfrac{1}{\pi} \int_{0}^{\pi} x \cos 2x \, dx = -\dfrac{1}{4\pi} \int_{-\pi}^{0} 2x \cos 2x (2 \, dx) + \dfrac{1}{4\pi} \int_{0}^{\pi} 2x \cos 2x (2 \, dx)$

$\quad = -\dfrac{1}{4\pi}(\cos 2x + 2x \sin 2x)\Big|_{-\pi}^{0} + \dfrac{1}{4\pi}(\cos 2x + 2x \sin 2x)\Big|_{0}^{\pi} = 0$

$$a_3 = \frac{1}{\pi} \int_{-\pi}^{0} -x \cos 3x \, dx + \frac{1}{\pi} \int_{0}^{\pi} x \cos 3x \, dx = -\frac{1}{9\pi} \int_{-\pi}^{0} 3x \cos 3x (3 \, dx) + \frac{1}{9\pi} \int_{0}^{\pi} 3x \cos 3x (3 \, dx)$$

$$= -\frac{1}{9\pi} (\cos 3x + 3x \sin 3x) \Big|_{-\pi}^{0} + \frac{1}{9\pi} (\cos 3x + 3x \sin 3x) \Big|_{0}^{\pi} = -\frac{4}{9\pi}$$

$$b_1 = \frac{1}{\pi} \int_{-\pi}^{0} -x \sin x \, dx + \frac{1}{\pi} \int_{0}^{\pi} x \sin x \, dx = -\frac{1}{\pi} (\sin x - x \cos x) \Big|_{-\pi}^{0} + \frac{1}{\pi} (\sin x - x \cos x) \Big|_{0}^{\pi} = -1 + 1 = 0$$

$$b_2 = \frac{1}{\pi} \int_{-\pi}^{0} -x \sin 2x \, dx + \frac{1}{\pi} \int_{0}^{\pi} x \sin 2x \, dx = -\frac{1}{4\pi} \int_{-\pi}^{0} 2x \sin 2x (2 \, dx) + \frac{1}{4\pi} \int_{0}^{\pi} 2x \sin 2x (2dx)$$

$$= -\frac{1}{4\pi} (\sin 2x - 2x \cos 2x) \Big|_{-\pi}^{0} + \frac{1}{4\pi} (\sin 2x - 2x \cos 2x) \Big|_{0}^{\pi} = \frac{1}{2} - \frac{1}{2} = 0$$

$b_n = 0$ for all values of n.

$$f(x) = \frac{\pi}{2} - \frac{4}{\pi} \cos x + 0 \cos 2x - \frac{4}{9\pi} \cos 3x + \cdots$$

$$= \frac{\pi}{2} - \frac{4}{\pi} \cos x - \frac{4}{9\pi} \cos 3x - \cdots$$

13. $y_1 = \frac{\pi}{4} - \left(\frac{2}{\pi}\right)(\cos x + 9^{-1} \cos 3x) + \sin x - 0.5 \sin 2x$

$$x_{\min} = -8, = x_{\max} = 8$$
$$y_{\min} = -1, = y_{\max} = 3$$

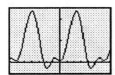

29.7   More On Fourier Series

1. $f(x) = \begin{cases} 5 & -3 \le x < 0 \\ 0 & 0 \le x < 3 \end{cases}$

from the graph $f(x)$ is
neither odd nor even.

5. $f(x) = |x|$ $-4 \le x < 4$

from the graph $f(x)$
is even.

9. $f(x) \begin{cases} 2 & -\pi \le x < -\dfrac{\pi}{2}, \dfrac{\pi}{2} \le x < \pi \\ 3 & -\dfrac{\pi}{2} \le x < \dfrac{\pi}{2} \end{cases}$

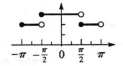

From Example 2, $f(x) = \dfrac{1}{2} + \dfrac{2}{\pi}\left(\cos x - \dfrac{\cos 3x}{3} + \dfrac{\cos 5x}{5} - \cdots\right) + 2.$

13. $f(x) \begin{cases} 5 & -3 \le x < 0 \\ 0 & 0 \le x < 3 \end{cases}$

period $= 6 = 2L$, $L = 3$

$a_0 = \dfrac{1}{2L}\int_{-L}^{L} f(x)\,dx = \dfrac{1}{6}\int_{-3}^{0} 5\,dx + \dfrac{1}{6}\int_{0}^{3} 0\cdot dx = \dfrac{5}{2}$

$a_n = \dfrac{1}{L}\int_{-L}^{L} f(x)\cos\dfrac{n\pi x}{L}\,dx = \dfrac{1}{3}\int_{-3}^{0} 5\cos\dfrac{n\pi x}{3}\,dx + \dfrac{1}{3}\int_{0}^{3} 0\cdot\cos\dfrac{n\pi x}{3}\,dx$

$a_n = \dfrac{5\sin(n\pi)}{n\pi} = 0, n = 1,2,3\cdots$

$b_n = \dfrac{1}{L}\int_{-L}^{L} f(x)\sin\dfrac{n\pi x}{L}\,dx = \dfrac{1}{3}\int_{-3}^{0} 5\sin\dfrac{n\pi x}{3}\,dx + \dfrac{1}{3}\int_{0}^{3} 0\cdot\sin\dfrac{2\pi x}{3}\,dx$

$b_n = \dfrac{5\cos(n\pi) - 5}{n\pi} = \dfrac{5}{\pi}\left(\dfrac{\cos(n\pi) - 1}{n}\right)$

n	b_n
1	$\dfrac{5}{\pi}\cdot(-2) = \dfrac{-10}{\pi}$
2	0
3	$\dfrac{5}{\pi}\left(-\dfrac{2}{3}\right) = \dfrac{-10}{3\pi}$
4	0
5	$\dfrac{5}{\pi}\left(-\dfrac{2}{5}\right) = \dfrac{-10}{5\pi}$

$f(x) = a_0 + a_1\cos\dfrac{\pi x}{L} + a_2\cos\dfrac{2\pi x}{L} + a_3\cos\dfrac{3\pi x}{L} + \cdots + b_1,\sin\dfrac{\pi x}{L} + b_2\dfrac{2\pi x}{L} + b_3\dfrac{3\pi x}{L} + \cdots$

$f(x) = \dfrac{5}{2} - \dfrac{10}{\pi}\left(\sin\dfrac{\pi x}{3} + \dfrac{1}{3}\sin\dfrac{3\pi x}{3} + \dfrac{1}{5}\sin\dfrac{5\pi x}{3} + \cdots\right)$

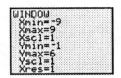  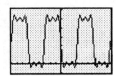

17. $f(x) = \begin{cases} -x & -4 \le x < 0 \\ x & 0 \le x < 4 \end{cases}$

$$a_0 = \frac{1}{8}\int_{-4}^{0} -x\,dx + \frac{1}{8}\int_{0}^{4} x\,dx = -\frac{1}{16}x^2\Big|_{-4}^{0} + \frac{1}{16}x^2\Big|_{0}^{4} = 2$$

$$a_n = \frac{1}{4}\int_{-4}^{0} -x\cos\frac{n\pi x}{4}\,dx + \frac{1}{4}\int x\cos\frac{n\pi x}{4}\,dx$$

$$= -\frac{1}{4}\frac{16}{(n\pi)^2}\int_{-4}^{0}\frac{n\pi}{4}x\cos\frac{n\pi}{4}x\frac{n\pi}{4}\,dx + \frac{1}{4}\frac{16}{(n\pi)^2}\int_{0}^{4}\frac{n\pi}{4}x\cos\frac{n\pi}{4}x\frac{n\pi}{4}\,dx$$

$$= -\frac{4}{(n\pi)^2}\left(\cos\frac{n\pi x}{4} + \frac{n\pi x}{4}\sin\frac{n\pi x}{4}\right)\Big|_{-4}^{0} + \frac{4}{(n\pi)^2}\left(\cos\frac{n\pi x}{4} + \frac{n\pi x}{4}\sin\frac{n\pi x}{4}\right)\Big|_{0}^{4}$$

$$= -\frac{4}{(n\pi)^2}(\cos 0 - [\cos(-n\pi) + n\pi\sin n\pi]) + \frac{4}{(n\pi)^2}(\cos n\pi + n\pi\sin n\pi - [\cos 0])$$

$$= -\frac{4}{(n\pi)^2}(1 - \cos n\pi - n\pi\sin n\pi) + \frac{4}{(n\pi)^2}(\cos n\pi + n\pi\sin n\pi - 1)$$

$$= -\frac{4}{(n\pi)^2}(1 - \cos n\pi - n\pi\sin n\pi - \cos n\pi - n\pi\sin n\pi + 1) = -\frac{4}{(n\pi)^2}(2 - 2\cos n\pi - 2n\pi\sin n\pi)$$

$$a_1 = -\frac{16}{\pi^2};\ a_2 = 0;\ a_3 = -\frac{16}{9\pi^2}$$

$$b_n = \frac{1}{4}\int_{-4}^{0} -x\sin\frac{n\pi x}{4}\,dx + \frac{1}{4}\int_{0}^{4} x\sin\frac{n\pi x}{4}\,dx$$

$$= -\frac{1}{4}\frac{16}{(n\pi)^2}\int_{-4}^{0}\frac{n\pi x}{4}\sin\frac{n\pi x}{4}\left(\frac{n\pi}{4}\,dx\right) + \frac{1}{4}\frac{16}{(n\pi)^2}\int_{0}^{4}\frac{n\pi x}{4}\sin\frac{n\pi x}{4}??\frac{n\pi\,dx}{4}$$

$$= -\frac{4}{(n\pi)^2}\left(\sin\frac{n\pi x}{4} - \frac{n\pi x}{4}\cos\frac{n\pi x}{4}\right)\Big|_{-4}^{0} + \frac{4}{(n\pi)^2}\left(\sin\frac{n\pi x}{4} - \frac{n\pi x}{4}\cos\frac{n\pi x}{4}\right)\Big|_{0}^{4}$$

$$= -\frac{4}{(n\pi)^2}\{-[\sin(-n\pi) + n\pi\cos(-n\pi)]\} + \frac{4}{(n\pi)^2}(\sin n\pi - n\pi\cos n\pi)$$

$$= -\frac{4}{(n\pi)^2}(\sin n\pi - n\pi\cos n\pi) + \frac{4}{(n\pi)^2}(\sin n\pi - n\pi\cos n\pi) = 0,\ \text{for all } n.$$

Therefore, $f(x) = 2 - \frac{16}{\pi^2}\cos\frac{\pi x}{4} - \frac{16}{9\pi^2}\cos\frac{3\pi x}{4} = 2 - \frac{16}{\pi^2}\left(\cos\frac{\pi x}{4} + \frac{1}{9}\cos\frac{3\pi x}{4} + \cdots\right)$

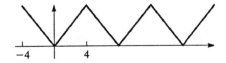

and comparing with calcuator,

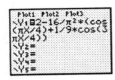 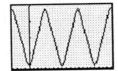

Chapter 29 Review Exercises

1. $f(x) = \dfrac{1}{1 + e^x} = (1 + e^x)^{-1}$

$f'(x) = -(1 + e^x)^{-2}(e^x) = e^x(1 + e^x)^{-2}$
$f''(x) = -(1 + e^x)^{-2}e^x + e^x(2)(1 + e^x)^{-3}(e^x)$
$f'''(x) = -(1 + e^x)^{-2}e^x + e^x(2)(1 + e^x)^{-3}(e^x) + 2e^{2x}(-3)(1 + e^x)^{-4}e^x + (1 + e^x)^{-3}(2e^{2x})(2)$

$f(0) = \dfrac{1}{1 + 1} = \dfrac{1}{2}$

$f'(0) = -1(2^{-2}) = -\dfrac{1}{4}$

$f''(0) = -\dfrac{1}{4} + \dfrac{1}{4} = 0$

$f'''(0) = -\dfrac{1}{4} + \dfrac{1}{4} - \dfrac{3}{8} + \dfrac{1}{2} = \dfrac{1}{8}$

$f(x) = \dfrac{1}{2} - \dfrac{1}{4}x + \dfrac{0x^2}{2!} + \left(\dfrac{1}{8}\right)\dfrac{x^3}{3!} + \cdots$

$\qquad = \dfrac{1}{2} - \dfrac{1}{4}x + \dfrac{1}{48}x^3 - \cdots$

5. $f(x) = (x + 1)^{1/3}$ $f(0) = 1$ $f(x) = 1 + \dfrac{1}{3}x - \dfrac{2x^2}{9(2)} + \cdots$

$f'(x) = \dfrac{1}{3}(x + 1)^{-2/3}$ $f'(0) = \dfrac{1}{3}$ $\qquad = 1 + \dfrac{1}{3}x - \dfrac{1}{9}x^2 + \cdots$

$f''(x) = -\dfrac{2}{9}(x + 1)^{-5/3}$ $f''(0) = -\dfrac{2}{9}$

9. In $e^x = 1 + x + \dfrac{x^2}{2} + \cdots$ Let $x = -0.2$

$e^{-0.2} = 1 - 0.2 + \dfrac{(0.2)^2}{2} + \cdots = 0.82$

13. Taylor Series: $f(x) = f(a) + f'(a)(x-a) + \dfrac{f''(a)(x-a)^2}{2!} + \cdots$

Let $f(x) = \dfrac{1}{x}$

$\qquad f'(x) = -\dfrac{1}{x^2}$

$\qquad f''(x) = \dfrac{2}{x^3}$

and $a = 1$, then

$$f(x) = 1 - (x-1) + 2 \cdot \dfrac{(x-1)^2}{2!} + \cdots \text{ from which}$$

$$f(1.086) = \dfrac{1}{1.086} = 1 - (1.086 - 1) + (1.086 - 1)^2$$

$$1.986^{-1} = 0.9214$$

17. $f(x) = \tan x; \; a = \dfrac{\pi}{4}$ $\qquad\qquad\qquad f\left(\dfrac{\pi}{4}\right) = 1$

$\qquad f'(x) = \sec^2 x = 1 + \tan^2 x \qquad\qquad f'\left(\dfrac{\pi}{4}\right) = 2$

$\qquad f''(x) = 2 \tan x \sec^2 x \qquad\qquad\quad f''\left(\dfrac{\pi}{4}\right) = 4$

$$f(x) = 1 + 2\left(x - \dfrac{\pi}{4}\right) + \dfrac{4\left(x - \dfrac{\pi}{4}\right)^2}{2!} + \cdots$$

$$\tan 43.62° = \tan(45° - 1.38°) = \tan\left(\dfrac{\pi}{4} - \dfrac{1.38\pi}{180}\right) = 1 + 2\left(\dfrac{\pi}{4} - \dfrac{1.38\pi}{180} - \dfrac{\pi}{4}\right) + 2\left(\dfrac{\pi}{4} - \dfrac{1.38\pi}{180} - \dfrac{\pi}{4}\right)^2 + \cdots$$

$$= 1 + 2(-0.0240855) + 2(0.0005801) = 0.953$$

21. $\displaystyle\int_{0.1}^{0.2} \dfrac{1 - \dfrac{x^2}{2} + \dfrac{x^4}{24} + \cdots}{\sqrt{x}} \, dx = \int_{0.1}^{0.2} \left(x^{-1/2} - \dfrac{x^{3/2}}{2} + \dfrac{x^{7/2}}{24} + \cdots\right) dx$

$$= \int_{0.1}^{0.2} x^{-1/2} dx - \dfrac{1}{2}\int_{0.1}^{0.2} x^{3/2} dx + \dfrac{1}{24}\int_{0.1}^{0.2} x^{7/2} dx$$

$$= 2x^{1/2} - \dfrac{1}{5}x^{5/2} + \dfrac{1}{108}x^{9/2} + \cdots \Big|_{0.1}^{0.2} = 0.259$$

25. $f(x) \begin{cases} 0 & -\pi \le x < 0 \\ x - 1 & 0 \le x < \pi \end{cases}$

is Example 2 of section 29-6 shifted down 1 unit. The Fourier series is

$$f(x) = -1 + \dfrac{2+\pi}{4} - \dfrac{2}{\pi}\cos x - \dfrac{2}{9\pi}\cos 3x - \cdots + \left(\dfrac{\pi - 2}{\pi}\right)\sin x - \dfrac{1}{2}\sin 2x + \cdots$$

29. From example 2, $f(x) = \dfrac{1}{2} + \dfrac{2}{\pi}\left(\cos x - \dfrac{1}{3}\cos 3x + \dfrac{1}{5}\cos 5x - \cdots\right)$

33. It is a geometric series for which $|r| < 1 = 0.75$. Therefore the series converges.

$$S = \dfrac{64}{1 - 0.75} = 256$$

37. $\ln(1+x)^4 = 4\ln(1+x) = 4\left(x - \dfrac{x^2}{2} + \dfrac{x^3}{3} - \dfrac{x^4}{4} + \cdots\right)$

$\qquad\qquad = 4x - 2x^2 + \dfrac{4x^3}{3} - x^4 + \cdots$

41. $\quad\begin{aligned}
f(x) &= \cos x & f(0) &= 1\\
f'(x) &= -\sin x & f'(0) &= 0\\
f''(x) &= -\cos x & f''(0) &= -1\\
f'''(x) &= \sin x & f'''(0) &= 0\\
f^{iv}(x) &= \cos x & f^{iv}(0) &= 1
\end{aligned}$

$\quad f(x) = 1 + 0x - \dfrac{1}{2}x^2 + \dfrac{0x^3}{3!} + \dfrac{1x^4}{4!} = 1 - \dfrac{1}{2}x^2 + \dfrac{1}{24}x^4 + \cdots$

$\quad \sec x = \dfrac{1}{\cos x} = \dfrac{1}{1 - \frac{1}{2}x^2 - \frac{1}{24}x^4} = 1 + \dfrac{1}{2}x^2 + \dfrac{5}{24}x^4 + \cdots$

The long division process is shown below:

$$
\begin{array}{r}
1 + \frac{1}{2}x^2 + \frac{5}{24}x^4 \\[4pt]
\hline
1 - \frac{1}{2}x^2 + \frac{1}{24}x^4\,\overline{)\,1 + 0\quad + 0\quad + 0\quad + 0} \\
\end{array}
$$

$$
\begin{array}{r}
1 - \frac{1}{2}x^2 + \frac{1}{24}x^4 \\
\hline
\frac{1}{2}x^2 - \frac{1}{24}x^4 \\
\frac{1}{2}x^2 - \frac{1}{4}x^4 \\
\hline
\frac{5}{24}x^4 + 0 \quad + \quad 0 \\
\frac{5}{24}x^4 - \frac{5}{48}x^2 + \frac{5}{576}x^4 \\
\hline
\end{array}
$$

45. $A = \displaystyle\int_{0.1}^{0.2} \dfrac{x - \sin x}{x^2}\,dx = \int_{0.1}^{0.2} \dfrac{x - \left(x - \frac{x^3}{3!} + \frac{x^5}{5!}\right)}{x^2}\,dx$

$\qquad = \displaystyle\int_{0.1}^{0.2} \dfrac{\frac{x^3}{3!} + \frac{x^5}{5!}}{x^2}\,dx = \int_{0.1}^{0.2} \left(\dfrac{x}{6} - \dfrac{x^3}{120}\right)\,dx$

$\qquad = \left(\dfrac{x^2}{12} - \dfrac{x^4}{480}\right)\Big|_{0.1}^{0.2} = \left(\dfrac{0.4}{12} - \dfrac{0.0016}{480}\right) - \left(\dfrac{0.01}{12} - \dfrac{0.0001}{480}\right)$

$\qquad = 0.0025$

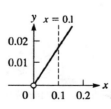

49. $N = N_0 e^{-\lambda t} \cdot N = N_0\left[1 + (-\lambda t) + \dfrac{(-\lambda t)^2}{2!} + \dfrac{(-\lambda t)3}{3!} + \cdots\right] = N_0\left[1 - \lambda t + \dfrac{\lambda^2 t^2}{2} - \dfrac{\lambda^3 t^3}{6} + \cdots\right]$

53.
$$\frac{N_0}{1 - e^{-kt}} = N_0 \left(\frac{1}{1 - e^{-kt}} \right) ; \text{ Let } x = e^{-kt}$$

$$N_0 \left(\frac{N_0}{1 - e^{-kt}} \right) = N_0 \left(\frac{1}{1 - x} \right)$$

The Maclaurin's expansion for $f(x) = \dfrac{1}{1 - x}$ is:

$$f(x) = \frac{1}{1 - x}; \qquad f(0) = 1$$

$$f'(x) = \frac{1}{(1 - x)^2}; \qquad f'(0) = 1$$

$$f''(x) = \frac{2}{(1 - x)^3}: \qquad f''(0) = 2$$

$$f(x) = 1 + x + \frac{2x^2}{2!} + \cdots = 1 + x + x^2 + \cdots$$

Substituting $e^{-k/t}$ for x; $f(x) = 1 + e^{-kt} + e^{-2kt} + \cdots$;

$$\therefore \frac{N_0}{1 - e^{-kt}} = N_0(1 + e^{-kt} + e^{-2kt} + \cdots)$$

57. Answers may vary. One way would be to use a Taylor Series and choosing an a-value from the known values which is close the angle for the particular trig function being computed.

DIFFERENTIAL EQUATIONS

30.1 Solutions of Differential Equations

1. $y = e^{-x^2}$; $\dfrac{dy}{dx} = -2xe^{-x^2}$

Substitute y and $\dfrac{dy}{dx}$ into the differential equation.

$\dfrac{dy}{dx} + 2xy = 0$, $-2xe^{-x^2} + 2x(e^{-x^2}) = 0$; $0 = 0$ particular solution. no c.

5. $\dfrac{dy}{dx} - y = 1$; $y = e^x - 1$; $\dfrac{dy}{dx} = e^x$

Substitute y and y'.

$e^x - (e^x - 1) = 1$; $e^x - e^x + 1 = 1$;
$1 = 1$ identity

9. $\dfrac{dy}{dx} = 2x$; $y = x^2 + 1$; $\dfrac{dy}{dx} = 2x$

Substitute. $2x = 2x$ identity

13. $y' + 2y = 2x$; $y = ce^{-2x} + x - \dfrac{1}{2}$;

$y' = -2ce^{-2x} + 1$

Substitute y and y'.

$-2ce^{-2x} + 1 + 2\left(ce^{-2x} + x - \dfrac{1}{2}\right) = 2x$

$\qquad -2ce^{-2x} + 1 + 2ce^{-2x} + 2x - 1 = 2x$

$2x = 2x$ identity

17. $x^2 y' + y^2 = 0$; $xy = cx + cy$; $y = \dfrac{cx}{x - c}$

$y' = \dfrac{(x - c)(c) - cx(1)}{(x - c)^2} = \dfrac{-c^2}{(c - x)^2}$

Substituting,

$x^2 y' + y^2 = x^2 \left[\dfrac{-c^2}{(x - c)^2}\right] + \left(\dfrac{cx}{x - c}\right)^2$

$= \dfrac{-c^2 x^2}{(x - c)^2} + \dfrac{c^2 x^2}{(x - c)^2} = 0$; $0 = 0$ identity

21. $y' + y = 2\cos x$;
$\qquad y = \sin x + \cos x - e^{-x}$
$\qquad y' = \cos x - \sin x - e^{-x}(-1) = \cos x - \sin x + e^{-x}$

Substitute,

$\qquad y' + y = \cos x - \sin x + e^{-x} + \sin x + \cos x - e^{-x} = 2\cos x$
$\qquad 2\cos x = 2\cos x$ identity

25. $\cos x \dfrac{dy}{dx} + \sin x = 1 - y$; $y = \dfrac{x + c}{\sec x + \tan x}$

$\dfrac{dy}{dx} = \dfrac{(\sec x + \tan x)(1) - (x + c)(\sec x \tan x + \sec^2 x)}{(\sec x + \tan x)^2} = \dfrac{(\sec x + \tan x) - (x + c)(\sec x)(\tan x + \sec x)}{(\sec x + \tan x)^2}$

$= \dfrac{(\sec x + \tan x)[1 - (x + c)(\sec x)]}{(\sec x + \tan x)(\sec x + \tan x)} = \dfrac{1 - (x + c)(\sec x)}{(\sec x + \tan x)}$

Substituting,

$$\cos x\frac{dy}{dx} + \sin x = \cos x\left[\frac{1-(x+c)(\sec x)}{\sec x + \tan x}\right] + \sin x = \frac{\cos x - (x+c)}{\sec x + \tan x} + \sin x$$

$$= \frac{\cos x - (x+c)}{\sec x + \tan x} + \frac{\sin x \sec x + \sin x \tan x}{\sec x + \tan x} = \frac{\cos x - x - c + \dfrac{\sin x}{\cos x} + \dfrac{\sin^2 x}{\cos x}}{\sec x + \tan x}$$

$$= \frac{\dfrac{\cos^2 x}{\cos x} + \dfrac{\sin^2 x}{\cos x} + \tan x - x - c}{\sec x + \tan x} = \frac{\dfrac{1}{\cos x} + \tan x - x - c}{\sec x + \tan x} = \frac{\sec x + \tan x - x - c}{\sec x + \tan x}$$

$$= 1 - \frac{(x+c)}{\sec x + \tan x} = 1 - y$$

30.2   Separation of Variables

1. $2x\,dx + dy = 0$; integrate. $x^2 + y = c$;
$y = c - x^2$

5. $\dfrac{dV}{dP} = \dfrac{-V}{p^2}$; $\dfrac{-dV}{V} = \dfrac{dP}{p^2}$; $\dfrac{dP}{p^2} + \dfrac{dV}{V} = 0$

$\displaystyle\int P^{-2}dP + \int \frac{dV}{V} = 0$; $\dfrac{P^{-1}}{-1} + \ln V = c$;

$\ln V = \dfrac{1}{P} + c$

9. $\qquad dy + \ln xy\,dx = (4x + \ln y)dx$;
$dy + (\ln x + \ln y)dx = 4x\,dx + \ln y\,dx$
$dy + \ln x\,dx + \ln y\,dy = 4x\,dx + \ln y\,dx$
$\qquad dy + \ln x\,dx = 4x\,dx$

Integrate: $y + x\ln x - x - 2x^2 = c$
$y = 2x^2 + x - x\ln x + c$

$\displaystyle\int dy + \int \ln x\,dx - \int 4x\,dx = 0$;

$y + x\ln x - x - 2x^2 = c$

13. $\quad e^{x+y}dx + dy = 0$
$\quad e^x e^y dx + dy = 0$

$\quad e^x dx + \dfrac{dy}{e^y} = 0$

$\quad e^x dx + e^{-y}dy = 0$

Integrate:

$\quad e^x - e^{-y} = c$

17. $x\dfrac{dy}{dx} = y^2 + y^2\ln x$; divide by xy^2 and multiply by dx

$\dfrac{dy}{y^2} = \dfrac{dx}{x} + \dfrac{1}{x}\ln x\,dx$; $\dfrac{dy}{y^2} = (1 + \ln x)\dfrac{dy}{dx}$; integrate

$-\dfrac{1}{y} = \dfrac{1}{2}(1 + \ln x)^2 + \dfrac{c}{2}$; $-2 = y(1 + \ln x)^2 + cy$

$y(1 + \ln x)^2 + cy + 2 = 0$

21. $yx^2\,dx = y\,dx - x^2\,dy$

Divide by y and by x^2; $dx = \dfrac{dx}{x^2} - \dfrac{dy}{y}$; integrate

$x + c = \dfrac{x^{-1}}{-1} - \ln y$

$x + \dfrac{1}{x} + \ln y = c$; $x^2 + 1 + x\ln y + cx = 0$

25. $2\ln t\,dt + t\,di = 0$;

$2\ln t\dfrac{dt}{t} + di = 0$; integrate

$\dfrac{2(\ln t)^2}{2} + i = 0$; $(\ln t)^2 + i = c$;

$i = c - (\ln t)^2$

29. $\dfrac{dy}{dx} + yx^2 = 0$; $\dfrac{dy}{y} + x^2\,dx = 0$

Integrate: $\ln y + \dfrac{x^3}{3} + c$

Substitute $x = 0, y = 1$; $\ln 1 = c$; $c = 0$

$3\ln y + x^3 = 0$

33. $\dfrac{dy}{dx} = (1 - y)\cos x$; $\cos x$; $x = \dfrac{\pi}{6}$ when $y = 0$

$dy = (1 - y)\cos x\,dx$

$\dfrac{1}{1 - y}\,dy = \cos x\,dx$; integrate

$-\ln(1 - y) = \sin x + c$; $\sin x + \ln(1 - y) = c$

Substitute $x = \dfrac{\pi}{6}$, $y = 0$; $\sin\dfrac{\pi}{6} + \ln 1 = c$; $c = \dfrac{1}{2}$

$\sin x + \ln(1 - y) = \dfrac{1}{2}$; $2\ln(1 - y) = 1 - 2\sin x$

30.3   Integrating Combinations

1. $x\,dy + y\,dx + x\,dx = 0$; $d(xy) + x\,dx = 0$

$xy + \dfrac{x^2}{2} = c$; $2xy + x^2 = c$

5. $A^3\,dr + A^2 r\,dA + r\,dA - A\,dr = 0$

$A\,dr + r\,dA - \left(\dfrac{A\,dr - r\,dA}{A^2}\right) = 0$

$Ar - \dfrac{r}{A} = c$; $A^2 r - r = cA$

9. $\sqrt{x^2 + y^2}\,dx - 2y\,dy = 2x\,dx$;
$(x^2 + y^2)\,dx = 2x\,dx + 2y\,dy$
$dx = (x^2 + y^2)^{-1/2}d(x^2 + y^2)$
$x = 2(x^2 + y^2)^{1/2} + c$
$2\sqrt{x^2 + y^2} = x + c$

13. $y\,dy - x\,dx + (y^2 - x^2)dx = 0$

$$\frac{y\,dy - x\,dx}{y^2 - x^2} + dx = 0$$

$$\frac{2y\,dy - 2x\,dx}{y^2 - x^2} + 2\,dx = 0$$

$$\ln(y^2 - x^2) + 2x = c$$

17. $2(x\,dy + y\,dx) + 3x^2 dx = 0$

$$x\,dy + y\,dx + \frac{3}{2}x^2 dx = 0$$

$$d(xy) - \frac{3}{2}x^2 dx = 0$$

$$xy + \frac{1}{2}x^3 + c = 0$$

$$2xy + x^3 + c = 0$$

Substituting $x = 1$, $y = 2$; $4 + 1 + c = 0$;
$c = -5$
$2xy + x^3 = 5$

30.4   The Linear Differential Equations of the First Order

1. $dy + y\,dx = e^{-x}dx$; $P = 1$,
$Q = e^{-x}$; $e^{\int dx} = e^x$

$$ye^x = \int e^{-x}e^x dx + c$$

$$ye^x = \int dx + c;$$

$$ye^x = x + c$$
$$y = e^{-x}(x + c)$$

5. $\dfrac{dy}{dx} - 2y = 4$; $dy - 2y\,dx = 4\,dx$

$P = -2$, $Q = 4$; $e^{\int -2dx} = e^{-2x}$

$$ye^{-2x} = \int 4e^{-2x}dx + c$$

$$ye^{-2x} = 4\left[-\frac{1}{2}\int e^{-2x}(-2\,dx)\right] + c$$

$$ye^{-2x} = -2e^{-2x} + c \quad \text{or} \quad y = -2 + ce^{2x}$$

9. $2x\,dy + y\,dx = 8x^3 dx$; $dy + \dfrac{1}{2x}y\,dx = 4x^2 dx$

$$P = \frac{1}{2x}, \ Q = 4x^2$$

$$e^{\int (1/2)x\,dx} = e^{(1/2)\ln x} = e^{\ln x^{1/2}} = x^{1/2}$$

$$yx^{1/2} = \int 4x^2(x^{1/2})dx + c = \int 4x^{5/2}dx + c = 4x^{7/2}\left(\frac{2}{7}\right) + c = \frac{8}{7}x^{7/2} + c$$

$$y = \frac{8}{7}x^3 + \frac{c}{\sqrt{x}}$$

13. $\qquad\qquad \sin x\dfrac{dy}{dx} = 1 - y\cos x$

$$\sin x\,dy + y\cos x\,dx = dx$$

$$dy + y\frac{\cos x}{\sin x}dx = \frac{dx}{\sin x}$$

$$dy + \cot x\,y\,dx = \csc x\,dx$$

$$P = \cot x, \ Q = \csc x$$

$$e^{\int \cot x\,dx} = e^{\ln \sin x} = \sin x$$

$$y\sin x = \int \csc x\sin x\,dx$$

$$y\sin x = \int dx + c = x + c$$

$$y = \frac{x + c}{\sin x} = (x + c)\csc x$$

17. $ds = (te^{4t} + 4s)dt$; $ds - 4s\,dt = te^{4t}dt$; $ds - 4s\,dt = te^{4t}dt$; $P = -4$; $Q = te^{4t}$; $e^{\int -4dt} = e^{-4t}$

$$se^{-4t} = \int te^{4t}e^{-4t}dt + c; \ se^{-4t} = \int t\,dt + c; \ se^{-4t} = \frac{t^2}{2} + c$$

21. $x\dfrac{dy}{dx} = y + (x^2 - 1)^2$; $x\dfrac{dy}{dx} = y + x^4 - 2x^2 + 1$; $dy = \dfrac{1}{x}y\,dx + \dfrac{1}{x}(x^4 - 2x^2 + 1)dx$;

$dy - \dfrac{1}{x}y\,dx = \left(x^3 - 2x + \dfrac{1}{x}\right)dx$; $P = -\dfrac{1}{x}$; $Q = \left(x^3 - 2x + \dfrac{1}{x}\right)$; $e^{\int(-1/x)dx} = e^{-\ln x} = e^{\ln x^{-1}} = \dfrac{1}{x}$

$\dfrac{y}{x} = \displaystyle\int \left(x^3 - 2x + \dfrac{1}{x}\right)\left(\dfrac{1}{x}\right)dx + c$

$\dfrac{y}{x} = \displaystyle\int (x^2 - 2 + x^{-2})dx + c = \dfrac{x^3}{3} - 2x - x^{-1} + c$

$y = \dfrac{x^4}{3} - 2x^2 - 1 + cx$; $3y = x^4 - 6x^2 - 3 + cx$

25. $y' = 2(1 - y)$; solve by separation of variables.

$\dfrac{dy}{1 - y} = 2\,dx$; $-\ln(1 - y) = 2x - \ln c$; $\ln\dfrac{c}{1 - y} = 2x$; $c = (1 - y)e^{2x}$; $1 - y = ce^{-2x}$; $y = 1 - ce^{-2x}$

Solve as a first order equation.

$dy = 2\,dx - 2y\,dx$; $dy + 2y\,dx = 2\,dx$

$ye^{\int 2dx} = \displaystyle\int e^{\int 2dx}\,dx$; $ye^{2x} = \displaystyle\int 2e^{2x}\,dx$

$ye^{2x} = e^{2x} + c$; $y = 1 + ce^{-2x}$

29. $\dfrac{dy}{dx} + 2y\cot x = 4\cos x$; $x = \dfrac{\pi}{2}$, $y = \dfrac{1}{3}$; $dy + 2y\cot x\,dx = 4\cos x\,dx$; $P = 2\cot x$; $Q = 4\cos x$

$e^{\int P dx} = e^{\int 2\cot x dx} = 2^{2\ln|\sin x|} = e^{\ln|\sin x|^2} = |\sin x|^2$

$y(\sin x)^2 = \displaystyle\int 4\cos x(\sin x)^2 dx + c = \dfrac{4(\sin x)^3}{3} + c$

$y = \dfrac{4}{3}\sin x + c(\csc^2 x)$

$x = \dfrac{\pi}{2}$ when $y = \dfrac{1}{3}$; $\dfrac{1}{3} = \dfrac{4}{3}\sin\dfrac{\pi}{2} + c$; $c = -1$

$y = \dfrac{4}{3}\sin x - \csc^2 x$

30.5   Elementary Applications

1. $\dfrac{dy}{dx} = \dfrac{2x}{y}$; $y\,dy = 2x\,dx$; $\dfrac{1}{2}y^2 = x^2 + c$

Substitute $x = 2$, $y = 3$; $\dfrac{1}{2}(9) = 4 + c$; $c = 0.5$

$\dfrac{1}{2}y^2 = x^2 + 0.5$; $y^2 = 2x^2 + 1$; $y = \pm\sqrt{2x^2 + 1}$

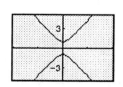

5. See Example 2; $\dfrac{dy}{dx} = ce^x$; $y = ce^x$; $c = \dfrac{y}{e^x}$

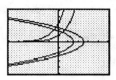

Substitute for c in the equation for the derivative.

$\dfrac{dy}{dx} = \dfrac{y}{e^x} e^x = y$; $\left.\dfrac{dy}{dx}\right|_{OT} = -\dfrac{1}{y}$; $y\,dy = -dx$

Integrating, $\dfrac{y^2}{2} = -x + \dfrac{c}{2}$; $y^2 = c - 2x$

9. See Example 3; $N = N_o e^{kt}$

Use the condition that half this isotope decays in 40.0 s.

$N = \dfrac{N_0}{2}$ when $t = 40.0$ s. $\dfrac{N_0}{2} = N_0 e^{40.0k}$; $0.5 = e^{40.0k}$; $0.5^{1/40.0} = e^k$; $0.5^{0.025} = e^k$; $N = N_0(0.5)^{0.025}$

Evaluating when $t = 60.0$ s,

$N = N_0(0.5)^{0.025(60.0)} = N_0(0.5)^{1.50} = 0.354N_0$

Therefore, 35.4% remains after 60.0 s.

13. $r\dfrac{dS}{dr} = 2(a - S)$; $r\,dS = 2a\,dr - 2S\,dr$; $dS + \dfrac{2}{r}S\,dx = 2a\dfrac{dr}{r}$; $y = S$, $x = r$, $P = \dfrac{2}{r}$, $Q = \dfrac{2a}{r}$

$e^{\int P dx} = e^{\int (2/r)dr} = e^{2\ln x} = e^{\ln r^2} = r^2$

$Sr^2 = \displaystyle\int \left(2a\dfrac{dr}{r}\right) - 2a\ln r + c = \int 2ar\,dr$

$Sr^2 = ar^2 + c$; $S = a + \dfrac{c}{r^2}$

17. $\dfrac{dT}{dt} = k(T - 80.0)$; $\dfrac{dT}{T - 80.0} = k\,dt$; $\ln(T - 80.0) = kt + \ln c$; $T = 80.0 + ce^{kt}$; $200 = 80.0 + c$;

$T = 80.0 + 120e^{kt}$; $140 = 80.0 + 120e^{5k}$; $e^{5k} = 0.5$; $e^k = (0.5)^{1/5}$; $T = 80.0 + 120(0.5)^{t/5}$;

$100 = 80.0 + 120(0.5)^{t/5}$; $0.5^{t/5} = \dfrac{1}{6.0}$; $t = 5\dfrac{\ln\frac{1}{6.0}}{\ln\frac{1}{2}} = 13$ min

21. See Example 4; $i = \dfrac{E}{R}(1 - e^{-(R/L)t})$

$\displaystyle\lim_{i \to \infty} \left(\dfrac{E}{R} - \dfrac{E}{R}e^{-(R/L)t}\right) = \dfrac{E}{R} - 0 = \dfrac{E}{R}$

25. $Ri + \dfrac{q}{C} = 0$; $i = \dfrac{dq}{dt}$; $R\dfrac{dq}{dt} + \dfrac{q}{C} = 0$; $\dfrac{dq}{q} + \dfrac{1}{RC}\,dt = 0$

Integrating, $\ln q = -\dfrac{1}{RC}t + c$

Let $q = q_0$ when $t = 0$; $\ln q_0 = c$; $\ln q = -\dfrac{1}{RC}t + \ln q_0$; $\ln q - \ln q_0 = -\dfrac{1}{RC}t$; $\ln\dfrac{q}{q_0} = -\dfrac{1}{RC}t$;

$\dfrac{q}{q_0} = e^{(-1/RC)t}$; $q = q_0 e^{-t/RC}$

29. $\dfrac{dv}{dt} = 32 - v$; $\dfrac{dv}{32 - v} = dt$; $-\dfrac{-dv}{32 - v} = dt$

Integrating, $(-1)\ln(32 - v) = t - \ln c$; $\ln\dfrac{32 - v}{c} = -t$; $32 - v = ce^{-t}$; $-v = -32 + ce^{-t}$; $v = 32 - ce^{-t}$

Starting from rest means $v = 0$ when $t = 0$; $0 = 32 - c$; $c = 32$; $v = 32 - 32e^{-t} = 32(1 - e^{-t})$

$\lim\limits_{t \to \infty} 32(1 - e^{-t}) = 32$

33. $\dfrac{dx}{dt} = 6t - 3t^2$; $dx = (6t - 3t^2)$; $x = \dfrac{6t^2}{2} - \dfrac{3t^3}{3} + c$; $x = 3t^2 - t^3 + c$; $x = 0$ for $t = 0$; $0 = c$; $x = 3t^2 - t^3$;
$y = 2(3t^2 - t^3) - (3t^2 - t^3)^2$; $y = -t^6 + 6t^5 - 9t^4 - 2t^3 + 6t^2$

37. $\dfrac{dv}{dt} = kv$; $\dfrac{dv}{v} = k\,dt$; integrating, $\ln v = kt + \ln c$

Let $v = 16\ 500$ when $t = 0$; $16\ 500 = c$; $\ln v = kt + 16500$; $\ln v - \ln 16\ 500 - kt$; $\ln\dfrac{v}{16\ 500} = kt$;

$\dfrac{v}{16\ 500} = e^{kt}$; $v = 16\ 500e^{kt}$

Let $v = 9850$ when $t = 3$; $9850 = 16\ 500e^{3k}$; $e^k = (0.597)^{1/3}$; $v = 16\ 500(0.597)^{t/3}$

After 11 years, $v = 16\ 500(0.597)^{11/3} = \$2490$

30.6   Higher-Order Homogeneous Equations

1. $D^2y - Dy - 6y = 0$; $m^2 - m - 6 = 0$; $(m - 3)(m + 2) = 0$; $m_1 = 3$ and $m_2 = -2$; $y = c_1e^{3x} + c_2e^{-2x}$

5. $D^2y - 3\,Dy = 0$; $m^2 - m = 0$; $m_1 = 0$ and $m_2 = 3$; $y = c_1e^0 + c_2e^{3x}$; $y = c_1 + c_2e^{3x}$

9. $3\,D^2y + 8\,Dy - 3y = 0$; $3m^2 + 8m - 3 = 0$; $(3m - 1)(m + 3) = 0$; $m_1 = \frac{1}{3}$ and $m_2 = -3$;
$y = c_1e^{x/3} + c_2e^{-3x}$

13. $2\dfrac{d^2y}{dx^2} - 4\dfrac{dy}{dx} + y = 0$; $2\,D^2y - 4\,Dy + y = 0$; $2m^2 - 4m + 1 = 0$

Quadratic formula: $m = \dfrac{4 \pm \sqrt{16 - 8}}{4}$; $m_1 = 1 + \dfrac{\sqrt{2}}{2}$, $m_2 = 1 - \dfrac{\sqrt{2}}{2}$

$y = c_1e^{(1 + (\sqrt{2}/2))x} + c_2e^{(1 - (\sqrt{2}/2))x}$; $y = c_1e^xe^{(\sqrt{2}/2)x} + c_2e^xe^{-(\sqrt{2}/2)x} = e^x\big(c_1e^{x(\sqrt{2}/2)} + c_2e^{-x(\sqrt{2}/2)}\big)$

17. $y'' = 3y' + y$; $D^2y - 3\,Dy - y = 0$; $m^2 - 3m - 1 = 0$

Quadratic formula: $m = \dfrac{3 \pm \sqrt{9 + 4}}{2}$; $m_1 = \dfrac{3}{2} + \dfrac{\sqrt{13}}{2}$; $m_2 = \dfrac{3}{2} - \dfrac{\sqrt{13}}{2}$

$y = c_1e^{((3/2) + (\sqrt{13}/2))x} + c_2e^{((3/2) - (\sqrt{13}/2))x}$; $y = e^{3x/2}\big(c_1e^{x(\sqrt{13}/2)} + c_2e^{-x(\sqrt{13}/2)}\big)$

21. $D^2y - 4\,Dy - 21y = 0$; $m^2 - 4m - 21 = 0$; $(m - 7)(m + 3) = 0$; $m_1 = 7$ and $m_2 = -3$; $y = c_1e^{7x} + c_2e^{-3x}$

The derivative of the general equation is: $Dy = 7c_1e^{7x} - 3c_2e^{-3x}$
Substituting $Dy = 0$, $y = 2$ when $x = 0$: $c_1 + c_2 = 2$ and $7c_1 - 3c_2 = 0$.
Solving this system: $c_1 = \dfrac{3}{5}$ and $c_2 = \dfrac{7}{5}$. Therefore, $y = \dfrac{3}{5}e^{7x} + \dfrac{7}{5}e^{-3x}$; $y = \dfrac{1}{5}(3e^{7x} + 7e^{-3x})$

25. $D^3y - 2\,D^2y - 3\,Dy = 0$; $m^3 - 2m^2 - 3m = 0$; $m(m + 1)(m - 3) = 0$; $m_1 = 0$, $m_2 = -1$, and $m_3 = 3$;
$y = c_1e^0 + c_2e^{-x} + c_3e^{3x}$; $y = c_1 + c_2e^{-x} + c_3e^{3x}$

30.7   Auxiliary Equations with Repeated or Complex Roots

1. $D^2y - 2Dy + y = 0$; $m^2 - 2m + 1 = 0$; $(m-1)^2 = 0$; $m = 1, 1$

   $y = e^x(c_1 + c_2x)$; $y = (c_1 + c_2x)e^x$

5. $D^2y + 9y = 0$; $m^2 + 9 = 0$; $m_1 = 3j$ and $m_2 = -3j$, $\alpha = 0$, $\beta = 3$

   $y = e^{0x}(c_1 \sin 3x + c_2 \cos 3x)$; $y = c_1 \sin 3x + c_2 \cos 3x$

9. $D^4y - y = 0$
   $m^4 - 1 = 0$
   $(m^2 - 1)(m^2 + 1) = 0$
   $(m-1)(m+1)(m^2 + 1) = 0$
   $m = \pm 1$, $m = \pm i$
   $y = c_1e^x + c_2e^{-x} + c_3 \sin x + c_4 \cos(-x)$
   $y = c_1e^x + c_2e^{-x} + c_3 \sin x + c_4 \cos x$

13. $16D^2y - 24Dy + 9y = 0$; $16m^2 - 24m + 9 = 0$; $(4m - 3)^2 = 0$; $m = \dfrac{3}{4}, \dfrac{3}{4}$

   $y = e^{3x/4}(c_1 + c_2x)$

17. $2D^2y + 5y = 4Dy$; $2D^2y + 5y - 4Dy = 0$; $2m^2 - 4m + 5 = 0$

   Quadratic formula: $m = \dfrac{4 \pm \sqrt{16 - 40}}{4} = \dfrac{4 \pm 2\sqrt{-6}}{4}$

   $m_1 = 1 + \dfrac{\sqrt{6}}{2}j$; $m_2 = 1 - \dfrac{\sqrt{6}}{2}j$; $\alpha = 1, \beta = \dfrac{1}{2}\sqrt{6}$

   $y = e^x\left(c_1 \cos \dfrac{1}{2}\sqrt{6}x + c_2 \sin \dfrac{1}{2}\sqrt{6}x\right)$

21. $2D^2y - 3Dy - y = 0$; $2m^2 - 3m - 1 = 0$

   By the quadratic formula, $m = \dfrac{3 \pm \sqrt{9 + 8}}{4}$

   $m_1 = \dfrac{3}{4} + \dfrac{\sqrt{17}}{4}$, $m_2 = \dfrac{3}{4} - \dfrac{\sqrt{17}}{4}$

   $y = c_1e^{((3/4)+(\sqrt{17}/4))x} + c_2e^{((3/4)+(\sqrt{17}/4))x}$; $y = e^{(3/4)x}\left(c_1e^{x(\sqrt{17}/4)} + c_2e^{-x(\sqrt{17}/4)}\right)$

25. $D^3y - 6D^2y + 12Dy - 8y = 0$
   $m^3 - 6m^2 + 12m - 8 = 0$
   $(m-2)(m^2 - 4m + 4) = 0$
   $(m-2)(m-2)(m-2) = 0$
   $m = 2, 2, 2$ repeated root
   $y = e^{2x}(c_1 + c_2x + c_3x^2)$

29. $D^2y + 2Dy + 10y = 0$; $m^2 + 2m + 10 = 0$

   By the quadratic formula, $m = \dfrac{-2 \pm \sqrt{4 - 40}}{2}$

   $m_1 = -1 + 3j$; $m_2 = -1 - 3j$, $\alpha = -1$, $\beta = 3$; $y = e^{-x}(c_1 \sin 3x + c_2 \cos 3x)$

Substituting $y = 0$ when $x = 0$; $0 = e^0(c_1 \sin 0 + c_2 \cos 0)$; $c_2 = 0$

Substituting $y = e^{-1}$, $x = \dfrac{\pi}{6}$; $e^{-1} = e^{-\pi/6}\left(c_1 \sin \dfrac{\pi}{2}\right)$; $e^{-1} = e^{-\pi/6}c_1$; $c_1 = e^{\pi/6-1}$

$y = e^{-x}[e^{(\pi/6-1)} \sin 3x] = e^{(\pi/6-1-x)} \sin 3x$

33. $y = c_1 e^{3x} + c_2 e^{-3x}$

$(m - 3)(m + 3) = 0$

$m^2 - 9 = 0$; $(D^2 - 9)y = 0$

30.8   Solutions of Nonhomogeneous Equations

1. $D^2y - Dy - 2y = 4$; $m^2 - m - 2 = 0$

$(m - 2)(m + 1) = 0$; $m_1 = 2$, $m_2 = -1$

$y_c = c_1 e^{2x} + c_2 e^{-x}$

$y_p = A$; $Dy_p = 0$; $D^2y_p = 0$

Substituting in diff. equation, $0 - 0 - 2A = 4$; $A = -2$

Therefore, $y_p = -2$; $y = c_1 e^{2x} + c_2 e^{-x} - 2$

5. $y'' - 3y' = 2e^x + xe^x$; $D^2y - 3\,Dy = 2e^x + xe^x$

$m^2 - 3m = 0$; $m(m - 3) = 0$; $m_1 = 0$, $m_2 = 3$

$y_c = c_1 e^0 + c_2 e^{3x} = c_1 + c_2 e^{3x}$; $y_p = Ae^x + Bxe^x$

$Dy_p = Ae^x + B(xe^x + e^x) = Ae^x + Bxe^x + Be^x$

$D^2y_p = Ae^x + Be^x + B(xe^x + e^x) = Ae^x + Be^x + Bxe^x + Be^x$

$D^2y_p = Ae^x + 2Be^x + Bxe^x$

Substituting in diff. equation:

$Ae^x + 2Be^x + Bxe^x - 3(Ae^x + Bxe^x + Be^x) = 2e^x + xe^x$

$Ae^x + 2Be^x + Bxe^x - 3Ae^x - 3Bxe^x - 3Be^x = 2e^x + xe^x$

$\qquad\qquad -2Ae^x - Be^x - 2Bxe^x = 2e^x + xe^x$

$\qquad\quad e^x(-2A - B) + xe^x(-2B) = 2e^x + xe^x$

$-2A - B = 2$; $-2A + \dfrac{1}{2} = 2$; $-2A = \dfrac{3}{2}$; $A = -\dfrac{3}{4}$; $-2B = 1$; $B = -\dfrac{1}{2}$

$y = c_1 + c_2 e^{3x} - \dfrac{3}{4}e^x - \dfrac{1}{2}xe^x$

9. $\dfrac{d^2y}{dx^2} - 2\dfrac{dy}{dx} + y = 2x + x^2 + \sin 3x$

$D^2y - 2\,Dy + y = 2x + x^2 + \sin 3x$

$m^2 - 2m + 1 = 0$; $(m - 1)^2 = 0$; $m = 1, 1$

$y_c = e^x(c_1 + c_2x)$; $y_p = A + Bx + Cx^2 + E \sin 3x + F \cos 3x$

$Dy_p = B + 3Cx + 3E \cos 3x - 3F \sin 3x$

$D^2y_p = 2c - 9E \sin 3x - 9F \cos 3x$

Substituting in diff. equation:

$2C - 9E\sin 3x - 9F\cos 3x - 2(B + 2Cx + 3E\cos 3x - 3F\sin 3x) + A + Bx + Cx^2 + E\sin 3x + F\cos 3x$
$\quad = 2x + x^2 + \sin 3x$

$2C - 9E\sin 3x - 9F\cos 3x - 2B - 4Cx - 6E\cos 3x + 6F\sin 3x + A + Bx + Cx^2 + E\sin 3x + F\cos 3x$
$\quad = 2x + x^2 + \sin 3x$

$2C - 2B + A - 8E\sin 3x + 6F\sin 3x - 8F\cos 3x - 6E\cos 3x - 4Cx + Bx + Cx^2$
$\quad = 2x + x^2 + \sin 3x$

$(2C - 2B + A) + \sin 3x(6F - 8E) + \cos 3x(-8F - 6E) + x(B - 4) + Cx^2$
$\quad = 2x + x^2 + \sin 3x$

$2C - 2B + A = 0; \ A = 10$

$6F - 8E = 1; \ -8F - 6E = 0; \ E = -\dfrac{2}{25}; \ F = \dfrac{3}{50}; \ B - 4 = 2; \ B = 6; \ C = 1$

$y = e^x(c_1 + c_2 x) + 10 + 6x + x^2 - \dfrac{2}{25}\sin 3x + \dfrac{3}{50}\cos 3x$

13. $D^2 y - Dy - 30y = 10; \ m^2 - m - 30 = 0; \ (m - 6)(m + 5) = 0; \ m_1 = -5; \ m_2 = 6$

$y_c = c_1 e^{-5x} + c_2 e^{6x}; \ y_p = A; \ Dy_p = 0; \ D^2 y_p = 0; \ 0 = 0 - 30A = 10; \ A = -\dfrac{1}{3}; \ y_p = -\dfrac{1}{3};$

$y = c_1 e^{-5x} + c_2 e^{6x} - \dfrac{1}{3}$

17. $D^2 y - 4y = \sin x + 2\cos x; \ m^2 - 4 = 0; \ m_1 = 2; \ m_2 = -2$

$y_c = c_1 e^{2x} + c_2 e^{-2x}$

$y_p = A\sin x + B\cos x; \ Dy_p = A\cos x - B\sin x$

$D^2 y_p = -A\sin x - B\cos x$

$-A\sin x - B\cos x - 4A\sin x - 4B\cos x = \sin x + 2\cos x$

$\qquad\qquad -5A\sin x - 5B\cos x = \sin x + 2\cos x$

$$-5A = 1; \ A = -\dfrac{1}{5}; \ -5B = 2; \ B = -\dfrac{2}{5}$$

$y_p = -\dfrac{1}{5}\sin x - \dfrac{2}{5}\cos x$

$y = c_1 e^{2x} + c_2 e^{-2x} - \dfrac{1}{5}\sin x - \dfrac{2}{5}\cos x$

21. $D^2 y + 5Dy + 4y = xe^x + 4; \ m^2 + 5m + 4 = 0; \ (m + 1)(m + 4) = 0; \ m_1 = -1, \ m_2 = -4$

$y_c = c_1 e^{-x} + c_2 e^{-4x}; \ y_p = Ae^x + Bxe^x + C$

$Dy_p = Ae^x + B(xe^x + e^x) = Ae^x + Bxe^x + Be^x$

$D^2 yp = Ae^x - B(xe^x + e^x) + Be^x = Ae^x + 2Be^x + Bxe^x$

$Ae^x + 2Be^x + Bxe^x + 5(Ae^x + Bxe^x + Be^x) + 4(Ae^x + Bxe^x + C) = xe^x + 4$

$(10A + 7B)e^x + 10Bxe^x + 4C = xe^x + 4$

$10A + 7B = 0; \ 10B = 1; \ B = \dfrac{1}{10}$

$4C = 4; \ C = 1; \ 10A + 7\left(\dfrac{1}{10}\right) = 0$

$A = -\dfrac{7}{100}; \ y_p = -\dfrac{7}{100}e^x + \dfrac{1}{10}e^x + 1$

$y = c_1 e^{-x} + c_2 e^{-4x} - \dfrac{7}{100}e^x + \dfrac{1}{10}xe^x + 1$

25. $D^2 y + y = \cos x$

$m^2 + 1 = 0$

$m = \pm i$

$y_c = c_1 \sin x + c_2 \cos x$

let $y_p = x(A \sin x + B \cos x)$

$Dy_p = x(A \cos x - B \sin x) + A \sin x + B \cos x$

$D^2 y_p = x(-A \sin x - B \cos x) + A \cos x - B \sin x + A \cos x - B \sin x$

$D^2 y_p + y_p = \cos x$

$-x(A \sin x + B \cos x) + 2A \cos x - 2B \sin x + x(A \sin x + B \cos x) = \cos x$

$$2A \cos x - 2B \sin x = \cos x$$

$2A = 1, B = 0$

$A = \dfrac{1}{2}$

$y_p = \dfrac{1}{2} x \sin x$

$y = y_c + y_p$

$y = c_1 \sin x + c_2 \cos x + \dfrac{1}{2} x \sin x$

29. $D^2 y - Dy - 6y = 5 - e^x; \; m^2 - m - 6 = 0; \; (m-3)(m+2) = 0; \; m_1 = 3, \; m_2 = -2$

$y_c = c_1 e^{3x} + c_2 e^{-2x}$

$y_p = A + Be^x; \; Dy_p = Be^x; \; D^2 y_p = Be^x$

$Be^x - Be^x - 6(A + Be^x) = 5 - e^x$

$-6A - 6Be^x = 5 - e^x; \; -6A = 5; \; A = -\dfrac{5}{6}; \; -6B = -1; \; B = \dfrac{1}{6}$

$y_p = -\dfrac{5}{6} + \dfrac{1}{6} e^x; \; y = c_1 e^{3x} + c_2 e^{-2x} + \dfrac{1}{6} e^x - \dfrac{5}{6}$

Substituting $x = 0$ when $y = 2$; $3c_1 + 3c_2 = 8$

$Dy = 3c_1 e^{3x} - 2c_2 e^{-2x} + \dfrac{1}{6} e^x$

Substituting, $Dy = 4$ when $x = 0$, $18c_1 - 12c_2 = 23$

Solving the two linear equations simultaneously, $c_1 = \dfrac{11}{6}, \; c_2 = \dfrac{5}{6}$

$y = \dfrac{11}{6} e^{3x} + \dfrac{5}{6} e^{-2x} + \dfrac{1}{6} e^x - \dfrac{5}{6} = \dfrac{1}{6}(11e^{3x} + 5e^{-2x} + e^x - 5)$

30.9   Applications of Higher-Order Equations

1. $D^2 \theta + \dfrac{g}{l} \theta = 0; \; g = 9.8 \text{ m/s}^2, \; l = 0.1 \text{ m}; \; D^2 \theta + 9.8\theta = 0; \; m^2 + 9.8 = 0$

$m_1 = \sqrt{9.8}j, \; m_2 = -\sqrt{9.8}j; \; \alpha = 0, \beta = \sqrt{9.8}$

$x = c_1 \sin \sqrt{9.8}t + c_2 \cos \sqrt{9.8}t$

Substituting $\theta = 0.1$ when $t = 0$; $0.1 = c_1 \sin 0 + c_2 \cos 0$; $0.1 = c_2$; $D_x = c_1 \cos \sqrt{9.8}t - c_2 \sin \sqrt{9.8}t$

Substituting $D\theta = 0$ when $t = 0$; $0 = c_1 \cos 0 - c_2 \sin 0$; $c_1 = 0$; $\theta = 0.1 \cos \sqrt{9.8}t = 0.1 \cos 3.1t$

$\sqrt{9.8}t$	t	$\cos \sqrt{9.8}t$	$0.1 \cos \sqrt{9.8}t$
0	0	1	0.1
$\frac{\pi}{2}$	0.50	0	0
π	1.00	-1	-0.1
$\frac{3\pi}{2}$	1.50	0	0
2π	2.00	1	0.1

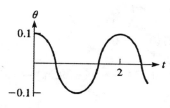

5. To get spring constant: $F = kx$; $4.00 = k(0.125)$; $k = 32.0$ lb/ft

To get mass of object: $F = ma$; $4.00 = m(32.0)$; $m = 0.125 \dfrac{\text{lb} \cdot \text{s}^2}{\text{ft}}$ (a slug)

Using Newton's Second Law: mass $\times$ accel. = restoring force

$0.125 \dfrac{d^2x}{dt^2} = -32.0x$; $D^2x + 256x = 0$; $m^2 + 256 = 0$; $m = \pm 16.0j$; $x = c_1 \sin 16.0t + c_2 \cos 16.0t$
$D_x = 16.0c_1 \cos 16.0t - 16.0c_2 \sin 16.0t$

Let $x = 0.250$ ft when $t = 0$; $0.250 = c_1 \sin 0 + c_2 \cos 0$; $c_2 = 0.250$
Let $Dx = 0$ when $t = 0$, $0 = 16.0c_1 \cos 0 - 16.0c_2 \sin 0$; $c_1 = 0$; $x = 0.250 \cos 16.0t$

9. $$L\frac{d^2q}{dt^2} + R\frac{dq}{dt} + \frac{q}{C} = E$$

$$0.200\frac{d^2\theta}{dt^2} + 8.00\frac{dq}{dt} + 10^6 q = 0$$

$$0.200m^2 + 8.00m + 10^6 = 0$$

$$m = \frac{-8.00 \pm \sqrt{64.0 - 0.800 \times 10^6}}{0.400}; \; m = -20.0 \pm 2240j; \; \alpha = -20.0, \; \beta = 2240$$

$q = e^{-20.0t}(c_1 \sin 2240t + c_2 \cos 2240t)$; $t = 0$, $q = 0$

Therefore, $c_2 = 0$

$\dfrac{dq}{dt} = e^{-20.0t}(2240c_1 \cos 2240t - 2240c_2 \sin 2240t) + (c_1 \sin 2240t + c_2 \cos 2240t)(-20.0e^{-20.0t})$

$t = 0$, $i = 0.500$; therefore, $c_1 = 2.24 \times 10^{-4}$; $q = e^{-20.0t}(2.24 \times 10^{-4}) \sin 2240t$
$q = 2.24 \times 10^{-4} e^{-20.0t} \sin 2240t$

13. $0.500D^2q + 10.0Dq + \dfrac{q}{200 \times 10^6} = 120 \sin 120\pi t$

$1.00D^2q + 20.0Dq + 10^4 q = 240 \sin 120\pi t$
$1.00m^2 + 20.0m + 10^4 = 0$

$m = \dfrac{-20.0 \pm \sqrt{400 - 4 \times 10^4}}{200} = -10.0 \pm 99.5j$; $q_c = e^{-10.0t}(c_1 \sin 99.5t + c_2 \cos 99.5t)$;
$q_p = A \sin 120\pi t + B \cos 120\pi t$

$$Dq_p = 120\pi A \cos 120\pi t - 120\pi B \sin 120\pi t$$
$$D^2 q_p = -142\,000\,A \sin 120\pi t - 142\,000 B \cos 120\pi t - 142\,000\,A120\pi t - 142\,000\,B \cos 120\pi t$$
$$\qquad + 7540A \cos 120\pi t - 7540B \sin 120\pi t + 10\,000A \sin 120\pi t + 10\,000B \cos 120\pi t$$
$$\qquad = 240 \sin 120\pi t$$
$$-132\,000A - 7540B = 240$$
$$-132\,000B + 7540A = 0$$
$$A = -1.81 \times 10^{-3},\, B = -1.03 \times 10^{-4}$$
$$q = e^{-10.0t}(c_1 \sin 99.5t + c_2 \cos 99.5t) - 1.81 \times 10^{-3} \sin 120\pi t - 1.03 \times 10^{-4} \cos 120\pi t$$

17. $1.00D^2 q + 5.00 Dq + \dfrac{q}{150 \times 10^{-6}} = 120 \sin 100t$

$$\qquad 1.00 D^2 q + 5.00 Dq + 6670 q = 120 \sin 100t$$

$$q_p = A \sin 100t + B \cos 100t$$
$$Dq_p = 100A \cos 100t - 100B \sin 100t$$
$$D^2 q_p = -10^4 A \sin 100t - 10^4 B \cos 100t$$
$$-10^4 A \sin 100t - 10^4 B \cos 100t + 5.00(100A \cos 100t - 100B \sin 100t) + 6670(A \sin 100t + B \cos 100t)$$
$$= 120 \sin 100t$$
$$(-10^4 A - 500B + 6670A) \sin 100t + (-10^4 B + 500A + 6670B) \cos 100t = 120 \sin 100t$$
$$3330A + 500B = -120$$
$$500A - 3330B = 0$$
$$A = -0.0352;\; B = 0.005\,28;\; q_p = -0.0352 \sin 100t + 0.005\,28 \cos 100t;$$
$$i_p = -3.52 \cos 100t + 0.528 \sin 100t$$

30.10   Laplace Transforms

1. $f(t) = 1;\; L(f) = L(t)$

$$F(s) = \int_0^\infty e^{-st}\,dt = -\frac{1}{s}\int_0^\infty e^{-st}(-s\,dt) = \lim_{c\to\infty}\left[-\frac{1}{s}\int_0^c e^{-st}(-s\,dt)\right] = \lim_{c\to\infty}\left[-\frac{1}{s}e^{-st}\right]\Big|_0^c$$

$$L(t) = \lim_{c\to\infty}\left[-\frac{1}{s}e^{-sc} + \frac{1}{s}\right] = 0 + \frac{1}{s} = \frac{1}{s}$$

5. $f(t) = e^{3t}$; from transform (3) of the table, $a = -3$; $L(3t) = \dfrac{1}{s-3}$

9. $f(t) = \cos 2t - \sin 2t;\; L(f) = L(\cos 2t) - L(\sin 2t)$

By transforms (5) and (6),

$$L(f) = \frac{s}{s^2 + 4} - \frac{2}{s^2 + 4};\; L(f) = \frac{s-2}{s^2 + 4}$$

13. $y'' + y';\; f(0) = 0;\; f'(0) = 0$
$$L[f''(y) + f'(y)] = L(f'') + L(f') = s^2 L(f) - sf(0) - f'(0) + sL(f) - f(0) = s^2 L(f)$$
$$\qquad - s(0) - 0 + sL(f) - 0$$
$$\qquad = s^2 L(f) + sL(f)$$

17. $L^{-1}(F) = L^{-1}\left(\dfrac{2}{s^3}\right) = 2L^{-1}\left(\dfrac{1}{s^3}\right);\ L^{-1}(F) = \dfrac{2t^2}{2} = t^2;$ transform (2)

21. $L^{-1}(F) = L^{-1}\dfrac{1}{(s+1)^3} = L^{-1}\dfrac{1}{2}\left[\dfrac{2}{(s+1)^3}\right] = \dfrac{1}{2}t^2 e^{-t};$ transform (12)

25. $F(s) = \dfrac{4s^2 - 8}{(s+1)(s-2)(s-3)} = \dfrac{-\frac{1}{3}}{s+1} + \dfrac{-\frac{8}{3}}{s-2} + \dfrac{7}{s-3}$

   $L^{-1}(F) = -\dfrac{1}{3}L^{-1}\left(\dfrac{1}{s+1}\right) - \dfrac{8}{3}L^{-1}\left(\dfrac{1}{s-2}\right) + 7L^{-1}\dfrac{1}{s-3}$

   $f(t) = -\dfrac{1}{3}e^{-t} - \dfrac{8}{3}e^{2t} + 7e^{3t}$

30.11   Solving Differential Equations by Laplace Transforms

1. $y' + y = 0;\ y(0) = 1;\ L(y') + L(y) = L(0);\ L(y') + L(y) = 0;\ sL(y) - y(0) + L(y) = 0$

   $sL(y) - 1 + L(y) = 0;\ (s+1)L(y) = 1;\ L(y) = \dfrac{1}{s+1};\ a = -1,$ transforms (3); $y = e^{-t}$

5. $y' + 3y = e^{-3t};\ y(0) = 1;\ L(y') + L(3y) = L(e^{-3t});\ L(y') + 3L(y) = L(3^{-3t});\ [sL(y-1] + 3L(y) = \dfrac{1}{s+3}$

   $(s+3)L(y) = \dfrac{1}{s+3} + 1;\ L(y) = \dfrac{1}{(s+3)^2} + \dfrac{1}{s+3}$

   The inverse is found from transforms (11) and (3).

   $y = te^{-3t} + e^{-3t} = (1+t)e^{-3t}$

9. $y'' + 2y' = 0;\ y(0) = 0,\ y'(0) = 2;\ L(y'') + L(2y') = 0;\ L(y'') + 2L(y') = 0$

   $[s^2 L(y) - 0 - 2] + 2sL(y) - 0 = 0;\ (s^2 + 2s)L(y) = 2;\ L(y) = \dfrac{2}{s^2 + 2s} = \dfrac{2}{s(s+2)}$

   By transform (4), $a = 2;\ y = 1 - e^{-2t}$

13. $y'' + y = 1;\ y(0) = 1;\ y'(0) = 1;\ L(y'') + L(y) = L(1);\ s^2 L(y) - s - 1 + L(y) = \dfrac{1}{s}$

   $(s^2 + 1)L(y) = \dfrac{1}{s} + s + 1;\ L(y) = \dfrac{1}{s(s^2 + 1)} + \dfrac{s}{s^2 + 1} + \dfrac{1}{s^2 + 1}$

   By transforms (7), (5), and (6), $y = 1 - \cos t + \cos t + \sin t;\ y = 1 + \sin t$

17. $y'' - 4y = 10e^{3t},\ y(0) = 5,\ y'(0) = 0;\ L(y'') - 4L(y) = 10 \cdot L(e^{3t})$

   $s^2 L(y) - s \cdot y(0) - y'(0) - 4L(y) = \dfrac{10}{s-3};\ s^2 L(y) - 5s - 0 - 4L(y) = \dfrac{10}{s-3}$

   $(s^2 - 4)L(y) = 5s + \dfrac{10}{s-3}$

$$L(y) = \frac{5s}{(s+2)(s-2)} + \frac{10}{(s+2)(s-2)(s-3)}$$

$$L(y) = \frac{\frac{5}{2}}{s+2} + \frac{\frac{5}{2}}{s-2} + \frac{\frac{1}{2}}{s+2} + \frac{-\frac{5}{2}}{s-2} + \frac{2}{s-3}$$

$$L(y) = \frac{3}{s+2} + \frac{2}{s-3}$$

$$y = 3e^{-2t} + 2e^{3t}$$

21. $2v' = 6 - v$; since the object starts from rest, $f(0) = 0$, $f'(0) = 0$

$$2L(v') + L(v) = 6L(1); \ 2sL(v) - 0 + L(v) = \frac{6}{s}; \ (2s+1)L(v) = \frac{6}{s}$$

$$L(v) = \frac{6}{s(2s+1)} = 6\left[\frac{\frac{1}{2}}{s\left(s+\frac{1}{2}\right)}\right]$$

By transforms (4), $v = 6(1 - e^{-t/2})$

25. $10\dfrac{d^2q}{dt^2} + \dfrac{q}{4 \times 10^{-5}} = 100\sin 50t$; $q(0) = 0$ and $q'(0) = 0$

$$10L(q'') + 2.5 \times 10^4 L(q) = L(100\sin 50t)$$
$$10s^2 L(q) + 2.5 \times 10^4 L(q) = L(100\sin 50t)$$

$$(s^2 + 2.5 \times 10^3)L(q) = 10\left(\frac{50}{s^2 + 50}\right); \ L(q) = \frac{500}{(s^2 + 50^2)^2}; \ L(q) = \frac{1}{500}\frac{2(50)^3}{(s^2+50^2)^2}$$

By transforms (15),

$$q = \frac{1}{500}(\sin 50t - 50t\cos 50t)$$

$$i = \frac{dq}{dt} = \frac{1}{500}\{50\cos 50t - [50t(-50\sin 50t) + (\cos 50t)50]\} = \frac{1}{500}(2500t\sin 50t) = 5t\sin 50t$$

29. $0.2Di + 10i = 50e^{-100t}$, $i(0) = 0$

$$0.2L(Di) + 10L(i) = 50L(e^{-100t}); \ 0.2(s \cdot L(i) - i(0)) + 10L(i) = 50 \cdot \frac{1}{s+100}; \ \frac{s}{5}L(i) + 10L(i) = \frac{50}{s+100};$$
$$sL(i) + 50L(i) = \frac{250}{s+100}; \ (s+50) \cdot L(i) = \frac{250}{s+100}$$

$$L(i) = \frac{250}{(s+50)(s+100)} = \frac{5}{s+50} + \frac{-5}{s+100}; \ i = 5 \cdot e^{-50t} - 5e^{-100t}$$

Chapter 30 Review Exercises

1. $4xy^3 dx + (x^2 + 1)dy = 0$; divide by y^3 and $x^2 + 1$

$$\frac{4x}{x^2+1}dx + \frac{dy}{y^3} = 0; \text{ integrating, } 2\ln(x^2+1) - \frac{1}{2y^2} = c$$

5. $2D^2y + Dy = 0$. The auxiliary equation is $2m^2 + m = 0$

$m(2m + 1) = 0$; $m_1 = 0$ and $m_2 = -\dfrac{1}{2}$

$y = c_1 e^0 + c_2 e^{-1/2x}$; $y = c_1 + c_2 e^{-x/2}$

9. $(x + y)dx + (x + y^3)dy = 0$; $x\,dx + y\,dx + x\,dy + y^3\,dy = 0$; $x\,dx + d(xy) + y^3\,dy = 0$

Integrating, $\dfrac{1}{2}x^2 + xy + \dfrac{1}{4}y^4 = c_1$; $2x^2 + 4xy + y^4 = c$

13. $dy = (2y + y^2)dx$; $\dfrac{dy}{(2y + y^2)} = dx$

$-\dfrac{1}{2}\ln\left(\dfrac{2+y}{y}\right) = x + \ln c_1$; $\ln\dfrac{2+y}{y} = -2x - \ln c_1^2$ $\left(\dfrac{2+y}{y}\right) = -2x$; $\dfrac{2+y}{y} = \dfrac{e^{-2x}}{c_1^2}$;

$y = c_1^2(y + 2)e^{2x}$; $y = c(y + 2)e^{2x}$

17. $y' + 4y = 2e^{-2x}$; $\dfrac{dy}{dx} + 4y = 2e^{-2x}$; $dy + 4 \cdot y\,dx = 2e^{-2x}\,dx$

$e^{\int 4dx} = e^{4x}$; $e^{4x}dy + 4ye^{4x}dx = 2e^{-2x} \cdot e^{4x}dx$; $d(ye^{4x}) = 2e^{2x}dx$; $\int d(ye^{4x}) = \int e^{2x} \cdot (2\,dx)$;

$ye^{4x} = e^{2x} + c$; $y = e^{-2x} + ce^{-4x}$

21. $2D^2y + Dy - 3y = 6$; $2m^2 + m - 3 = 0$; $(m - 1)(2m + 3) = 0$; $m_1 = 1$, $m_2 = -\dfrac{3}{2}$

$y_c = c_1 e^x + c_2 e^{-3x/2}$; $y_p = A$; $y_p' = 0$; $y_p'' = 0$

Substituting into the differential equation, $2(0) + 0 - 3A = 6$; $A = -2$; $y_p = -2$.
$y = c_1 e^x + c_2 e^{-3x/2} - 2$

25. $9D^2y - 18Dy + 8y = 16 + 4x$

$D^2y - 2Dy + \dfrac{8}{9}y = \dfrac{16}{9} + \dfrac{4}{9}x$

$$m^2 - 2m + \dfrac{8}{9} = 0;\ m = \dfrac{2 \pm \sqrt{4 - 4\left(\frac{8}{9}\right)}}{2};\ m_1 = \dfrac{2}{3};\ m_2 = \dfrac{4}{3}$$
$$y_c = c_1 e^{2x/3} + c_2 e^{4x/3};\ y_p = A + Bx;\ y_p' = B;\ y_p'' = 0$$

Substituting into the differential equation,

$$0 - 2B + \dfrac{8}{9}(A + Bx) = \dfrac{16}{9} + \dfrac{4}{9}x$$
$$\left(-2B + \dfrac{8}{9}A\right) + \dfrac{8}{9}Bx = \dfrac{16}{9} + \dfrac{4}{9}x;\ -2B + \dfrac{8}{9}A = \dfrac{16}{9};$$
$$\dfrac{8}{9}B = \dfrac{4}{9};\ B = \dfrac{1}{2};\ -2\left(\dfrac{1}{2}\right) + \dfrac{8}{9}A = \dfrac{16}{9};\ A = \dfrac{25}{8};\ y_p = \dfrac{1}{2}x + \dfrac{25}{8}$$
$$y = c_1 e^{2x/3} + c_2 e^{4x/3} + \dfrac{1}{2}x + \dfrac{25}{8}$$

29.
$$y'' - 7y' - 8y = 2e^{-x}$$
$$D^2 y_c - 7Dy_c - 8y_c = 0$$
$$m^2 - 7m - 8 = 0$$
$$(m+1)(m-8) = 0$$
$$m = -1, \; m = 8$$
$$y_c = c_1 e^{-x} + c_2 e^{8x}$$

Let $y_p = Axe^{-x}$, $y_p' = Ae^{-x}(1-x)$, $y_p'' = Ae^{-x}(x-2)$

$$Ae^{-x}(x-2) - 7Ae^{-x}(1-x) - 8Axe^{-x} = 2e^{-x}$$
$$A(x-2) - 7A(1-x) - 8Ax = 2$$
$$-9A = 2$$
$$A = \frac{-2}{9}$$

$$y_p = \frac{-2}{9} xe^{-x}$$

$$y = y_c + y_p$$

$$y = c_1 e^{-x} + c_2 e^{8x} + \frac{-2}{9} xe^{-x}$$

33. $3y' = 2y \cot x$; $\dfrac{dy}{dx} = \dfrac{2}{3} y \cot x$; $\dfrac{dy}{y} = \dfrac{2}{3} \cot x \, dx$; integrating,

$$\ln y = \frac{2}{3} \ln \sin x + \ln c; \;\; \ln y - \ln \sin^{2/3} x = \ln c;$$

$$\ln \frac{y}{\sin^{2/3} x} = \ln c; \; \frac{y}{\sin^{2/3} x} = c; \; y = c \sin^{2/3} x$$

Substituting $y = 2$ when $x = \dfrac{\pi}{2}$, $2 = c \sin^{2/3} \dfrac{\pi}{2}$; $c = 2$

Therefore $y = 2 \sin^{2/3} x = 2\sqrt[3]{\sin^2 x}$; $y^3 = 8 \sin^2 x$

37. $D^2 y + Dy + 4y = 0$. $m^2 + m + 4 = 0$; $m = \dfrac{-1 \pm \sqrt{-15}}{2}$

$$m_1 = -\frac{1}{2} + \frac{\sqrt{15}}{2} j; \; m_2 = -\frac{1}{2} - \frac{\sqrt{15}}{2} j; \; \alpha = -\frac{1}{2}, \; \beta = \frac{\sqrt{15}}{2}; \text{ by Eq. (30-17)}$$

$$y = e^{-x/2} \left(c_1 \sin \frac{\sqrt{15}}{2} x + c_2 \cos \frac{\sqrt{15}}{2} x \right)$$

$$Dy = e^{-x/2} \left(\frac{\sqrt{15}}{2} c_1 \cos \frac{\sqrt{15}}{2} x - c_2 \sin \frac{\sqrt{15}}{2} x \right) + \left(c_1 \sin \frac{\sqrt{15}}{2} x + c_2 \cos \frac{\sqrt{15}}{2} x \right) \left(-\frac{1}{2} e^{-x} \right)$$

Substituting $Dy = \sqrt{15}$, $y = 0$ when $x = 0$; $c_2 = 0$; $c_1 = 2$.

$$y = 2 e^{-x/2} \sin \left(\frac{1}{2} \sqrt{15} x \right)$$

41. $L(4y') - L(y) = 0$; $4L(y') - L(y) = 0$; $4sL(y) - 1 - L(y) = 0$, by Eq. (30-24)

$$(4s - 1)L(y) = 1; \; L(y) = \frac{1}{4s - 1} = \frac{\frac{1}{4}}{s - \frac{1}{4}}$$

By transforms (3), $y = e^{t/4}$.

45. $L(y'') + L(y) = 0$; $y(0) = 0$, $y'(0) = -4$, Eq. (30-25)

$s^2 L(y) + 4 + L(y) = 0$; $(s^2 + 1)L(y) = -4$; $L(y) = -4\left(\dfrac{1}{s^2+1}\right)$

By transforms (6), $y = -4\sin t$.

49. $\boxed{1}\ v = \pi\dfrac{4}{3}r^3$; $\boxed{2}\ A = 4\pi r^2$; $\dfrac{dv}{dt} = kA$

From $\boxed{1}$, $\dfrac{\left(d\frac{4}{3}\pi r^3\right)}{dt} = k(4\pi r^2)$; $4\pi r^2\dfrac{dr}{dt} = 4\pi k r^2$; $\dfrac{dr}{dt} = k$; $r = kt + C$; $r_0 = 0 + C$; $C = r_0$; $r = r_0 + kt$

53. $N = N_0 e^{-kt}$

$\dfrac{N_0}{2} = N_0 e^{-kt(1.28\times10^9)}$

$e^{k(1.28\times10^9)} = 2$

$k(1.28\times10^9) = \ln 2$

$$k = \frac{\ln 2}{1.28\times10^9}$$

$$N = N_0 e^{-\frac{\ln 2}{1.28\times10^9}\cdot t}$$

$$0.75N_0 = N_0 e^{\frac{-\ln 2}{1.28\times10^9}\cdot t}$$

$$e^{\frac{\ln 2}{1.28\times10^9}\cdot t} = \frac{4}{3}$$

$$\frac{\ln 2}{1.28\times10^9}\cdot t = \ln\frac{4}{3}$$

$$t = \frac{\ln\frac{4}{3}}{\frac{\ln 2}{1.28\times10^9}}$$

$$t = 5.31\times10^8 \text{ years.}$$

57. $y = cx^5$; $c = \dfrac{y}{x^5}$; $\dfrac{dy}{dx} = 5cx^4$; $\dfrac{dy}{dx} = 5\left(\dfrac{y}{x^5}\right)x^4 = \dfrac{5y}{x}$; $\dfrac{dy}{dx}\Big|_{OT} = -\dfrac{x}{5y}$; $5y\,dy = -x\,dx$

Integrating, $\dfrac{5y^2}{2} = -\dfrac{x^2}{2}$; $5y^2 + x^2 = c$

61. $F = kx$; $40 = 0.50k$; $k = 80$ N/m; $m = 4\,kg$; $4\dfrac{d^2x}{dt^2} = -16\dfrac{dx}{dt} - 80x$; $4D^2x + 16Dx + 80 = 0$;

$D^2x + 4Dx + 20 = 0$; $4m^2 + 16m + 80 = 0$; $m_1 = -2 + 4j$; $m_2 = -2 - 4j$; $x = e^{-2t}(c_1 \sin 4t + c_2 \cos 4t)$

Let $x = 0.50$ when $t = 0$; $0.50 = c_2$

$Dx = e^{-2t}(4c_1 \cos 4t - 4c_2 \sin 4t) + (c_1 \sin 4t + c_2 \cos 4t)(-2)e^{-2t}$

Let $Dx = 0$ when $t = 0$; $0 = 4c_1 - 2c_2$; $0 = 4c_1 - 1$; $c_1 = 0.25$

$x = e^{-2t}(0.25 \sin 4t + 0.5 \cos 4t) = 0.25e^{-2t}(\sin 4t + 2\cos 4t)$ underdamped

65. $R = 20\ \Omega,\ L = 4\ \text{H},\ C = 10^{-4}\ \text{F};\ V = 100\ \text{V};\ q(0) = 10^{-2}\ \text{C};\ 4q'' + 20q' + 10^4 q = 100;$

$$\frac{1}{25}q'' + \frac{1}{5}q' + 100q = 1$$

$$\frac{1}{25}[s^2 L(q) - s(10^{-2})] + \frac{1}{5}[sL(q) - 10^{-2}] + 100L(q) = L(1)$$

$$\frac{1}{25}s^2 L(q) - \frac{1}{2500}s + \frac{1}{5}sL(q) - \frac{1}{500} + 100L(q) = \frac{1}{s}$$

$$L(q)\left(\frac{1}{25}s^2 + \frac{1}{5}s + 100\right) = \frac{1}{s} + \frac{1}{2500}s + \frac{1}{500}$$

$$L(q)\left(\frac{s^2 + 5s + 2500}{25}\right) = \frac{2500 + s^2 + 5s}{2500s}$$

$$L(q) = \frac{1}{100s};\ q = \frac{1}{100};\ i = \frac{dq}{dt} = 0$$

69. $0.25\dfrac{d^2 q}{dt^2} + \dfrac{4.0\,dq}{dt} + \dfrac{q}{10^{-4}} = 0;\ q(0) = 400\ \mu C = 4.0 \times 10^{-4}\ C$

$i = \dfrac{dq}{dt} = q'(0);\ 0.25L(q'') + 4.0L(q') + 10^4 L(q) = 0;\ L(q'') + 16L(q') + 4.0 \times 10^4 L(q) = 0$

$s^2 L(q) - 4.0 \times 10^{-4}s - 0 + 16sL(q) - 4.0 \times 10^{-4} + 4.0 \times 10^4 L(q) = 0$

$(s^2 + 16s + 4.0 \times 10^{-4})L(q) = 4.0 \times 10^{-4}s + 64 \times 10^{-4}$

$$L(q) = 4.0 \times 10^{-4}\left[\frac{s + 16}{s^2 + 16s + 64 + 4 \times 10^4}\right]$$

$$= 4.0 \times 10^{-4}\left[\frac{s + 8}{(s + 8)^2 + 200^2} + \frac{8}{(s + 8)^2 + 200^2}\right]$$

$$= 4.0 \times 10^{-4}\left[\frac{s + 8}{(s + 8)^2 + 200^2} + \frac{8}{200} \times \frac{200}{(s + 8)^2 + 200^2}\right]$$

$$q = 4.0 \times 10^{-4}(e^{-8t}\cos 200t + 0.04e^{-8t}\sin 200t)$$

$$q = 10^{-4}e^{-8t}(4.0\cos 200t + 0.16\sin 200t)$$

73. $\dfrac{d^2 y}{dx^2} = \dfrac{1}{EI}m = \dfrac{1}{EI}(2000x - 40x^2);$ integrating

$$\frac{dy}{dx} = \frac{1}{EI}\left[\left(2000\frac{x^2}{2} - \frac{40x^3}{3}\right) + c_1\right]$$

$$\frac{dy}{dx} = \frac{1}{EI}\left(1000x^2 - \frac{40x^3}{3} + c_1\right);\ \text{integrating}$$

$$y = \frac{1}{EI}\left(\frac{1000x^3}{3} - \frac{10x^4}{3} + c_1 x + c_2\right) = \frac{10}{3EI}(100x^3 - x^4) + c_1 x + c_2$$

$y = 0$ when $x = 0;\ 0 = c_2;\ y = \dfrac{10}{3EI}(100x^3 - x^4) + c_1 x$

$y = 0$ when $x = L;\ 0 = \dfrac{10}{3EI}(100L^3 - L^4) + c_1 L$

$$c_1 L = -\frac{10}{3EI}(100L^3 - L^4); \; c_1 = -\frac{10}{3EI}(100L^2 - L^3)$$

$$y = \frac{10}{EI}(100x^3 - x^4) - \frac{10}{3EI}(100L^2 - L^3)x = \frac{10}{3EI}[100x^3 - x^4 - (100L^2 - L^3)x]$$

$$= \frac{10}{3EI}[100x^3 - x^4 + xL^2(L - 100)]$$

77.

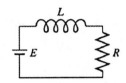

$$L \cdot \frac{di}{dt} + iR = E$$

$$i(0) = 0$$

(a) separation of variables.

$$L\,di + iR\,dt = E\,dt$$
$$L\,di = (E - iR)\,dt$$

$$\int \frac{L}{E - iR}\,di = \int dt$$

$$-\frac{L}{R} \int \frac{-R\,di}{E - iR} = \int dt$$

$$-\frac{L}{R}\ln(E - iR) = t + \ln C$$

$$-\frac{L}{R}\ln(E - 0 \cdot R) = 0 + \ln C$$

$$\ln E^{-L/R} = \ln C$$
$$C = E^{-L/R}$$

$$-\frac{L}{R} \cdot \ln(E - iR) = t + \ln E^{-L/R}$$

$$\ln(E - iR)^{L/R} + \ln E^{-L/R} = -t$$
$$\ln[(E - iR)^{L/R} \cdot E^{-L/R}] = -t$$
$$(E - iR)^{L/R} \cdot E^{-L/R} = e^{-t}$$
$$(E - iR)^{L/R} = E^{L/R} \cdot e^{-t}$$
$$E - iR = E \cdot e^{-R/L \cdot t}$$
$$iR = E - E \cdot e^{-R/L \cdot t}$$

$$i = \frac{E}{R}(1 - e^{-R/L \cdot t})$$

(b) linear differential equation of first order, use integrating factor

$$L\,di + iR\,dt = E\,dt$$

$$di + \frac{R}{L}\cdot i\,dt = \frac{E}{L}\,dt$$

$$e^{\int R/L\,dt} = e^{(R/L)t}$$

$$e^{R/Lt}\cdot di + \frac{R}{L}\cdot e^{(R/L)t}\cdot i\,dt = \frac{E}{L}e^{(R/L)t}dt$$

$$\int d(ie^{(R/L)t}) = \int \frac{E}{L}e^{(R/L)t}dt = \frac{E}{R}\int e^{(R/L)t}\cdot\frac{R}{L}dt$$

$$i\cdot e^{(R/L)t} = \frac{E}{R}e^{(R/L)t} + C$$

$$0\cdot e^{R/L\cdot 0} = \frac{E}{R}e^{R/L\cdot 0} + C$$

$$C = \frac{-E}{R}$$

$$i\cdot e^{(R/L)t} = \frac{E}{R}e^{R/Lt} - \frac{E}{R}$$

$$i = \frac{E}{R}(1 - e^{-(R/L)t})$$

(c) Laplace transforms: use L for inductance and $\mathcal{L}$ for the Laplace transform

$$L\frac{di}{dt} + iR = E$$

$$L\mathcal{L}(i') + R\mathcal{L}(i) = \mathcal{L}(E)$$

$$L\cdot(s\mathcal{L}(i) - i(0)) + R\mathcal{L}(i) = \frac{E}{s}$$

$$Ls\mathcal{L}(i) + R\mathcal{L}(i) = \frac{E}{s}$$

$$\mathcal{L}(i)(Ls + R) = \frac{E}{s}$$

$$\mathcal{L}(i) = \frac{E}{s(Ls + R)} = \frac{E}{Ls\cdot\left(s + \frac{R}{L}\right)} = \frac{\frac{E}{L}}{s\left(s + \frac{R}{L}\right)}$$

$$\mathcal{L}(i) = \frac{\frac{E}{R}}{s} - \frac{\frac{E}{R}}{s + \frac{R}{L}}$$

$$i = \frac{E}{R}(1 - e^{-(R/L)t})$$